PROCEEDINGS OF THE 2ND INTERNATIONAL SYMPOSIUM ON FIELD
MEASUREMENTS IN GEOMECHANICS / KOBE / 6-9 APRIL / 1987

FIELD MEASUREMENTS IN GEOMECHANICS

Edited by
S.SAKURAI
Kobe University, Japan

VOLUME TWO
4 *Underground openings*
5 *Waste disposal problems*
6 *Interpretation of field measurement results*

A.A.BALKEMA / ROTTERDAM / BROOKFIELD / 1988

The texts of the various papers in this volume were set individually by typists under the supervision of each of the authors concerned.

Published by

A.A.Balkema, P.O.Box 1675, 3000 BR Rotterdam, Netherlands

A.A.Balkema Publishers, Old Post Road, Brookfield, VT 05036, USA

For the complete set of two volumes ISBN 90 6191 778 6

For volume 1: ISBN 90 6191 750 6

For volume 2: ISBN 90 6191 751 4

© 1988 A.A.Balkema, Rotterdam

Printed in the Netherlands

Table of contents

5 Waste disposal problems

4 Underground openings

2nd International Symposium on Field Measurements in Geomechanics, Sakurai (ed.)
© 1988 Balkema, Rotterdam. ISBN 90 6191 778 6

The understanding of measured changes in rock structure, in situ stress and water flow caused by underground excavation

J.A.Hudson
Department of Mineral Resources Engineering, Imperial College of Science and Technology, London, UK

1 INTRODUCTION

Throughout recorded history, underground openings have been created as protection from the elements for the living, and for the dead. Man has also searched relentlessly for minerals. More recently, underground openings have been created and considered for a wide variety of civil engineering purposes, including transport and more exotic activities such as superconductive magnetic energy storage and radioactive waste disposal. But it is only over the last few decades that sophisticated field measurements supported by advanced analysis techniques have been used to provide assistance in the design and performance of underground openings. Thus, the history of engineered underground openings has been one of thousands of years of experience and just a few decades of scientific measurement and theory.

Currently, it is possible to make extremely accurate field measurements; and, similarly, it is possible to conduct very detailed and accurate computer analyses of the ground response to the excavation disturbance. The integration of these two capabilities, the understanding of the phenomena observed, and the crucial need to identify interactive effects are the subject of this paper.

Can the observations and the theory be matched? Are the reasons for the measurement values understood? Can the 'pure' and the 'interactive' causal components be separated? Are underground openings really designed on the basis of measurements and/or theory? Discussion in the paper is concentrated on the effect of underground excavation on the rock structure (especially the discontinuities), the stress distribution and groundwater flow.

2 THE EFFECT OF EXCAVATION

2.1 Primary effects

The creation of an excavation in the gravity loaded rock mass can a dramatic effect on the rock mass structure, the stress field and the water flow. This is because the removal of the rock to create the excavation has the following effects:

a) the resistance to deformation is removed, i.e. the original elastic modulus of the rock has been replaced by the zero modulus of the excavation opening;

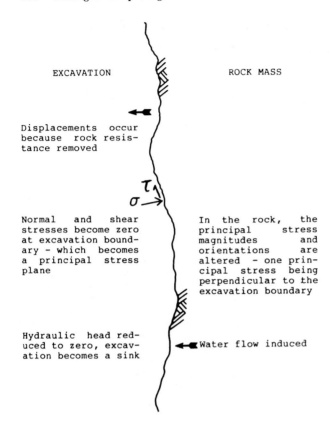

EXCAVATION   ROCK MASS

Displacements occur
because rock resis-
tance removed

Normal and shear   In the rock, the
stresses become zero   principal stress
at excavation bound-   magnitudes and
ary - which becomes   orientations are
a principal stress   altered - one prin-
plane   cipal stress being
                          perpendicular to the
                          excavation boundary

Hydraulic head red-
uced to zero, excav-   Water flow induced
ation becomes a sink

Figure 1. The effects of underground rock excavation on deformation
resistance, local stress field and water flow.

   b) the local stress field in the surrounding rock has been perturbed
because the walls of the excavation are now principal stress planes
(there are no shear stresses on the excavation walls) and in particular
the stress field has been rotated to have one principal component perpendicular and the other two principal components parallel to the excavation surface;
   c) the hydraulic head is reduced to zero and so the excavation acts as
a local sink for groundwater flow.
   These effects are represented in Figure 1 where the excavation opening
is on the left of the diagram and the remaining rock on the right side
of the diagram.
   Additionally, these changes do not occur instantly as is often assumed
in modelling, but progressively as excavation proceeds.

2.2 Secondary or interactive effects

There has been a tendency in the past to consider the effects a) - c)
above separately and to solve specific problems for the individual
parameter variations. In the case of the rock mass structure, the

removal of the resistance to deformation can cause the excavation walls to move as a continuum in an elastic, viscoelastic, plastic or more complex constitutive mode. Alternatively, if the rock mass is moving under structural control, the discontinuities will dictate the type of block movement that can occur. The development of rock mechanics has always included study of the former case; in latter years, considerable progress has been made on block geometry and movements (e.g. Goodman and Shi 1985). There is also, of course, the case of combined responses where both the intact material and the discontinuities are playing their parts.

In a similar way, the effect of excavation on the stress field in the surrounding rock has been extensively studied and, to a lesser extent, the effect of excavation on water flow. In both these cases and where the discontinuities are producing significant effects, there is a move to consider the near-field by analysis methods which can consider the specific effects of discontinuities (either with deterministic proper- ties or as statistically generated) and the far field as an equivalent continuum where the effect of the discontinuities is incorporated into the analysis via equivalent material properties (Lorig and Brady 1984). The concept of the Representative Elemental Volume (REV) is very impor- tant in this context as the continuum modelling can only be validly conducted for volumes containing a statistically representative sample of discontinuities.

However, very little work has been conducted on the interaction bet- ween the primary effects, and it is these interactive effects that one encounters in practice. Given the three factors described relating to the rock mass structure, in situ stress and water flow, there are three pairs of complementary interaction terms:

a) the effect of rock mass structure on in situ stress (the structure influencing the stress field), and the complementary but different effect of the in situ stress on the rock mass structure (by inducing deformation and failure);

b) the effect of the rock mass structure (especially the discontinuity characteristics) on the water flow, and the converse effect of the water flow on the rock mass structure (through degradation);

c) the effect of the in situ stress on water flow (because the discon- tinuity permeability is very sensitive to any stress induced changes in aperture, and the well known effect of water on the in situ stress as developed via the effective stress concept.

Now as the excavation proceeds and, as explained in 2.1, there are significant effects on the primary factors of rock mass structure, in situ stress and water flow, so also the secondary or interactive factors described above come into play. The major difficulties in excavation are often associated with extremes of deformation, high stresses and high water flows but these extremes will be localized through the secon- dary interactions. For example, other factors being equal, the greatest amount of water flow into an excavation will occur in regions of tensile stress where the rock mass permeability is highest.

The ramifications of the existence of the interactive effects are that analysis methods must be 'coupled' to provide realistic and useful modelling and that interpretations of measured changes in the values of deformation, stress and water flow must be conducted within the frame- work of knowledge containing the interactive effects. Indeed, the design of a field investigation programme should take into account these interactive effects so that the different types of tranducers are opt- imally located to allow for interpretation of the primary and secondary effects.

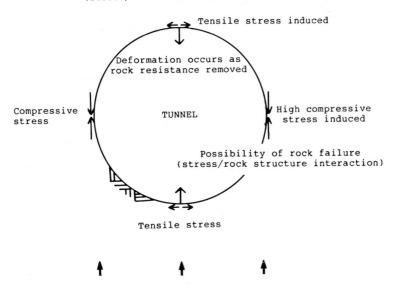

High
vertical
stress

Region of increased permeability
(stress/water flow interaction)

Tensile stress induced

Deformation occurs as
rock resistance removed

Compressive
stress

TUNNEL

High compressive
stress induced

Possibility of rock failure
(stress/rock structure interaction)

Tensile stress

Figure 2.   Effects of excavation for a circular tunnel in an unfractured
rock mass – the three primary effects and two interactive secondary
effects.

3 ILLUSTRATION OF THE PRIMARY AND SECONDARY EFFECTS FOR A ROCK MASS WITH AND WITHOUT JOINTS

The discussion above holds for all excavations but for convenience the
ideas will be illustrated, firstly, with reference to a circular tunnel
in a solid rock mass and, secondly, with reference to an arbitrary
excavation in a discontinuous rock mass.

3.1 Primary and secondary effects in an unfractured rock mass

The circumstances in an unfractured rock mass are relatively simple and
provide a useful starting point for considering the interactions as
illustrated in Figure 2.

   In the case where there is a high vertical stress and a relatively low
horizontal stress, we would expect a tensile stress to be present in the
crown and invert and for high compressive stresses to be induced in the
sidewalls at axis level.   As shown in Figure 1, the three primary
effects will occur: a squeezing inward of the crown and invert, principal stresses parallel to the tunnel periphery, and the potential for
increased groundwater flow.   The two main interactive terms are that the
rock permeability will be increased in the regions of tensile stress and

that the rock at the crown and invert could fail due to the tensile stress - and, similarly, the rock at axis level could fail as a result of the increased compressive stress. If the rock failed in a brittle manner, there would be an alteration to the rock structure.

These are the main effects. However, if the rock were anisotropic or we considered the long term effects of water flowing through the newly fractured high permeability zones, then there would be further secondary and even tertiary effects to consider.

3.2 Primary and secondary effects in a discontinuous rock mass

When discontinuities are considered within this context of primary and secondary or interactive effects, there are far more factors to take into account than for an unfractured rock mass, however complex the constitutive mode of the intact rock. The discontinuites play a major role in rock deformation, the natural and induced rock stress field and groundwater flow. Thus, the excavation behaviour can be dominated by the direct primary effects of the discontinuities and the interactions outlined in items a) to c) in Section 2.2.

The interactive effects are illustrated and listed in Figure 3 which demonstrates that, even in this simple case (with no time, temperature or other complicating factors), the understanding of measured field values can easily be fraught with difficulties. Unless these interactions are identified and coherently interpreted, it is very unlikely that the behaviour of the excavation would be correctly understood.

A brief description of each interaction term is given below to indicate that these effects can be very significant, both in terms of actual excavation behaviour and the interpretation of measured field values.

Interaction 1: (Rock structure/stress) - Local stress field affected by discontinuities. Both the pre-existing rock stress field and the induced stress field around the excavation can be strongly affected by the presence of discontinuities (Crouch & Starfield 1983; Hyett et al. 1986). It is also helpful if the natural stress field can be interpreted in the structural geology context as highlighted in, for example, Chapter 3 of Uemura and Mizutani (1984). In Figure 3, if the discontinuites had very little shear resistance, the principal field stresses could well be sub-parallel and sub-perpendicular to the discontinuites, invalidating any assumption that the principal stresses were vertical and horizontal. Similarly, adjacent to the excavation the stresses could be completely different from those calculated for an unfractured rock mass.

Interaction 2: (Rock structure/water flow) - Water flows along discontinuities. The permeability of a discontinuous rock mass is generally governed by flow through the discontinuities. Thus, the geometrical and hydrogeological characteristics of the rock mass mainly govern the permeability. With the parallel plate model, the permeability of a discontinuity is proportional to the cube of the aperture, with the result that the permeability is highly sensitive to changes in aperture caused by changes to the rock structure during excavation. The influence of rock structure characteristics on water flow has been the subject of recent research, (e.g. Rouleau and Gale 1985).

Interaction 3: (Stress/water flow) - High normal stresses reduce the discontinuity permeability. Because the discontinuity permeability depends, inter alia, on the discontinuity aperture, changes in the local stress field will affect the aperture and hence the permeability. In particular, if any tensile stress is applied across a discontinuity, the permeability can change by several orders of magnitude. Thus, the

DISCONTINUOUS ROCK MASS

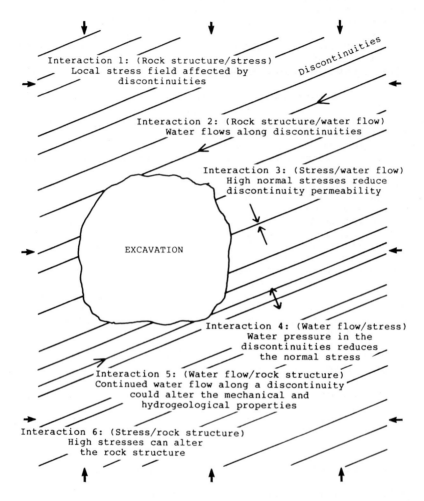

Interaction 1: (Rock structure/stress)
Local stress field affected by
discontinuities

Discontinuities

Interaction 2: (Rock structure/water flow)
Water flows along discontinuities

Interaction 3: (Stress/water flow)
High normal stresses reduce
discontinuity permeability

EXCAVATION

Interaction 4: (Water flow/stress)
Water pressure in the
discontinuities reduces
the normal stress

Interaction 5: (Water flow/rock structure)
Continued water flow along a discontinuity
could alter the mechanical and
hydrogeological properties

Interaction 6: (Stress/rock structure)
High stresses can alter
the rock structure

Figure 3. Effects of excavation in a discontinuous rock mass: the three primary effects occur plus the six interactive effects indicated.

complete analysis of the quantity of water flowing into an excavation and the location of greatest flow around the excavation boundaries is a complex problem.

Interaction 4: (Water flow/stress) – Water pressure in the discont-inuities reduces the normal stress. This is, of course, the effective stress concept with all its ramifications. Whilst considerable work has been conducted on effective stresses in soils, the work in rock masses has not led to such clear conclusions. This is because both the intact rock and the discontinuities contribute to the permeability, and the effective stress ought not to be considered constant through the preexcavation fractured rock mass. Needless to say, after excavation the effective stresses are even more difficult with the varying water pressure around the excavation.

Interaction 5: (Water flow/rock structure) – Continued water flow along a discontinuity could alter the mechanical and hydrogeological rock mass properties. This is a very important but much neglected subject. Any discontinuity permeability values, either measured or assumed, are highly likely to change fairly rapidly with time following the major perturbation to the rock system caused by excavation. Scouring, or other types of erosion caused by enhanced water flow, will increase discontinuity permeability. Conversely, where silt-filled fissures are present, the whole hydrogeological system could eventually become almost impermeable due to the development of silt plugs in key hydrological nodes and pathways.

Interaction 6: (Stress/rock structure) – High stresses can alter the rock structure. We have seen in the simpler case in Figure 2 that stress could cause fracturing around a circular tunnel. In the general case illustrated in Figure 3, the effect of stress creating new fractures would have all kinds of effects. The stress field itself would be redistributed; the local permeability could be significantly altered; the ground displacements would increase; and so on. Currently, there are very accurate methods of measuring ground movement profiles (Kovari & Amstad 1984) so that the alteration to the rock structure could be studied in detail from site measurements.

4 DISCUSSION AND CONCLUSIONS

The above description has shown that interpreting the effects of excavating in a gravity-loaded discontinuous rock mass is far from simple. The main effects, together with binary, tertiary and further effects, have to be identified if the rock mass response to excavation is to be understood. Ideally, this knowledge should guide the choice of instrumentation, its location, methods of data reduction and final interpretation. The approach will therefore enable the planning of field instrumentation programmes to be extended following the type of system set out by Franklin (1977). Also, such knowledge will assist in the interpretation of all field measurement data relating to underground excavation. A wealth of such data is available in Saari (1986). Moreover, analytical models developed to predict and interpret the rock mass behaviour must be 'coupled' so that the interactive effects are taken into account.

The author is currently developing a rock mechanics interaction matrix approach to the problem of identifying and characterizing the interactions (Hudson 1987). Concurrently, considerable progress is being made on coupled models as the existing analytical models are further developed and enhanced by the inclusion of the interactive effects (e.g. Hart and St. John 1986).

REFERENCES

Crouch, S.L. & A.M. Starfield 1983. Boundary element methods in solid mechanics. London: Allen & Unwin.
Franklin, J.A. 1977. Some practical considerations in the planning of field instrumentation. In K. Kovari (ed.) Proc. of the Int. Symp.

Field measurements in rock mechanics held in Zurich, p.3-13, Rotterdam: Balkema.

Goodman, R.E. & G.H. Shi 1985. Block theory and its application to rock engineering. New Jersey: Prentice-Hall.

Hart, R.D. & C.M. St. John 1986. Formulation of a fully coupled thermal-mechanical-fluid flow model for non-linear geologic systems. Int. J. Rock Mech. Min. Sci. & Geomech. Absts. 23: 213-224.

Hudson, J.A. 1987. Rock mechanics interaction matrices: Part I – the concept. Int. J. Rock Mech. Min. Sci. & Geomech. Absts. 24: in press.

Hyett, A.J.,C.G. Dyke & J.A. Hudson 1986. A critical examination of basic concepts associated with the existence and measurement of in situ stress. In O. Stephansson (ed.) Proc. Int. Symp. on Rock stress and rock stress measurements, 387-396. Stockholm: Centek.

Kovari, K. & Ch. Amstad 1984. Fundamentals of deformation measurements. In K. Kovari (ed.) Proc. Int. Symp. Field measurements in geomechanics, 219-239. Rotterdam: Balkema.

Lorig, L.J. & B.G.H. Brady 1984. A hybrid computational scheme for excavation and support design in jointed rock media. In E.T. Brown and J.A. Hudson (eds.) Proc. ISRM Symp.   Design and performance of underground openings, 105-112. London: Thomas Telford.

Rouleau, A. & J.E. Gale 1985. Statistical characterization of the fracture system in the Stripa granite, Sweden. Int. J.   Rock Mech. Min. Sci. & Geomech. Absts. 22: 353-367.

Saari, K. 1986. (Ed.) Large rock caverns. Proc. Int. Symp. Oxford: Pergamon   Press.

Uemura, T. & S. Mizutani 1984. Geological structures. New York: John Wiley.

2nd International Symposium on Field Measurements in Geomechanics, Sakurai (ed.)
© 1988 Balkema, Rotterdam. ISBN 90 6191 778 6

Integration of field instrumentation and computer simulation: Development and application of the SPDR method

Shosei Serata
Serata Geomechanics, Inc., Richmond, Calif., USA

1 INTRODUCTION

In spite of the recent advancements made in instrumentation and analysis techniques, geomechanics is still left very much in the primitive stage of empirical rules and has not yet achieved the status of a fully quantitative science with regard to analysis, design, and construction of earth structures, especially in complex ground. In order to further advance geomechanics from this stage, a new quantitative method of integrating field instrumentation work and computer simulation analysis has been developed. The new method, known as the SPDR method, has been validated through its wide applications in solving underground mining problems, and thus making the problem solving approach truly quantitative for the first time.

The SPDR method is the final outcome of our past two decades of studies aimed at quantifying geomechanics. The method encompasses the comprehensive use of the following:

1) <u>Constitutive model:</u> A constitutive model of generalized earth materials to represent time-dependent behaviors of all possible earth materials in a unified numerical form, known as REM.

2) <u>Finite element program</u>: A finite element computer simulation program (R) based on the REM model for realistic computer simulation of earth structures in complex ground.

3) <u>Comprehensive field instrumentation</u>: A computerized hardware system capable of measuring the three basic in situ behaviors of stress states (S), material properties (P), and ground deformation (D) in complex ground.

4) <u>Field calibration</u>: A sequence of scientific steps for field calibration of the REM computer model against SPD field measurements.

5) <u>Implementation</u>: Field implementation of the SPDR method with a number of successful applications.

The SPDR method was first developed to meet the immediate need of solving the severe ground failure problems for the potash mining industry in Saskatchewan, Canada. The method is now being successfully applied to benefit coal mining in the U.S., Canada and Australia. In these applications, the new approach has been able to succesfully eliminate the ground failure problems, which are inherent to the conventional method of mining. The application of the SPDR method and its significant accomplishments in geomechanical work are discussed below, using an example of a salt mine in the Michigan salt basin.

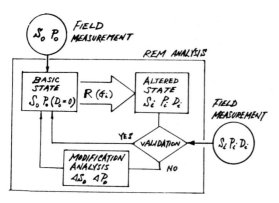

FIG. 1. SPDR Method of Integrating Field Measurements, (S_i, P_i, D_i) and REM Computer Simulation $(R[G_i])$. Initial Ground State (S_o, P_o) is Established Prior to Analyzing Design Geometry (G_i) Using REM Computer Program (R).

2 SPDR METHOD

The SPDR method integrates the two separate operations of field instrumentation and computer analysis for the purpose of engineering applications in geomechanics. The operating scheme, validation/-calibration, as well as the specific approach in regard to new and pre-existing structures, are discussed below:

Operating Scheme

The operating scheme of the SPDR method is shown in Fig. 1, illustrating the relationship between the field measurements and the computer simulation analysis. Here, S, P and D represent the three basic characteristics of ground media: stress state (S), material properties (P), and ground deformation (D). S_o, P_o, and D_o represent the initial condition of the ground with initial ground geometry (G_o), prior to any geometric alteration (G_i). S_i, P_i, and D_i represent the condition of the altered state (G_i).

In an ideal case of computer simulation, S_o and P_o can be measured, while D_o is considered to be zero to represent the initial ground state. The $S_i P_i D_i$ behaviors of the constructed structure (i.e., of the altered geometry, G_i) can then be computed by conducting simulation analysis, using the program R, with the initial state parameters S_o, P_o as input.

Validation/Calibration

Validity of the basic ground computer model is tested by comparing the computed and measured values of S_i, P_i, D_i. If computer projections do not match well with in situ measurements, then the ground model needs to be calibrated. The calibration is accomplished when proper values of S_o and P_o are found for which the differences between the simulated and measured values of S_i, P_i, D_i fall within a narrow range, defined by the required accuracy for the specific design application. After the model calibration, the program R can be applied, with calibrated S_o and P_o as input parameters, for evaluating any number of design possibilities $(G_i, i=1,2,...n)$.

New Construction

In application to new construction, where the new structure (G_i) is to be designed, field measurements are required only to determine in situ S_o and P_o. Following the field measurements, a finite element model of the ground is constructed with initially determined S_o and P_o. Using the basic ground model so developed, the computer program (R) can be used to evaluate various design geometries (G_i) by analyzing the simulated S_i, P_i, D_i behaviors for $i=1,2,...n$, to search for the optimum design parameters.

In case of insufficient field data on S_o and P_o, only a preliminary computer model of the ground is constructed, using estimated S_o and P_o as valuable parameters. The model is then applied to analyze the effects of S_o and P_o upon the proposed design G_i. Based on the results of this analysis, the decision to go ahead with the proposed design is made with known calculated uncertainties, if any.

Pre-Existing Structures

In application to pre-existing structures, the conditions of S_i, P_i, D_i are measured in situ and then back analysis is performed using the computer program (R) to determine the initial ground condition of S_o and P_o. Such analysis can be performed effectively since a large amount of actual field data, S_i, P_i, and D_i, are usually obtainable by measurement. For example, in the case of ground deformation D_i alone, we can easily obtain accurate in situ deformation ΔD and deformation rate $D=\Delta D/\Delta t$ in relation to time and space. It is indeed this large amount of accurate in situ S_i, P_i, D_i field data obtainable through the computerized data acquisition system as well as the ability to back analyze the data, which made the development of the SPDR method a reality.

3 COMPONENTS OF SPDR METHOD

The SPDR method encompasses the comprehensive use of four distinct proprietary tools: in situ stress measurement system (S), in situ property measurement system (P), deformation measurement system (D), and the REM computer simulation program (R). These tools are briefly described below.

In Situ Stress Measurement System (S)

A new type of stress measuring probe (S-100) has been developed especially for application with the SPDR method. The probe is made of a highly expandable plastic tubing, capable of loading a borehole to induce a set of two mutually perpendicular fracture planes without allowing the loading liquid to contact the borehole boundary. The stress state (S) is determined from the diametral deformation of the borehole with a number of transducers emplaced within the plastic tubing.

As illustrated in Fig. 2, the probe enables us to observe the initiation of the two fracture planes at B_1 and B_2. Free separation points of the double fracture planes are detected in the reloading cycles at E_1 and E_2 where the subscripts 1 and 2 denote the primary and secondary fracture planes, respectively. From the pressure-

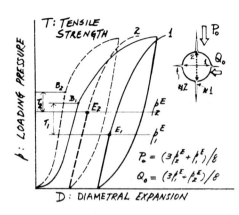

FIG. 2. Double Fracture Method of Determining Maximum and Minimum Stresses P_o, Q_o from Free Separation Pressures p_1^E and p_2^E of Primary (#1) and Secondary (#2) Fracture Planes Created by Hydraulic Loading Using Soft Plastic Tube of S-100 Stressmeter.

deformation (p-D) curves, free separation pressure readings, p_1^E and p_2^E, are obtained, and the stress state (S) is determined in terms of maximum and minimum stresses, P_o and Q_o by solving the following simultaneous equations:

$$P_o = (3p_2^E + p_1^E)/8$$
$$Q_o = (3p_1^E + p_2^E)/8$$

(1)

The stress orientation is determined from the diametral deformation pattern of the borehole cross-section.

This double fracture method with the S-100 probe is simple, fast, and economical and all it requires is a single borehole. The probe application is not limited by the depth of measurement, and the stress states are determined in real time.

The probe has been proven to perform well in both brittle fractured rocks and ductile yielding media [1]. It can also be used in highly stratified ground with mixed layers of different materials. Further details of the stressmeter probe and its application are described in another paper presented at this conference [2].

In Situ Property Measurement System (P)

The property measurement probe (P-100) is another device specially developed for the SPDR method. The probe is used with the same computerized stress/property measuring system which is used for the stress measuring probe, S-100. The probe is capable of conducting a set of 8 simultaneous penetration tests, with the 8 penetrometer pistons deployed symmetrically around the probe. Several material properties are determined from the P-100 data, including uniaxial failure, triaxial failure, Young's modulus, deformation modulus, plastic strength, plastic viscosity, non-uniformity, and anisotropy.

The most important use of P-100 measurements is to determine point properties and their spatial distribution in the ground. The P-100 is extremely effective in detecting anomalous properties which would never be obtained in conventional laboratory core testing, such as cracks, micro-fractures, clay seams, sand layers, and other small scale geologic weaknesses.

The point properties determined by the P-100 probe are compared with the mass media properties determined by the S-100 measurements. These properties include tensile strength, compressive strength, Young's modulus, deformation modulus and creep properties. The material properties determined from the two probes, S-100 and P-100, are always uniquely different due to the scaling effect. The scaling effect is taken into consideration by conducting dimensional analysis on the properties (P) for input to the computer program (R), for which they must be adjusted to specific, individual finite element sizes. Further details of the device and its application are described in another paper presented at this conference [3].

Deformation Measurement System (D)

A set of six basic types of deformation sensing devices was developed especially for the SPDR method. The set is required to verify balance of ground movement as illustrated in the case of the underground opening in Fig. 3. The function of each deformation measuring instrument, in relation to its application, is depicted in the figure and summarized in Table 1.

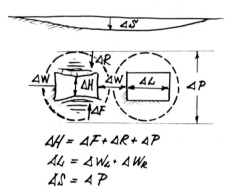

FIG. 3. Basic Balance Relationships Among Deformation Measurements of Room Closure (ΔH), Floor Heave (ΔF), Pillar Squeeze (ΔP), Lateral Room Closure (ΔL), Pillar Width Expansion (ΔW) and Surface Subsidence (ΔS).

$$\Delta H = \Delta F + \Delta R + \Delta P$$
$$\Delta L = \Delta W_L + \Delta W_R$$
$$\Delta S = \Delta P$$

Table 1. Ground Deformation (D) Measurement System

Type	Instrument	Feature	Application
Displacement	Creepmeter (C)	Wire connection	$\Delta H, \Delta L$
Strain	Extensometer (E)	Single and multiple with anchored wires	$\Delta R, \Delta F, \Delta W$
Room closure rate alarm	Closure Rate Meter (CRM)	Multiple pogo sticks, safety monitor alarm	$\dot{H}$
Creep rate	Microcreepmeter	Super resolution (20 μ)	$\dot{H}, \dot{R}, \dot{F}, \dot{W}$
Surface subsidence	Microlevelmeter	High resolution (0.1 μ)	$\Delta S, \dot{S}$
Data acquisition computer	Data collection	Totally computerized data acquis. system	Data coll. & process.

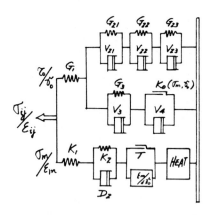

FIG. 4.   REM Constitutive
Model of Generalized Earth
Materials, Capable of
Simulating Wide Range of
Material Types, Including
Fluids, Soils, Sand,
Brittle-Fractured Rocks, and
Yielding Media.

Data from these different probes are collected and analyzed manually
as well as by computer. The data collection computer is capable of
assigning memory to individual transducers so that the computer can
collect data from individual transducers as needed. This eliminates
the need for long distance wiring and frequent data collection trips.
The data-accumulating memory of the transducers is so programmed by
the computer that they can be left alone till the desired time of
data collection.

REM Computer Simulation Program (R)

A special finite element computer simulation program (R) was
developed for the application of the SPDR method [4,5]. The essence
of the program is the REM constitutive model which is schematically
illustrated in Fig. 4. The constitutive model is specially designed
to represent all types of earth materials, including fluids, soils,
sands, soft rocks, hard rocks, ductile-brittle media, brittle
fractured masses and their combination, by properly selecting a set
of property coefficient values for individual geological materials.
ror a given material type, the REM model is capable of simulating
material behaviors, including elasticity, viscoelasticity,
viscoplasticity, strength deterioration, strain hardening,
temperature effect, and volume expansion. The generalization of the
material properties is instrumental in simulating large complex
ground structures using only a small amount of computer time, as
illustrated in the following field application example.

4 SPDR FIELD APPLICATION EXAMPLE IN UNDERGROUND SALT MINE

The SPDR method was applied to eliminate a severe roof failure
problem at the Sifto Salt mine, Goderich, Ontario. The mine had been
operating in a 30 m thick A-2 salt bed in the Michigan salt basin, at
a depth of 600 m under Lake Huron. The conventional room and pillar
method of mining with 15-20 m wide rooms and 55-75 m wide pillars
had been used since the start of mining in 1959. The mine layout as
of 1984 is shown in Fig. 5. After the start of mining, the roof
started deteriorating slowly but steadily. This deterioration
accelerated with the expansion of mining, and by 1982 the roof
failure became very difficult to control by conventional means of

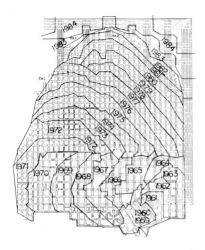

FIG. 5. Plan View of Entire
Layout of Sifto Salt Mine,
Showing Transition of Mining
Method from Room and Pillar to
4-Room Stress Control.

support. At this time, the SPDR method was applied to analyze the
cause of the failure, devise a new design, and to implement it to
eliminate the roof failure problem. Major development stages of the
work performed are summarized below.

Application Scheme of SPDR Method

As a permanent solution to the problem, a 4-room stress-controlled
entry was proposed to entirely replace the traditional room and
pillar method of mining. Implementation of the SPDR method involved
the following 10 steps:

Step	Measure
1	Constructing 4-room test entry, G_1, as shown in Fig. 6;
2	Measuring in situ behavior (S_1,P_1,D_1) of the test entry G_1;
3	Constructing REM computer model using the estimated basic ground condition S_0,P_0;
4	Conducting REM simulation analysis $R(G_1)$ to test in situ measured behavior (S_1,P_1,D_1) against computer projection;
5	Validating the basic ground condition S_0,P_0 through the scheme outlined in Fig. 1;
6	Constructing computer mesh for the proposed design, G_2, for future mining;
7	Conducting REM simulation analysis of $R(G_2)$ to compute S_2,P_2,D_2;
8	Repeating REM analysis of Step 7 with trial geometries $G_3,G_4,...G_n$ until design G_i with optimum design parameters was determined;
9	Implementing the design G_i in the mine;
10	Monitoring the behavior S_i,P_i,D_i of the constructed structure G_i.

Calibration of Basic Ground Condition (S_0,P_0)

As Steps 1 and 2, the test entry G_1 was constructed and the ground
conditions S_1,P_1,D_1 were measured in five test holes, H1, H2, H3, V1,
and V2, as shown in Fig. 6. At the same time, a computer model of the

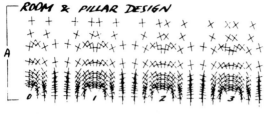

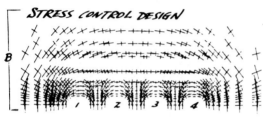

FIG. 6. 4-Room Test Entry Used for In Situ Measurement of Stress States (S_1), Material Properties (P_1), and Deformation (D_1) for SPDR Method of Design Analysis at Sifto Salt Mine

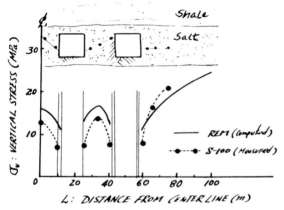

FIG. 7. Comparison of Measured and Computed Distribution of Vertical Stresses in Pillars of 4-Room Test Entry in Sifto Salt Mine.

test entry G_1 was constructed and its simulated behavior (S_1,P_1,D_1) was projected[1] as Steps 3 and 4. The measured S_1 values of the vertical stresses across the pillars and D_1 values of the lateral expansion of the pillar were compared with the corresponding computer predicted values, as shown in Figs. 7 and 8, respectively. In both cases of S_1 and D_1, the agreements between the measured and computer values were found[1] to be sufficiently close to accept the validation of S_0 and P_0 values as Step 5.

New Mine Design Analysis

A new mine design, G_2 was proposed, based on the above study, and constructed as Step 6.[2] Its behavior (S_2,P_2,D_2) was simulated as Step 7. The design was then analyzed through a series of sensitivity analyses with the trial geometries $G_3,G_4,...G_n$. The sensitivity analysis led to the optimum design G_i as Step 8. The optimum design for renewed development of the mine was found to be a 4-room stress controlled entry with 130 m wide abutment pillars. Implementation of

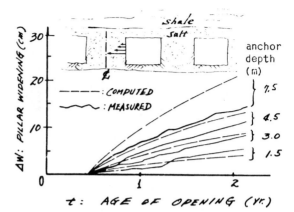

FIG. 8. Comparison of
Field Measurement vs.
Computer Simulation.
Projection of Yield
Pillar Expansion.

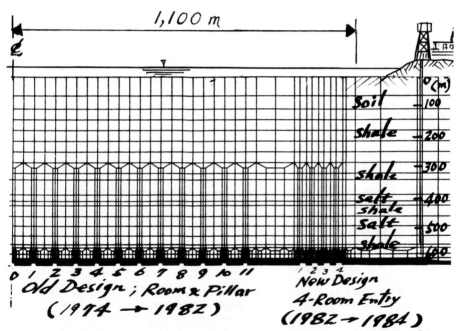

FIG. 9. Finite Element Profile of Sifto Salt Mine Simulating 10
Years of Progressive Excavation (Stress Control Mining Replaces
Conventional Mining in 1982).

the design in plan as well as in profile is shown in Figs. 5 and 9,
respectively.

For the REM simulation analysis $R(G_i)$, the stress distribution
around the old entries of the room-and-pillar design were obtained
and are plotted as the global distribution of principal stresses in
Fig. 10. An arc of stress concentration is formed in the immediate
roof formation over every old room, in the case of the conventional
room-and-pillar design (Fig. 10A). However, the concentrated

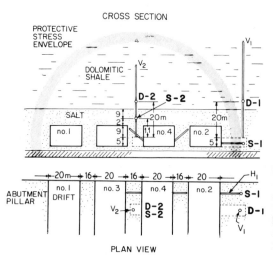

FIG. 10. Comparison of Stress Distribution Patterns Between Old and New Designs Illustrating Mechanism Natural Stabilization of Roof in Stress Control Design.

stresses disappear from the immediate roof over the new 4-room entry designed by the Stress Control Method, as shown in Fig. 10B. Naturally, the reduction of stress means a self-stabilization of the roof strata, especially over the two center rooms. Thus the Stress Control Method helped to enhance the stability of the immediate roof without the need of any artificial roof support system.

5 DISCUSSION AND CONCLUSIONS

Based on the results of the Sifto field study, as well as many other similar studies conducted with the SPDR method, it can be concluded that the method has been very successful in achieving the goals set forth. Some of the salient features of the method are discussed in detail as follows:

Utility of the SPDR Method

The SPDR method is shown to provide a new quantitative means for analyzing, designing, and quality controlling construction of earth structures in complex ground. This method is very useful as it is simple, scientific, and economical. The magnitude of benefit derived from the method is significant, as illustrated by the salt mine example. The design improvement shown in the example is typical of this method, and similar improvements are now being made in other mining industries worldwide.

Applicability to Geotechnical Areas

Although the SPDR method was originated and perfected in the mining industry, it can be applied equally well in solving general geotechnical problems where stricter quality control is desirable. Recently, the SPDR method has been used in a wide range of geotechnical projects, including tunnelling, foundation, slope stability and dam analysis, with promising results.

Computerization of the SPDR Method

The time involved in analyzing the behavior of earth structures using the SPDR method is heavily dependent on the extent of computerization of the analytical techniques. Since the beginning, through early stages of its development, the method has been very time-consuming due to lack of such computerization. Even with partial computerization of those techniques, the work involved in the salt mine example described above took nearly two years to complete. More recently, with rapid computerization and automation of analysis, similar work in coal mines can now be completed in 2-3 months. This indicates that with further improvement in speed and automation of the methodology, it will be possible to apply the SPDR method even in ground where the basic conditions of S_o and P_o are rapidly changing.

Method of Interpretation

The key to the success of the SPDR method is the integration of the four different techniques of stress measurements (S), property measurements (P), deformation measurements (D), and the REM computer program (R). Integration of the above techniques is accomplished by using the field measurement data for calibrating and validating the model of the ground. As a result, a fine-tuned ground model is established which can be used repeatedly to evaluate the responses of various design modifications.

Accuracy of the Method

The accuracy of the S,P,D projections using conventional quantitative methods has always been questionable and therefore earth structures have always been designed with a high factor for safety. The SPDR method promises relatively high accuracy, and its present level of attainable accuracy is found adequate for most engineering applications. Nevertheless, attempts are being made to further improve its accuracy through ongoing research at the SGI laboratory in Richmond, California.

REFERENCES

1. Serata, S., & Kikuchi, S., 1986, "A diametral deformation method for in situ stress and rock property measurement," <u>Int'l Journal of Mining and Geological Engineering</u>, vol. 4, pp. 15-38.
2. Kikuchi, S., Nakamura, T., & Serata, S., 1987, "Double fracture method of stress measurement in hard rock formations," paper to be presented at 2nd Int'l Conf. on Field Measurements in Geomechanics, Kobe, Japan.
3. Chern, S-J, & Serata, S., 1987, "Computerized system for measuring material properties as a function of ground scale," paper to be presented at 2nd Int'l Conf. on Field Measurements in Geomechanics, Kobe, Japan.
4. Sakurai, S., 1966, "Time-dependent behavior of circular cylindrical cavity in continuous medium of brittle aggregate," Ph.D. dissertation, Department of Civil Engineering, Michigan State University, East Lansing.

5. Adachi, T., 1970, "Construction of continuum theories describing behaviors of rock aggregate based on tensor testing," Ph.D. dissertation, College of Engineering, University of California, Berkeley.
6. Serata, S., 1982, "Stress Control methods: Quantitative approach to stabilizing mine openings in weak ground," Proc., 1st Int'l Conf. on Stability in Underground Mining, Vancouver.

ACKNOWLEDGMENTS

The author wishes to express his appreciation to Mssrs. Don Dickie and Al Hamilton of the Sifto Salt mine of Domtar Chemical, who planned and managed the mine work with vision and vigor, to final completion. The author is also indebted to Mssrs. Shinji Kikuchi and Tetsuye Nakamura of Japan Development Corp. (Kokudo Kaihatsu K.K.) who carried out the stress and property measurement portion of the Sifto mine study. The author also wishes to acknowledge the innovative contributions of Professors Shunske Sakurai and Toshihisa Adachi during the original development of the REM constitutive model while working on their Ph.D.s at Michigan State University and the University of California, respectively.

2nd International Symposium on Field Measurements in Geomechanics, Sakurai (ed.)
© 1988 Balkema, Rotterdam. ISBN 90 6191 778 6

Field study of rock damage around a gallery

P.Egger
Rock Mechanics Laboratory, Swiss Federal Institute of Technology, Lausanne

1. INTRODUCTION

Unavoidably, underground excavations mean stress relief and more or less marked rock damage near the wall. Whereas in civil works, increased deformations and reduced stability are the main concern, for the disposal of toxic or radioactive waste, the interest is focused on the permeability of the damaged rock zone which may cause preferential seepage channels.

Intensity and depth of the rock damage depend on parameters such as geological and hydrogeological conditions, virgin stresses and shape of the cavity as well as excavation and support methods.

The present paper describes a field study performed in the Grimsel Underground Rock Laboratory owned by the Swiss National Agency for the Disposal of Radioactive Waste.

2. THE GRIMSEL UNDERGROUND ROCK LABORATORY

This Laboratory is situated in Central Switzerland, in compact crystalline rocks such as granites and granodiorites, at a depth of 450 m. Generally the discontinuities of the rock mass are widely spaced, but there exist more or less mylonitized shear zones.

Figure 1 shows the system of galleries forming the Underground Laboratory, 3.50 m in diameter and excavated by a TBM. These galleries start from the access tunnel to the Grimsel II Underground Power Station [Müller-Salzburg and Egger, 1982] and permit a comprehensive test programme to be carried out [Lieb, 1985].

3. STUDY OF THE ROCK DAMAGE

3.1 Geological Conditions

The Rock Mechanics Laboratory of the Swiss Federal Institute of Technology in Lausanne was in charge of the design and execution of the test programme aimed at defining and quantifying the rock damage caused by a mechanical tunnel excavation. The test zone is entirely situated in granodiorite (Fig. 1 and Fig. 2) with a uniaxial compressive strength of approximately 120 MPa. The rock mass structure is characterized by a subvertical schistosity, nearly normal to the axis of the test drift

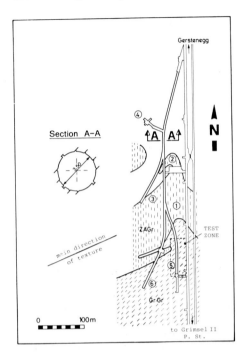

Fig. 1. Layout of the Grimsel
Underground Rock Laboratory

1.. Migration Test
2.. Offices and Shops
3.. Heater Test
4.. Radial Flow Test
5.. Rock Damage Study
6.. Ventilation Test
ZaGr.. Granite
GrGr.. Granodiorite

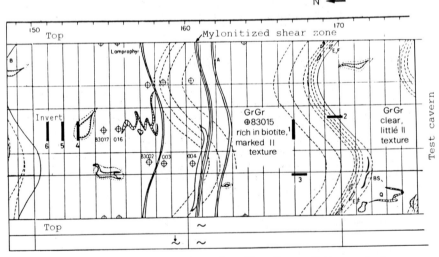

Fig. 2. Geological map of the tunnel wall – Zone of the rock damage
study

and locally concentrated in shear zones. The hydrogeological conditions
were studied in advance by means of a series of 100 m long horizontal
boreholes. In the test zone, the water pressure was 1.0 MPa and its
flow rate 0.17 ℓ/min or 1.7 cm^3/min per open joint.

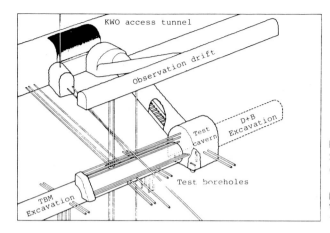

Fig. 3.
Schematic layout
of the boreholes
(parallel and
perpendicular to
the test drift)
for the field tests

3.2   Testing Concept

Due to the sound rock conditions and the moderate overburden, the deformations and rock damage were anticipated to be very limited. Therefore, the test methods had to be extremely precise and reliable. In
addition, it was decided to perform the tests **before** and **after** the
excavation of the test drift whenever possible, in order to be able to
distinguish between local geological variations and the effect of the
tunnel excavation. This concept required the execution of a complete
pattern of boreholes before the test drift excavation (Fig. 3).

The second basic idea for the test programme was the study of the
scale effect, e.g. for the evaluation of the rock deformability the
size of the tested samples was varied from a small laboratory sample to
the rock mass around the test galleries [Egger, 1976].

In order to enable the interpretation of the field measurements, it
was also necessary to determine the virgin rock stresses.

3.3   Virgin Stresses in the Rock Mass

The evaluation of the virgin rock stresses was done at three different
scales : overcoring, stress relief and recompression of 1 square meter
slots and monitoring the displacements caused by the excavation of the
test drift. The results obtained from these three methods show surpri-
singly good agreement, namely a maximum stress thas was subhorizontal
and in the NW-SE direction, of approximately double the overburden
pressure. They confirm earlier measurements performed for the Grimsel
II Power Station [Descoeudres and Egger, 1977]. Perpendicularly to the
test drift, the K_0-value is about 1.5. Figure 4 compares the measured
displacements with the theoretical predictions, showing the best agreement for $K_0 = 1.5$ and $E/q_v = 3'000$ which means $E \simeq 35$ GPa.

3.4   Displacements

The displacements caused by the tunnel excavation were systematically
measured by means of a Distometer device, generally showing values of
about 1 mm. These values agree with those predicted for a homogeneous
elastic rock mass, thus proving the absence of any noticeable rock
damage.

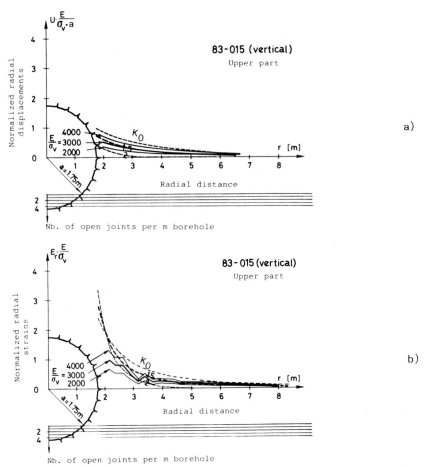

Fig. 4. Vertical borehole 83-015 : Measured (———) and forecast (----) values of the displacements, as function of K_0 and of E/σ_v

In order to know the displacements within the rock mass as well, a number of boreholes crossing the planned test drift were equipped with measuring pads for the "Sliding Micrometer". Figures 5 and 6 show the displacements along a vertical and a subhorizontal borehole at several stages of the tunnel excavation. The number of open joints per meter of borehole is also shown in these figures. In the upper part of the vertical borehole (compact rock, no joints), the maximum displacement is 0.3 mm at the tunnel wall; the lower part of this borehole crosses a shear zone, and the displacements come to a stop after the crossing of the TBM (stages C, D of Fig. 5).

The subhorizontal borehole (Fig. 6) is situated in the immediate neighbourhood of this shear zone, crossing it along several meters at both sides of the test drift. In addition, it crosses the highly laminated zone found in the drift at Sta. 170, at a depth of between 10 and 15 m.

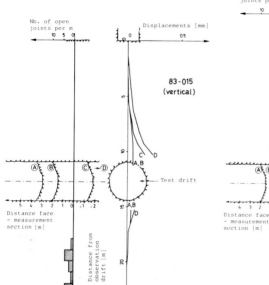

Fig. 5. Vertical displacements-
"Sliding Micrometer" results
(borehole 83-015)

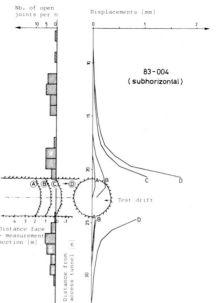

Fig. 6. Subhorizontal displacements-
"Sliding Micrometer" results
(borehole 83-004)

The measurements show a sharply increased mobility of the rock mass
situated between the laminated and the shear zone, e.g. 1.8 mm displacements at the East wall of the drift, compared with only 0.8 mm at the
opposite side.

3.5 Velocity of elastic waves

In order to characterize the rock mass, the P-wave velocities (v_p)
were measured systematically using the Cross-Hole-Method between parallel boreholes spaced at 60 cm or 1.0 m. This method allows detection of anticipated variations of v_p along a distance of 10 cm near
the test drift. The measurements were performed with the help of specially conceived watertight probes containing air jacks aligned with
the transducers which pushed the latter firmly against the borehole
wall.

The main measurements were carried out before and after the tunnel
excavation; they were completed by additional measurements between
short boreholes drilled from the test drift.

The P-wave velocity of the virgin rock mass has been found to vary
between 4.6 and 5.7 km/s, but lies essentially between 5.0 and 5.3 km/s
(Fig. 7, continuous line). The lower values hold for the shear or laminated zones. The rock texture causes a slight anisotropy of v_p, not
exceeding 7 percent.

The tunnel excavation results in the decrease of v_p, particularly
in the shear zones (Fig. 7, dashed line). The reduction of v_p reaches
1 km/s or 20% of the original value; the effect of the tunnel excavation extends to a depth of 1.20 m.

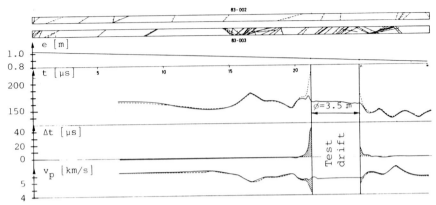

Fig. 7. P-wave velocity vp between the subhorizontal boreholes
83-002/3 (perpendicular to the test drift)

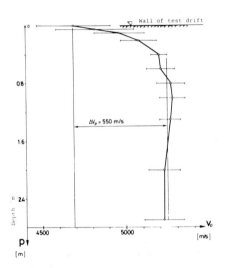

Fig. 8. P-wave velocity vp near
the wall of the test drift.
Average of 8 pairs of boreholes

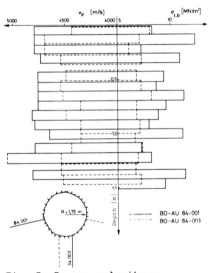

Fig. 9. P-wave velocity vp
of unloaded samples and indirect
tensile strength $\sigma_{t,b}$ vs. depth

    Away from the shear zones, this influence is limited to about 60 cm
(Fig. 8), the reduction of vp at the tunnel wall being 0.55 km/s or
10.5% on the average.
    Laboratory measurements of the P-wave velocities of core samples
yielded two important results : no influence of the depth, either on
vp in the unloaded state, or on the indirect tensile strength (Fig.
9). At first sight, these results seem to be in contradiction with the
field measurements (e.g. Fig. 8), but vp has been found to increase
noticeably with the stresses applied to the samples. From this, it may
be concluded that the reduction of vp near the tunnel wall is caused
by the stress redistributions following the excavation rather than by
damage of the rock.

3.6   Rock Deformability

The deformability measurements performed on samples of various sizes
showed a negligible scale effect, thus confirming the soundness of the
rock mass.
    The study of the rock damage caused by the tunnel excavation was done
by two methods. The first method was by means of a borehole dilatometer, before and after the tunnel excavation. The tests were performed
in several boreholes, at a maximum distance of 1 meter from the tunnel
wall. They were run by repeated loading-unloading cycles and yielded
average E-moduli of 29 GPa at loading and 39 GPa at unloading, for the
intact rock. The tunnel excavation reduced both moduli by approximately
15%. The second method determined the rock mass deformability by means
of Large Flat Jacks (A $\simeq$ 1 square meter). This method enabled the evaluation of the E-moduli depending on the depth and on the orientation
with respect to the rock texture. For the loading steps from 0.5 to
8 MPa (E_1) and from 8 to 14 MPa (E_2) the following deformation moduli
were found (in GPa) :

Depth [m]	$\perp$ to the texture (slot 1)		‖ to the texture (slot 2)	
	E_1	E_2	E_1	E_2
± 0	--	--	19.5	21.7
0.37	21.4	30	25.0	41.6
0.74	31.6	40.4	29.0	48.0

    In addition to a slight anisotropy which decreased with depth, a con-
siderable reduction of the E-moduli near the surface (= tunnel wall)
was measured; these results are in good agreement with the reduction of
v_P (Fig. 8) in the same zone.
    The deformability of the rock mass at a large scale was evaluated by
the comparison between the measured displacements around the test drift
and the theoretical forecast (Fig. 4); as it was pointed out in § 3.3,
the best fit was found for E values of approximately 35 GPa.

3.7   Permeability

For repositories of radioactive waste it is of the utmost importance to
know whether or not underground excavations necessarily cause an increase in rock permeability. In the technical literature, papers can be
found upholding this assumption [e.g. Kelsall et al., 1984] as well as
expressing contrary opinions [Wilson et al., 1983]. To the best of the
author's knowledge, all described cases dealt with cavities excavated
by drill and blast, whereas the galleries of the Grimsel Laboratory had
been driven by a TBM.
    The rock mass permeability was evaluated using a double approach :
    • The **joint permeability** was determined by borehole water tests (Lu-
geon tests) before and after the excavation of the test drift. Because
of the small absorption, a special device permitting the measurement of
flow rates on the order of 1 cm³/min had to be conceived.

The tests performed in a borehole parallel to the drift at 60 cm from its wall generally yielded flow rates of 2 to 4 cm^3/min (for 1 MPa water pressure) before and 4 to 5.5 cm^3/min after the tunnel excavation. A theoretical flow model showed that this increase could be explained by the modified border conditions (tunnel wall = zero potential surface) alone and did not require a higher permeable rock zone near the excavation. Only the tests executed in the shear zone near Sta. 160 caused the fines to be washed out, entailing a considerable increase of the flow rate.

For a detailed study of the immediate neighbourhood of the gallery, particularly short mechanical packers were used instead of the inflatable packers used elsewhere. This study confirmed the absence of an increased permeability of the joints near the drift with the exception of the shear zones.

The flow rates absorbed during the tests agree remarkably well with the flow rates of the investigation borehole (1.7 cm^3/min per joint) situated in the test zone. This proves the good reliability of the measuring device used.

• The **permeability of the intact rock** was determined, on the one hand, in the laboratory on samples taken at various depths. Both radial flow tests (hollow samples) as well as axial ones (disks) showed similar permeabilities on the order of $6 \cdot 10^{-12}$ m/s, independent of depth. On the other hand, a field test was performed between three parallel slots of about 1 m^2 area (Fig. 10). These slots were executed by means of a circular diamond saw (S4 to S6, see also Fig. 2) at 1 m mutual distances. A concrete slab was poured on the surface and measures were taken to prevent any seepage at the rock-concrete interface.

Water pressure was applied stepwise up to a maximum value of 0.45 MPa, to the central slot S5 and seepage took place towards the open lateral slots S4 and S6. A combined pressure controlling and flow rate metering system enabled the tests to be carried out. Figure 11 shows an excerpt from the test results which yielded a permeability of the 2 m^3 rock mass in question of $k = 3.3 \cdot 10^{-12}$ m/s when considering the real curved seepage paths (for an isotropic medium) or $k = 5 \cdot 10^{-12}$ m/s for

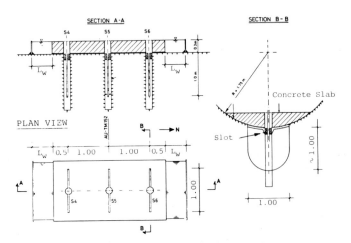

Fig. 10. Layout of the water test between 3 parallel slots (S4-S5-S6)

an idealised parallel flow. In addition, the flow rate vs. pressure curve did not pass through the origin of the coordinate system, but crossed the pressure axis at a negative value of −40 kPa. Moreover, the water level in the standpipes installed at the lateral slots decreased steadily, in spite of the flow arriving from the central slot. It was found that evaporation at the tunnel wall, combined with capillary forces, was responsible for the observed phenomenon. When artificial water ponds of a variable length, L_w, were created at both sides of the concrete slab (see Figs. 10 and 11), the flow rate leaving the standpipes at the lateral slots diminished accordingly, whilst the flow from the central slot remained unaffected. The same phenomenon had also been observed in the laboratory tests where the evaporation acted like a negative pressure of −35 to −50 kPa.

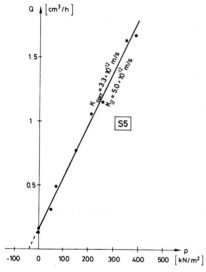

Fig. 11. Water test between parallel slots - flow rates vs. pressure and vs. time; influence of the length L_w of the lateral ponds

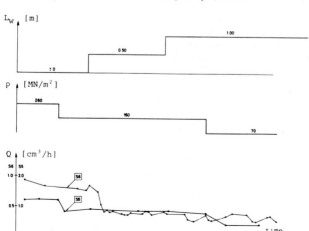

4.   CONCLUSIONS

The field study of the rock damage caused by the excavation of a
gallery, described above, allows the following conclusions to be
drawn :
  • Mechanical rock damage which could have jeopardized the safety of
the workers or the tunnel stability never did occur. Only 10 m of the
800 m total drift required a light support with rock bolts and wire
mesh; this was at the crossing of a narrowly jointed zone under a very
acute angle.
  • Rock damage in a larger sense was detected in the shear zones, particularly near Sta. 160 : with a 1.8 mm, the displacements of the East
wall of the drift are clearly higher than elsewhere. It appears that
these displacements and the erosion of fines (see Section 3.7) are
linked to heavily tectonised zones. In these particular zones, certain
rock damage has doubtlessly occurred.
  • However, aside from these disturbed rock zones, the field and laboratory tests have shown that neither the strength nor the permeability
of the rock were affected by the excavation of the test gallery. The
measured reductions of the P-wave velocity and of the deformation modulus, some 10 to 20% at the tunnel wall, can be explained by the
stress redistributions. In particular, the water tests performed in
boreholes, between slots and in the laboratory have yielded remarkably
constant values before and after the tunnel excavation and over the
investigated depths, respectively.
  Nevertheless the rock damage observed in the rare shear zones and the
lack of knowledge of a possible connection between these systems require a considerable increase of the scale of the tests in order to
determine the preferential seepage paths within the rock mass. For
example, the rock volume under investigation could be considerably
increased by an "Hydraulic Chamber" such as currently used for the design of pressure tunnels. Thus, the water tests performed for the present field study could be completed, to greater advantage, by such a
system.

5.   BIBLIOGRAPHIC REFERENCES

Descoeudres F. and Egger P. : "Monitoring System for Large Underground
  Openings - Experiences from the Grimsel-Oberaar Scheme". Proc. Int.
  Symp. Field Meas. in Rock Mech., Balkema, pp. 535-549, Zurich, 1977.
Egger P. : "Le rôle des mesures et auscultations dans la construction
  souterraine". Doc. SIA No 12, pp. 33-43, Zurich, 1976.
Kelsall P.C., Case J.B. and Chabannes C.R. : "Evaluation of Excavation
  - induced Changes in Rock Permeability". Int. J. Rock Mech. Min. Sci.
  (21), No 3, pp. 123-135, 1984.
Lieb R.W. : "Grimsel Test Site. Overview and Test Programs". Nagra
  Technical Report 85-46, Baden, 1985.
Müller L. und Egger P. : "Standsicherheit und Bau von Kavernen in Ab-
  hängigkeit von den geologischen Bedingungen". Proc. Symp. ISRM Rock
  Mech., Caverns and Pressure Shafts, vol. 3, pp. 1179-1191, Aachen,
  1982.
Wilson C.R. et al. : "Large-scale Hydraulic Conductivity Measurements
  in Fractured Granite". Int. J. Rock Mech. Min. Sci. (20), No 6, pp.
  269-276, 1983.

Performance and results of geomechanical measurements in a rock salt gallery

M.Haupt & O.Natau
Institut für Bodenmechanik und Felsmechanik, Universität Karlsruhe, FR Germany

1 INTRODUCTION

In a German rock salt mine a gallery was drifted in order to investigate the viscous and the time independent stress-strain-behaviour in the near-field of the opening and to determine the qualification of field measurement methods for the application in rock salt. The gallery is situated in flat embedded rock salt in a depth of about 700 m. Due to the great distance to the excavation districts influences by large-scale stress redistributions could be excluded. The length of the gallery amounted to 110 m.

Since there is anhydrite embedded beyond the rock salt and updomes were observed at other sites of the mine it was taken into account that anhydrite could hurt the drifting of the gallery or the measurement results. Really at a gallery length of 75 m anhydrite was encountered at the heading face and consequently the direction of driving was changed. In some boreholes directed perpendicular to the gallery axis anhydrite was met, too.

Driving of the gallery took place by blasting. First by an advance per round of about 7 m a cross-section of 3m*6m was drifted and subsequently this part was enlarged by smooth blasting to a final shape of 4m*7m.

2 INSTALLATION OF MEASURING DEVICES

Before the beginning of drifting a short transverse gallery was excavated. In this transverse gallery 8 horizontal boreholes with a length of 15 m were made parallel to the measuring gallery being planned (fig. 1). At the bottom of each borehole 2 hydraulic pressure cells directed perpendicular to each other were fitted to measure tangential and radial stresses. Afterwards the borehole was filled with a special rock-salt-like mortar. By this method an entire contact arised between the hard-inclusion cells and the surrounding rock masses in time due to the viscous properties of the rock salt.

Furthermore before the beginning of drifting five 60 m long horizontal boreholes were made with distances of 2 m to the top and to the side walls of the measuring gallery (fig. 1). At the bottom of the boreholes 3o m long multi-point deflectometers were placed to measure radial displacements.

In the gallery 4 convergence measuring profiles were disposed with 10 measuring points connected by 18 measuring distances, respectively. With

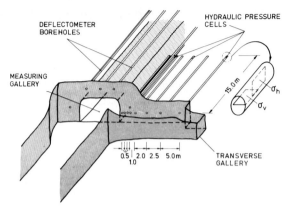

Fig. 1: Transverse gallery with deflectometer boreholes for stress
        measurements

the purpose of determination of over- and underprofile and for control
of rock fall profile measurements were performed at the 4 convergence
measuring sections using a laser beam profile measuring instrument.
Moreover at 2 measuring profiles 8 extensometer stations were installed,
respectively. Each extensometer station consists of a mini-extensometer
with measuring points at 0.5/1.0/2.0 and 4.0 m and a rod extensometer
with measuring points at 8.0 and 16.0 m. The arrangement of these
measuring devices not only allowed the determination of deformations at
the gallery surface but also the investigation of the displacement field
in the near-field of the gallery.

The loosening zone and its temporal and spatial change was
investigated by different methods. At 3 measuring profiles each
consisting of 16 boreholes being 5 m long and disposed at the floor, the
top, the side walls and the corners of the cross section optical
probings were carried out using a special rigid and collapsible optical
sonde with a diameter of 22 mm. At 2 measuring profiles each consisting
of six 5 m long boreholes placed at the side walls and the floor of the
gallery permeability tests with air were executed with a self-
constructed device. For this purpose a 0.5 m long segment of the
borehole was sealed off by a double-packer sonde and supplied with a
pneumatic pressure of 1.0 bar. Subsequently the air intake was stopped
and the decline in pressure was monitored for a time of 30 minutes. This
process was carried out at different borehole depths and repeated in
time. At one measuring profile dilatometer tests were performed using a
LNEC-borehole dilatometer. By this equipment deformations could be
measured in 4 directions radial to the borehole axis.

3 EVALUATION AND RESULTS

3.1 Stress measurements

Stress measurements are devided in 2 stages: First the spontaneous, time
independent stress redistributions in the surroundings of the heading
face caused by drifting of the gallery are investigated and afterwards
the viscous changes of stress distribution in the near-field of the
gallery will be considered.

In the first part viscous effects are not taken into account. It was observed that the enlargement of the gallery by smooth blasting did not cause any considerable stress redistributions. This leads to the conclusion that the zone taken off by the enlargement blasting must have been weakened and disturbed and hence had not been able to be stressed. By this result the positive effects of enlargement by smooth blasting are confirmed: A disturbed zone is taken off without a debiliation of the rock mass caused by repeated stress redistributions.

Fig. 2 shows the distribution of tangential and radial stresses at the side wall depending on the distance from the heading face. In contrast to reality the heading face is fix and the measuring profiles vary in the depiction. The stress is plotted dimensionless as σ/σ_0 where σ_0 is the "primary stress" measured before beginning of drifting.

It results from the depiction that neither the radial (horizontal) stresses nor the tangential (vertical) stresses at a distance of about 7.5 m behind the heading face are influenced noteworthy by driving the gallery. Even close by the heading face radial stresses hardly show any changes. Obviously stress redistributions only occur in the closer surroundings of the heading face. No change of radial stresses can be noticed more than 7 m in front of the heading face. Almost the same concerns to tangential stresses. Sponteaneous stress redistributions only can be observed within a region of about 1 gallery diameter to all sides of the heading face.

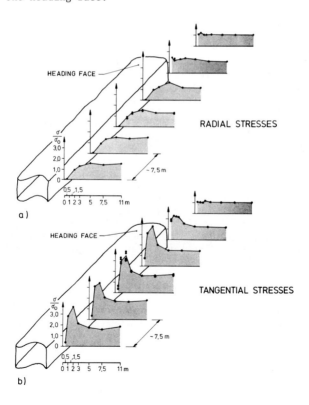

Fig. 2: Radial (a) and tangential (b) stresses at the side wall

Considering the stress distribution dependent on the radial distance from the side wall it can be noticed that the radial stresses behave quite normal. They increase continuously and reach the initial value of $\sigma/\sigma_0 = 1$ at a distance from the side walls of 2-3 m already. The tangential stresses exhibit a clear maximum at a distance of 2 m. It is a remarkable observation that the tangential stresses nearby the side walls are very low. This is an indication of the presence of a small disturbed zone without a load-bearing capacity though most of the loosening zone had already been taken off by the smooth enlargement blasting.

The second part of the evaluations deals with the time-dependent stress redistributions within the observation time of more than 1300 days. Though the results are not as clearly as those of the spontaneous stress redistributions it could be ascertained that the maximum tangential stress nearby the gallery decreases in time. The horizontal stresses remained constant in the near-field and grew insignificantly at a greater distance from the side wall. This increase of radial stresses was due to the fact that the contact between the pressure cells, the mortar and the rock masses slowly grows stronger in time if there is no considerable rise of stress. Both increase and decrease of stresses occur with decreasing stress rates which means that stresses nearly were constant at the end of the observation period.

3.2 Displacement measurements

Due to blasting the extensometer rods got out of place and hence the measuring results hardly could be taken in to account. Though the producer repeated had installed multi-point deflectometers on similar conditions without any complaint it must be concluded that this kind of deflectometers is unsuitable for an application in the near-field of a gallery in rock salt drifted by blasting.

Due to their low accuracy of $\pm$ 1 cm profile measurements are not able to substitute precision measurements as for example convergence measurements. Firstly they were carried out to compare the ideal and the actual cross-section of the gallery whereby the ideal profile was a 4m*7m-rectangular with rounded corners. As could be proved the ideal profile can be reached very good by smooth enlargement blasting.

The second purpose of the profile measurements was to control rock fall. In fig. 3 one measuring profile is shown dependent on time. Each single profile is composed of the initial and the actual profile. The bottom of the gallery was covered by loose salt and therefore often changed its shape. It should be ignored in the depiction. In this measuring profile a few rock falls occured at the top (accentuated by hatching in the depiction). No dependency between the intensity of rock fall and time could be observed. Hence in an unsupported gallery in rock salt rock falls may happen at any time.

Convergence measuring curves are represented in a linear and in a logarithmic scale in fig. 4. It can be perceived that the convergence rates are very high at the beginning. The convergence rate decreases continuously without reaching a constant value. Obviously convergence proceeds with a power law

(1) $$e_c = (t/t_0)^a$$

whereby e_c: convergence,
t, t_0:   time and
a: constant, $0 < a < 1$

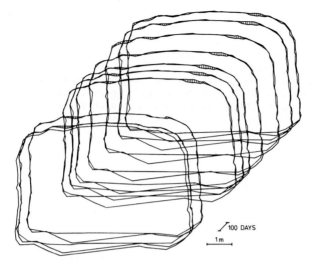

Fig. 3: Profile measurements

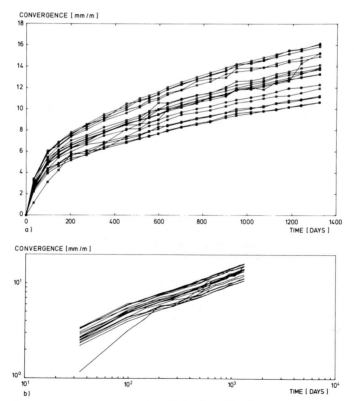

Fig. 4: Convergence measurements: Relative displacements
in a linear (a) and a logarithmical (b) scale

and will not stop in a not-too-distant future. However, this equation is not valid for laboratory creep tests because it includes relaxation.

The arrangement of the measuring points and the measuring distances in connection with survey measurements permits the determination of displacements of each measuring point within the measuring plane dependent on time. The result is shown in fig. 5 for one of four measuring profiles. Essentially the directions of the displacements are radial. Remarkable horizontal components can be observed at the exterior measuring points at the bottom. They indicate buckling or arching of the bottom. Really this occurrence was found in other galleries of the mine, too.

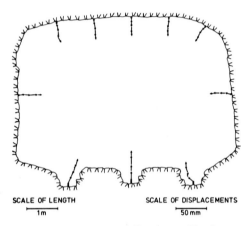

SCALE OF LENGTH   SCALE OF DISPLACEMENTS
1m   50 mm

Fig. 5: Convergence measurements: Absolute displacements after
        50, 200, 400 and 1200 days

The measurements also demonstrated the influence of the gallery shape on the distribution of displacements. Rounded corners cause displacements dispersed almost uniformly over the shape. Using simplifying assumptions the convergence of plane can be estimated as

$$(2) \qquad\qquad (\Delta A/A_0) = 2e_c$$

whereby $\Delta A/A_0$: convergence of plane and
        e_c: convergence (see equ. 1).

By means of extensometer measurements relative displacements between the head of the extensometer at the skin of the gallery and the extensometer measuring points fixed at different depths in the borehole are measured. Absolute displacements can be determined by the assumption that the farest extensometer points at 16.0 m do not change their position. This assumption was confirmed in connection with convergence and survey measurements.

Displacements at the gallery skin unavoidably are very similar to those determined by the aid of convergence measurements. As it is depicted in fig. 6 in the near-field of the gallery time dependent radial displacements all develop similar. A stationary state is not reached.

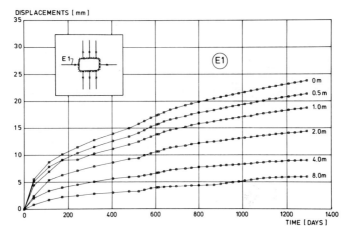

Fig.: 6: Extensometer measurements

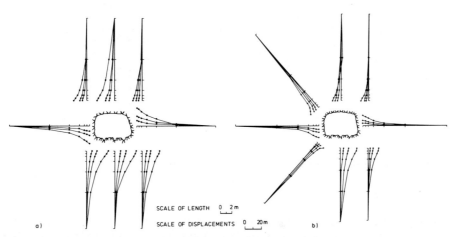

Fig. 7: Extensometer measurements: Displacements after
50, 200 and 800 days

The spatial propagation of radial displacements is represented in fig.
7. On account of better clearness the extensometers are displaced
somewhat in the depiction. Radial displacements decrease considerably
with increasing distance from the gallery but increase in time. On the
basis of fig. 7 lines of identical radial displacements can be
constructed as depicted in fig. 8. These lines demonstrate the time
dependent development of the displacement field in the near-field of the
gallery. Though there is an approximately uniform distribution of the
displacements at the gallery skin as a result of the convergence
measurements fig. 8 shows a nonuniform distribution in the rock masses.
The displacement gradient in the corners of the gallery is considerably
higher than at the sides though the corners are rounded. Hence the shape
of the gallery distinctly takes effect on the displacement field in the
near-field.

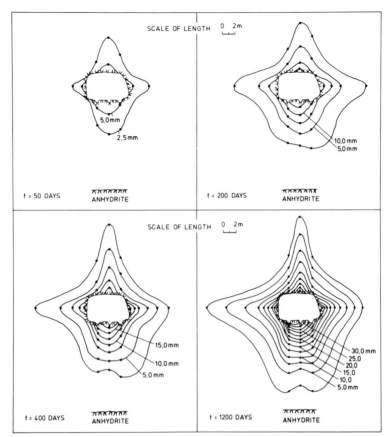

Fig. 8: Extensometer measurements: Lines of identical displacements

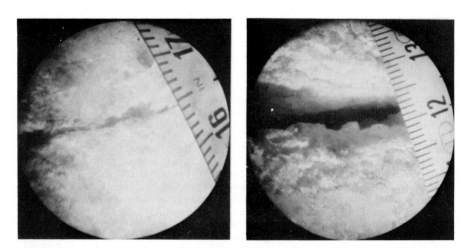

Fig. 9: Cracks observed in boreholes

3.3 Investigation of the loosening zone

The propagation of cracks indicating the intensity of the loosening zone
was investigated by optical probings. Cracks exclusively occurred in
tangential direction. The width was between a fraction of millimeters
and some millimeters (fig. 9). No cracks were observed more than 1 m
away from the skin of the gallery within the observation time. Cracks
were distributed uniformly around the profile. It could be prooved that
most of the cracks emerged in time and hence were not caused directly by
blasting. Rather they propagate due to time-dependent deformations and
stress redistributions.

Other indications of a loosening zone were expected by permeability
tests. Essentially permeability and dilatometer measurements described
following are not used to determinate characteristic values of the rock
salt but to gain comparison values for the investigation of the spatial
and the time dependent propagation of a loosening zone. Permeability
tests mainly were performed in borehole-depths of 1.2 and 4.0 m. As
expected permeability decreases with increasing distance from the
gallery and on principle there was no change in time. Fig. 10 shows the
time dependent increase of permeability measured within 6 boreholes in a
depth of 1.2 m. It is a remarkable observation that the permeability
increased in time not only nearby the gallery but at a distance of 4.0
m, too. That means that the loosening zone both intensifies and spreads
continuously in time. On account of the results of the extensometer
measurements it must be deduced that the expanding displacement field is
responsible for the emergence of very small cracks which can not be
detected with the naked eye but considerably enhance the permeability.

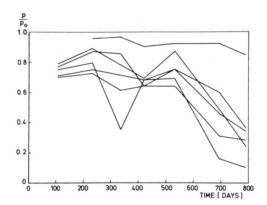

Fig. 10: Permeability measurements in different boreholes at a
borehole-depth of 1,2 m;
$P_o = p(t = 0) = 1$ bar, $p = p(t = 30$ min$)$

Since dilatometer tests only were performed at few points of time no
time dependency of the stress-deformation-behaviour could be determined.
It was found out that the deformability in longitudinal direction of the
gallery is lower than tangential to the cross-section. There was no
dependency of deformability on the distance to the gallery skin and no
dependency on the stress level. Another result is shown in fig. 11. As
other characteristic values the deformation moduli in the depiction

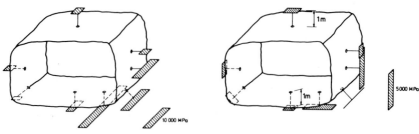

a) MODULI OF DEFORMATION PARALLEL TO GALLERY AXIS

b) MODULI OF DEFORMATION IN TANGENTIAL DIRECTION

Fig. 11: Dilatometer measurements: Moduli of deformation

should not been considered as material parameters but as comparison values. There is a remarkable decrease of deformability – which is an increase of the deformation modulus – at the lower corner on the right side of the profile. In this region anhydrite comes close to the gallery and it must be supposed that anhydrite intercalations reducing the deformability are responsible for this observation.

4 CONCLUSIONS

The measurement results show the importance of displacement measurements not only at the skin of the gallery but also in the near-field. By knowledge of the displacement field in the surroundings of the gallery numeric calculations can be verified much better than with the aid of convergence measurement curves only. A broad investigation of the loosening zone is possible by means of optical probings and permeability measurements by inclusion of stress measurements. These measurents allow conclusions on the time dependent permeability around the gallery profile which for example is important for underground dams. In any case extensive results can be achieved not by using single measurement methods like convergence measurements but only by evalution of various different measurement techniques.

2nd International Symposium on Field Measurements in Geomechanics, Sakurai (ed.)
© 1988 Balkema, Rotterdam. ISBN 90 6191 778 6

Study on the permeability change of rock mass due to underground excavation

I.Motojima
Central Research Institute of Electric Power Industry, Japan

1 INTRODUCTION

Clarification of the loosening zone in a rock mass around large-scale caverns is an important theme from the viewpoints of safety during the actual excavation process, and the long-term stability of the cavern. The permeability variation caused by this loosening is also an important factor in the design of underground drainage systems around the cavern. Also, since permeability depends on cracks and joints in the rock mass, it gives useful information on the local conditions of the rock mass, in terms of continuity and the opening of crack or joints.

In this paper, during the excavation of caverns for underground power stations, the permeability of the rock mass around the cavern was measured, and studies on loosening zone of the rock mass were then made based on the results of these measurements.

2 MEASUREMENT METHOD AND EQUIPMENT

Measurements were made using the equipment developed by our organization to investigate underground water. The purpose of using this equipment was to measure the permeability and pressure of underground water in bore-holes rapidly and accurately. The main properties are shown in Table 1.

The permeability was measured by the constant pressure method, and the maximum pressure was made at 1 - 2 kg/cm^2 to prevent destruction of the rock mass around the bore-holes.

Permeability measurements were made in bore-holes at intervals of 2 m The initial measurement was performed before the excavation, and repeated 4 - 6 times throughout the period of excavation.

Permeability is expressed using Lugeon conversion values. Its overall accuracy is ±10%.

3 RESULTS OF MEASUREMENTS

3.1 Case of Tk Power Station Site

The Tk Power Station is an underground facility of the water pumping

Table 1. Properties of Groundwater Measuring Equipment

Composition	Properties	
Bore-hole Probe	Maximum Outer Diameter:	55 mmø
	Packer Form:	Air Packers made of gum tube
	Length:	1 m
	Maximum Pressure:	12 kg/cm^2
Pressure-Transducer	Maximum Measuring Range:	10 kg/cm^2
	Thermal Characteristics at 0°C:	±0.001% F.S./°C
	Thermal Sensitivity Characteristics:	±0.0015% F.S./°C
	Hysteresis:	±0.035% F.S.
	Measurement Error:	±0.020% F.S.
Electric Flow Meter	Transducing Type:	Float and Differential Trans Type
	Range of Flow Rate:	10 - 150 cc/min
		100 - 1500 cc/min
	Accuracy:	±0.1% F.S.
Self-Recorder	Recording Type:	Point Marking
		Marking Time Intervals 3 sec.
	Paper Speed:	10 - 600 cm/hr
	Indication Accuracy:	0.3% F.S.
Digital Tester	Maximum Figure:	19999
	Maximum Disintegration:	1 μV (DCV)
	Measuring Accuracy:	±0.025% of rdg
	Input Impedance:	10 MΩ

type and has a capacity of 1,280 MW. This facility was constructed inside an underground cavern with a height of 55 m, width of 27 m and length of 165 m.

(1) General description of the measurements

Measurements of permeability were made utilizing two horizontal bore-holes bored from adit to middle height of the cavern as shown in Fig. 1. Measurements were made 6 times during the excavation of the cavern, which covered a period of 1 year and 5 months.

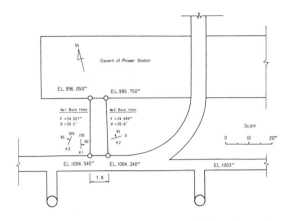

Fig. 1   Measurement Bore-Hole Layout

Table 2.   Appearance of Bore-Holes before Cavern Excavation

	Hole No. 1	Hole No. 2
Number of Joints	92	83
Joint Frequency	4.0 (Jst/m)	3.6 (Jst/m)
Average Permeability	0.79 (Lugeon)	0.23 (Lugeon)

   The rock mass at the site of the measurements was fresh and hard granite.   Appearances of bore-holes before excavation of the cavern are shown in Table 2.

(2) Variations of permeability

Before excavation started, the average permeability was 0.79 Lugeon at Hole No. 1, and 0.23 Lugeon at Hole No. 2.   After completion of the excavation, however, permeability was 28.8 Lugeon at Hole No. 1 and 18.1 Lugeon at Hole No. 2.   Near the walls of the cavern, permeability increased from 0.03 - 0.1 Lugeon (before excavation) to 30 - 55 Lugeon (after completion).   Fig. 2 shows typical results of the measurements.
   The ratio of permeability after excavation to before is 36.4 times for Hole No. 1 and 78.8 times for Hole No. 2.   At the zone of approximately 5 m from the cavern wall, the ratio is extremely large, reaching as high as 100 times.   However, it decreases as the distance from the cavern increases, and at a distance of approximately 20 m it is about 10 times.

(3) Relationship between permeability variations and excavation

As the excavation lift reaches the same elevation as the
measuring hole, the permeability of the zone 5 m from the cavern wall
begins to increase abruptly.   Afterward, permeability at a more distant

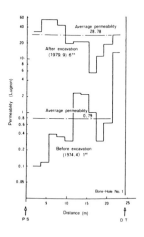

Fig. 2.1
Result of Permeability
Measurement

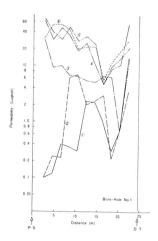

Fig. 2.2
Result of Permeability
Measurement

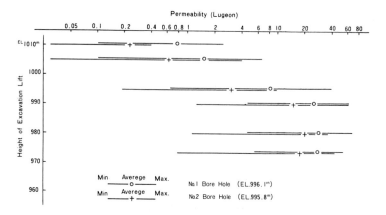

Fig. 3   Relationship Between Permeability Variation and
         Excavation of Cavern

zone also gradually increased as the excavation progressed. This
phenomenon indicates that loosening of the rock mass gradually increased to a more distant zone as the excavation work progressed. The
present measurements, however, did not clarify how far from the cavern
the loosening zone of the rock mass would advance.

Next, as shown in Fig. 3, the variation of permeability was extremely
large during the period of excavation from the elevation of the measuring
hole to 5 m below. And after that, almost no change was detected. In
other words, it can be supposed that considerable loosening of the rock
mass at the floor of the excavation reached 5 m below, thus considerable
caution is required in order to maintain safety.

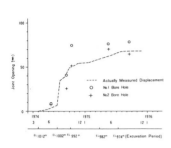

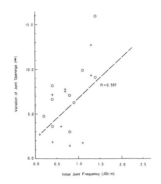

Fig. 4
Comparison of Variation
in Crack Opening to
Actually Measured Displacement

Fig. 5
Relationship Between
Variation of Joint Openings
and Initial Joint Frequency

(4) Variation in crack openings before and after excavation

Based on Bore Hole T.V observations and permeability measurements, the
following equation has been obtained to express the relationship
between the internal width of cracks and permeability.

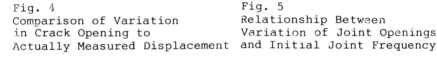

$$\Delta D = 0.774\left\{ (N_A^2\, L_{UA})^{1/3} - (N_B^2 L_{UB})^{1/3} \right\}$$

where,

L_U : permeability (Lugeon)

N : joint frequency (jts/m)

A and B are the conditions of the cavern excavation.

ΔD : variations in crack openings from cavern excavation condition A to B (mm/m).

The variations in crack openings calculated with the above equation
agreed comparatively well with the actually measured displacements, and
also corresponded fairly closely with the results of observations made
using a Bore Hole T.V. Fig. 4 shows the calculated variations in crack
openings.
    The above results were used to make studies on variations of crack
openings in rock mass. As shown in Fig. 5, the results indicated a
high degree of correlation between the variation of crack openings and
joint frequency (R = 0.60), and that loosening of around the rock mass due
to excavation depends largely on the initial joint frequency.

3.2 Case of M Power Station Site

The M Power Station is an underground facility of the water pumping
type and has a capacity of 1,200 MW. This facility was constructed

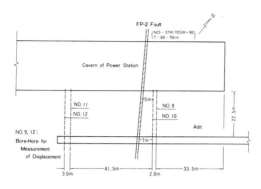

Fig. 6   Measurement Bore-Hole Layout

Table 3.   Appearance of Bore-Holes before Cavern Excavation

	Hole No. 10	Hole No. 11
Number of Joints	42	33
Joint Frequency	2.0 (Jts/m)	1.4 (Jts/m)
Average Permeability	0.34 (Lugeon)	0.0 (Lugeon)

inside an underground cavern with a height of 46.3 m, width of 23.5 m and length of 155.5 m.

(1) General description of the measurements

Measurements of permeability were made utilizing two horizontal bore-holes bored from adit to the crown base of the cavern as shown in Fig. 6. Measurements were made 5 times during the excavation of the cavern, which covered a period of 2 years and 2 months.
   The rock mass at the site of the measurements was fresh and hard granite with local porphyrite and aplite. The appearance of the bore-holes before excavation of the cavern are shown in Table 3.

(2) Variations in permeability

Before excavation started,the average permeability was 0.34 Lugeon at Hole No. 10 and less than the minimum measured value (0.025 Lugeon) for the entire space at Hole No. 11.   After the excavation was completed, however, these values had changed to 3.52 Lugeon at Hole No. 10 and 21.20 Lugeon at Hole No. 11.
   The increase was remarkable for both holes at a zone of 10 - 12 m from the cavern wall.   The permeability in this zone was higher at Hole No. 11 but the permeability variation was less.

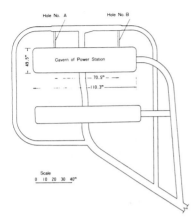

Fig. 7   Measurement Bore-Hole Layout

Table 4.   Appearance of Bore-Holes Before Cavern Excavation

	Hole No. A	Hole No. B
Number of Joints	19	20
Joint Frequency	1.10 (Jst/m)	1.16 (Jst/m)
Average Permeability	-	-

(3) Relationship between permeability variations and excavation

The relationship between excavation lift and permeability variation
indicates that the variation is greatest when the excavation lift
reaches the same elevation as the measuring hole. After that, almost
no change was detected at Hole No. 10, but, there was a great deal of
change at Hole No. 11. In addition, the quality around the rock mass
at Hole No. 11 was better, and there were fewer initial joints.

3.3 Case of Tm Power Station Site

The Tm Power Station is an underground facility of the water pumping
type and has a capacity of 1,200 MW. This facility was constructed
inside an underground cavern with a height of 49.5 m, width of 26.2 m
and length of 116.3 m.

(1) General description of the measurements

Measurements of permeability were made utilizing two horizontal boreholes bored from adit to middle height of the cavern as shown in Fig. 7.
Measurements were made 4 times during the excavation of the cavern,
which covered a period of 5 months.
    The rock mass at the site of the measurements was Lipuritic tuff-

breccia and Conglomerate in Tertiary Period, and each was separated
by a fault. The rock mass was of the non statification type and of
good quality. Hole No. B was located near the fault. The appearance
of bore-holes before excavation of the cavern is shown in Table 4.

(2) Variations of permeability

Both before and after excavation, permeability was almost zero, except
for local zones; Hole A showed permeability of only 0.1 - 0.2 Lugeon
at the zone 13.3 - 15.3 m from the cavern wall, and there were no
changes during the excavation. At Hole B, permeability varied from
0.047 to 0.083 Lugeon at the zone 9.2 - 11.3 m from the excavation
wall, and permeability at zone 3.3 - 5.3 m from the cavern wall increased to 1.57 Lugeon after excavation.

(3) Relationship between the variations of permeability and cracks

Normally, permeability of rock mass depends on the quantity of cracks
in the rock mass, but permeability is zero unless the cracks are
connected. When this fact is taken into consideration, of the cracks
confirmed by means of boring cores, only 4 at Hole A and 2 at Hole B
are thought to indicate the presence of cracks before the commencement
of the excavation. Most were thought to be unconnected.
   5 at Hole A and 3 at Hole B showed variations of crack openings due
to the excavation (including the openings of fine cracks) as was
confirmed by a survey made using the bore-hole T.V. Of these, however,
only 2 at Hole B were found to be continuous, while only 2 from Hole A
were thought to have this possibility. It was determined that the
number of crack openings thought to be continuous or to have enlarged
as a result of the excavation was extremely small. Rock displacement
was also very small at 2 - 3 mm at the locations where permeability was
measured.
   From the above results, the conclusion was that continuous cracks
which could influence the stability of the cavern will not be generated
or become enlarged as a result of the underground cavern excavation.

3.4 Case of I Power Station Site

The I Power Station is an underground facility of the water pumping
type and has a capacity of 1,050 MW. This facility was constructed
inside an underground cavern with a height of 51.0 m, width of 33.5 m
and length of 160.0 m. The shape of the cavern at the cross-section is
like an egg-shell, and on the excavation of each stage, the wall of the cavern
is secured by coating the rock with shotcrete and setting prestressed
anchors.
   As for the design of the cavern's cross-section, this type was introduced for the first time in Japan.

(1) General description of the measurements

The locations of the bore-holes are shown in Fig. 8. During the excavation of the cavern, which covered a period of 2 years and 4 months,
measurements were made 5 times at 9 bore-holes. The rock mass at the
site of the measurements was hard siliceous sanstone, breccia, and
claystone. The appearance of the bore-holes before excavation of the
cavern are shown in Table 5.

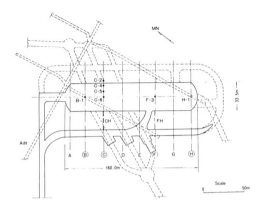

Fig. 8   Measurement Bore-Hole Layout

Table 5.   Appearance of Bore-Hole before Cavern Excavation

	Siliceous Sandstone	breccia	claystone
Joint Frequency	2.0 (Jst/m)	0.5 (Jst/m)	0.7 (Jst/m)
Average Permeability	0.79 (Lugeon)	0.30 (Lugeon)	0.0 (Lugeon)

(2) Variations of permeability

The average permeability near the crown of the cavern before the excavation was 0.47 Lugeon. Due to excavation of the crown, permeability was 0.74 Lugeon, but increased to 2.71 Lugeon due to excavation near the floor of the generator compartment. Almost no change was seen after succeeding excavations of the cavern. Variation in permeability was found at the upper zone, 12 m from the cavern. Fig. 9 shows a typical example of measurements made near the crown section.

Permeability near the height of the generator compartiment's floor was almost zero before the excavation but, after the excavation, the average permeability had become 5.31 Lugeon. Permeability was relatively large (10 - 40 Lugeon) at the zone 5 - 6 m from the cavern wall, but gradually decreased as the distance from the cavern wall increased, becoming zero at a distance of about 10 m. Fig. 10 shows a typical example of measurements made near the wall.

Permeability was very low in the area around the rock mass and, as shown in Fig. 11, variations mainly took place at zones where the Luge·n unit was not zero before the excavation. This means that permeability variations depend on the number and/or size of the openings of pre-existing cracks. In addition, no appreciable difference was seen in the measurements of various types of rocks around the cavern excavated.

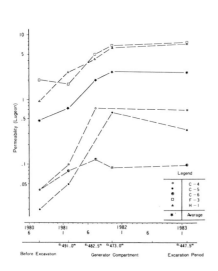

Fig. 9
Result of Permeability
Measurement
(Around Crown of Cavern)

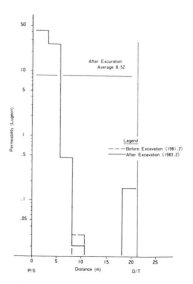

Fig. 10
Result of Permeability
Measurement
(Around Wall of Cavern)

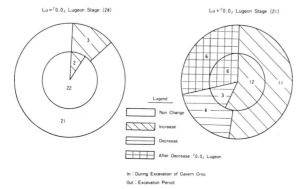

Fig. 11 Variation of Permeability Legend

(3) Relationship between the variations of permeability and the
    displacement of the rock mass

It has been thought that the displacement of rock mass, in part, could
be due to the deformation in rocks themselves, but the conclusion is that,
overall, this displacement is caused by the openings of existing cracks and
joints. There are also very few openings of cracks and joints in rock
mass, or new cracks, caused by excavation, thus in conclusion,
excavation near the floor of the generator compartment is stable.
    Variation in crack openings was calculated as being 2 - 3 mm (ℓ = 20 m)

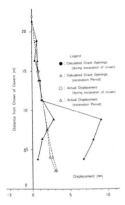

Fig. 12   Relationship Between Calculated Crack Openings
          and Actual Displacement Around the Crown of Cavern

near the crown of the cavern and 10 - 20 mm (ℓ = 20 m) near the height
of the floor of the generator compartment using the equation in the
last section. These values agreed very closely with the displacements
actually measured (Fig. 12).

  A low permeability zone, thought to be due to excavation, formed
near the crown section. There is a possibility that a relatively high
permeability zone exists in the deep part (Fig. 12).

  The above results indicate that there is very small variation of
permeability or rock displacement due to the excavation of the cavern,and
that there are no continuous cracks or expansion which would influence
the stability of the cavern. These results are thought to be due to
the use of an egg-shell like cross section.

4 CONCLUSIONS

Permeability measurements were conducted throughout the excavation
period of the caverns for underground electric power stations, and
clarified the previously unclear variations in permeability due to
underground excavations. The relationship between permeability and the
variations in cracks openings in the rock mass has been obtained
through measurement and studies of the variations.

  As a result, by measuring the permeability around the rock mass
of the cavern, it was possible to understand the looseness of the
rock mass caused by the excavation, and to confirm that it is
possible to contribute to safety control during the process of excavating caverns.

Río Grande I tailrace tunnel monitoring

Roberto I.Cravero, Raúl E.Sarra Pistone & Juan C.del Río
Agua y Energía Eléctrica. Insp., Río Grande I, Argentina

1 INTRODUCTION

The Río Grande I hydroelectric complex is a pumped-storage scheme with underground powerhouse, located in the province of Córdoba, in the center of the Argentine Republic.

The 5.5 km-long tailrace tunnel links the powerhouse with the lower reservoir. It has a vaulted section 12 m wide and 18 m high with an area of 206 m2. The rock overburden oscillates between 28 and 180 m. It will be under pressure (max. 0.4 MPa) during the whole of the useful lifetime of the complex. Nevertheless, only 20% of its length was concrete or shotcrete lined.

The tunnel's function is to conduct turbined water to the lower reservoir (max. 500 m3/sec) and back to the upper one during pumping operation (max. 360 m3/sec).

In this paper, the monitoring system used during construction, design criteria as well as global interpretation of results are discussed.

In order to control excavation and support performance and to infer the degree of stability accomplished, critical fault zones and special structures, like portals, were instrumented. Result interpretation was based on systematic in-situ observations, displacement vs. time-relationship analysis, mean deformation calculations, structural models and standard section instrumentation.

2 GEOLOGY AND GEOTECHNICS

A total length of 4500 m of the tunnel was excavated in massive, scarcely jointed-tonalitic gneiss characteristic of the C°Pelado geological environment. The remainder 1000 m traversed weathered-schistous gneiss typical of the Contraembalse environment (fig. 1).

The following three groups of faults strongly influenced the tunnel: FN, 45°/60°; FP, 60°/25°; FV, 35°/315° (dip/dip direction), (Sarra Pistone, del Río, 1982).

Monitoring sections were located in correspondence with the regional fault belts FTS, FT20 and FN1140.

In general, faults were of the reverse type with claily gauged shear seams and did not surpass 1 m in thickness. Fault belts were of variable thickness but could achieve a maximun of 150 m.

A geological-geotechnical zonification of the tunnel was carried out

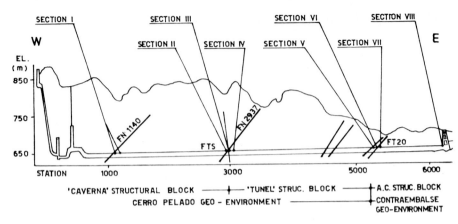

Figure 1. Tailrace tunnel longitudinal profile.

Table 1. Rock mass geotechnical parameters

Geo-environ. ment	Deformability		Strength		Reduction factors	
	E (MPa)	ν (-)	Uniaxial (MPa)	Joints	m (-)	s (-)
C°Pelado	60.000	0.20	110	Ø=33° c=0.4MPa	12.3– 7.06	0.2– 0.07
Contraembalse	30.000	0.20	110		1.77– 0.26	0.004– 0.0001

in scale 1:200, based on available information for design and adjusted during the entire 1st. stage excavation. It was used for the 2nd. stage design, as support for classification of different cases of instability and for corresponding selection of suitable excavation and support methodology.

Geotechnical parameters of Río Grande rock mass and tests carried out were published elsewhere (del Río, Sarra Pistone, Cravero, 1983a), (Dorso, del Río, de la Torre, Sarra Pistone, 1982). For the reader's illustration, the most relevant parameters of rock are listed in table 1, referred to the mentioned geological environments.

At the cavern site, in-situ stress measurements were carried out. Values obtained are summarized in table 2. For the cavern structural block (fig. 1) limited by regional faults, that included measurement sites, it was possible to assume a ratio $k = \sigma h/\sigma v = \sigma 1/\sigma 2 = 2.2$. The same did not occur with the other structural blocks as in the tunnel block and A. C. block. Nevertheless, for mathematical structural simulation of the section, a study of sensitivity on this variable was carried out and the

most unfavorable situations among the feasible ones, were taken into
account for the final design of the monitoring sections.

3 EXCAVATION AND SUPPORT

On account of the size of the tunnel section, it was necessary to determine the excavation in two stages: one in vault 8 m high and another in
benching 10 m high. However, in fault zones, more conservative methodo-
logies were adopted.

First stage through FTS and FN1140 faults was excavated using a central pilot tunnel 6 m wide and the second one in FTS and FT20 was attac-
ked in two horizontal benches, 5 m high each.

Second stage in Contraembalse poor quality rock was excavated with
a full-height-central trench and subsequent completion by removing the
remaining rock from the walls using a smooth-blasting technique (fig. 2)

Controlled-blasting technique was employed, alternatively pre-split-
ting or smooth-blasting, in correspondence with rock quality. In high
risk sites, controlled-vibration technique was used too, in order to
minimize rock damage.

Support consisted basically in mechanically-anchored rock bolts,
pretensioned up to half of bar failure load capacity and grouted with
cement. It was dimensioned according to rock mass-self-support-contri-
bution criterion, to provide stability to rock blocks that could fail
by gravity, sliding or toppling.

Time of installation agreed with the permissible estimated-stand-up
time for each section (del Río, Sarra Pistone, Cravero, 1983b).

4 MONITORING PROJECT

4.1 Design criterion

In the original design, no provision was made for a monitoring system
for the tailrace tunnel. The project discussed in this paper resulted
from the necessity of structural behaviour control during construction.

As a consequence, the control criterion was not unique. It was improved as construction went ahead and excavation methodology was adjusted to rock mass conditions (del Río, Sarra Pistone, Cravero, 1986).

At the first stage, only the vault was instrumented at stations 5135
and 1120, for monitoring a large gravity wedge and a threshold section

Table 2. In-situ stress measurement results.

	Mean value (MPa)	Direction	Observations
σ_1	9.2	Horizontal	$\perp$ tunnel axis
σ_2	4.2	Vertical	
σ_3	3.1	Horizontal	$\parallel$ tunnel axis

of a high risk fault zone, respectively.

Before starting the second stage, all the excavation methodology was reviewed. As a result, among other things, monitoring was included as part of the methodology itself in order to:
-optimize the construction technique,
-dimension required support,
-decide the time of installation,
-verify some hypotheses about structural response, and
-infer the degree of stability achieved.

Basically, displacements were measured using multiple-position-rod extensometers with deflectometers of 0.01 mm.

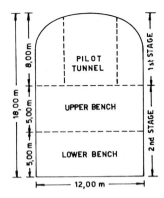

Figure 2. Tailrace tunnel.
Excavation stages.

Monitoring sections were selected according to the following objectives:
a. to control fault zones,
b. to control good quality rock mass as a reference and link with available mathematical model results,
c. to control special sections as portals, tunnel intersections, etc.

The positioning of extensometers at the base of the 1st-stage walls responded to the expectation of maximun displacements at those points once benching was carried out, resulting from studies made using methods of finite elements (Agua y Energía E.,1976) and boundary elements (del Río, Sarra Pistone, Cravero, 1983a), for an in-situ stress factor k=2.2. Besides, such a location could allow the detection of large unstable rock masses at walls that could move towards the tunnel because of the brusque stress relief that should provoke the 10 m-high benching.

Another point considered to be critical for monitoring purposes, was the tunnel crown in weak rock zones since, for $k > 1$, high stress concentration was foreseen and shear failures could start, and for $k < 1$ risk could arise from loosening of unstable gravity blocks. Despite access difficulties some extensometers were set up and measured in such points.

4.2 Monitoring sections

Twenty two extensometers were distributed in eight sections as shown in table 3 and in figure 1. Most sections were selected in order to control fault zones, objective 4.1.a. According to objective 4.1.b one section was instrumented. As a special section, the downstream-access portal was instrumented. Results were published by del Río and Cravero 1986, in a detailed report. Being a singular structure, any attempt to correlate its behaviour with that of the tunnel proper lacked sense. Therefore it is not dealt with in this paper.

Tunnel instability problems were classified in function of discontinuity types, their interaction with the tunnel, block kinematism, type of rock, weathering degree and groundwater seepage. Such a classification was based on the geological-geotechnical zonification mentioned

in 2. (del Río, Sarra Pistone, Cravero, 1983a).
    In correspondence with that classification, different methodologies
of excavation, support and monitoring were adopted.

5 ANALYSIS OF RESULTS

For interpretation of monitoring data, available, systematically- collected information related to tunnel sections was analyzed. The geo-zonifi-
cation and classification of instability situations constituted the fundamental frame for such an analysis. Knowledge about mechanical behaviour
of materials, was based on the laboratory and in-situ test program carried out all over the site. Structural behaviour was studied by means of
the mathematical models mentioned in 4.1, being aware of their limited
representativity since the simulated rock mass was continuous, elastic,
homogeneous and isotropic clearly the opposite to the actual one in fault
zones. With the aim of increasing their modelling capacity, a sound rock
section was instrumented as a full-scale physical model. The structural
response allowed the models to be adjusted and so simulate geologically-
more-severe situations introducing rock parameters characteristic of
fault zones.
    The resource most utilized for monitoring interpretation was that of
analysis of displacements vs. time curves, including the relationship
with the excavation rate (figs.3, 4, 5, b) as well as graphs of total
displacements referred to the deepest point, the slopes of which permitted an apreciation of deformations within each segment stretched between
two consecutive anchorages (fig.3, 4, 5, a).
    Different excavation methodologies, using controlled blasting and con-
trolled vibration technique, as well as differences in the application of
support, in quantity and time, among others, are variables that make the
result classification task difficult.
    On the other hand, available space does not permit further discussion
of details about data that would show the influence of every variable.
    However, some common features that represent standard behaviour could
be identified, being of great utility for any monitoring interpretation.
    In figures 3, 4 and 5 the monitoring results are summarized, groupped
in four characteristic arrays or dispositions, elaborated in function of
the mentioned classification of instability cases for a better understanding of the global behaviour of the tunnel:
    A: Instruments that monitored sliding bloks over fault planes in the
       sidewalls.

Table 3. Monitoring sections.

Sections	I	II	III	IV	V	VI	VII	VIII
Northern sidewall	E1095N E1110N	E2912N E2921N	E2933N	E2993N	E5185N	E5220N	E5234N	
Southern sidewall		E2921S	E2933S	E3007S	E5185S	E5220S	E5234S	
Crown	E1120C		E2933C	E2993C	E5185C E5183C	E5220C		E6198C E6199C

B: Instruments that monitored toppling fenomena on the sidewalls.
C: Instruments that monitored the crown.
D: Instruments installed in standard sections, as references.

5.1 Disposition A (fig.3)

This groups instruments that controlled sliding blocks on claily-gauged-
faults. The following extensometers were included in this group (E: extensometer, number: station, N: northern wall, S: southern wall, C:
crown): E1095N and E1110N at the FN1140 fault, E2912N and E2921N at the
FTS fault and E3007S at the FP3010 fault.
    Characteristic features: little sensitivity of the rock mass response
to the approach of the excavation front and the occurrence of great deformations (0.1%) in correspondence with the traverse point, in which
the extensometer station was achieved, particularly at the mid-segment
(3 to 8 m deep) when intersecting the fault (see E1095N, E1110N, E3007S),
which registered the major strains compared with those of other dispositions.
    Influence of concrete lining stiffness was shown by contrast between
superficial point displacement of E1095N (δ = 4.3 mm) and that of
E1110N (δ = 1.6 mm) in the lined stretch.
    Both extensometers monitored the same sliding wedge and in spite of
being 15 m apart, reacted simultaneously to the excavation advance, confirming the expected behaviour in block of the sliding mass.
    Finally, it must be pointed out that a curious shortening of deep
segments (8 to 17 m deep), occurred when the full section had been completed, (see E1095N, E1110N, E2921N). The greatest deformations of
E1095N corresponded to full-bench excavation.

5.2 Disposition B (fig.4)

Instruments that monitored toppling movements at the sidewalls, which did
not exclude block sliding kinematism, but the difference with the disposition A was that the principal fault ran behind the block and did not
work as a sliding plane. The main interest was knowing how deep a non-
sliding fault could significantly influence the excavation in such cases.
    B-extensometers registered the greatest displacements at segments
that traversed the main fault: deep segment (8 to 18 m), E2921S, E2933S,
and E5185S. Mid-segment (3 to 8 m), E5220S and superficial segment (0 to
3m), E2993N. Related to other dispositions, these instruments registered
the greatest deformations (0.01%) of the deepest segments, and showed
the most sensitive response to the advance of the excavation front, even
when it was 20 m away from the traverse point.
    E5220S and E5185S registered important displacements (aprox. 5 mm) at
the surface point installed on the concrete lining. Contrarily, the
E2933S showed a superficial segment shortening as a consequence of the
stiffness of the intensive rock bolting as well as that of the steel-
reinforced-concrete lining applied there.

5.3 Disposition C (fig.5)

Includes extensomenters set in the crown. E5183C and E1120C were instal-

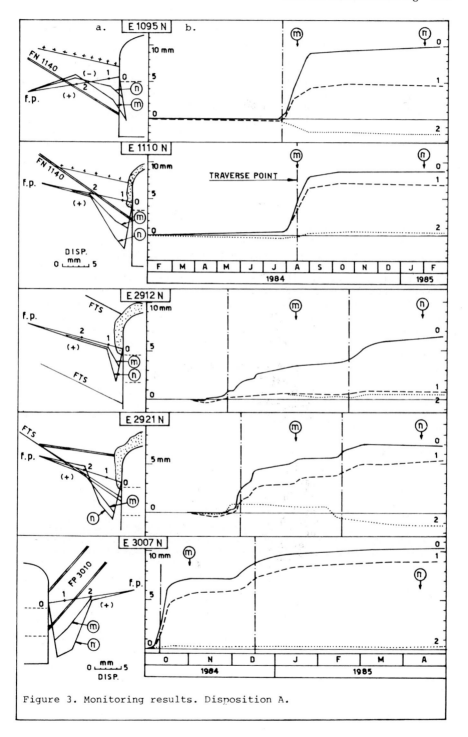

Figure 3. Monitoring results. Disposition A.

led in the first stage while E2933C, E5185C and E5220C were installed
inmediately before starting the second stage.

Data corresponding to E5183C and E5185C gathered together, constituted
the only complete record of the crown deformation during both stages.The
first one controlled a gravity wedge of great dimensions (450 m3). A marked descent of the crown related to 1st-stage blastings, was registered
once the top heading had   achieved the FN5170 fault.

Having been already installed the adequate support, displacement increments became zero when the front was at least 2-diameter away from
the traverse point.

On the other hand, superficial segment deformations, absence of wedge movements and a little descent of concrete arches (δ = 1.4 mm),
were observed during the 2nd. stage, once 1st-stage concrete lining had
been constructed.

E5220C controlled a weathered rock mass between faults FN 5170 and
FN 5220. Displacements at the deep segment were registered. Inflection
point of displacement vs. time curves were delayed with respect to the
traverse point. A sensitive response to a 35 m-far-benching of the same
weathered rock mass was shown one year later. All these elements indicated that the mentioned fault belt acted rather like a structural piece,
so that the extensometer responded to changes of excavation geometry
whenever these occurred within that belt.

The E2933C checked the behaviour of the poor-quality rock mass of the
FTS fault, excavated with controlled vibration technique in two horizontal benches. A small descent of the crown was observed as a consequence
of the upper benching. In contrast, a marked ascent was registered once
the lower bench had been excavated, in agreement with mathematical predictions.

The rock mass relative insensitiveness and delaying of the inflection
point with respect to the traverse point during the lower benching, indicated that the crown reacted more to geometrical changes than to blasting vibrations, neutralized, particularly in this case, by using careful techniques.

5.4 Disposition D

D-extensometers monitored the standard, nearly homogeneous, isotropic
and moderately-jointed rock mass. From two instruments installed, the
E5235N was the only that functioned during the entire bench excavation.

Measured displacements were 0.32 mm at the surface point and 0.25 mm
at the depth of 3 m. Calculated ones were 0.47 mm and 0.42 mm respectively. Taking into account that these were absolute displacements and
that measured ones were referred to a deep point assumed to be fix,
though calculations showed that the displacement there amounted almost
a 40% of that at the surface point, so it can be concluded that both
measured and calculated values were practically coincident.

Such a verification confirmed the usefulness of structural analysis
in elastic field, as a reference for studying more complex situations.
Simulating poorer rock mass conditions, greater displacements were obtained and points where elastic failure occurred were identified providing a guide for structural interpretation.

6 REMARKS

- A sigmoidea configuration with inflection point coincident with the

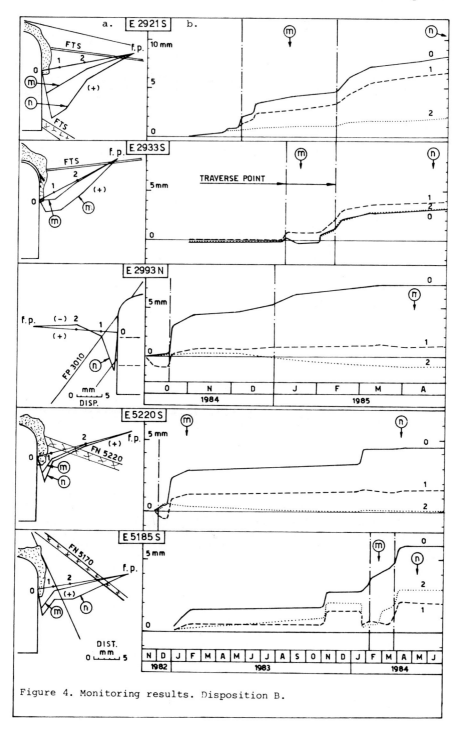

Figure 4. Monitoring results. Disposition B.

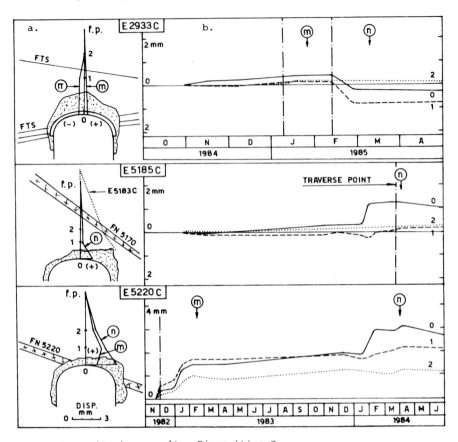

Figure 5. Monitoring results. Disposition C.

traverse point was characteristic of displacement vs. time curves corresponding to sidewall extensometers. One sigmoidea configuration corresponded to each semi-bench excavation.
- The inflection point did not coincide with the traversing point in
those cases in which advance direction was opposite to that of sliding,
favoring displacement before the traversing station (e.g. E3007S), and
in those cases in which displacements were related mainly to geometrical
changes of sections (e.g. E2933C, E5220C).
- Deformations were directly related to blasting which had a double influence on the behaviour of sections. On one hand, geometrical changes
arose. On the other hand, vibrations that could induce sliding through
discontinuities, were produced.
    Time-dependant deformations were negligible.
- Where careful excavation technique was employed (FTS fault), significant reduction in horizontal extension of sigmoidea curves, was achieved.
Minimun value was obtained at FTS zone: 5 m to each side of the traverse
point.
- Sliding block structures distinguished by reacting notoriously at tra-

verse points, varying the amount of response in function of support stiffness and stabilizing shortly after the necessary support had been installed.
- Toppling structures turned out to be more important than was expected with regard to deformation depth. The oustanding features of such structures were: occurrence of deep movements, shown far from the traversing point, and the contrasting fact that section stabilization only required cortical support.
- Finally, the importance of systematic control during underground construction is emphasized. Monitoring is recognized as an useful tool for the team in charge of the technical direction of works. This must be considered indispensable for the making of decisions, for the evaluation of finished excavation stability conditions and, as a reference, for later controls during project operation.

7 BIBLIOGRAPHY

Agua y Energía Eléctrica, 1976. Análisis de la estabilidad de la sección del túnel de restitución por el método de elementos finitos. Internal report.
del Río, J.C., Sarra Pistone, R. E. & Cravero, R.I. 1983a. Estudio de la estabilidad del túnel de restitución del complejo hidroeléctrico Río Grande N°1, Córdoba. Proc. II Symp. of ASAGAI. Vol II, 49-67.
del Río, J.C., Sarra Pistone, R.E. & Cravero, R.I. 1983b. Rock bolting used in the underground excavations of Río Grande I. Argentina. Lisbon. Proc. Int. Symp. of IAEG on Eng. Geol. and Underground Construction. Vol. I, III.9-21.
del Río, J.C. & Cravero, R.I. 1986. Monitoring of a large portal excavated in rock. Buenos Aires. Proc. V Int. Congress of IAEG. Vol. I.
del Río, J.C., Sarra Pistone, R.E. & Cravero, R.I. 1986. Construction of a large section tunnel in weak rock. Buenos Aires. Proc. V Int. Congress of IAEG. Vol. II.
Dorso, R., del Río, J.C., de la Torre, D.G. & Sarra Pistone, R.E. 1982. Powehouse cavern of the hydroelectric complex 'Río Grande N°1'. Aachen. Proc. ISRM Symp. on Rock Mechanics, Cavern and Pressure Shafts. Vol. I, 221-238.
Sarra Pistone, R.E. & del Río, J.C. 1982. Excavation and treatment of the principal faults in the tailrace tunnel of the Río Grande I hydroel. complex. New Delhi. Proc. IV Int. Congress of IAEG. Vol.IV, 139-152.

2nd International Symposium on Field Measurements in Geomechanics, Sakurai (ed.)
© 1988 Balkema, Rotterdam. ISBN 90 6191 778 6

Río Grande I powerhouse cavern monitoring

R. Sarra Pistone & R. Cravero
Agua y Energía Eléctrica, Insp., Río Grande I, Argentina
S. Sakurai
Kobe University, Japan

1 INTRODUCTION

Río Grande I is a pumped-storage hydroelectric scheme whose powerhouse is in a cavern excavated in massive metamorphic gneiss at a depth of 150 m.

Main chamber dimensions are: length 105 m, height 50 m and maximum width 27 m. For its construction, 110,000 m^3 of rock were excavated by blasting in five descendent stages, from 1979 to 1981 (Dorso, del Río, de la Torre, Sarra Pistone, 1982).

This paper deals specifically with the monitoring project of the vault, its implementation and interpretation of the results. In general, five sections were instrumented in correspondence to the main geological features surveyed during the pilot tunnel excavation, using multiple point-rod extensometers for direct measurement of displacements, and load cells for rock bolts for the indirect record of bar elongations.

2 CAVERN: GEOLOGY AND GEOTECHNICS

The cavern was integrally excavated in tonalitic-massive, fresh gneiss, with rough and tight discontinuities affected by some principal faults shown in figure 1 and described in table 1.

The mechanical parameters of the rock were determined by means of laboratory and in-situ tests. Deformability: core tests, micro-seismic measurements and plate bearing tests in two perpendicular directions were carried out, showing an elastic (E=60,000 MPa, ν =0.20), fairly isotropic and homogeneous rock mass. Intact rock strength: c=24 MPa, $\emptyset$=45°. Discontinuities shear strength: c=0-0.35 MPa, $\emptyset$=33° for σ_N varying from 0.2 to 4.0 MPa. The latter results can be considered as residual ones because the tested joints had sliding signals.

With respect to water circulation, the rock mass is highly impervious, and seepage is restricted only to some joints at the cavern walls. One must bear in mind that the main 100 m-deep reservoir was completely filled two years ago and its bottom is only 300 m over the cavern roof.

Theoretical studies of displacements and stresses induced around the excavation were made using a bi-dimensional boundary element model, simulating different construction stages for the two different sections corresponding to those of turbines and draft tubes (fig. 2).

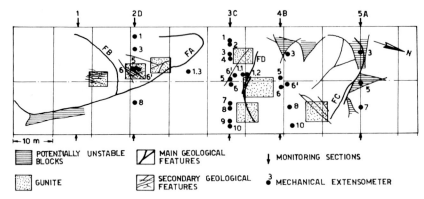

Figure 1. Plane view of the cavern roof. Main geological features and monitoring sections.

Rock mass was assumed to be continuous, elastic, isotropic and homogeneous. Displacements calculated with such a method were used as reference for the interpretation of measured ones.

On the other hand, structural singularities like the intersections of principal discontinuities with roof and walls, and several large rock blocks limited by two or three joints with proved possibilities of falling down, were carefully controlled (fig. 1).

3   EXCAVATION AND SUPPORT OF THE CAVERN VAULT

Detailed information about the excavation methodology of the cavern was published by Dorso, et al. in 1982. Here, only the first stage of excavation is summarily discussed.

Vault excavation started with a central pilot tunnel (fig. 3,A). After that, a cross-shaped enlargement was excavated at the mid-section (B), and finally four front sections were blasted alternatively from the center to each extreme (C, D).

Table 1. Characteristics of main discontinuities that affect the cavern vault.

Name	Dip Direction	Dip	Category	Comments
FA	230°	40°	I	Category I: Faults; plane and
FB	155°	40°	I	smooth surfaces of great area
FC	180°	50°	I	and small thickness. Asso-
FD	100°	75°	I	ciated to intensively jointed
				belts. Generally contain
				detritic/clayey fills or wea-
				thered milonite fragments.

Category II: Non-planar, discontinuous and rough surfaces, sometimes with lateral micro-jointed rock.

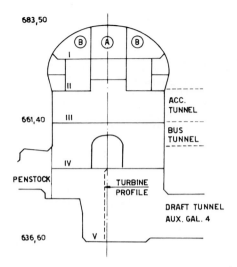

683,50

661,40

636,60

ACC.
TUNNEL

BUS
TUNNEL

TURBINE
PROFILE

PENSTOCK

DRAFT TUNNEL
AUX. GAL. 4

Figure 2. Cavern excavation
stages.

Mechanically anchored rock bolts
were used as the main support of
the vault. The 25 mm and 19 mm
diameter bars were pretensioned
at 14 tons and 9 tons respectively and injected with cement
grouting (del Río, Sarra Pistone,
Cravero, 1983).

The support design was based
on a structural criterion, so it
was dimensioned and located in a
function of geological discontinuities that could potentially
arise in unstable rock blocks,
assuming gravity and sliding
mechanisms as the characteristic
types of failures.

Rock bolts were installed for-
lowing the excavation rate, although the final adjustment of
the system was accomplished after
the stage had been completed, by
retensioning and grouting bolts
already set and adding some reinforcement bolts as a secondary support system.

Mechanical extensometers and load cells were installed immediately
after blasting as near as possible to the excavation front. Load cells
and their cable connections resulted in serious damage near blastings
and should be repaired. As a consequence, no systematic reading was
recorded until the cavern was completed.

4 MONITORING SECTIONS, INSTRUMENTS AND PERFORMANCE

The final five-monitoring section arrangement was designed by adapting
the original layout of four sections to the real geological features
uncovered with the pilot tunnel excavation. Some instruments were set
apart from them to check particular features in the roof, as well as in
the walls.

The rock mass monolithic structure, the low level of stresses induced
by excavation, and the good behaviour of rock observed during the vault
construction permitted us to foresee a satisfactory response from the
walls. Therefore, it was concluded to monitor only singular places of
geological weakness.

Mechanical extensometers were installed and data recording was maintained for just the time necessary to assess the structure behaviour

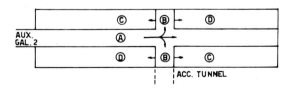

AUX.
GAL. 2

ACC. TUNNEL

Figure 3. Excavation scheme of the cavern's first stage.

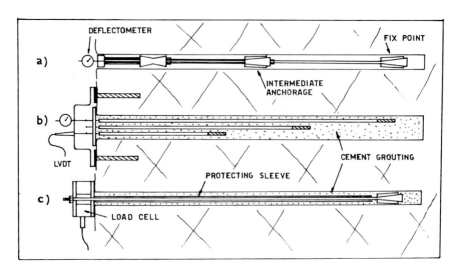

Figure 4. Instruments used to monitor the cavern: a) mechanical extensometer, b) electrical extensometer, c) load cells for rock bolts.

together with the support performance, as far as they were influenced
by blastings. After stability had been assured, the instruments were
removed.

Displacements were measured at the roof, using two types of borehole
extensometers of multiple points. During the first stage of excavation,
two-position mechanically-anchored rod extensometers were set. Readings were made by means of 0.01 mm deflectometers (fig. 4,a).

These instruments turned out to be simple and robust enough to resist
severe underground conditions and yielded reliable information. Hence,
the interpretation of the excavation behaviour was based fundamentally
on this data.

During the excavation of the 4th stage, three and four-position
electrical extensometers were installed. Sleeved rods were introduced
and grouted in an NX borehole, and referenced to an anchored head that
allowed the utilization of both manual deflectometers and linear variation transducers for remote readings (fig. 4,b).

In September of 1980, they were ready for manual measuring but access
was interrupted because scaffolds were severely damaged by blasting.
Once repaired, electrical connections were undertaken and just completed
in 1982. However, the automatic system had several problems due to de-
ficiencies in the electrical supply and electronic failures, therefore,
today's available readings were considered unreliable for being taken
into account for quantitative analyses.

Indirect measurements of vault deformation were realized by means of
vibrating wire load cells for rock bolts (fig. 4,c). In spite of being
reliable and strong instruments, several problems occurred frequently
when cables and protection boxes were broken by nearby blasting operations.

5 ANALYSIS OF RESULTS

As it was said, only mechanical extensometer data was considered for
quantitative analysis. Electrical extensometer data allowed for only
qualitative interpretations about the general trend of vault stability
after the excavation was concluded.

Mechanical extensometers on the other hand, were accessible only
during the construction period. Nevertheless, conclusions obtained
from their data were considered to be highly representative of the
actual deformation state, since the rock mass should behave elastically
for the stress level imposed by the opening. Viscous strains were
proved to be negligible in field rock tests and in measurements carried
out during the access tunnel construction.

For result analysis, measured displacements were compared to elastic
ones calculated with the boundary element method. Differences in excess were assumed to be displacements that occurred along discontinuities, in other words, they were interpreted as being related to structural singularities of the rock mass.

Furthermore, it was admitted that some abnormal displacements could
be explained by the sliding of mechanical anchorages caused by vibrations related to nearby blasting (e.g. ext. EC7 in fig. 5).

Results obtained from some instruments considered as the most representative of major structural behaviour, are summarized in figures 5,
6 and 7. Outstanding observations from each section are discussed in
the next paragraphs.

5.1 Section 3

Section 3 is considered to be the most representative for the following
reasons:
    a. It was the first section instrumented, and therefore, it resulted
in the most complete record of displacements available.
    b. It is the section least affected by the tri-dimensional effect
(Martins, 1985). Other sections are, at most, little more than one
diameter from each head-wall.
    c. It is not superficially affected by any fault.
The upstream mid-section is influenced in depth by the FA fault.
Extensometers EC3, EC4, EC5 and E1.1 (fig. 5) registered deep displacements between 1.90 m and 10 m, due to movements along the fault.
Approximately 65% of the total displacements occurred at the 1st stage
before definitive support has been installed.

The downstream mid-section is not affected by category I structures.
Displacements were basically superificial (fig. 5, EC7). Extensometers
EC7 and EC8 registered important movements before stage 3 started, pre-
sumably influenced by excavation works for the access to that level.

5.2 Section 2

FA and FB fault behaviour is controlled (fig. 1). ED5, ED6, ED6'
(fig. 6) indicated deep plastic movements due to the descent of block
limited for those faults, fundamentally in the first stage, achieving
a relative stabilization when definitive support was installed at its
completion.

The ED8 was set at the end of the first stage and recorded movements

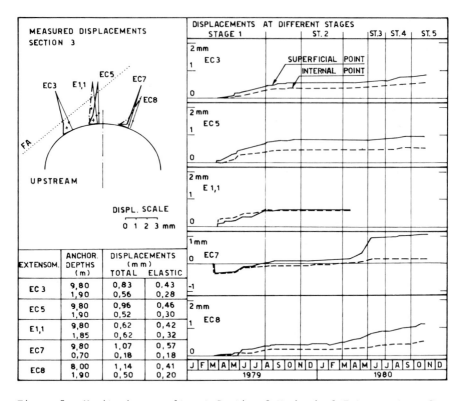

Figure 5.   Monitoring results at Section 3-Mechanical Extensometers C.

of the superficial crown during the excavation of the second stage and the beginning of the third, due to displacements occurring along the FA fault.

5.3   Section 4

Instruments registered purely elastic displacements, most of them occurring at the first stage. Some instruments, EB8 and EB10, indicated superficial inelastic movements during the excavation of accesses and large benches at the third and fourth stages.

5.4   Section 5

This section controls potentially unstable blocks (fig. 7). Extensometers were installed after stage 1 had been finished. 80% of the displacements were produced before the start of the second stage, due to the influence of nearby blasting and before the installation of the definitive support. The excavation became stable after the blasting was completed.

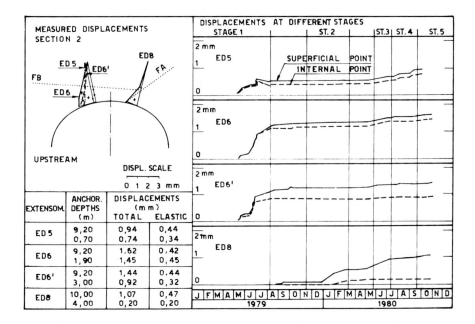

Figure 6. Monitoring results at Section 2-Mechanical extensometers D

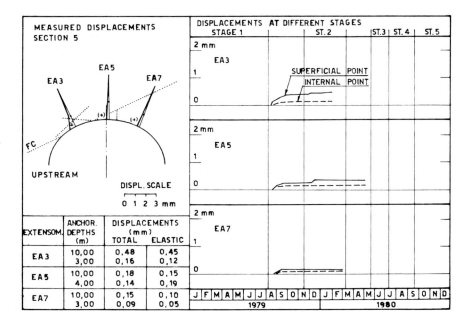

Figure 7. Monitoring results at Section 5-Mechanical extensometers A

6   INTERPRETATION OF RESULTS ON THE BASIS OF STRAIN

A method for monitoring the stability of underground caverns on the basis of strain has been proposed by Sakurai (1981). This method has been named the Direct Strain Evaluation Technique (DSET), and is a fundamental concept of a comparison between the strain occurring around a cavern, and the critical strain of geomaterials in the ground. If the occurring strain is smaller than the critical strain, stability of the cavern is secured. The DSET has been applied for evaluation of the stability of the Rio Grande I powerhouse cavern, and the results are shown below.

Strain and displacements are related by kinematics. Thus, when measuring displacements, one can immediately calculate strain through use of the following equation, without knowing either the initial stress or the mechanical properties of the ground.

$$\{ \varepsilon \} = [B] \{ u \} \tag{1}$$

where [B] is a matrix only depending on the location of the measuring points. $\{u\}$ is a measured displacement vector. However, the disadvantage of this equation is that the results of strain distribution become less reliable when the amount of measured displacement data is limited. (Such is the case with the Rio Grande I powerhouse cavern.) In order to overcome this disadvantage, a so-called back analysis may be introduced. Namely, the initial stress and mechanical constants are firstly determined from the measured displacements, such that the following value becomes minimal, i.e.,

$$R = \sum_{i}^{N} (u_i - u_i^*)^2 \rightarrow min. \tag{2}$$

where u_i and u_i^* are calculated and measured displacements, respectively. The strain distribution is then calculated by using these as input data for an ordinary analysis such as FEM, BEM, etc.

Sakurai and Shinji (1983) proposed a method based on FEM performing the above mentioned back analysis to obtain strain distribution from measured displacements. Since the method has been presented elsewhere, only a brief description is given here.

Assuming the ground consists of a homogeneous isotropic elastic material, the following equation can be derived.

$$[A] \{ \sigma_0 \} = \{ u \} \tag{3}$$

where [A] is a matrix expressed as a function of Poisson's ratio, ν is a measured displacement vector, and $\{\sigma_0\}$ is a vector defined in a two dimensional state as

$$\{ \sigma_0 \} = \{ \sigma_{x0}/E \quad \sigma_{y0}/E \quad \tau_{xy0}/E \}^T \tag{4}$$

in which σ_{x0}, σ_{y0}, and τ_{xy0} are components of the initial stress acting in the ground prior to excavation of the cavern. E denotes Young's modulus.

The amount of measured displacement data is, in general, taken to be greater than three. Thus, a least squares method can be used for solving equ. (3) to determine $\{\sigma_0\}$. Poisson's ratio is usually assumed to be an appropriate value.

Assuming the vertical component of initial stress is equal to the overburden pressure, i.e.,

$$\sigma_{y0} = \gamma \cdot H \quad (\gamma : \text{unit weight, H : height of overburden}) \tag{5}$$

yields all the components of the initial stress and Young's modulus by splitting the normalized initial stress. The results of back analysis are shown in table 2.

It appears that in section 3 a loosened zone occurs in the vicinity of the cavern crown. Thus, the reduced Young's modulus is used for the elements near the crown in this section. The reduction ratio is $Er/E = 0.3$. This is determined so as to minimize R given in equ. (2). The comparison between the measured and calculated displacements proves the accuracy of the back analysis. The results are shown in figs. 8, 9, and 10 for each section. They all indicate a good coincidence between the two distributions, and therefore, the accuracy of the back analysis is demonstrated.

The strain distribution is calculated and shown as a coloured graph on the CRT of the microcomputer. One example is shown in photo 1. The maximum shear strain distribution obtained after the final stage of excavation is presented in fig. 11 for each section. Comparing this strain with the critical strain of the material, one can immediately assess the stability of the cavern. The largest value of the maximum shear strain occurs near the crown at section 3, and its value is approximately 0.4%, while the critical shear strain under a uniaxial condition is approximately 0.6%.

Table 2. Results of back analysis (Poisson's ratio : 0.3)

Section	Normalized Initial Stress	Initial stress (MPa)	Young's modulus (MPa)
2	$\sigma_{xo}/E = -0.179\times10^{-3}$ $\sigma_{yo}/E = -0.109\times10^{-3}$ $\tau_{xyo}/E = -0.156\times10^{-4}$	$\sigma_{xo} = -6.28$ $\sigma_{yo} = -3.82$ $\tau_{xyo} = -0.55$	35089.
3	$\sigma_{xo}/E = -0.677\times10^{-4}$ $\sigma_{yo}/E = -0.916\times10^{-4}$ $\tau_{xyo}/E = -0.344\times10^{-4}$	$\sigma_{xo} = -2.82$ $\sigma_{yo} = -3.82$ $\tau_{xyo} = -1.44$	41754.
5	$\sigma_{xo}/E = -0.601\times10^{-4}$ $\sigma_{yo}/E = -0.262\times10^{-4}$ $\tau_{xyo}/E = 0.212\times10^{-4}$	$\sigma_{xo} = -8.76$ $\sigma_{yo} = -3.82$ $\tau_{xyo} = 3.09$	145802.

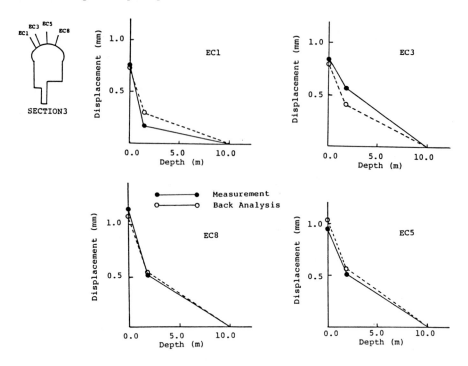

Figure 8. Comparison between measured and calculated displacements
(Section 3)

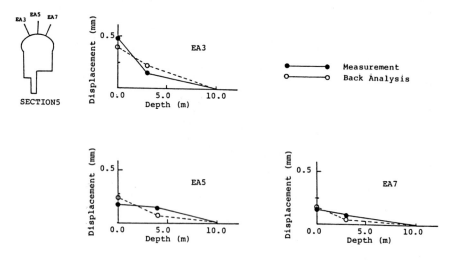

Figure 9. Comparison between measured and calculated displacements
(Section 5)

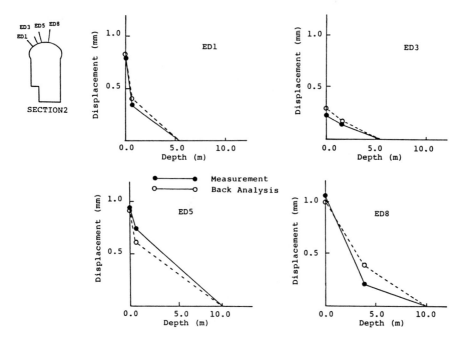

Figure 10. Comparison between measured and calculated displacements (Section 2)

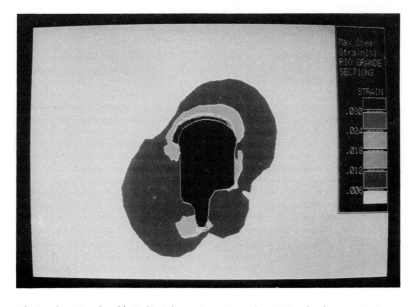

Photo 1. Strain distribution shown on the CRT of microcomputer

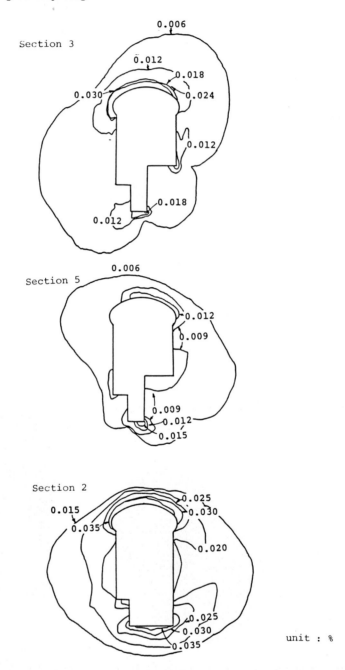

Figure 11. Maximum shear strain distribution after the final stage
of excavation

7 REMARKS

The security level required for the Rio Grande I powerhouse cavern
design implied detailed rock mechanics and structural behaviour studies,
strict geological-geotechnical control during construction, immediate
installation of structural supports, as well as additional and special
supports to minimize rock block movements (or collapse), to avoid blasted rock relaxation and to foresee any defects in high stress concentration points (pillars, galery intersections, etc.) respectively.
  In addition, a monitoring system based on multiple position-rod
extensometers and load cells was designed to check rock mass response
during and after construction.
  Mechanical extensometers registered displacements of similar magnitude to those calculated by bi-dimensional mathematical models on a
standard rock mass. Differences in excess were interpreted as being
related to movements along discontinuities induced by the opening and/or
by blasting vibrations. The Direct Strain Evaluation Technique (DSET)
was applied for analyzing the measured displacements, and the stability
of the cavern was fully secured.
  The cavern's deformation state after construction was considered to
be representative of the actual one, since the viscous behaviour turned
out to be negligible.
  The general trend of the observed strain was desacceleration until
stabilization at construction completion, and load cells, definitively
installed in 1982, showed no significant variations in initial loading
(between 13 and 15 tons).
  Some instruments clearly confirmed the influence of support in strain
limitation and structural stabilization.
  Electrical extensometers did not yield reliable readings due to elec-
tronical problems in the readout unit.
  Sidewall monitoring in the original design was reduced according to
a good performance expectation. Only structural singularities were
controlled. The instruments were removed after stability had been
accomplished.

BIBLIOGRAPHY

del Rio, J.C., Sarra Pistone, R.E. & Cravero, R.I. 1983. Rock bolting
    used in the underground excavations of Rio Grande I-Argentina. Proc.
    IAEG Symp. on Underground Construction. Lisbon. Vol. 1.
Dorso, R., del Rio, J.C., de la Torre, D.G. & Sarra Pistone, R.E. 1982.
    Powerhouse cavern of the hydroelectric complex 'Rio Grande I'. Proc.
    ISRM Symp. on Caverns and Pressure Shafts. Aachen. Vol. 1.
Martins, C. de S. 1985. Contribuicao para o estudo de estruturas sub-
    terraneas associadas a empreendimentos hidroelectricos. Tese de
    Especialista. Lisbon. LNEC.
Sakurai, S. 1981. Direct strain evaluation technique in construction
    of underground opening. Proc. 22nd US Symposium on Rock Mechanics,
    MIT.
Sakurai, S., & Shinji, M. 1983. A monitoring system for the excavation
    of underground openings based on microcomputers. Proc. ISRM Symposium.
    Design and Performance of Underground Excavations. Cambridge.

Stability of a large scale egg-shaped cavern

S.Hibino & M.Motojima
Central Research Institute of Electric Power Industry, Abiko-shi, Chiba-ken, Japan

1 INTRODUCTION

Many types of underground structures have been constructed in these past few decades, such as tunnels for high-ways or rail-roads, and caverns for underground power stations or crude oil storage. The demands for construction of underground structures may increase more and more in the near future, because underground structures have some advantageous points; they are at a constant temperature and humidity, are isolated, earthquake-proof and fire-proof etc. In addition to these points, underground developments do not destroy the natural environment at the surface, and the effective use of the underground may bring more subaerial room in such a narrow and restricted country, like Japan. In the near future, caverns for several purposes, such as radioactive waste disposal, superconducting magnetic energy storage, compressed air energy storage and underground dams or nuclear power plants etc., will be constructed.

In this paper we will present the mechanisms of rock behaviour during excavation of a large scale cavern, which was clarified through the investigation of about 20 sites of underground power plants. We also describe and discuss the stability of an egg-shaped cavern in terms of numerical analyses and *in situ* measurements.

2 MECHANISM OF ROCK BEHAVIOUR DURING EXCAVATION OF LARGE SCALE CAVERNS

In the past twenty years, about 25 large caverns for hydraulic power plants have been constructed in Japan. Numerical analyses and measurements on rock behaviour during excavation at these sites were carried out in order to construct caverns safely (Hayashi 1970). Through the analyses and measurements, the following were revealed (Hibino 1983) :
(1) The total subsidence of ceiling rock mass was mostly generated during excavation of the arch part, and any conspicuous subsidence did not take place or did not increase during the following excavation of the main part. On the contrary, we recognized a recovery of subsidence (Fig. 1). (2) The horizontal displacements of the walls increased continuously during each stage of the main part of the excavation. (3) Stresses in the arched concrete lining depended on the amount of convergence of the caverns (Fig. 2).

These features of rock behaviour mentioned above greatly differ from those of tunnel excavation. Ordinally, tunnels have a circular section, and under a state of hydro-static pressure, horizontal external

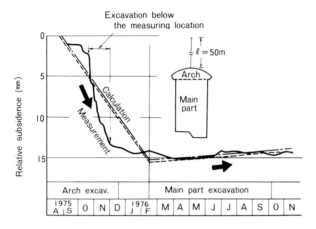

Figure 1. Time history of the subsidence showing the typical trend of the ceiling rock due to excavation

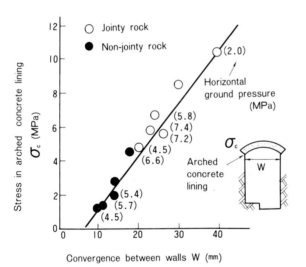

Figure 2. Relationship between stresses and convergencies

loads due to excavation are equal to the vertical ones. In contrast, caverns for hydraulic power plants have quasi-rectangular shapes in cross-section; the height of a cavern is nearly twice the size of its width. This double size means not only a double amount of horizontal external loads due to excavation, but also a magnifying of rock discontinuity effects. Rock discontinuity expressed by joints, bedding planes and faults plays a major role in increasing rock displacements or in the stability degradation of caverns; the larger in size the rock excavation surface becomes, the easier the opening of the joints becomes, which results in the enlargement of the relaxed zones of a cavern. The largeness of the excavation surface, therefore, has a great influence on the amount of displacement or on the stability of the cavern. Furthermore, ground pressure partly due to the tectonic stress in the Japanese Islands has a trend, in which the horizontal component of the pressure exceeds the vertical one (Fig. 3 and Table 1); an average ratio of the components is 1.36 (Kanagawa 1986). This value means that the external forces along the horizontal direction will be 2.7 times larger than those of the vertical direction. Under these conditions, we should note the behaviour of cavern walls. When a tunnel is to be excavated, we should care for both the ceiling and wall, or at least more care should be paid to the ceiling. Figure 4 shows the different mechanisms of rock behaviour of a tunnel and of a flat-shaped cavern under excavation.

Table 1. Results of *in situ* stress measurements in Japan using the over-coring method (Kanagawa 1986)

Site no.	Kind of rock	Density (t/m³)	Young's modulus (GPa)	Elevation (m)	Depth (m)	Principal stress σ_1 (MPa)	σ_2 (MPa)	σ_3 (MPa)	Horizontal stress σHmax (MPa)	σHmin (MPa)	Vertical stress component σ_v (MPa)	$\dfrac{\sigma\text{Hmax}}{\sigma_v}$
1	Granite	2.5	18	1054	250	10.8	6.4	0	10.6	0.6	6.0	1.77
2	Shale	2.6	8	386	214	-	-	-	9.0	4.6	7.3	1.24
3	Granite	2.5	20	320	280	9.6	7.5	4.9	7.9	4.9	9.2	0.86
4	Black schist	2.6	10	580	270	11.1	5.4	3.7	8.7	4.4	7.2	1.22
5	Granite	2.5	24	185	370	23.4	13.2	7.2	20.2	11.1	12.5	1.61
6	Mudstone	1.7	0.8	-40	70	1.24	1.08	1.07	1.16	1.09	1.17	0.99
7	Green schist	2.5	5	5	30	0.89	0.66	0.46	0.77	0.53	0.71	1.10
8	Rhyolite	2.5	12	260	165	4.2	3.3	2.5	4.0	3.0	2.9	1.37
9	Granite	2.5	30	130	510	15.8	11.1	6.3	15.5	6.4	11.2	1.39
10	Schalstein	2.5	7	140	210	6.2	4.8	4.7	6.2	4.8	4.8	1.29
11	Siliceous sandstone	2.5	26	540	420	15.7	10.6	7.8	14.7	8.8	10.6	1.39
12	Breccia	2.5	27	540	395	12.1	8.5	7.6	11.4	7.6	9.1	1.24
13	Conglomerate	2.5	14	601	270	8.2	5.5	4.9	8.1	5.0	5.5	1.48
14	Conglomerate	2.5	2.6	-16	22	1.06	0.72	0.41	0.92	0.71	0.55	1.68
15	Quartz diorite	2.5	-	550	15	-	-	-	7.4	2.6	2.8	2.62
16	Rhyolite	2.6	16	460	335	9.0	6.2	4.6	8.9	5.6	5.2	1.72
17	Mudstone	2.0	1.2	-20	30	-	-	-	0.49	0.45	0.55	0.89
18	Granite	2.5	12	-47	71	5.5	4.6	4.1	5.5	4.5	4.2	1.30
19	Rhyolite	2.6	10	663	192	5.1	4.3	1.7	4.4	1.7	5.0	0.88
20	Tuff breccia	2.6	7	664	241	5.0	3.7	2.9	4.1	2.9	4.6	0.89
21	Porphyrite	2.5	20	358	285	10.4	7.0	4.1	8.7	7.0	5.8	1.51
22	Porphyrite	2.5	20	358	285	8.9	5.9	3.0	6.4	5.6	6.0	1.07
23	Slate	2.5	11	674	316	12.1	7.9	5.5	12.1	6.7	6.9	1.76

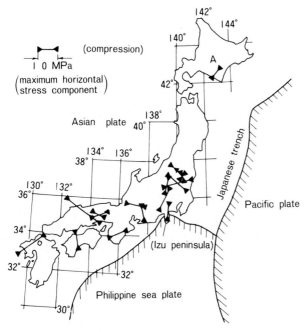

Figure 3. Directions of maximum horizontal compressive stresses σHmax (Kanagawa 1986)

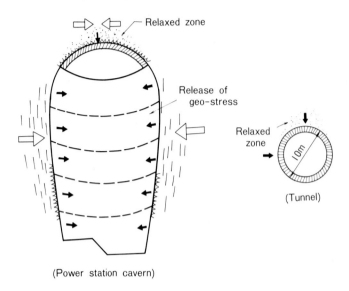

(Power station cavern)

Figure 4. Differences of rock behaviour around a
cavern and a tunnel

3 STABILITY COMPARISON BETWEEN AN EGG-SHAPED CAVERN AND A MUSHROOM-
  SHAPED CAVERN

A mushroom-shaped cavern is efficient for the construction of power
plants built near the surface, because the arched concrete lining is more
effective in sustaining vertical external loads which are exerted as a
main pressure due to the overburden weight. In the construction of deeply
sited plants (>100m), however, main external loads are not vertical, but
horizontal. Thus, a mushroom-shaped cavern would not be very effective in
this case. As noted below, stability of both the egg-shaped and mushroom-shaped caverns are compared by numerical analyses (Hibino 1978).
   In the numerical excavation analyses, the mushroom-shaped cavern has
an arched concrete lining, and the egg-shaped cavern has a shotcrete
lining. Analyses were carried out for both types caverns under the
same conditions (Fig. 5 and Table 2), and relaxed zones or displacements were calculated (Figs. 6 and 7). In this paper, the term relaxed zones
means the zones where Poisson's ratio becomes greater than 0.45. These
zones need stability reinforcement by pre-stressed anchors or rock bolts.
The area of relaxed zones and the amount of displacements in the egg-
shaped cavern are smaller and less than those of the mushroom-shaped one.
Stability of the egg-shaped cavern exceeds that of the mushroom-shaped
cavern.
   The reason for this stability gap is ascribed to the shape of the
arch part of the mushroom-shaped cavern. In order to place the arched
concrete lining, a notched shape excavation is necessary in the upper
part of the cavern (Fig. 8), and the notched corner produces a weak
point of rock masses, from which the relaxed zones are enlarged (part A
in Fig. 8). In contrast, the notched shape excavation is not necessary
for the egg-shaped cavern. The stress trajectories around the cavern
are so smooth that the relaxed zones and displacements are smaller and
less than those of the mushroom type cavern.

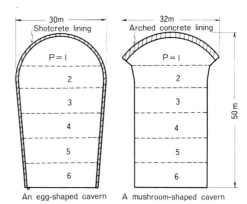

Figure 5. Cavern shapes and excavation stages (p) in the numerical analyses

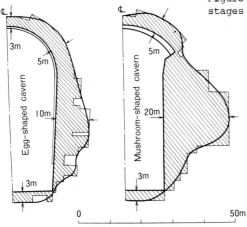

Figure 6. Relaxed zones due to excavation (numerical analyses)

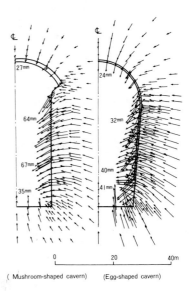

Figure 7. Comparison of displacements due to excavation (numerical analyses)

688 *Underground openings*

Table 2. Input data of the numerical analyses

Ground pressure	Horizontal stress	σ_{xo} (MPa)	7.14
	Vertical stress	σ_{yo} (MPa)	10.2
	Shearing stress	τ_{xyo} (MPa)	0.
Properties of rock mass	Deformability	Do (GPa)	10.2
	Poisson's ratio	ν_o	0.28
	Creep coefficient	α	1.0
	Shearing strength	τ_r (MPa)	1.02

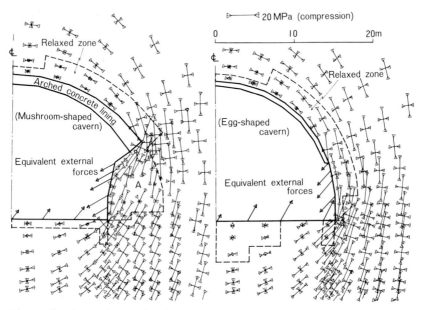

Figure 8. Comparison of relaxed zones and stresses at the upper part excavation (numerical analyses)

4. CONSTRUCTION OF AN EGG-SHAPED CAVERN AND *IN SITU* MEASUREMENTS

The Tokyo Electric Power Company had planned to construct a cavern as the site for a hydraulic underground power station, and we have carried out analyses on cavern shapes (Motojima 1982). The results indicated that the egg type cavern was superior to the mushroom type one. The company decided to build the egg type cavern and completed excavation of the cavern by 1982. Some noticeable results were obtained by measuring the displacements of rock masses (Motojima 1984).

Geology around the egg-shaped cavern is shown in Figure 9. Rock masses consist of siliceous sandstone, volcanic breccia and alternation of slate and sandstone. The cavern has a shape of 33.5 m in width, 51 m in height and 160 m in length. Each excavation stage and reinforcement patterns are shown in Figures 10 and 11. Rock masses were reinforced with pre-stressed rock anchors (length; 15 m and 10 m, respectively) and rock bolts (length; 5 m).

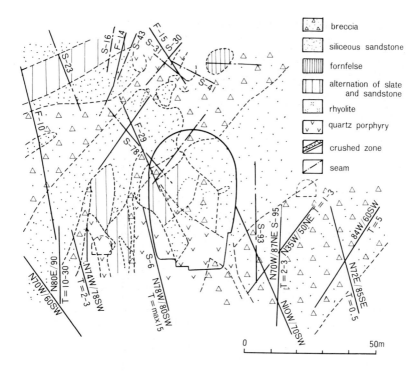

Figure 9. Geological map of the vertical section

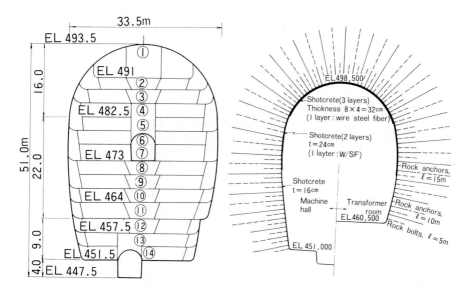

Figure 10. Excavation stages of the cavern

Figure 11. Pattern of the reinforcements

The ceiling part of the cavern surface was covered with a 32 cm thick shotcrete lining (16 cm; plain concrete, 8 cm; steel fiber concrete). One example of measurement layouts is shown in Figure 12, where extensometers were used for displacements. Extensometers were set close to each other near the excavation surface to detect an extent of relaxed zones.

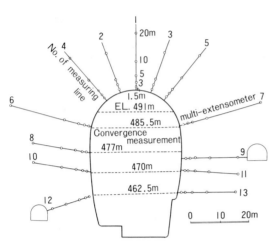

Figure 12. Layout of the displacement measurement of the section C

The result of displacement measurements is shown in Figure 13. The general trend of the variation during the excavation stage is the same as the results obtained at the other sites (see, section 2). The distribution of displacements measured along each line shows that the displacement value decreased rapidly as the distance from the excavation surface increased.

This tendency differs from those obtained in the other sites where mushroom type caverns were excavated. The trend may originate from smooth trajectories of ground pressure around the egg-shaped cavern. Displacements in the penstock side, where the rock mass is volcanic

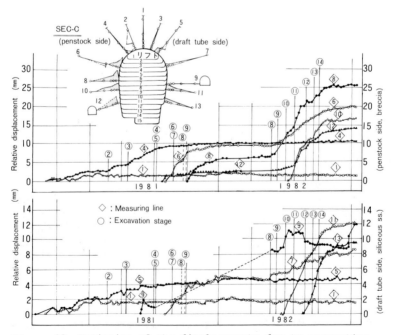

Figure 13. Variation of the displacements due to excavation

breccia, are greater than those in the draft tube side, where the rock mass is siliceous sandstone. Geo-stresses act the cavern nearly symmetrically, so the difference in displacements may depend on the difference in rock types. Figure 14 shows a distribution of strains in the rock masses, which were calculated from the measured displacements. Sharp changes in strain distribution may show us a boundary of relaxed zones during excavation, which are shown by the dotted line. In section C of Figure 14, the extent of relaxed zones is about 5 m or so, and the strain magnitude in the area exceeds 0.2%. An estimation of the extent of the relaxed zones is rather difficult because the measuring method for the relaxed zones has not been established yet. However, the size of the relaxed zones in this section C may be less than those of other mushroom-shaped caverns constructed in the past.

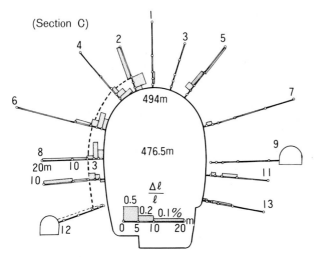

Figure 14. Strain distributions and estimation of the relaxed zone (section C)

5 CONCLUSION

The stability of large scale caverns depends on many factors, such as the properties of rock masses, initial state of geo-stresses, shape of the cavern, excavation methods or reinforcing methods etc. In this paper, we have clarified the effects of cavern shapes. The main results of our studies are as follows;
(1) Caverns for hydraulic power stations have a flat face in a vertical cross-section and the height is twice that of the width. This flatness causes instability of the walls of the caverns, which is due to the increase in horizontal equivalent external loads due to excavation. The flatness also magnifies the effect of rock discontinuity.
(2) The egg-shaped cavern is superior in stability to the mushroom-shaped cavern. The mushroom type cavern has a notched shape at the upper part of the cavern in order to place the arched concrete lining. This shape produces a weak point in rock masses; such findings were obtained through numerical excavation analyses.

(3) The egg type cavern was constructed by the Tokyo Electric Power Company for the underground power station. By this study carried out during excavation, characteristic behaviour of rock displacements and the extent of relaxed zones were revealed. The stability of the egg-shaped cavern was confirmed.

REFERENCES

Hayashi, M. and Hibino, S. 1970. Visco-plastic analysis on progressive relaxation of underground excavation works. Proc. of the 2nd Conf. on Rock Mechanics, 2 : No. 4-25
Hibino, S. 1978. Stability of a large scale cavern during excavation and effects of cavern shape. J. Butsuri Tanko-Geophysical Prospecting, 31, 4 : 73-83
Hibino, S., Motojima, M. and Kanagawa, T. 1983. Behaviour of rocks around large caverns during excavation. Proc. of the 5th Int. Congress on Rock Mechanics. 2 : D199-202
Kanagawa, T. and et al. 1986. In situ stress measurements in the Japanese Islands : Over-coring results from a multi-element gauge used at 23 sites. Int. J. Rock Mech. Min. Sci. & Geomech. Abstr. 23, 1 : 29-39
Motojima, M., Hibino, S. and Kanagawa, T. 1982. Numerical analyses on stability of an egg-shaped cavern. Technical report of the Central Research Institute of Electric Power Industry (in abb. CRIEPI). No. 382019
Motojima, M. and Hibino, S. 1984. Rock behaviour of the egg-shaped cavern during excavation. Technical report of CRIEPI. No. 383057

2nd International Symposium on Field Measurements in Geomechanics, Sakurai (ed.)
© 1988 Balkema, Rotterdam. ISBN 90 6191 778 6   693

Construction stage instrumentation for the underground openings of Tehri dam complex

V.K.Mehrotra
U.P. Irrigation Research Institute, Roorkee, India

1 INTRODUCTION

Tehri dam complex of 8 x 250 MW installed capacity shall be the largest underground cavern to be constructed in Himalayan rocks. The complex is proposed to be developed in two stages. Four conventional type machines of 250 MW each will be installed in Ist stage, while four reversible pump turbines of 250 MW each will be ereceted in 2nd stage. Geological exploration revealed that no suitable site for surface power house is available because of deep seated overburden occupying the hill terrain involving huge excavations and costly treatments required for stabilising the slopes. Therefore a proposal has been made to construct an underground power station. Each of the two power house complex shall have three major cavities aligned parallel to each other. The machine hall cavity shall house the main power generating equipments, the transformer cavity for installing the main transformers and collection gallery cum expansion chamber shall serve as downstream surge tank.

The maximum size of the cavity in stage-1 complex is proposed to be 184 m long x 21.5 m wide x 56 m deep, while the machine hall cavity of stage-2 shall be 194 m long x 22 m wide x 62 m deep. There will come up many other cavities of smaller sizes inter-connected with the main cavities to serve as water conductor system, cable galleries etc. The power house shall be approachable by two adits, one at the crown level and the other at eraction bay level.

The proposed Tehri dam because of its complex geology and underground power house system is a challenge to civil engineers and geologists. The important component of the dam complex is its underground system of cavities passing through complex geological formations and different initial insitu stress conditions. The excavation of the underground cavities is of particular interest because uncommonly large sections are to be dealt with under most unfavourable geological conditions.

Preconstruction investigations have already been done in several drifts but these are not sufficient to provide conclusive decisions about the rockmass behaviour and may not be representative while working with the prototype cavern. It was therefore felt necessary to perform more tests in the tunnels already under construction in order to generate more data. Detailed geotechnical investigations are going on and a unique instrumentation scheme has been planned for implementation in the power house cavern. The process of instrumentation and data collection shall continue from construction stage to completion stage of the project. Dissemination of the data so obtained shall result in betterment of already

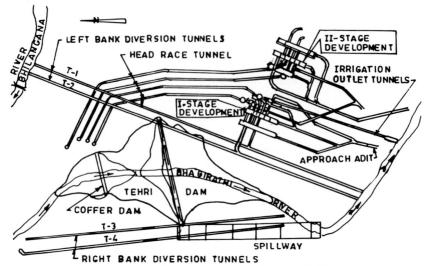

Figure 1. General plan of proposed Tehri Dam Complex showing main features.

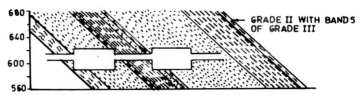

Figure 2. Geological Section along the Long Axis of Machine hall of Tehri Dam Complex.

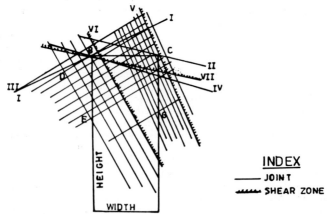

Figure 3. Cross Section along the short Axis of Tehri Dam Complex.

recommended design parameters. It is further planned to prepare photo-elastic models of the above cavities in order to determine boundary stresses in laboratory. Such studies were carried out by Bhargava et al. (1985) in order to determine the boundary stress pattern along the roof vault of machine hall cavity for different rise-span ratios. A corelation of bound-ary stresses between laboratory and field observations shall be stablished after sufficient data have been obtained through instrumentation. But because of the limits of applicability, design cannot be based wholly on photo-elastic observations, and therefore final selection of design parameters, size and shape of cavities would mainly depend upon the results of observa-tions obtained during construction stage instrumentation of the project.

In the field of geomechanics particularly relevant to river valley projects, despite tremendous advancement in technology, a full understanding of natural forces and phenomenon is still not possible because in view of the discontinuities, hetrogenities, weathering and many other factors, involved in rock masses, a number of unknown features have to be catered for. The important question still remains - what may be the actual behaviour of the ground during underground excavation ? A properly programmed and conducted geological and geotechnical investigation is necessary to understand the influence of geological conditions and the resulting states of stresses due to excavation and insitu stress condition on the ultimate behaviour of the underground openings. The author believes that the expe-rience, gained during execution of this tremendous work, shall certainly intensify the present knowledge of engineers the world over.

2 GEOLOGICAL FEATURES

The rock formations in the Tehri dam complex area belong to phyllites of Chandpur series. On the basis of their physical condition, the argilla-ceous and arenaceous materials present and the varying magnitude of tectonic deformations suffered by them, the rocks have been broadly graded as grade I, grade II and grade III. Grade III being the worst and generally weathered. The power house complex is being located mainly in grade I and grade II with localized thin bands of inferior rock formations. There are a number of shear zones passing in the power house area forming planes of weakness in the cavities. The general plan of the proposed Tehri dam complex and geological sections along the long and short axis of the machine hall have been respectively shown in figure 1, 2, and 3. In the design of underground caverns, consideration is to be given to the ability of ground either to support itself or be supported by new structural means. In every case, the engineering properties of rock matrix need be determined in order to select the most probable values of the design parameters. As construction at the site shall proceed and geotechnical conditions observed, the designer shall critically judge his assumed para-meters and change them where and when necessary. Observations during construction stage are not only necessary but a must for a rational design process, and instrumentation during this stage shall positively provide a better insight in the selection of design parameters. The present paper highlights the details of proposed instrumentation for the underground structures of the dam complex mainly the tunnels and the power house.

3 PROPOSALS FOR INSTRUMENTATION

Mainly there are two types of cavities in the proposed dam complex, the tunnels and the underground power house. In order to understand the

rock mass behaviour, and to generate design data, and also to have a feel of the most probable location of an instrument following studies have already been completed:
1. Shear strength parameters
2. Deformation modulus of the rock types
3. Groutability and permeability tests
4. Laboratory tests on rock material
5. Monitoring of movements in tunnels
6. Rock load measurements on tunnel supports
7. Determination of insitu state of stress

It is proposed to provide extensive instrumentation during construction of the project and observe the behaviour of structure during all stages. The data obtained from instrumentation and through different approaches shall provide valuable results and it will be possible to assess the most real values of design parameters. It is planned to conduct many more tests at different locations in head race tunnels, approach adits, experimental drifts constructed within the power house complex. The locations of the proposed instruments shall be such that all important rock types and their variations are covered statistically. These instruments shall measure the deformations, loads and stresses in rocks, supports and linings.

3.1 Measurement of rock movements during construction stage of dam

The following instruments are installed for measuring the rock movements at different places.

1. Measuring bolts:
With the advance of underground excavation, ground rock mass will get thrusted with higher loads resulting in convergence of side walls. Measuring bolts shall be installed for observing convergency measurements with the help of tape extensometer. These shall be a set of simple rock bolts used to measure convergence in rock along the length of the bolt. A group of bolts shall be installed in diametrically opposite pairs across the underground openings. It is highly convenient to measure in horizontal and vertical directions and the two diameters inclined to them at 45°. Depending upon the behaviour of nodal discs, exact location of the measurements shall however be decided by the field engineer. The measuring bolts shall be of non corrosive steel of 40 mm dia. from 2 to 4 m long depending upon the site conditions. These bolts shall be anchored at the bottom of drill holes. A stainless steel ball silver soldered to the other end of these bolts shall form the actual reference surface. The various diameters and chords shall be measured using tape extensometer from the start of excavation. The observations shall be recorded till the lining of underground structure is complete and the stresses are finally stabilised. These observations shall be of immense use in ascertaining the wall displacements during various stages of excavation and shall also indicate whether or not to provide suitable anchors to control the convergence. After vigilant inspection and study of the geology of the area, author feels that time dependent convergence shall be more dominating for the safety of tunnels and the main power house cavity, than the convergence on account of gravitational forces. Convergence shall increase with time firstly because of creep property of a nodal disc and secondly because of the advance of tunnel face. Critical situations may sometimes occur at places where the ground is soft and observations shall have to be taken more frequently at those places. Although it is not possible to pin point the location of measuring bolts, yet following guidelines are given:
1. Main test sections shall be constructed in all underground caverns.

A minimum of two such sections shall be provided in the main power house cavity of the proposed complex. In other cavities the number of test sections shall depend upon the geology of the area and the method and sequence of excavation. The location of the test section shall be decided in consultation with the field engineers and the geologist. One test section shall be built near the major rock defect.

2. Besides the main test sections which will help to judge the behaviour of nodal discs, convergency measurements shall be performed routinely in distance of 50 m i.e. at every 50 m of tunnel advance. This distance can vary depending upon the geological variation in rock formation. Closure observations shall also be taken at the time of benching to know the effect of benching on the rock mass behaviour.

2. Borehole extensometer:

Rock mass is blocky in nature. During excavation of an underground cavern, the whole system gets disturbed and a readjustment of blocks takes place initiating movements inside the rock mass. Specially designed borehole extensometers shall be used to measure displacement of rock surrounding the excavation providing a precise means for studying resultant displacements deep inside the rock mass along the axis of a borehole. The data so obtained shall provide information on the behaviour and extent of the zone of loosened and fractured rock around the openings. It is proposed to instal multipoint borehole extensometers preferably five nos. at a particular test section (figure 4). The data so obtained shall help to locate failure planes, localised slips, joint movements, relaxation of rock around the openings, etc. The author feels it necessary that all important discontinuities are located by using borehole television camera before executing the installation program of extensometers. It is also essential that in all underground monitoring operations, extensometers are installed near the face of excavation (preferably within 3 m) to record the displacements occurring immediately after excavation or as the next advance is excavated. The sequence of installation shall be such that the whole work is completed within 12-16 hours and all readings are taken within the excavation cycle.

It is to be kept in mind that because of the blocky nature, displacements occurring in rock mass seldom react to travel long distances and in most cases, it remains a local phenomenon. Under the most captious orientations nodal discs get finally active and it becomes a dangerous situation for the stability of structure. When such conditions occur displacements are accelerating and it is an indication of a more wide-spread, serious and positive condition resulting in the instability of underground structure. The idea is that in rock mass the stability criteria should never be based totally on a displacement magnitude but rates of displacement should be observed carefully. All the borehole extensometers installed at the project site shall therefore be provided with arrangements for observing continuous axial measurements of rocks. Site engineer shall be careful to observe that sudden increments in the rate of rock movements are surely an early indication of local unstable conditions. It is further to point that high rates of displacements that are not on account of excavation or that continue after construction has advanced well beyond the extensometer location, are sure indication of overall unstable condition within the rock mass. Therefore plots of not only the displacement versus time but rate of displacement versus time shall also be recorded at each test section.

It is recommended that multipoint borehole extensometers developed by Central Mining Research Station, Dhanbad, India are installed at site (Sharma and Joshi, 1982).

1. Their unique design provides maximum information from a single

① FOUR POINT ROD EXTENSOMETERS
② LOAD CELLS FOR ROCK ANCHORS (TIE RODS)
③ CONVERGENCE MEASURING DEVICE

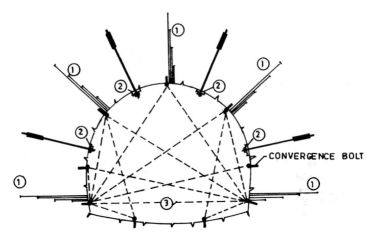

A – TUNNEL WITHOUT LINING OR BEFORE LINING

① FOUR POINT ROD EXTENSOMETERS
②③ HYDRAULIC CELLS FOR STRESS IN CONCRETE (RADIAL AND TENGENTIAL)
④ LOAD CELLS FOR ROCK ANCHORS (TIERODS)
⑤ LOAD CELLS FOR STEEL LINING.
⑥ CONVERGENCE MEASURING DEVICE

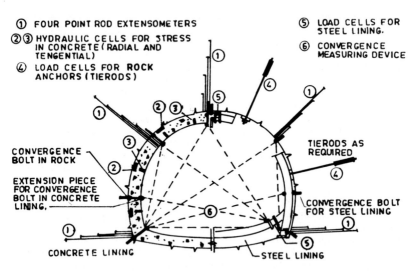

B – TUNNEL WITH CONCRETE LINING OR STEEL LINING

Figure 4. Proposed instrumentation for tunnels in Tehri dam complex.

borehole and thus reduces the cost otherwise on drilling operations. These prove therefore to be economical.

2. Installation is possible internally or externally when the overburden is not too deep.

3. Best suited for longterm measurements.

4. Can be modified to give an alarm when displacements exceed a desired value.

5. Active fracture zone or a nodal disc penetrated by the instrumented borehole can be located.

3. Borehole deflectometers:

Borehole deflectometers shall be installed to measure rock deformations at right angles to the axis of the borehole. These need be installed for the measurement of movements in shear zones and lateral displacements of tunnels. The instrument shall essentially consist of two flexibly connected parts, the deflection arm and the guide housing. There shall be a pivoted joint between the deflection arm head and the guide housing allowing the arm to vary its position as the instrument moves up and down the cased borehole. The two set of rollers provide longitudinal stability for the guide housing while the deflection arm has only one set. An anchored and tensioned steel wire passes through a knife edge orifice plate in the guide housing. Deflection of borehole results a corresponding bending of the wire and the angular deviation of wire is measured by electrical transducers.

4. Slope indicators:

Slope indicators need be installed in order to measure stability of natural and cut slopes, lateral movements of rocks and stability of shafts, tunnels and underground constructions. The rock in the vicinity of portal is generally unstable and extra loads due to sliding of rock mass come on the tunnel supports near the portals. Author recommends to instal accelerometer type slope indicators having a closed loop servo accelerometer circuit.

3.2 Measurement of loads on supports and linings

Measurement of loads and stresses actually transferred on tunnel or cavity supports enable their comparison with those assumed in the design. It is necessary to observe the actual loads and stresses coming on the supports. The spacing of supports can be better determined using the measurements made at the beginning of an underground excavation project.

1. Load cells:

Load cells shall be used for measuring the thrust in steel sections used for supporting the tunnel or the underground cavity. These cells shall be of special type capable of operating satisfactorily under submerged conditions and subjected to blasting shocks. Load cells of mechanical type containing either an elastic cup spring or a lever system which deforms during load application are recommended. In a measuring section at least three consecutive rows of load cells shall be installed. The location of cells in general shall be so dispersed as to cover statistically all the rock types found in different underground openings. It is also suggested to instal a few electrical type of load cells at specific locations so that it is possible to monitor them during post excavation stage of the project. The load cells shall be installed, in general at the springing, at the crown and under the foot of legs of the horse shoe section. This will determine the horizontal and the vertical components of loads coming on the supports.

2. Pressure cells:

During underground construction it shall be necessary to observe distribution of stress on a rib, in concrete, at the interface between concrete and rock or between rock and steel supports. These stresses shall be observed by installing stressmeters on supporting structure. The observed data shall provide a quantitative measure about the development of rock loads during construction and shall be of immense utility in critical evaluation of the rock support interaction because ground movements and support stresses being interdependent are also effected by the method and sequence of excavation. Cells shall be held in position by non-corrosive pins driven in rock or tied in place with steel reinforcement before placement of concrete. Cells placed at the concrete rock interface shall be embedded in a thin pad of lean mortar. Several groups of cells shall be installed around an instrumented section to observe distribution of pressure on and within the lining of the structure.

3. Rockbolt dynamometers:

Rockbolt dynamometers shall be installed to observe the changes of bolt load with time so that bolts may be tightened again whenever it becomes necessary. These shall be indicative of the excessive bolt loads and hence the need of additional support, if any.

3.3 Measurement of stresses in support and linings

Nothing better if the stresses in an underground excavation are measured directly. Direct measurement of stresses is of significant importance. As most of the design methods are based on limiting streses, the observation of actual stresses developed in the rock mass at the contact face of rock and structure is necessary for the assessment of the structural behaviour, improvement in design methods or to the control of safety. The stresses shall be measured directly by means of stressmeters.

4 OTHER MEASUREMENTS

Depending upon the specific requirement of a structure or under a special circumstance, following equipments may be installed.

1. Temperature measuring devices:

Measurement of surface temperature, interior and the ambient temperature is quite important for structures like intake towers, exposed partly to hot sun and partly submerged in water. Thermal stresses may be very high under such conditions. Various types of specially designed thermometers shall be used for this purpose.

2. Permanent warning devices:

In order to avoid any casualty at the construction site, it is strongly proposed to instal some of such instruments especially those installed underground.

3. Dynamic measurements:

During underground construction, it may at times be necessary to consider the effect of blasting on nearby completed structures. Instrumentation may be very helpful in arriving at safe charge for such locations.

4. Pore water pressure measurements:

It is essential to observe pore water pressure measurements during and after completion of the construction. In a number of constructions such as underground power houses, tunnels and dam foundations etc., it is imperative due to presence of entrapped water pocket or water table or reservoir filling to observe pore water pressure in the rock joints and shear zones.

5. LOCATICN CF INSTRUMENTS

The locations of instruments are difficult to pinpoint in advance. However, the most critical locations shall be decided in consultation with the field engineers and the geologist. The following criteria may be observed.

1. All nodal discs shall be instrumented thoroughly.
2. A minimum of three discs that may represent typical behaviour shall be instrumented.
3 Specific locations shall be intrumented in consultation with the geologist and the field engineer during construction stage of the project.
4. Though uneconomical yet it is recommended to have two systems of independent measurements.

6. MONITORING

If the measurements indicate a constant increase in deformation, it is necessary that measurements are taken daily. The results should be plotted as quickly as possible and shown to the field engineer. Weekly measurements shall be sufficient, if the results show a distinct tendency to stabilization once the tunnel face has passed the test section by a distance of atleast two tunnel diameters (Bieniawski 1984).

7. SUMMARY AND CONCLUSIONS

It is well known that no laboratory or field test is complete in itself. However knowledge about the during construction and post construction behaviour of rock has to be gathered with whatever tools available at hand. The geology at the dam site is quite complex and is regarded poor for supporting cavity spans. The grade II rock mostly present in these cavities is conspicuously banded due to the rapid alternations of arenaceous and argillaceous material. In physical quality and competence this unit is considered next to grade I phyllite. The rock in this unit is considerably impregnated with quartz veins, both along and across the foliation planes. Attempt has been made as given in the preceding pages to assess the likely deformations, loads and insitu stresses in such rock formation with the help of instruments available and developed in the country. The earlier investigations done for determining the various rock parameters have also been mentioned. It is hoped that the observations with the help of proposed instrumentation scheme would be useful for designing this major and important national project. A number of similar projects are proposed to be constructed in lower Himalayan region confronting the same complex geology. Therefore it is expected that the results of these observations will also help in assessing the design parameters for those project structures. Success of the proposed scheme would however depend upon the Vigilant Inspection and Instrumentation of Nodal Discs that may be existing in the area under consideration. The paper is presented for any constructive criticisms and suggestions which would be considered with a sense of gratitude for better understanding of the rock behaviour to engineering community.

8.   ACKNOWLEDGEMENT

The author wishes to acknowledge with thanks the guidance provided by Dr. P.S. Nigam, Engineer-in-Chief, Mr. R.K. Agarwal, Engineer-in-Chief (Design & Research) and Mr. D.N. Bhargava, Director, Research Institute, U.P. Government Irrigation Department, for finalisation of the instrumentation plan for the project. Author is also thankful to Dr. B.V.K.Lavania, Professor, University of Roorkee for his suggestions and critical review of the paper.

REFERENCES

Bieniawski, Z.T. 1984, "Rock Mechanics Design in Mining and Tunneling", 149. Rotterdam: Balkema.
Bhargava, D.N., V.K. Mehrotra & S. Mitra, "Use of photoelasticity in design of underground power house openings", Proceedings of III Symposium on Rock Mechanics, 21-27.Roorkee, India, Nov. 16-18, 1985.
Design report, Tehri dam project (Power House Complex), Irrigation Design Organisation, Roorkee (India), October, 1984.
Mehrotra, V.K. & S. Mitra, 1985, "2-D Photoelastic Studies of Power House (machine hall) Cavity for Various Stages of Excavations". Technical memorandum no.74, 1-6, U.P. Irrigation Research Institute, Roorkee, India.
Mehrotra, V.K. & S. Mitra, 1986, "Effect of boundary stresses around machine hall cavity in the vicinity of other cavaties". Technical memorandum no. 126, 1-5, U.P. Irrigation Research Institute, Roorkee, India.
Sharma, V.M. & A.B. Joshi, 1982, "The need for Instrumentation in Underground Excavations", Symposium on Design and Construction of Diversion Tunnels, Outlet Tunnels, Gate Structures and Intake Structures, No.159, Vol. I, 165-166. New Delhi: CBIP.

2nd International Symposium on Field Measurements in Geomechanics, Sakurai (ed.)
© *1988 Balkema, Rotterdam. ISBN 90 6191 778 6*

Field measurement of Washuzan Tunnel

Yasuji Saeki & Shinichi Ooe
Kojima Construction Office, Second Construction Bureau, Honshu-Shikoku Bridge Authority, Japan
Kakuo Takeuchi
Fukuyama Consultant Co. Ltd
Previously Kijima Construction Office, Second Construction Bureau, Honshu-Shikoku Bridge Authority, Japan

1. Introduction

The Washuzan Tunnel is a four bore tunnel situated at the northern end of the strait portion of the Kojima-Sakaide Route of the Honshu-Shikoku Bridges.

The southern mouth of the Washuzan Tunnel is adjacent to the Shimotsui-Seto Bridge, a suspension bridge for combined highway and railway use, and as the structural standard of the highway and railway is strict there is no option in the plane and vertical alignment, necessitating the sectional composition of the suspension bridge to have to be extended into the tunnel almost as it is. Therefore, the tunnel had to be constructed in the two levels, approaching up and down and right and left, which is unparalleled in the world. (Fig-1)

In this paper the field measurement that has particular significance from the viewpoint of construction management of the Washuzan Tunnel is reported below:

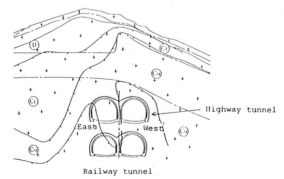

Fig-1   Cross-section of Washuzan Tunnel

2. Summary of Washuzan Tunnel

1) Topography and geology

The Washu mountain is located on the Tip of the peninsular that projects into the Seto Inland Sea and is a hill 137 m above sea level
expanding to about 0.6 Km in the north-south direction and to about
1.5 Km in the east-west direction.   On the slope, both north and
south, shallow valleys are scattered, the ridgeline extends in the
east-west direction in a zigzag line and the east end sinks into the
Shimotsui Strait.   Fig-2 shows the vertical section of the geology at
the center of the route.   With the exception of part of the top soil,
most of the geology consist of igneous rock, the main portion of which
is black micagranite.   Also the basic salt rocks represented by
diorite can be seen.   From the distribution of the granite rocks and
the diorite rocks, it is assumed that the granite rocks penetrated the
basic salt rocks and intruded into it afterward, but the entire
geological construction becomes complicated as it shows diversified
contact form.   With regard to the joint group, the dip is almost
perpendicular and the strike is predominantly 45° - 90° to the tunnel
axis.   However, there are some that have nearly horizontal dip and it
may be said to form a complete block.   Also, the considerable cracks
have developed in the fresh and rigid bedrock in the deeper layer,
making it a significant characteristic of the geological construction.

2) Specification of the tunnel

Fig-3 shows the typical section and Table-1 the specification of the
tunnel.

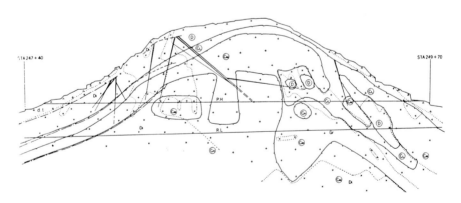

Fig-2   Geological profile

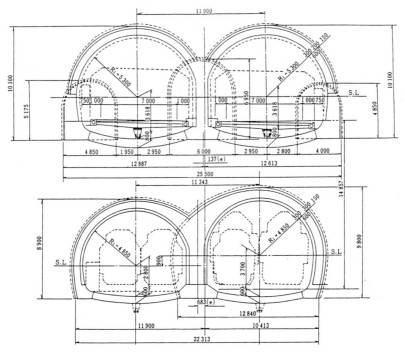

Fig-3   Cross-section of highway and railway tunnel

Table-1.   Specification of the tunnel

		Railway Tunnel		Highway Tunnel		
		East (conventional train)	West (Shinkansen)	East (down traffic)	West (up traffic)	
Length (m)		230		205		
Inclination		1% (Honshu side ↗ Shikoku side)		0.852% (Honshu side ↘ Shikoku side)		
Cross-section area of excavation (m²)		81	123	119	119	
Finish internal cross-section area (m²)		59	69	75	75	
Shotcrete (cm)		20		20		
Steel support		H-150		H-150		
Rockbelt (per section)	Main tunnel	SN bolt (TD-24) ℓ = 3m x 6 ℓ = 4m x 7 ℓ = 6m x 2	SN bolt (TD-24) ℓ = 3m x 10 ℓ = 4m x 5 ℓ = 6m x 2 Glass fibre ℓ = 3m x 6	Deformed steel bar (SD 35) ℓ = 3m x 13	Deformed steel bar (SD 35) ℓ = 3m x 13	
	Pilot tunnel			Deformed steel bar (SD 35) ℓ = 2m x 5 Glass fibre ℓ = 2m x 6	Deformed steel bar (SD 35) ℓ = 3m x 3 Glass fibre ℓ = 2m x 10	Deformed steel bar (SD 35) ℓ = 2m x 2 Glass fibre ℓ = 2m x 5
Lining (cm)		60 (double reinforcement)		60 (semi-reinforcement)		
Decorative lining * (cm)		30		30		

* to be constructed as required.

3) Method of construction

As a result of examination of dynamic stability, workability, economy, etc. the sequence of construction of the Washuzan Tunnel was determined and the completion of the lower railway tunnel up to the lining was followed by the excavation of the upper highway tunnel. For both the highway and the railway tunnel, the tunnel on the west side preceded that of the east side tunnel, and the excavation of the east side tunnel commenced after the lining concrete of the west side tunnel had hardened.

As the four tunnels were to be constructed in proximity to each other in this construction work, considerations had to be given so as to control the damage of the natural ground to a minimum and to reduce ill-effects among the tunnels themselves as much as possible. Therefore the NATM method of construction was adopted throughout the entire length.

3. Field Measurement

1) Objective of field measurement

The objectives of field measurement of this construction work are given below:
① To grasp the behaviors of the peripheral natural ground of the tunnel and the effect of each support member, ensuring the safety of tunnel excavation and improving the economy of construction.
② To examine the mutual effect of four tunnels.
③ To use the results as input data in the various analyses which have been performed in the construction of the tunnel.
① The above refers to the field measurement that is incorporated in the ordinary NATM system element and 2 and 3 are unique to this tunnel.
② The above can be classified into two types. One is to judge the effect of excavation of the preceding tunnel on the natural ground and the loosening domain, etc. and use them as reference for the tunnel excavation that follows. To meet this, the measurements of subsidence of ground surface and underground displacement (outside the tunnel) and an elastic wave survey in the tunnel were performed. The other is to judge the effect of the excavation that follows on the preceding tunnel that has been completed and use them as reference to take necessary and adequate action. For this purpose, the measurements of convergence, underground displacement (in the tunnel) and lining stress on the completed tunnel were performed.

③ The above is to establish the initial stress of natural ground and the bedrock properties values so that the measured values of the convergence of the preceding tunnel, the settlement of the tunnel crown and the underground displacement (both in the tunnel and outside the tunnel) match the calculated values, aiming at improving the accuracy of the analysis.
   Table-2 Summarizes these.

Table-2 Classification of field measurement by objective

Objective	Items of measurement
To judge stability of natural ground in the vicinity of the tunnels and to secure the safety of excavation as well as improving economy.	Convergence, crown settlement, underground displacement (in the tunnel), shotcrete stress, visual observation of shotcrete surface, subsidence of ground surface, rockbolt axial force.
To judge the extent and scope of the effect of excavation of the preceding tunnel and to use them as reference for excavation of the tunnel following.	Subsidence of ground surface Underground displacement (outside the tunnel) Elastic wave survey in the tunnel
To judge the effect of excavation of the tunnel following on the excavated tunnel and to use them as reference for actions to be taken.	Convergence underground Underground displacement (in the tunnel) Lining stress
To use results of measurement as input data to improve the accuracy of analysis.	Convergence Crown settlement Underground displacement (in the tunnel) Underground displacement (outside the tunnel)

2) Plan of field measurement

Table-3 shows the summary of field measurement, Fig-4 the location of measurements performed at the time of excavation of highway tunnel and Fig-5 the typical section of the tunnel.

Table-3 Summary of field measurement

No.	Item			Number	Frequency of measurement	Accuracy	Instrument used
I	Survey of face			About 1,000 faces	5 m in front and rear of the section of the measurement of convergence joint dressing is performed at each face.		Clinometer Tape measure
II	Convergence			8 cross-sections 96 measuring lines West side 6 x 8 = 48 east side 4 x 8 = 32 after lining 1 x 8 x 2 = 16	Once a day before and after the passing of face, thereafter reduce frequency, e.g. once a week, once a month, etc.	Higher than 2/100 mm	Distometer Tape extensometer (spare)
III	Crown settlement			8 cross-sections 64 measuring lines West side 1 x 8 = 8 east side 1 x 8 = 8 afterlining 3 x 8 x 2 = 48	Same as above	Higher than 1/10 mm	1st class leveling instrument 1st class staff
IV	Lining stress			7 cross-sections 154 measuring points West side 12 x 7 = 84 east side 10 x 7 = 70	Same as above	Smaller than minimum reading value 19 kgf/cm^2	Carlson type bar gage Carlson meter
V	Underground displacement (in the tunnel)			8 cross-sections 96 bars West side 3 x 8 = 24 east side 3 x 8 = 24 after lining 3 x 8 x 2 = 48	Same as above	Higher than 2/100 mm	Multiple point bedrock displacement gauge
VI	Observation of shotcrete and lining concrete and measurement of cracks			—	Constantly	Higher than 1/10 mm on cracks	Caliper (contact gauge)
VII	Rockbolt axial force			3 cross-sections 27 bolts West side 5 x 3 = 15 east side 4 x 3 = 12	Once a day before and after the passing of face, thereafter reduce. Frequency, e.g. once a week, once a month, etc.	Higher than 1/100 mm	Mechanical anchor
VIII	Elastic wave survey in the tunnel			3 cross-sections 15 measuring lines 600 m West side 3 x 3 = 9 lines 40 m/measuring lines east side 3 x 2 = 6 measuring lines	—	—	Amplifier field graph, vibration receiver take out cable, data recorder, blaster, etc.
IX	Subsidence of ground surface			5 cross-sections 42 measuring points (8 + 8 + 8 + 9 + 9)	2 times/month	Higher than 1/10 mm	1st class leveling instrument 1st class staff
X	Underground displacement (outside the tunnel)	i	(one component) B	ℓ = 165 m 1 hole	Same as above	Higher than 1/100 mm	Horizontal inclinometer
		ii	(one component)	3 cross-sections 6 holes 181 m	Once a day before and after the passing of face, thereafter reduce the frequency, e.g. once a week, once a month, etc.	Higher than 1/1,000 m	Sliding micrometer
		iii	(two components) B	1 cross-section 2 holes 110 m West side 56 m east side 54 m	Same as above	Higher than 1/100 mm	High-accuracy inclinometer in the hole
		iv	(three components)	2 cross-sections 4 holes 258 m West side 68 + 53 = 121 east side 79 + 58 = 137	Same as above	Horizontal - higher than 1/500 mm Vertical - higher than 1/1,000 mm	Vertical inclinometer

The special features of the field measurement of this construction work are given below:

① Items of measurement are diversified
② Many measurements are to be taken from outside
③ Accuracy of measurement is high
④ Many measurements was performed after the secondary lining

① The above has the objective to make as correct a judgement as possible by mutually supplementing each measurement data and by comprehensive evaluation. For example, by comparing the axial force distribution of rockbolts and the results of measurements of underground displacement in the tunnel and the elastic wave survey in the tunnel, it is possible to accurately assume the loosened domain of the bedrock, etc.

② The above is to grasp the behaviors of the whole of the natural ground and obtain the average behaviors of the natural ground and the scope of the effect of excavation. It is possible to grasp the displacement of the ground before the face is reached, thereby obtaining total displacement.

③ The above is due to the fact that the bedrock in this region is relatively sound, the breaking strain is small, the displacement was assumed to be small from the analysis performed in advance and it was feared that with an ordinary instruments check the changes in the measured values by the time elapsed could not be identified, making delicate management impossible and as the cracks developed it was considered desirable for the accuracy of instruments to be as high as possible so as to be able to perform the comprehensive examination in consideration of the effect of these cracks based on the results of measurements, etc.

④ This is considered to be an important and controlling factor to grasp the mutual effect among the tunnels themselves for the safe and secure construction because of the peculiar nature of this construction work.

On the other hand, it is important to prepare in advance the methodology of how to reflect the results of measurement on the design and construction and the specific management criteria. In the construction of this tunnel it was considered questionable to apply the conventional general management criteria to the construction work due to the peculiar nature of the four tunnels which had to be excavated in close proximity. Therefore the construction work was managed which the management criteria were being established according to the following procedure:

First, from the results of measurements on the preceding tunnel the values of the initial stress of natural ground and bedrock proportion were assumed, based on which the displacement of natural ground and the shearing strain distribution of the following steps were assumed by analysis. The calculated values and the measured values were compared at each stage, correcting the initial stress of natural ground and the bedrock properties values and again an analysis was performed. In other words, the latest measured values were constantly reflected in the analysis using the calculated values as criteria for the construction management.

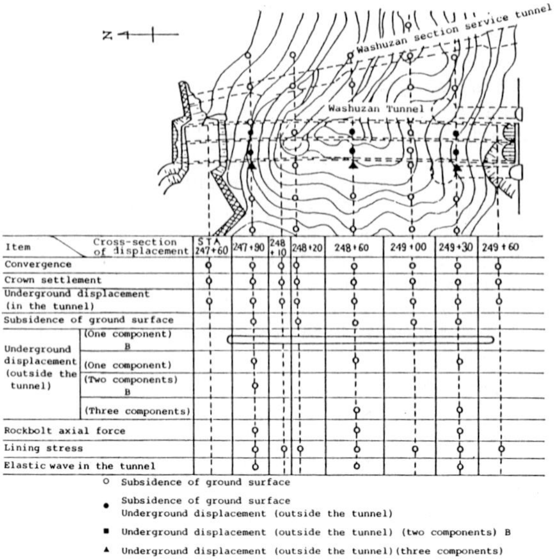

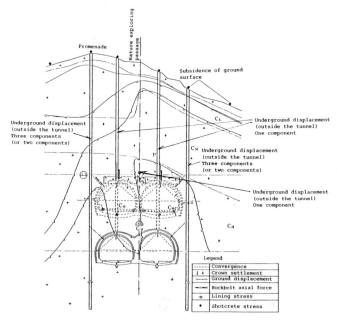

Item	Cross-section of displacement	STA 247+60	247+90	248+10	248+20	248+60	249+00	249+30	249+60
Convergence		○	○	○		○	○	○	○
Crown settlement		○	○	○		○	○	○	
Underground displacement (in the tunnel)		○	○	○	○	○	○	○	○
Subsidence of ground surface			○		○	○	○	○	○
Underground displacement (outside the tunnel)	(One component) B								
	(One component)		○			○		○	
	(Two components) B		○						
	(Three components)					○		○	
Rockbolt axial force			○			○		○	
Lining stress			○	○	○	○	○	○	○
Elastic wave in the tunnel			○			○		○	

○ Subsidence of ground surface

● Subsidence of ground surface
Underground displacement (outside the tunnel)

■ Underground displacement (outside the tunnel) (two components) B

▲ Underground displacement (outside the tunnel)(three components)

Fig-4 Location of field measurement

Fig-5 Model field measurement section

3) Results of measurements

In this report, of the main measured cross-sections (STA 247+90, STA 248+60, STA 249+30), the results of measurements from the viewpoint of mutual effect of tunnels on the west side tunnel of the railway tunnel STA 249+30, where the distance between the highway tunnel and the railway tunnel is the smallest, are shown.

Fig-6 shows the axial strain distribution in the periphery of the face that was assumed from the results of measurement taken of the components of the underground displacement (outside the tunnel) at the location immediately above the west side tunnel of the railway tunnel. From the results it can be seen that the axial elongation strain increased above the tunnel as a result of the passing of the face of the railway tunnel, the value of which is small, i.e.   about 0.1% maximum.

Fig-7 shows the convergence after the lining of the railway tunnel has been completed.   It appears that no significant effect was shown as a result of the passing of the highway tunnel face.

From these results, it can be seen that the four bore tunnel could be excavated using the present method of construction.

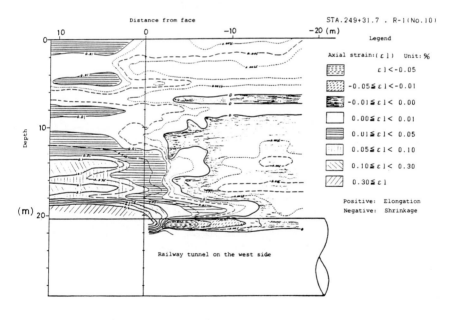

Fig-6   Axial strain distribution in the periphery of face using a sliding micrometer (No. 10)

Railway tunnel (STA249+30)

(+) Elongation
(-) Shrinkage

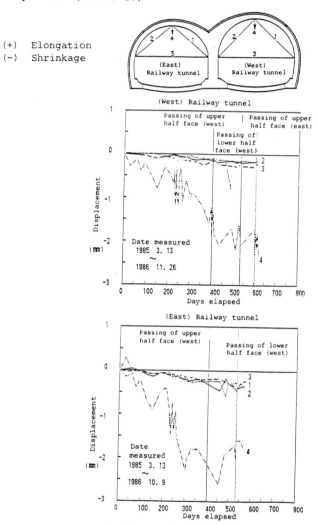

Fig-7 Transition of convergence

4. Conclusion

In the construction of the Washuzan Tunnel the field measurement was extremely effective as a means of grasping the mutual effect of the tunnels and monitoring the construction due to the peculiar nature of the construction work in which the tunnels had to be excavated in proximity to each other, the summary of which is reported herein.

The main construction work of the Washuzan Tunnel has almost been completed and only appurtenant works remain to be completed and is expected to be put to use in the spring of next year.

2nd International Symposium on Field Measurements in Geomechanics, Sakurai (ed.)
© 1988 Balkema, Rotterdam. ISBN 90 6191 778 6

Deformabilities of rock mass around an underground mine roadway by in-situ measurements

Yoshihiro Ogata & Yasuo Tomishima
National Research Institute for Pollution and Resources, Tsukuba, Japan
Yuichi Nishimatsu & Seisuke Ohkubo
Tokyo University, Japan

1 INTRODUCTION

When excavating a civil tunnel and underground mine roadway, the rock mass around these caverns is disturbed mechanically and transforms up to recovering a mechanical equilibrium again.

These deformations depend not only on the geological and physical properties of the rock mass, but also on the type and stiffness of support, such as rock bolts, steel arches, and concrete-linings, etc.

As the deformabilities of rock mass effect the design and setting of the supports, if it is possible to make a model of the deformabilities of the rock mass and estimate the pattern and value of deformation before excavating the caverns, then it is possible to plan an efficient and effective support system.

The deformabilities of a rock mass are generally estimated by means of the analysis of a mathematical model, such as the finite element method, etc., based on the data obtained by geological prospecting and physical properties of the boring core sample. In some cases, however, the phenomenon of deformation obtained by means of model experiments and in-situ measurements is analyzed for this purpose. In-situ measurement is a valid technique for estimating the practical deformations of rock mass around caverns, and is especially adopted for monitoring the behavior of a rock mass when excavating the tunnel by the New Austrian Tunnelling Method.

In-situ measurement is generally carried out after the excavation of caverns because of the limits on the underground structure. There are few examples of the behavior of rock mass being measured before excavating the roadway.

However, as the deformation of a rock mass practically occurs with the approach of the caverns, it is important to consider the deformability of the rock mass, including the deformation which is induced before excavating the caverns.

In order to make a model of deformability for the rock mass with respect to standardization of the design and setting of the supports for the underground roadway and civil tunnel, authors measured the deformation of the rock mass around the roadway before and after excavating the caverns in two underground mines.

2 MEASURING FIELD

2.1 Hiraki Mine

Hiraki Mine is located in the central part of Hyogo Prefecture and is mined for hydrothermal kaolinite-pyrophyllite by the underground mining method. The entire ore deposit is a massive type, whose size is about 250m in strike length and 40m to 50m in thickness, with about a 30 degree dipping. Main minerals that compose the ore are with 40% kaolinite (Al_2O_3 $2SiO_2$ $2H_2O$) and 60% quartz (SiO_2).
    The roadway which was selected as the object of the in-situ measurement was a horizontal exploratory roadway (drift) being driven about 80m under the ground level. The rock mass around the exploratory roadway is ore, and the mechanical properties of the rock samples obtained from the boring hole for setting the instruments are shown in Table 1. As shown in this table, the rock mass is relatively hard, so that this exploratory roadway is not supported.
    Fig. 1 shows the underground structure near the exploratory roadway and the setting pattern of the instruments in Hiraki Mine.

Table 1. Mechanical properties of rock mass (average, kgf/cm^2)

Mine	Hiraki	Matsumine
Rock mass	Ore (kaolinite)	Rock (tuff, gypsum, tuff mixing gypsum)
State of sample	Saturated	Natural
Uni-axial compressive strength	380	165
Tensile strength (Brazilian test)	19	12
Shear strength	105	–
Young's modulus	11.3×10^4	0.75×10^4

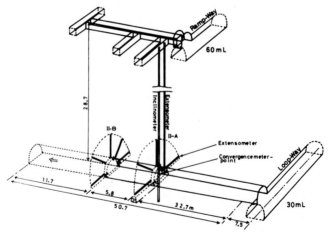

Fig. 1 Setting pattern of instruments in Hiraki Mine

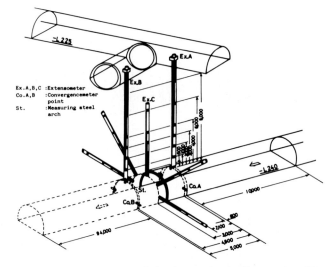

Fig. 2 Setting pattern of instruments in Matsumine Mine

2.2 Matsumine Mine

Mastumine Mine is located in the northern part of Akita Prefecture and Kuroko (black ore) is mined by underhand cut and fill with artificial roofing. Kuroko is composed of kuroko, yellow ore, and iron sulphide. The ore deposit is generally of the bedded type whose size is about 800m long of north-south and about 600m long east-west with an irregular thickness of 300 to 400m under the ground level.

The roadway which was selected as the object of the in-situ measurement was horizontal a exploratory roadway (drift) being driven about 300m under the ground level. The rock mass around the exploratory roadway is mainly composed of tuff, gypsum, and tuff including gypsum and the mechanical properties of the rock samples are shown in Table 1. As shown in this table, the rock mass is very weak. Thus, this exploratory roadway is supported by means of type 115 steel arches at an equal distance of 0.6m.

Fig. 2 shows the underground structure near the exploratory roadway and the setting pattern of the instruments in Matsumine Mine.

3 INSTRUMENTS AND MEASURING METHOD

The instruments and measuring methods are shown in Table 2. The main object of this measurement was to measure the deformation of the rock mass from before driving the exploratory roadway and just after blasting.

4 RESULTS

In these proceedings, the results measured with some extensometers, which were set in the vertical borehole bored in the roof of the exploratory roadway from the floor of upper level roadway, are indicated.

Table 2. Measuring instruments and methods

Mine	Hiraki		Matsumine	
Instruments setting				
Place	Roof and side of expected roadway (30mL)	Around the roadway (30mL)	Roof of expected roadway (-L240)	Around the roadway (-L240)
Method	From the floor of upper roadway (60mL)	From the roof and wall of the roadway (30mL)	From the floor of upper roadway (-L225)	From the roof and wall of the roadway (-L240)
Time	Before the face passes the measuring section	After the face passing	Before the face passing	After the face passing
Instrumentation				
Displacement in rockmass (extensor)	(1) Multi-extensometer with hydraulic anchoring * 6-anchors * 1-set x 1-section (2) Movable type inclinometer * 1-set x 1-section	(3) Multi-extensometer with hydraulic anchoring * 3-anchors * 4-sets x 1-section (4) Single-extensometer with cement-mortar anchoring * 5-sets x 1-section	(6) Multi-extensometer with hydraulic anchoring * 6-anchors * 1-set x 2-sections	(7) Multi-extensometer with hydraulic anchoring * 6-anchors * 5-sets x 1-section
Deformation of roadway (convergence)		(5) Steel tape convergencemeter * 3-points x 2-sections		(8) Steel tape convergencemeter * 3-points x 2-sections
Deformation of steel arch				(9) Strain gauge cemented on the side of steel arch * 10-positions x 1-set (10) Photograph of reflection tape cemented on the side of steel arch * 10-positions x 1-set

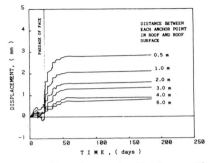

Fig. 3 The change in displacement
with time lapse in Hiraki
Mine

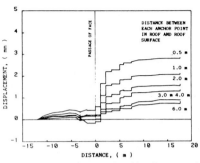

Fig. 4 The change in displacement
with advancing of the face
in Hiraki Mine

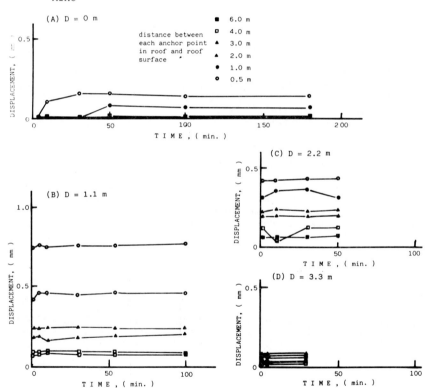

Fig. 5 The change in displacement with time lapse just after blasting
(D: distance from the face to the measuring section)

4.1 Hiraki Mine

Figs. 3 and 4 indicate the change in vertical displacement of the rock
mass in the roof of the exploratory roadway (30mL) measured from the
floor of upper level roadway (60mL) with time lapse and with advancing
of the roadway face (face).

Each anchor point in the roof begins to move according to the approach of the face to the measuring section, and is steeply displaced when the face passes the measuring section. But, the displacement rate decreaces gradually as the face passes away from the measuring section, and dwindles extremely when the face is located about 5m from the measuring section (same distance as the width of roadway).

Fig. 5 indicates the change in the vertical displacement of the roof in the exploratory roadway with time lapse just after each blasting. It is seen in Fig. 5 that the elastic displacement is instantaneously induced in the roof at the same time as the blasting, but the displacement with time lapse is scarcely induced. Namely, the deformation of this rock mass is the accumulation of the instantaneous elastic deformations which were induced by the occurrence of caverns, that is by the change of the face position with time lapse.

Because this rock mass is relatively hard and the roadway is relatively shallow, the displacement is very small and is about 3mm at a point of 1m in the roof.

4.2 Mastumine Mine

Figs. 6 and 7 indicate the change in the vertical displacement of the rock mass in the roof of the exploratory roadway (-L240) measured from the floor of the upper level roadway (-L225) with time lapse and with advancing of the roadway face (face).

Each anchor point in both roofs of two measuring sections A and B, begins to move when the face approaches within 5~6m of the measuring section, and the greatest displacement of these points is observed when the face is located between about 2m before and 4m behind the measuring section. But each anchor point in the roof continues to displace after the face passes about 5m away from the measuring section. Therefore, the displacement rate was calculated each time from a set of data. In this case, the inclination of simple equation was calculated from the adjacent 5 data by the least squares methods and was adopted as a displacement rate on the time when the middle datum was measured.

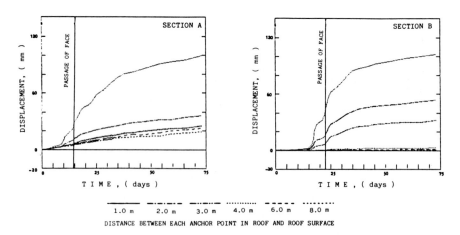

Fig. 6 The change in displacement with time lapse in Matsumine Mine

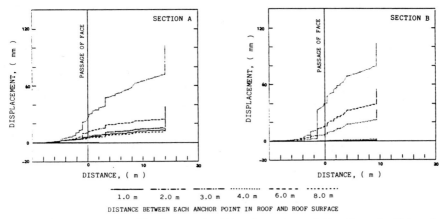

Fig. 7 The change in displacement with advancing of the face in Matsumine Mine

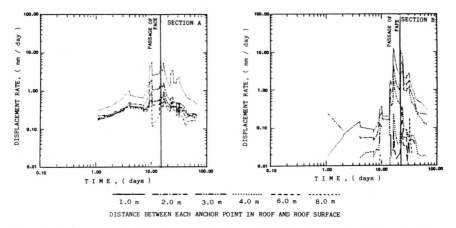

Fig. 8 The change in displacement rate with time lapse in Matsumine Mine

Fig. 8 indicates the change in the displacement rate with time lapse. It is seen in Fig. 8 that the displacement rate increases as the face approaches the measuring section and the three greatest peaks of displacement rate are observed of the time, before, and after the face passed the measuring section. In these cases, the maximum displacement rate is 8mm/day at section A and 12mm/day at section B.

Fig. 9 indicates the change in displacement of a 1m point in the roof with time lapse just after each blasting after the face passed the measuring section. It is seen in Fig. 9 that the displacement is not elastically induced just after the blasting but changes parabolically with time lapse. Namely, as is seen in Figs. 5 and 9 that the deformabilities of the rock mass in Matsumine Mine are very different from the deformabilities of the rock mass in Hiraki Mine, and the displacement rate calculated by the least squares method changes with time lapse in Matsumine Mine. Fig. 10 indicates the change in the displacement rate

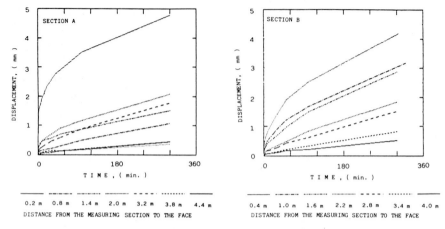

Fig. 9 The change in displacement of a 1m anchor point in the roof
with time lapse just after each blasting in Matsumine Mine

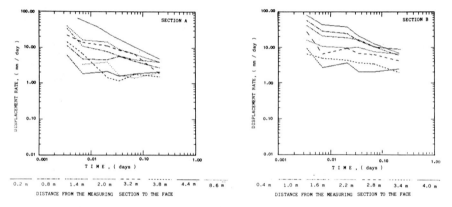

Fig. 10 The change in displacement rate of a 1m anchor point in the roof
with time lapse just after each blasting in Matsumine Mine

with time   lapse on both logarithmic graphs, and it is seen in these
figures that the displacement rate of this rock mass decreases lineally
with time lapse on both logarithmic graphs. The displacement rate also
decreases with the face location from the measuring section, but the
degree in reduction of the displacement rate is generally great when the
face is located close to the measuring section.
     To   estimate   the long-term change in the displacement rate from the
short   term change is an important factor for estimating the long-term
deformation and displacement of rock mass. It is in Fig. 11 that the
displacement rate of a 1m point in the roof was plotted with time lapse
on both logarithmic graphs. The displacement data which were measured
between a blasting and the next blasting was arranged as a group and, 3
groups at section A and 2 groups at section B were selected from many
groups. These 5 groups of displacement rates were calculated by the
following equation.

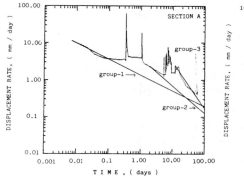
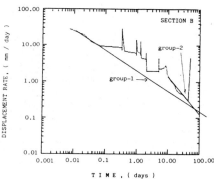

Fig. 11 The change in displacement rate of a 1m anchor point in the roof with time lapse after the face passed the measuring section in Matsumine Mine
(Straight lines indicate experimental equations)

Table 3. Coefficients of experimental equations for the five groups and the long-term displacement rate estimated from the experimental equations

Measuring section (A)				(B)	
Group number	(1)	(2)	(3)	(1)	(2)
Measuring date	11/28	11/29 ~12/03	12/14 ~1/17	12/05	12/14 ~1/17
Coefficient					
A	1.71	4.53	33.5	17.0	11.6
b	−0.474	−0.785	−1.14	−1.03	−0.976
Displacement rate (mm/day)					
after 100days	0.193	0.122	0.177	0.149	0.129
after 500days	0.090	0.034	0.028	0.028	0.027
after 1000days	0.065	0.020	0.013	0.014	0.014

$$\dot{d} = A\, t^b \tag{1}$$

where $\dot{d}$: displacement rate (mm/day)
    t: time (day)
    A, b: coefficients

The experimental equations are shown as lines in Fig. 11 and the coefficient, A and b, of the experimental equations and the long-term displacement rates estimated from these equations are indicated in Table 3. Because all the rock mass around the measured area was in almost the same geological and mechanical states, the long-term displacement rate which was estimated by the experimental equations, agrees relatively well for four out of the five groups.

5 CONCLUSION

With respect to the deformational measurement of rock mass around an exploratory roadway carried out in two mines, the measuring methods and a part of the obtained results have been indicated in this paper.

The conclusions obtained are as follows.

(1) the rock mass in Hiraki Mine indicates mainly elastic deformation.

(2) the rock mass in Matsumine Mine indicates in-elastic deformation which transforms as time lapses.

From these results, it is evident that it is possible to estimate the long-term displacement rate of rock mass around caverns from the short-term displacement rate measured just after excavation of the caverns.

2nd International Symposium on Field Measurements in Geomechanics, Sakurai (ed.)
© 1988 Balkema, Rotterdam. ISBN 90 6191 778 6

New developments in stress measurements for the design of underground openings

B.C.Haimson

University of Wisconsin, Madison, USA

ABSTRACT

This paper reviews the most recent improvements in the
hydrofracturing method with respect to testing technique and data
interpretation. It also reports on advances in the triaxial strain
cell and on the development of completely new methods of stress
measurement. The second half of the paper is a review of some recent
case histories emphasizing new applications of in situ stress tests
in the design of mine openings, hydroelectric schemes, and nuclear
waste repositories.

1 INTRODUCTION

Knowledge of the state of in situ stress is prerequisite in the
rational design of underground openings. However, until a few years
ago stress measurements were very restricted both in scope as well as
in methods employed. Typically, stress tests were conducted in large
hydro projects involving subsurface power plants (Dolcetta 1971), and
in potentially unstable mine openings (Leeman 1958). The methods
used involved overcoring and, hence, were limited in depth of
operation. Consequently, only post excavation measurements were
feasible, thus rendering the initial, preexcavation, design
tentative.

The advent of the hydrofracturing technique of stress
determination removed the depth obstacle and enabled the use of
exploration drill holes for evaluating the in situ stress condition
in the preexcavation stage of design (Haimson 1977, 1978). This was
a dramatic improvement since it provided the means of preventing
costly design modifications after the construction of underground
openings. Such 'midcourse' design changes had sometimes been
necessary when no stress tests could be conducted in advance of
excavation.

In the last several years additional improvements in the methods
of stress measurement and the increased appreciation of the benefits
resulting from direct determination of the stress regime have
contributed to a considerable broadening of the aspects of
underground opening design requiring such measurements. In this
paper we discuss some of the major advances in in situ stress

measurement, and review (with the help of case histories) several new
areas of design for which stress tests are carried out.

2   IMPROVEMENTS IN HYDROFRACTURING

The introduction of the hydrofracturing technique for stress
measurements lifted the depth barrier, since the method did not
require concentric overcoring or/and under-water strain gage bonding
(Haimson 1974). However, the initial method had some definite
drawbacks. For example it required heavy equipment such as drill rig
and drill rods for lowering the probe downhole, rendering the effort
rather expensive and cumbersome. Also, the interpretation of the
pressure-time records was ambiguous at times, and with some
exceptions the determination of the horizontal principal stresses was
possible only if the hydrofractures were along the axis of vertical
testholes. These drawbacks considerably restricted the use of the
method.
    In the last five years or so, however, major breakthroughs have
been registered in the use of the method, and with them the number of
practitioners and of applications have dramatically increased. Some
of the hydrofracturing improvements are described below.

2.1   Testing Techniques

The first major improvement in the conventional technique came with
the introduction of a flexible high pressure hose solely for the
inflation of the downhole packers. The hose was lowered strapped to
the outside of the drill rod and was instrumental in making
hydrofracturing a 'continuous' method (i.e. tests could be run
sequentially without the need to retrieve the packers after each
test, Haimson 1978). This improvement considerably decreased the
time spent on each test, lowered the cost of measurement and made it
more affordable.
    A subsequent improvement which dramatically contributed to
enhancing the cost efficiency of the method was the introduction of
wireline hydrofracturing (Rummel et al 1983; Haimson and Lee 1984;
Enever and Chopra 1986; Bjarnason et al 1986). This new system
consists of two major downhole assemblies, a wireline straddle packer
for hydrofracturing and a wireline impression - orienting tool for
the delineation and orientation of the hydrofracture. Each is
tripped into the testhole on a multiple conductor wireline operated
by a hoist mounted in a geophysical logging truck (Figure 1). By
replacing the conventional drillrig with a light weight tripod and a
portable hoist, and the drill rods with slim continuous high-pressure
hose and wireline, the new system is considerably lighter, easier to
operate, substantially faster, and thus more economical. The
wireline conductors and downhole instrumentation attached to the
straddle packer assembly provide hydrofracturing and packer pressure
readings taken right at the depth of testing. These are more
accurate then extrapolations based on the surface readings typically
employed in conventional testing. The self-contained wireline method
is particularly adaptable to relatively short hole testing such as
encountered in exploratory holes for the design of underground
openings, or in holes drilled from caverns and tunnels. Wireline

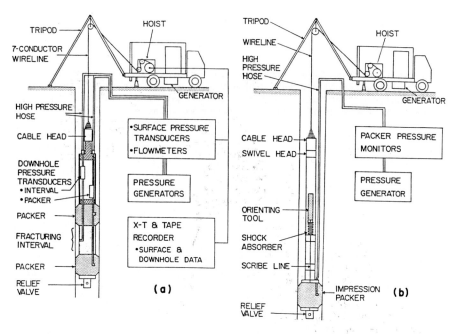

Figure 1 Schematics of the University of Wisconsin wireline hydrofracture assembly (a) and impression orienting tool (b).

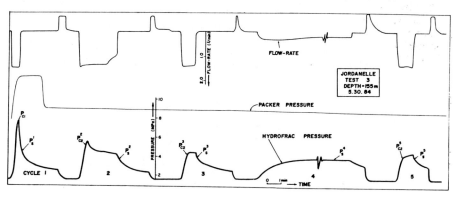

Figure 2 Typical hydrofracturing test pressure-time, flowrate-time analog record, showing P_s and P_{c2} for different cycles. Note that neither P_s, nor P_{c2} are obvious directly from this plot.

hydrofracturing has been used extensively in Europe and the United States at depths reaching 1000 m. However, it is generally not recommended in holes that may contain loose rock or other imperfections that may hinder the tool advance, since unlike drillrods the wireline is capable of neither pushing nor twisting to unjam the testing probe.

2.2   Interpretation

Clearly, the major contributor to the recent progress in
hydrofracturing data interpretation has been the introduction of
digital recording and processing. The unprecedented flexibility and
speed with which digitized field data can now be analyzed have opened
up techniques that previously were not technically feasible. For
example in the case of correctly identifying the critical pressure
values (the shut-in, P_s, and the refrac, P_{c2}) on a typical pressure-
time curve recorded during a test (Figure 2), the procedures have
varied with the practitioners and were all rather subjective (Zoback
and Haimson, 1982). The shut-in pressure directly yields the value
of the least horizontal principal stress S_h (in typical vertical
boreholes and axial hydrofractures) and is also an important
ingredient in the calculation of the maximum horizontal stress S_H.
The correct determination of P_s is thus a prerequisite to the
successful interpretation of hydrofracturing tests. It is now
possible to determine P_s unambiguously by invoking Muskat's (1937)
exponential model of the pressure decay in the test interval (Aamodt
and Kuriyagawa, 1983), and applying a nonlinear regression analysis
program to the recorded digital pressure data. The first segment of
the pressure-time curve following shut-in usually does not fit the
exponential model because the induced hydrofracture is still open.
The analysis discards in a systematic way those points not belonging
to the exponential decay until a best fit for the remainder of the
pressure-time curve is determined (Figure 3). The shut-in pressure
is then the value of the fitted exponential curve extrapolated to the
point in time where shut-in was initiated. Recently conducted
laboratory tests verify that the P_s thus obtained is within ±5% of S_h
(Figure 3).
     An alternative method of determining and/or confirming the shut-in
pressure is the use of pressure-flow rate (P-Q) plot obtained by
stepwise pressurization of the test interval after initial
hydrofracturing and pressure venting. The hydrofractured test

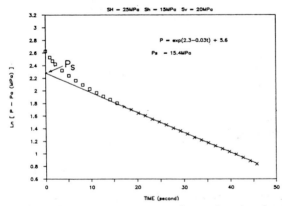

Figure   3   Shut-in pressure determination by fitting an exponential
decay curve to digitally recorded data. The square symbols are data
rejected by the non-linear regression program. Note the closeness of
P_s to applied stress S_h.

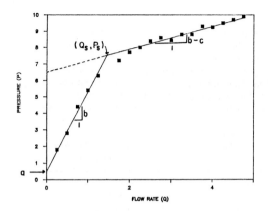

Figure 4 Pressure-flowrate data points, their bilinear regression fit, and the resultant P_s.

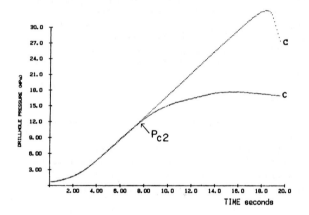

Figure 5 Determination of P_{c2} by picking out the point where the ascending pressure-time slope in cycle c departs from that in cycle a.

interval is first repressurized at a low constant flow rate until the pressure stabilizes. The procedure is then continued in a stepwise manner at increased levels of constant flow rates and corresponding constant pressures. A typical pressure-flow rate plot yields experimental points that roughly follow a nonlinear curve having two dominant slopes separated by a transition zone (Figure 4). The interpretation of this curve is that at pressures below P_s the hydrofracture remains closed and the pressure increases with the flow rate at a considerably higher gradient than that achieved after the fracture reopens. The point at which the change in slope of the P-Q fitted curve occurs is interpreted as being equal to the shut-in pressure P_s.

However, the selection of P_s within the transition zone on the P-Q curve is often difficult because the change in slope is not

sufficiently abrupt. To minimize the subjectivity involved in
pinpointing P_s, we recently introduced a statistical program which
employs a least square nonlinear regression technique to determine
the bilinear curve that best fits the digitized experimental
results. Thus, the transition point at the intersection of the two
straight lines can be determined unambiguously. One great advantage
of this analysis is that in the transition zone formed by the
experimental P-Q points, it determines uniquely which of the points
belong to the lower line and which to the upper one based on the best
statistical fit, thus avoiding any subjectivity (Figure 4).

The pressure at which a hydrofracture reopens (refrac pressure
P_{c2}) is not the peak pressure but the 'knee' that is often observed
during secondary cycles in the ascending pressure time curve (Figure
1). This 'knee' reflects the loss of some of the injected fluid to
the reopening fracture, which causes the rate of drillhole
pressurization to decline and forms the bend in the pressure-time
curve. Zoback and Haimson (1982) suggested that, since the 'knee' is
not always distinct, P_{c2} be picked out as the point at which the
slope of the ascending P-t curve in the second cycle departs from the
slope established by the first cycle (both cycles conducted at
identical flow rates). As shown in Figure 1 the bend in the curve
can be quite gradual and as in the case of P_s it appears that
pinpointing P_{c2} can become arbitrary. Using the digital data,
however, we can follow the change in slope of the secondary cycle and
continuously compare it to the change, if any, in the first cycle at
very close time intervals (say 10 points per second). The onset of
deviation of the secondary cycle from the first can be uniquely
determined as P_{c2} (Figure 5).

The proper relationship of P_{c2} to the largest horizontal stress S_H
is not quite clear. Hubbert and Willis' (1957) assumption that no
borehole fluid penetrates into the surrounding rock during
hydrofracturing pressurization, and that the pore pressure throughout
the rock remains at its background level P_o is quite unrealistic even
in tight rocks such as granites. Nevertheless, the Hubbert and
Willis equation is still widely used because of its mathematical
simplicity and because the calculated S_H has been found in some cases
to compare satisfactorily with results from other measurement
techniques (Haimson, 1981).

A complementary or alternative method to conventional
hydrofracturing has recently been suggested (Cornet and Valette,
1984), which eliminates the need for P_{c2} in determining the
horizontal in situ stresses. This new approach makes the reasonable
assumption that the vertical stress is a principal component and that
each of the non-zero components of the in situ stress tensor varies
linearly with depth. To calculate the stress tensor a minimum of six
(but preferably seven or more to reduce uncertainities)
hydrofracturing tests are required over the depth range of interest,
from which reliable P_s and fracture orientation values should be
obtained. These yield at least 6 values of normal stress (not
necessarily S_h) magnitudes and directions from which the principal
stresses can be calculated similarly to the strain rosette
analysis. The method actually works best in rocks for which the
induced hydrofractures are not always perpendicular to the direction
of S_h. In many crystalline rocks, such as granites and quartzites,
various weaknesses in the rock may in fact induce fractures which are
not well aligned with the principal stresses.

The suggested method could also be used to advantage in hydrofracturing tests conducted in inclined holes. In such cases one can at most obtain from conventional hydrofracturing interpretation the least horizontal stress (S_h) and its variation with depth. Using the new technique the complete stress tensor could theoretically be obtained in inclined holes provided sufficient tests are conducted which do not all result in identical hydrofracture orientations.

3 IMPROVEMENTS IN OVERCORING TECHNIQUES

In the last several years the triaxial strain cells have gained increased acceptability and utilization in post excavation design and to some extend in preexcavation design of underground openings. This development has been in part the direct result of two major improvements in the technique. Worotnicki and Walton (1976) modified Leeman's (1971) design by incorporating a solid cylindrical plastic body in which the strain rosettes are encapsulated. Gage bonding to rock is accomplished by filling with epoxy cement the annulus between the plastic cylinder and the borehole wall. This design makes the triaxial cell less dependent on borehole conditions and is completely water tight, allowing continuous recording of rosette strains during the process of overcoring. These features, which were not available in the original design, have increased the rate of success and the scope of utilization of the triaxial cell.

A different improvement in the original triaxial strain cell which renders it capable of being used at considerably greater depths than other overcoring methods (has been tested down to 500 m) was introduced by the Swedish Power Board, SPB (Hiltscher et al, 1979). This is a completely new design of the cell and inserting tool, and only the general principle has been preserved that nine properly oriented gages bonded to the borehole and overcored can be used to yield the complete strain tensor. There are major design changes in the way the rosettes are attached to the wall, in the orientation of the gages with respect to magnetic north, and most importantly in the glue used to adhere the gages to rock. The glue performs well in dry and wet conditions.

The great advantage of the SPB tool over other overcoring methods is that it can be used in exploratory holes, not unlike hydrofracturing, in the preliminary stages of excavation design. The revised triaxial cell is employed exclusively by the Swedish Power Board, which reports successful tests in a number of underground projects (Hallbjorn 1986).

4 NEW METHODS OF STRESS DETERMINATION

The last few years have seen the introduction of several new and innovative methods of estimating partially or totally the in situ stress tensor. The availability of these additional methods enhances the variety of choices open to the project engineer in terms of testing costs, suitability and preference, and is bound to increase the use of stress measurements in design.

One such innovative technique, which has already found use in several underground project designs is based on the natural phenomenon of borehole breakouts or borehole cross-sectional

elongation resulting from spalling (Figure 6). Based on a simple
linear elastic solution the maximum compressive stresses around a
vertical hole concentrate around the two points aligned with the
direction of S_h. The assumption that breakouts are the result of
compressive (shear) failure reached at or near the borehole wall in
S_h direction has been demonstrated both in the field (Gough and Bell
1982) and in the laboratory (Haimson and Herrick 1986). The
increasingly popular borehole sonic televiewer, a logging tool for
studying the state of fracturing around the borehole, can be used for
discerning zones of borehole breakouts (if any) and their
directions. Obviously, breakouts will only occur in rocks that are
sufficiently weak and/or in stress regimes that are appreciably
high. Presently, breakouts can only be utilized to determine the
directions of the major principal horizontal stresses. Stress
directions are a very important site characteristic required in
laying out the orientation of large caverns. However, the borehole
televiewer can also map the extent and shape of breakouts. Analysis
by Zoback et al (1985) and laboratory testing by Haimson and Herrick
(1986) show that the potential exists for using breakout dimensions
to also estimate the magnitudes of S_H and S_h (Figure 7). Thus,
borehole breakout logging which already yields essential, if
incomplete, stress information for design of underground openings
(Paillet 1985, Stock et al, 1985), promises to become an even more
important tool when proven techniques for estimating stress
magnitudes become available.

Another novel method of in situ stress estimation is by conducting
a differential strain curve analysis of oriented core extracted from
the depth of interest (Dey and Brown 1986). Cubical samples are
prepared from oriented core and a minimum of six strain gages are
bonded to their outside surfaces in predetermined directions. Each
sample is jacketed and loaded hydrostatically to a pressure of about
200 MPa so as to eliminate any crack porosity. From the pressure-
strain curve obtained for every gage the contribution of crack
closure to the recorded strain can be established. Using the six or
more gages, the three principal crack strains and their directions
can be determined. The principal stresses are then calculated using
measured rock elastic constants. Under the assumption that most of
the microcracks in the sample are due to in situ stress relief
occuring during the cutting of the core, and hence that in the
undisturbed field condition the number of open microcracks is
negligible, the established crack strain tensor can be interpreted in
terms of the in situ state stress. The method has been used on
several occasions, yielding reasonable results (Dey and Brown 1986;
Ren and Roegiers 1983), but it is still largely developmental.
However, it may become valuable in measurements of high in situ
stresses for which the assumption of microcrack closure is generally
correct.

5   NEW ASPECTS OF UNDERGROUND OPENING DESIGN UTILIZING STRESS MEASUREMENTS

5.1   Hydro projects

For several decades the design of large underground powerhouses in
hydroelectric schemes has incorporated the measurement of in situ

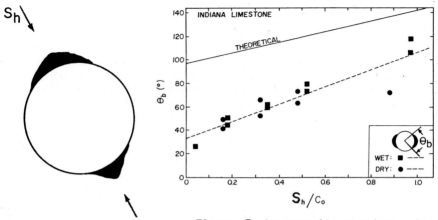

S_h

Figure 6 Typical borehole breakout as observed on a borehole televiewer log.

Figure 7 Apparent linear relationship between breakout span (θ_b) and normalized S_h (divided by the compressive strength, C_o).

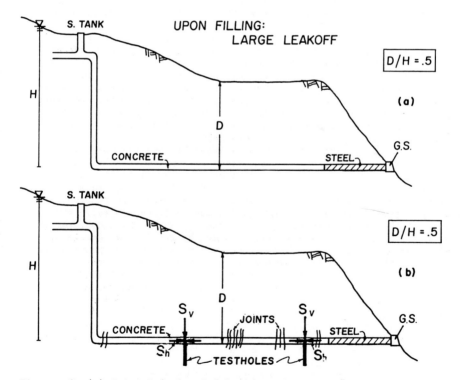

Figure 8 (a) Initial 'rule of 'thumb' condition was met in this case history but large leakoff was experienced upon water filling. (b) Measurements helped explain the leakoff: S_h was smaller than S_v and was perpendicular to a joint set crossing the tunnel.

stress. This came as a result of the realization that the rational
design of such large openings, in which each dimension is at least
several tens of meters long, requires reliable information of the
local field stresses. Until the mid seventies this information was
obtained using overcoring techniques (Benson et al 1971; Martinetti
1977; Anderson et al 1977). Hence, these measurements, conducted
after excavation, were restricted to confirming previous assumptions
made on the expected state of stress and used in the initial design
of the openings. A direct quote from Benson et al (1971) on the
design and construction of the Churchill Falls hydro project clearly
describes the situation then prevailing: "At the time of design, the
magnitude and orientation of the principal field stresses in the area
of the underground powerhouse were unkown as access had not yet been
gained to the underground area. Thus, for loading conditions for the
model, it was necessary to assume conditions which could be expected
to encompass the possible range of in situ stresses...". Haimson
(1977) reported on the first known stress measurements conducted
prior to excavation of underground powerhouses (Helms Pumped Storage
Project and Bad Creek Pumped Storage Project). The hydrofracturing
technique used in these projects was not only a reliable way of
estimating stresses (Haimson 1981) but also an unprecedented means of
obtaining stress information prior to excavation. These
characteristics gained hydrofracturing a growing acceptance as an
indispensable tool in the preexcavation design of large underground
openings (Haimson 1982, Haimson et al 1986).
   In the last several years direct stress measurements have been
increasingly used in other aspects of hydroelectric project design
such as penstocks and pressure shafts and tunnels, as well as in dam
foundation studies. One of the most critical decisions is the
determination of the required length of steel lining which, although
the most expensive tunnel reinforcement, is the only one to reliably
guard against profuse leakage into the surrounding rock mass. The
remainder of the tunnel is usually left unlined or concrete lined.
Measurements of hydrofracturing both before (Barton 1983; Haimson
1982), and after excavation (Haimson 1986; Vik and Tunbridge 1986;
Marulanda et al 1986) have been devised to replace the old 'rule of
thumb' empirical criteria (Selmer-Olsen 1970; Broch 1982) with a
measured-stress based analysis. Figure 8 depicts a case in which the
rule of thumb was first used to determine the extent of the concrete
lined pressure tunnel in a high head pumped storage project. The
rule of thumb: overburden (D)/water head in tunnel (H) > 0.4, where
both are measured in meters, was well met by the initial design of
the openings. However, profuse leakage was experienced upon filling
of the tunnel and shaft. Subsequent hydrofracturing stress
measurements conducted from the tunnel revealed that although the
premise of the empirical formula (D/H > 0.4) was wrong, since the
least principal stress was not vertical, no hydrofracturing
initiation by the tunnel pressurized water was to be expected.
However, since the least principal stress was horizontal and aligned
with the tunnel axis, vertical joints traversing the tunnel such as
those discovered at several points along the excavation could indeed
open up (hydraulic splitting) during or after tunnel filling. The
determination of the minimum stress magnitude and direction could
only be carried out by direct measurement. As a result of the
measurements it was decided that since the sources of leakage were
localized high pressure grouting would be sufficient to stabilize the

leakage problem.

Vik and Tunbridge (1986) discuss other new specific uses of post-excavation stress measurements to solve problems of underground hydroelectric plant design. In one case, the Kvilldal power plant in Norway, a water curtain was to be constructed above a leaking air-cushion surge chamber. To estimate the maximum allowable pressure in the water curtain so that inadvertent hydraulic fracturing could not be induced, a series of overcoring and hydrofracturing tests were undertaken. The water curtain maximum pressure was subsequently designed to be less than the measured minimum principal stress, and the air leakage was almost totally stopped.

Another application of stress measurements in tunnels of any kind has been the determination of stress in linings. A typical objective here is to monitor the increase in lining loading due to rock time-dependent deformation around the tunnel. Barla and Rossi (1984) report on two methods used: an overcoring technique preferred in concrete lining, and a flat jack test adoptable to brick masonry lining.

A different type of hydro project, for which stress measurements are not routinely conducted because of the shallow depth is the design and construction of dam foundations. In specific cases where high near-surface stresses are suspected, measurements at dam sites have recently been undertaken. Bock (1986) reports on a series of overcoring, hydrofracturing and borehole slotting stress tests in the river bed rock adjacent to the Burdekin Falls dam, Australia. Indications of high horizontal stresses, demonstrated by rock popping during foundation work were cause for concern. The purpose of the tests was to determine the near-surface stress regime in the river bed rock in order to reivew the stability of the dam and evaluate potential problems in grouting.

5.2 Underground mining

Stability problems in hard rock underground mines were the initial impetus for the development of modern techniques of stress measurement (Hast 1958; Leeman 1958). Over the last thirty years mining engineers have progressively realized that many aspects of mine design depend on direct stress determination. Accordingly, a sharp increase in underground stress measurements has occurred in the last decade. Major problems in underground mining that have required direct stress measurements have been rock bursts, pillar recovery and general cavern stability (see for example Del Greco et al, 1984).

Stress tests in coal mines have become more common only in recent years. A revealing case history in the Huntly coal mines of New Zealand has been reported by Mills et al (1986). There, with the introduction of modern mining techniques the depth of mining has gradually increased. However, at greater depths problems with roof stability were encountered. The initial tendency in design was to follow the australian coal mining experience (see for example Gale 1986) as well as the stress-depth relationship compiled by Hoek and Brown (1980). Both indicated horizontal stresses several times higher than overburden pressure at the depths mined at Huntly (no deeper than 300 m). These assumptions, however, led to unsatisfactory results. Local stress measurements were then undertaken which yielded horizontal to vertical stress ratios of 0.3

to 1.5. These were considerably lower than anticipated. It was, thus, determined that by modifying the initial dimensions of the underground roadways from a width/height ratio of 2:1 (following the australian model) to a narrower but taller opening, the local stress conditions were better met, and significant improvements in stability were achieved. In addition, it was concluded from the measurements that the two horizontal principal stresses were highly differential, and that to maintain a ratio close to unity between the far-field horizontal stress acting transversally to the roadways and the vertical stress, the axes of the roadways had to be aligned with the minor horizontal stress. Thus, although the australian experience with respect to mining direction was adopted, the actual orientation of the roadways was inverted because in australian coal mines the minor (and not the major) horizontal stress is closer to the overburden (Gale 1986). The significant design improvements over the original model that the Huntly mines were first trying to follow were made possible only as a result of undertaking their own stress test program.

The major stress measuring methods in underground mines are still the various overcoring techniques which are typically employed after a problem related to stress has been identified. However, progress in the direction of preexcavation rational design is being made. Enever and Wooltorton (1983), for example, used hydrofracturing to conduct an extensive series of in situ stress measurements from the surface to depths reaching 500 m in coal basins for the early planning of colliery layouts and for the selection of optimum sites for major entries. In various new mining ventures in the U.S. plans are to conduct stress measurements in advance of excavation, and this trend appears to have spread to other countries as well.

5.3 Nuclear waste repositories

Underground nuclear waste repositories are engineering rock structures that will be subjected to unprecedented requirements. A unique requirement is the significantly longer service life than would be expected for common civil structures. The long term stability requirement makes the knowledge of local in situ stress conditions an essential element in establishing the feasibility of a site for nuclear waste disposal and in rationally designing the appropriate subterranean openings.

In the United State extensive stress measurements by hydrofracturing have been conducted at the major potential sites for future repositories such as Hanford, Washington and the Nevada Test Site (Haimson 1987, Stock et al, 1985). Our measurements at Hanford, conducted well ahead of shaft sinking and emplacement room excavation, were used to establish several crucial aspects of design:
   a. Site suitability: the measured state of stress at Hanford was used to assess the stability of the repository host rock unit. A potentially unstable formation may not be suitable as a repository site because of the uncertainty regarding the safe isolation of radionuclides. To study the potential for instability a Mohr diagram was used in which were plotted the circle representing the stresses in the plane containing the measured maximum and minimum principal stresses and the lines depicting the criterion of failure of the rock mass and the criterion of frictional sliding along

natural fractures or joints. The circle appeared to be safely within the stable zone (entirely under the lines representing the failure criteria).

b. Preferred orientation of caverns: The original design of the underground nuclear waste repository at Hanford was based on an assumed ratio of 1:1 between any two principal stresses. Such an isotropic state of stress would be ideal since it would impose no constraints on the orientation of the required rock caverns. However, our measurements at the proposed repository horizon clearly indicated that such was not the case. Consequently, the design was changed to accomodate the measured stress situation which may be summarized as a ratio of approximately 2:1 between the largest horizontal stress and either of the other two principal components, and a roughly 1:1 ratio between the least horizontal and the vertical stresses (Figure 9). The "storage holes", being the most crucial caverns since they will house the radioactive waste canisters, will be oriented in the direction of the largest horizontal stress for maximum stability. The "emplacement room" which leads to the storage holes will by necessity have to be alligned with the minimum horizontal stress (Rockwell Hanford Operations, 1982).

c. Shape of rock caverns: The shape of the repository caverns should minimize stress concentrations around the openings so as to optimize stability. Again, the prerequisite to the rational selection of the least destablizing shape is knowledge of the in situ stresses. A properly conceived excavation shape can prevent large investments in reinforcement, and lower the risk of rock failure around the opening. At Hanford the storage holes would have a circular cross-section (Figure 9), which is the desired shape for the excavation method (boring), for the function of these holes (to store cylindrical canisters), and for least stress concentration, resulting in maximum stability (the axes of the holes are perpendicular to both the least horizontal and the vertical stresses which are approximately equal in magnitude). The emplacement room orientation is dictated by that of the storage holes and hence is less favorable. The suggested design calls for an elongated horse shoe (quasi-elliptical) shape with the width about twice the maximum height, to correspond with the 2:1 stress ratio (between the maximum horizontal and the vertical stresses) for optimum stress concentration around the room (Rockwell Hanford Operations 1982).

d. Support: The measured in situ stresses at Hanford provide the basic field data required for estimating needed corrective action, support and reinforcement of rock caverns. For example, the large diameter vertical shaft necessary for access to the repository could experience rock spalling, based on the highly differential horizontal stresses that would develop near the shaft wall. A suggested high density drilling fluid would provide temporary support during shaft sinking and, thus, eliminate or reduce the amount of spalling. Similarly, theoretical analyses of the rock stresses around the planned caverns based on our initial measurements, suggest that rock support may be required to improve stability in some of the horizontal openings.

Another precedent setting case of extensive in situ stress measurements were conducted at the future Forsmark nuclear power station and repository in Sweden (Carlsson and Christiansson 1986).

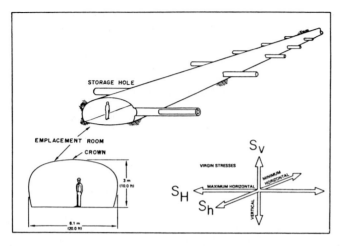

Figure 9 Proposed layout of emplacement room and waste canisters storage holes at Hanford, and its relation to the measured in situ stresses.

The Leeman door stopper and triaxial strain cell were used in eleven boreholes reaching 500 m depth in order to determine the state of stress at the site. The results indicated no apparent difference in stresses between various crystalline rock types at Forsmark. The measurements also revealed that the stresses just below a fractured and crushed zone at 300 m depth in one of the boreholes were anomalously high, and gave rise to core discing. An important conclusion was that large geologic structures such as faults appeared to influence the directions of the measured principal stresses. These relevant findings, which could not have been deduced without actual measurements, were closely considered in the design process of the Forsmark nuclear project.

6 SUMMARY AND CONCLUSIONS

This paper is an attempt to review new developments in stress measurements for the design of underground openings. Emphasis was placed on innovations in hydrofracturing and triaxial strain cell methods and on several new techniques such as borehole breakouts and differential strain curve analysis. In the second part of the paper recent case histories were used to demonstrate the growing importance of in situ stress in three major categories of design: mine openings, tunnels and pressure shafts, and nuclear waste repositories.

The major conclusion to be drawn is that stress measuring methods and their applications in design of underground openings are in a very dynamic state. While new and better techniques are being developed, the importance of in situ stress in rational design of openings is growing. We expect that in the not too distant future every sizeable underground project in both the mining and civil areas

will require preexcavation stress tests for initial decisions with
respect to site suitability, cavern orientations and shapes, and
amount of support, followed by post excavation tests for fine-tuning
the preliminary design.

7 REFERENCES

Aamodt R. and M. Kuriyagawa 1983. Measurement of instantaneous shut-
in pressure in crystalline rock, in Hydraulic Fracturing Stress
Measurements, eds. M. Zoback and B. Haimson, National Academy
Press, Washington, D.C., 139-142.
Anderson, J. G. C., J. Arthur and D. B. Powell 1977. The engineering
geology of the Dinorwic underground complex and its approach
tunnels, in Rock Engineering, University of Newcastle Upon Tyne
Press, 491-510.
Barla, G. and P. D. Rossi 1984. Stress measurements in tunnel
linings, in Field Measurements in Geomechanics, ed. K. Kovari,
Balkema, 987-998.
Barton, N. 1983. Hydraulic fracturing to estimate minimum stress and
rock mass stability at a pumped hydro project, in Hydraulic
Fracturing Stress Measurements, eds. M. Zoback and B. Haimson,
National Academy Press, Washington, DC 61-67.
Benson, R. P., R. J. Coulon, A. H. Merrit, P. Joli-Coeur and D. U.
Deere 1971. Rock Mechanics at Churchill Falls, in Underground Rock
Chambers, Am. Soc. Civil Engr., 407-486.
Bjarnason, B. O., O. Stephansson, A. Torikka, and K. Bergstrom
1986. Four years of hydrofracturing rock stress measurements in
Sweden, in Rock Stress, ed. O. Stephansson, CENTEK Publ., Lulea,
421-428.
Bock, H. 1986. In situ validation of the borehole slotting
stressmeter, in Rock Stress, ed. O. Stephansson, CENTEK Publ.
Lulea, 261-270.
Broch, E. 1982. The development of unlined pressure shafts and
tunnels in Norway, in Rock Mechanics: Caverns and Pressure Shafts,
ed. W. Wittke, A. A. Balkema, 545-554.
Carlsson, A. and R. Christiansson 1986. Rock stresses and geological
structures in the Forsmark area, in Rock Stress, ed. O.
Stephansson, CENTEK Publ., Lulea, 457-466.
Cornet, F. H. and B. Valette 1984. In situ stress determination from
hydraulic injection test data, J. Geophys. Res., 89B: 11527-11537.
Del Greco, O., G. Iabichino, N. Innaurato, and L. Stragiotti 1984.
Application of stress measurements for the solution of mining
problems in Italy, in Field Measurements in Geomechanics, ed. K.
Kovari, Balkema, vol. 2, 1007-1019.
Dey, T. N., and D. W. Brown 1986. Stress measurements in a deep
granitic rock mass using hydraulic fracturing and differential
strain curve analysis, in Rock Stress, ed. O. Stephansson, CENTEK
publ., Lulea, 351-358.
Dolcetta, M. 1971. Problems with large underground stations in
Italy, in Underground Rock Chambers, Am. Soc. Civil Engr., 243-286.
Enever, J. R., and P. N. Chopra 1986. Experience with hydraulic
fracture stress measurements in granite, in Rock Stress, ed. O.
Stephansson, CENTEK Publ., Lulea, 411-420.
Enever, J. R. and B. A. Wooltorton 1983. Experience with hydraulic
fracturing as a means of estimating in situ stress in australian

coal basin sediments, in Hydraulic Fracturing Stress Measurements, eds. M. Zoback and B. Haimson, National Academy Press, Washington, D.C., 28-43.

Gale, W. J. 1986. The application of stress measurements to the optimization of coal mine roadway driveage in the Illawarra coal measures, in Rock Stress, ed. O. Stephansson, CENTEK Publ., Lulea, 551-560.

Gough, D. I. and J. S. Bell 1982. Stress orientations from borehole wall fractures with examples from Colorado, east Texas and northern Canada. Canad. J. of Earth Sci. 19: 1358-1370.

Haimson, B. C. 1974. A simple method of estimating in situ stresses at great depths, in Field Testing and Instrumentation of Rock, Am. Soc. Testing and Materials (ASTM) Special Tech. Publ. 554, 156-182.

Haimson, B. C. 1977. Design of underground powerhouses and the importance of preexcavation stress measurements, in Design Methods in Rock Mechanics, eds. C. Fairhurst and S. L. Crouch, Am. Soc. Civil Engr., 197-204.

Haimson, B. C. 1978. The hydrofracturing stress measuring method and recent field results, Int. J. Rock mech. Min. Sci. and Geomech, Abstr., 15: 167-178.

Haimson, B. C. 1980. Near surface and deep hydrofracturing stress measurements in the Waterloo quartzite, Int. J. Rock Mech. Min. Sci. and Geomech. Abstr., 17: 81-88.

Haimson, B. C. 1981. Confirmation of hydrofracturing results through comparisons with other stress measurements, in Proceedings of the 22nd U.S. Symp. on Rock Mechanics, MIT Press, 379-385.

Haimson, B. C. 1982. Deephole preexcavation stress measurements for the design of underground powerhouses in the Sierra Nevada Mountains, in Rock Mechanics: Caverns and Pressure Shafts, ed. W. Wittke, Balkema, vol. 1, 31-40.

Haimson, B. C. 1986. Joints, in situ stress and leak-offs in hydroelectric pressure tunnels, in Large Underground Openings, University of Florence Press, vol. 2, 198-204.

Haimson, B. C. 1987. Stress measurements at Hanford, Washington for the design of a nuclear waste repository facility, Proc. 6th Int. Congress on Rock Mech., Montreal (in press).

Haimson, B. C. and C. G. Herrick 1986. Borehole breakouts - a new tool for estimating in situ stress?, in Rock Stress, ed. O. Stephansson, CENTEK Publ., Lulea, 271-280.

Haimson, B. C. and M. Y. Lee 1984. Development of a wireline hydrofracturing technique and its use at a site of induced seismicity, in Rock Mechanics in Productivity and Protection, eds. C. Dowding and M. Singh, Soc. Mining Engr. of AIME, 194-203.

Haimson, B. C., C. F. Lee, J. H. S. Huang 1986. High horizontal stresses at Niagara Falls, their measurement, and the design of a new hydroelectric plant, in Rock Stress, ed. O. Stephansson, CENTEK Publ., Lulea, 615-624.

Hallbjorn, L. 1986. Rock stress measurements performed by Swedish State Power Board, in Rock Stress, ed. O. Stephansson, CENTEK Publ., Lulea, 197-206.

Hast, N. 1958. The measurement of rock pressure in mines, Sveriges Geol. Undersokn. Arsbok, Ser. C, Auhandl. Uppsat., 52: 1-183.

Hiltscher R., J. Martna and L. Strindell 1979. The measurement of triaxial rock stresses in deep boreholes, in Proc. 4th Int. Congress on Rock Mech., Montreaux, vol. 2, 227.

Hoek, E. and E. T. Brown 1980. Underground excavation in rock,

Institute of Mining and Metallurgy, London.

Hubbert, M. K. and D. G. Willis 1957. Mechanics of hydraulic fracturing, Trans. AIME, 210: 153-166.

Leeman, E. R. 1958. The measurement of stress in the ground surrounding mining excavations, Ass. Min. Mngrs. S. Afr. Pap. and Disc., vol. 1958/9, 331-356.

Leeman, E. R. 1971. The CSIR 'doorstopper' and triaxial rock stress measuring instruments, Rock Mechanics, 3: 25-50.

Martinetti, S. 1977. Experience in field measuremnts for underground power stations in Italy, in Field Measurements in Rock Mechanics, ed. K. Kovari, Balkema, 509-534.

Marulanda, A., C. Ortiz, and R. Gutiereez 1986. Definition of the use of steel liners based on hydraulic fracturing tests; A case history, in Rock Stress, ed. O. Stephansson, CENTEK Publ., Lulea, 599-604.

Mills, K. W., M. J. Pender, and D. Depledge 1986. Measurement of in situ stress in coal, in Rock Stress, ed. O. Stephansson, CENTEK Publ., Lulea, 543-549.

Muskat, M. 1937. Use of data on the build-up of bottom-hole pressures, Trans. AIME, 123: 44-48.

Paillet, F. L. 1985. Acoustic televiewer and acoustic waveform logs used to characterize deeply buried basalt flows, Hanford Site, Washington, Open-file report 85-419, U.S. Geological Survey, Denver, Colorado.

Ren, N. K. and J. C. Roegiers 1983. Differential strain curve analysis, Proc. 5th Int. Congress Rock Mech., Melbourne, F117-F128.

Rockwell Hanford Operations 1982. Site characterization for the basalt waste isolation project. Report DOE/RL 82/3, Rockwell International, Richland, Washington.

Rummel, F., J. Baumgartner and H. J. Alheid 1983. Hydraulic fracturing stress measurements along the eastern boundary of the SW-german block, in Hydraulic Fracturing Stress Measurements, eds. M. Zoback and B. Haimson, National Academy Press, Washington, D.C., 3-17.

Selmer-Olsen, R. 1970. Experience with unlined pressure shafts in Norway, In Large Permanent Underground Openings, Universitetsforlaget, Oslo, 327-332.

Stock, J. M., J. H. Healy, S. H. Hickman and M. O. Zoback 1985. Hydraulic fracturing stress measurements at Yucca Mountain, Nevada. J. of Geophys. Res., 90:8691-8706.

Vik, G. and L. Tundbridge 1986. Hydraulic fracturing – a simple tool for controlling the safety of unlined high pressure shafts, in Rock Stress, ed. O. Stephansson, CENTEK Publ., Lulea, 591-598.

Worotnicki, G. and R. J. Walton 1976. Triaxial 'hollow inclusion' gauges for determination of rock stresses in situ, Supplement to ISRM Symp. on Investigation of Stress in Rock, the Institution of Engineers, Australia, 1-8.

Zoback, M. D. and B. C. Haimson 1982. Status of the hydraulic fracturing method for in situ stress measurements, in Issues in Rock Mechanics: Proc. 23rd U. S. Symposium on Rock Mechanics, Soc. of Mining Engineers of AIME, New York, 143-156.

Zoback, M. D., D. Moos, L. Martin, and R. N. Anderson 1985. Wellbore breakouts and in situ stress, J. Geophys. Res. 90: 5523-5530.

2nd International Symposium on Field Measurements in Geomechanics, Sakurai (ed.)
© 1988 Balkema, Rotterdam. ISBN 90 6191 778 6

Possibilities for prognosticating state of stress around underground openings on the basis of automated mapping

R.Parashkevov, E.Andonov & U.Dimitrov
Higher Institute of Mining and Geology, Sofia, Bulgaria

The results obtained in analytical and experimental studies of the stressed state of rocks around underground openings are more informative and well-grounded when a preliminary geomechanical prognostication mapping of the respective areas and volumes of the rock mass is carried out. The geomechanical prognostication and mapping of conditions are being intensively developed recently and find more and more wide application in physical and geomechanical grounding of various design and technological decisions (Bukrinskii,1985; Davis,1977; Kendall,Moran, 1963; Parashkevov et al.,1983 etc.).

The geomechanical mapping of a complex of indices for rock and mass properties allows for grounding new approaches for selection of proper areas and volumes of the rock mass where experimental-analytical measurements and assessments of the stress-strain state around underground openings and structures should be carried out.Among the great number of physical and geomechanical parameters of rocks, it is possible to select such a combination of factors, which can be used for prognosticating with an adequate precision the most typical processes and phenomena occuring within certain area or volume of the rock mass, depending on the alterations of the stress-strain state of the rock mass.

An example of geomechanical prognostication mapping of some basic physico-mechanical parameters in a certain area of Bobov Dol coal deposit is given in the present study (Fig.1), which includes: compressive strength R_n , modulus of elasticity E , Poisson's ratio ν , indices of rheological creep α/β, rock stability k at certain depth from the surface H and maximum tangential stresses in the rock mass τ_{max} .

Supposing the initial natural field of stresses in the rock mass is formed under the action of the gravity forces

$$\sigma_z = \gamma \cdot H \,; \quad \sigma_x = \sigma_y = \lambda \gamma H \,; \quad \lambda = \frac{\nu}{1-\nu} \tag{1}$$

Then as a basic indicator of the stress-strain state of the rocks around underground openings is analyzed the so called index of rock stability

$$k = \frac{\gamma \cdot H}{R_n} \,, \tag{2}$$

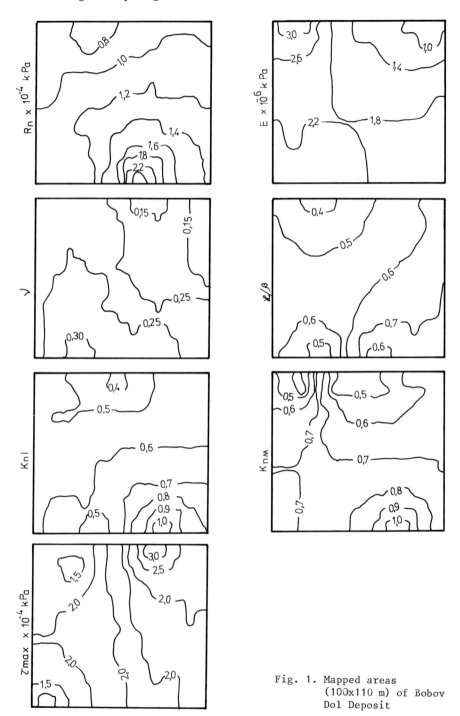

Fig. 1. Mapped areas
(100x110 m) of Bobov
Dol Deposit

where:

R_n – uniaxial compressive strength of rock varieties, Mpa;
γ – average-balanced volumetric mass of overburden, kH/m^3;
H – depth from the surface, m.

The compressive strength R_n of the host rocks is reduced to the conditions of the rock mass with taking into consideration the structural disturbance resulting from different factors.

Fig. 1 shows the results of mapping of rock stability index k , by using the compressive strength values R_{nl} obtained in laboratory conditions and those ones, calculated with taking into consideration the structural disturbance and reduced to the rock mass R_{nM}. The differences between the two maps allow a more correct prognostication of the expected alterations in the stress-strain state of rocks around underground openings, which are to be built in the mapped area. An additional information is obtained as well as by mapping the values of the maximum tangential stresses in the rock mass, according to the values of Poisson's ratio ν and the depth H (Fig.1).

$$\tau_{max} = \frac{1-2\nu}{1-\nu} \cdot \frac{\gamma H}{2} \qquad (3)$$

In order to make a proper selection of the points for experimental measurements of the stresses, it is necessary to make in the process of mapping an estimation of the reliability of the maps and to determine the optimum density of measurements network in the different points.

The optimum selection and layout of the control points (in the perspective measuring stations) are connected with the variability of the mapped geomechanical index k , as well as with the character and the requirements of the measurements to be made.

The authors use an approach, according to which for each point T_j of the uniform network a weighted mean is calculated on the basis of observations in n closest control points T_1, T_2, ...T_n by the distances to them $D_1 \leqslant D_2 \leqslant ... \leqslant D_n$. The values of the index in the control points are indicated by $k_1, k_2, ... \ k_n$. The distance of the point T_j to a control point T_i is

$$D_{ij} = \sqrt{(x_j - x_i)^2 + (y_j - y_i)^2}$$

where (x_j, y_j) are the co-ordinates of the point T_j and (x_i, y_i) of point T_i.

After having calculated D_{ij} according to all n control points, we define

$$k(x_j, y_j) = \frac{\sum\limits_{i=1}^{n} k_i / D_{ij}}{\sum\limits_{i=1}^{n} 1 / D_{ij}} \qquad (4)$$

The function $k(x_j, y_j)$ is obtained, which is defined in the points of the uniform network.

In order to estimate the reliability of realization in geometric representation of the index by a map, it is necessary to render an account of the density in the control points, as well as of the surface slope. On the basis of the experience it has been found out that the reliability is smaller in an area with a denser network of control

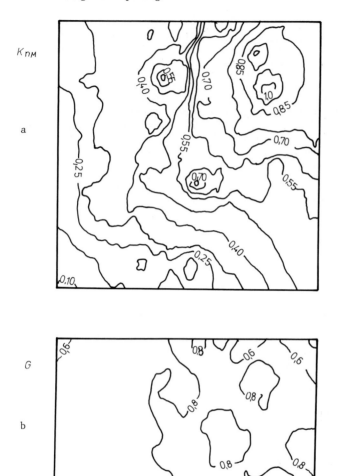

Fig. 2 - a,b. Mapped area (2000x2000 m)
    of Babino mine (39 p.)

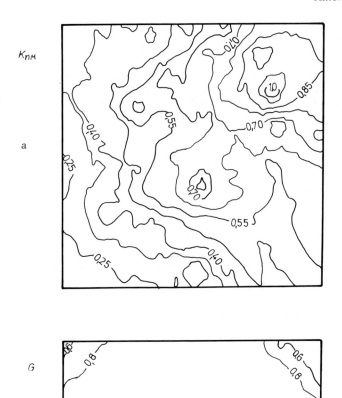

Fig. 3 - a,b. Mapped area (2000x2000 m)
of Babino mine (55 p.)

points because of the greater variability of the index.
  This question can be solved by a proper mathematical model.
  The term "relative reliability at a point of the represented surface"
means the control points density within a part of the surface whose
projection is a circular area around a point of the uniform network.
  Before making the reliability map it is necessary to standardize
the surface $k = k(x,y)$. This is realized by a steric transformation
of $Oxyk$ into $Ouvw$:

$$u = \frac{1}{\sqrt{S_o}} \cdot x \qquad v = \frac{1}{\sqrt{S_o}} \cdot y \qquad w = \frac{1}{k_s} \cdot k \qquad (5)$$

where S_o is the area of the mapped orthogonal region;
       k_s is a mean value of the index on the basis of all uniform nodes.
  The control points are unevenly located, i.e. can be considered as
accidentally taken. It means that they have a Poisson's distribution
in the plane Ouv. Let us choose the control points in such a way that
     $D_1 \le D_2 \le \ldots \le D_n$. Then the relative reliability can
be specified as a ratio between the n control points and the area S
of the part of the surface with projection the circle Q ,containing
these points. When making the numerical determination of the reliability
at a point of the uniform network, S will be defined as the area of
that part of the tangential plane, taken at this point with projection
the circle Q .
  After turning to the old co-ordinates $Oxyk$ we obtain the formula for
quantitative estimation of the relative geometrical reliability in a
point,

$$G_j = \frac{Ln \cdot k_s \cdot S_o \cdot \left(\sum_{i=1}^{n} \frac{1}{D_i} \right)^2}{\sqrt{k_s^2 + S_o k_x'^2 + S_o k_y'^2}} \qquad (6)$$

where: $Ln = \left(\sum_{i=1}^{n} \frac{2}{3} \frac{4}{5} \ldots \frac{2(i-1)}{2i-1} \right)^{-2}$

     k_x' , k_y' are the partial derivatives of $k = k(x,y)$,
taken at a concrete node (x_j, y_j).
  The derived formula can be used for comparison of two maps, representing the respective geomechanical index in two successive steps of the
mapping before and after adding new control points.
  An example about the different elements of the above suggested approach is shown in the Figs. 2 and 3. The parameter under consideration
is the rock stability index $k = \gamma H/R_n$ for the IV layer (roof) in Babino
mine, where:
  2.a - map of the index according to 39 control points;
  2.b - map of the reliability (about 1/3 of the mapped region has
smaller reliability);
  3.a - map of the index after adding new 16 control points (surface
shape has negligible alterations);
  3.b - map of the reliability after 39 + 16 = 55 points (after adding
the new control points, a higher reliability of the whole mapped area
is obtained).

The approach put forward for determination of the relative reliability at each point of geomechanical prognostication maps is an
effective means for a well-grounded prediction of the places of measuring the stressed state of rocks. In order to find the optimum number of measurements necessary for obtaining a preliminarily given
degree of reliability in prognostication mapping of geomechanical parameters, an algorithm of the process is made.

REFERENCES

Bukrinskii, V.A. 1985. Geometria nedr. M.: Nedra.
Davys,J. 1977. Statistika i analiz geologicheskikh dannikh. M.: Mir.
Kendall, M.G., Moran, P.A. 1963. Geometrical probability. New York:
  Hofner Publ.Co.
Parashkevov, R.D., Andonov, E.G., Dimitrov, J.L. 1983. Nauchni osnovi
  za sazdavane na novi technologii za razrabotvane na polezni izkopaemi
  v slozhni minno-geolozhki usloviya.Investigation report. Records of
  the Higher Institute of Mining & Geology. Sofia.

Rock stress and AE activity in Miike Coal Mine

Katsuhiko Sugawara, Katsuhiko Kaneko & Yuzo Obara
Faculty of Engineering, Kumamoto University, Japan

1. INTRODUCTION

Acoustic emission (AE) is a seismic phenomenon associated with mechanical instability within a material (Hardy, 1973). Relatively little is
known about the basic mechanisms responsible for seismic activity in
geological materials. As discussed in the present paper, however, AE
monitoring is one of the most useful means presently available for
evaluating global rock stress conditions and determining the stability
of mining excavations.

In this paper, the applicability of AE monitoring to the prediction
of coal burst on a longwall face is discussed. Coal burst is a very
difficult long standing problem in underground coal mining, and is
still fundamentally unsolved even today. This is also a seismic event
occurring close to the longwall face, causing damage to the working
place (Sugawara and co-workers, 1987a).

Two monitoring systems are examined in the multi-layered coal mining
field. One is high frequency AE monitoring available to investigate the
local AE activity, and the other is low frequency AE monitoring, applicable to the global investigation through all the mining fields. With
two case histories, a successful procedure for forecasting the hazardous district by use of a combination of in-situ stress measurements and
numerical stress analyses is demonstrated, and a practical rating
scheme to detect and locate potential regions of instability near the
longwall face are presented and discussed by analyzing the AE activity.

2. FIELD DESCRIPTION

AE monitoring has been performed in the deepest section of the Miike
Coal Mine (Japan). In this district, there are three workable coal
seams dipping about 5 degrees, namely in descending order of the upper
coal seam, the lower coal seam and the main coal seam as shown in
Fig.1. The working depth ranges from 500m to 700m below sea level. The
whole immediate roof and floor formations consist of a very rigid
sandstone of 70~95% in RQD (Kimura and co-workers, 1982).

Under ordinary circumstances, the upper coal seam (1.7~2.5m thick)
is firstly extracted by the retreating longwall method, and the main
coal seam (5~7m thick) is mined out by dividing it into two slices in

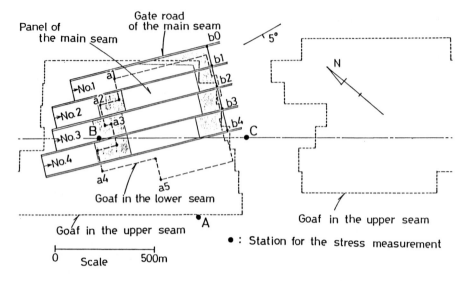

Fig.2 The longwall panels for the main coal seam in Yotsuyama No.35 district and the forecasted hazardous areas painted.

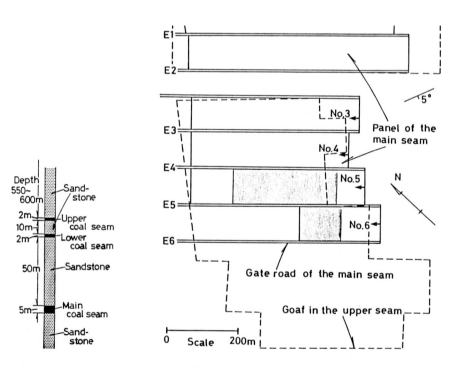

Fig.1 Geological columnar section.

Fig.3 The longwall panels for the main coal seam in Yotsuyama No.60 district and the forecasted hazardous areas painted.

order from the upper slice under the goaf of the upper coal seam. The lower coal seam, which is the thinnest one in this district, is partially extracted when it is workable, before the extraction of the main coal seam.

In this district, coal burst has occurred on the longwall face in the upper coal seam and also in the upper slice of the main coal seam. It has occurred repeatedly within the regions characterized by the frequent occurrence of audible rock noises and a comparatively small convergence of gate roads. They are pure coal bursts with no gas and no fracture of roof and floor sandstone, and belong to a transversal type of coal bursts. It can be noted that the high rigidity of the immediate roof and floor sandstone plays a very important role in their occurrence, and the damage results in dynamic ground movement induced by a large scale shearing fracture between the roof sandstone and the coal seam and/or between the coal seam and the floor sandstone (Sugawara and co-workers, 1987a). Such a boundary fracture can initiate in a highly stressed area in front of the face, which is named the abutment. In general, there is a zone of de-stressed coal seam between the abutment and the face. Such a zone is named the failure zone in this paper.

There is no problem when the boundary fracture propagates in a stable manner having a very low speed. The problem arises from its dynamic propagation. Unfortunately, very limited information is available on the stable-unstable transition of the present boundary fracture. However, it can be noted that a large scale boundary fracture in the immediate vicinity of the longwall face causes damage to the working place, while that at a distance results in only rock noise. In other words, it is necessary that the failure zone always has a sufficient

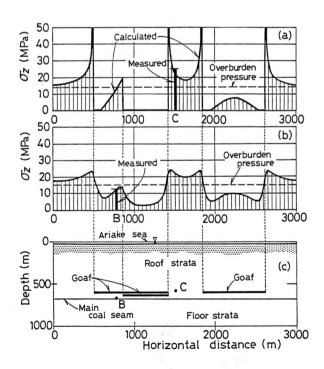

Fig.4
Calculated vertical stress σz in the BC section, in Fig.2, prior to the extraction of the main coal seam, comparing with the results obtained by the in-situ stress measurements.

width to prevent the coal burst, even if a large scale boundary fracture occurs in the abutment.

AE monitoring has been performed in the two districts adjoining each
other, called Yotsuyama No.35 and No.60 respectively, to detect and
locate potential regions of instability in front of the face and to
check the effectiveness of prevention measures. In the No.35 district,
Fig.2, a high frequency AE monitoring system has been adopted, while a
low frequency system has been installed in the No.60 district, Fig.3.

The subject of the present AE monitoring is the longwall mining in
the main coal seam, where the field stress is strongly disturbed by the
overlying goaf. Fig.4 shows an example of the field stress distribution
prior to the commencement of the longwall working in the main coal
seam. This is evaluated from a combination of in-situ stress measurements by the over-coring method and quasi-three dimensional field
stress analyses by the finite differential method (FDM). In regard to
the main coal seam, the stress distributions around the longwall excavation have been analyzed by the FDM for each face position, taking
the closing up of excavation into account, as well as the deformation
of strata. After that, the hazard of coal burst on the face has been
examined on the basis of the mean value of the vertical stress distributed along the face and/or the strain energy release rate due to the
face advance (Sugawara and Kaneko, 1986), and the painted regions in
Figs. 2 and 3 have finally been forecasted to be the hazardous area in
the main coal seam.

3. HIGH FREQUENCY AE MONITORING IN THE NO.35 DISTRICT

A high frequency AE monitoring system installed in the No.35 district
consists of an AE detector, amplifiers with filter, a discriminator and
a counter. The total frequency response is nearly flat in the frequency
range from 300Hz to 3000Hz. The detector is set at the bottom of the
roof borehole of 2m long drilled from the headgate road with a distance
of 100m from the face, and the AE activity is monitored at a distance
ranging from 20m to 100m from the face. After that, the detector is
shifted to a forward station successively.

AE is clearly active, as shown in Fig.5, during the cutting and
loading operation. During periods of down-time, the frequency of the AE
event decreases gradually as time proceeds. The cumulative number of
AE events is a practical parameter to express the AE activity. However, a decision made according of the total number of events will
possibily cause a misunderstanding of the AE activity in the abutment,
because AE in the abutment represents some of the AE events observed.
For an accurate decision, therefore, it is necessary to classify the AE
events observed. From the mechanism of the wave attenuation, the classification by the ringdown counts is considered to be applicable to
this case.

For this purpose, the AE event of a ringdown count greater than 10
has been selected, and the quantity of them per unit face advance has
been investigated. This is considered to be a datum related to the
intensity of the AE activity in the abutment currently questioned, but
influenced significantly by the wave attenuation depending upon the
propagation distance and so on. Therefore, this has been expressed as
$N(L)$, namely, a function of the dist⸱ ⸱e between the face and the
detector. Then the relative intensit⸱, κ, of the AE activity in the
abutment is approximately defined by

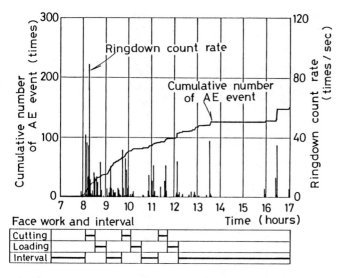

Fig.5 Time dependence of high frequency AE in district No.35.

$$K = N(L)/C(L), \qquad (1)$$

where $C(L)$ is a function for distance correction. By assuming that K is a constant and $C(L) = 1.0$ at $L = 50$m, eq.(1) can be rewritten as follows,

$$C(L) = N(L)/N(L=50\text{m}). \qquad (2)$$

According to eq.(2), the function $C(L)$ applicable to the present district is statistically estimated as shown in Fig.6, that is

$$C(L) = 3.03\exp(-L/40), \quad L \text{ in m.} \qquad (3)$$

By substituting eq.(3) into eq.(1), the relationship between the event number $N(L)$ and the relative intensity K has been determined as follows,

$$K = N(L) \cdot 0.33\exp(L/40), \quad L \text{ in m.} \qquad (4)$$

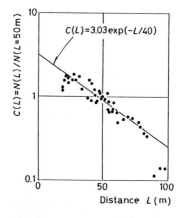

Fig.6
Estimation of the function $C(L)$ on the basis of eq.(2).

Fig.7 shows the changes of the K value with the face advance, comparing them with the mean vertical stress on the face evaluated by the FDM. It is noticeable that the K value has rapidly increased at some distance from the forecasted hazardous area and fluctuated having a high amplitude. The discrepancy between the AE active area and the forecasted hazardous area suggests the fact that the AE active area has moved forward with the face advance, keeping a sufficient distance from the face. Because, in Fig.7, K is plotted at the current face posi-

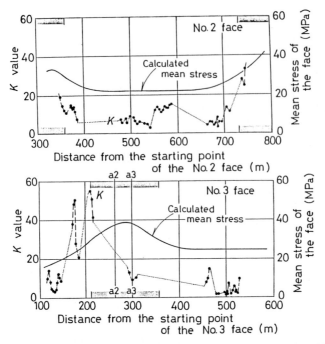

Fig.7 Changes of *K* value with the face advance in district No.35
and the forecasted hazard areas painted.

tion, not at the AE active point. Finally, the longwall face has gone
through these forecasted hazardous areas with no problem. From the
high frequency AE monitoring, it can be empirically noted that the **face**
is rather dangerous when it is at the bottom of *K*'s fluctuating curve.

4. LOW FREQUENCY AE MONITORING IN THE NO.60 DISTRICT

4.1 Source location of AE

In district No.60, a twelve channel source location system has been
installed to monitor the AE activity of the entire area. This system
consists of twelve seismometers, amplifiers, filters, A-D converters
and a mini-computer, and the total frequency response is flat in the
frequency range from 10Hz to 60Hz. For an accurate source location and
seismic energy evaluation, the seismometers have been arranged at
twelve fixed points in rock entries surrounding district No.60, and the
necessary values of primary wave velocity have been corrected successively by means of test-blasting on the face for every working shift.
In this system, the radiated seismic energy is expressed by a relative
value estimated according to the empirical formulae among the maximum
trace amplitude, the propagation distance and the seismic energy radiated (Sato and co-workers, 1986).
    Fig.8(a) is the projection of all the AE events on the coal seam
plane. Each event is represented by a circle having a diameter proportional to its seismic energy. The broken lines indicate the outline of

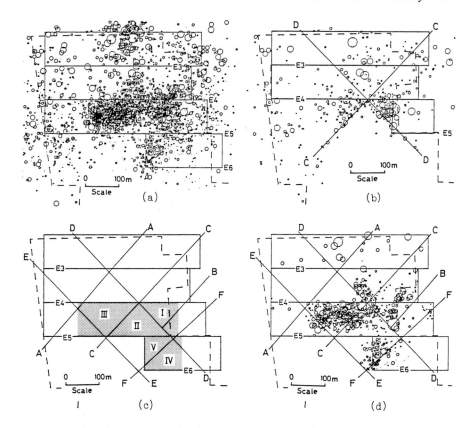

Fig.8 Source location of low frequency AE in district No.60.
(a) projection of all the AE events on the coal seam plane,
(b) distribution of the O-type events observed during a 100 days working,
(c) geometrical relations between the structural discontinuities A-A ~ F-F and the forecasted hazardous area painted,
(d) distribution of the N-type events.

the upper-lying goaf, and the solid lines show the longwall panels in the main coal seam. In each panel, the longwall face has advanced from the right to the left in the figure, and then, the working has moved to the adjacent dip-side panel successively.

In order to examine the AE activity in the abutment, the AE events must be classified into two types, named N-type and O-type respectively, according to the geometrical relation between the epicenter and the face. The former is the AE event located in the square region ranging from 40m in front of the face to 10m in rear of the face within the working panel. The others belong to the later. The N-type events are available to investigate the AE activity near the face, and the O-type events are considered to present abundant informations associated with the geological structure throughout the area.

Source locations of the O-type events are available for geological prospecting. Fig.8(b) shows the distribution of the O-type events observed during a 100 days working. This suggests the existence of some linear structures such as lines C-C and D-D crossing each other in front of the face. From a similar examination of the O-type events distribution, a total of six lines have been found out, as summarized in Fig.8(c). Among them, lines C-C and F-F are confirmed to coincide with the two dominant faults dipping about 80~90 degrees respectively. Therefore, the other lines are presumed to correspond to active discontinuities respectively, but they may be of small displacement. These discontinuities divide the forecasted hazardous area into five regions, from I to V in Fig.8(c).

Fig.8(d) shows all of the N-type events. It is noticeable that the N-type events are concentrated in regions I, III and V. This is considered to suggest a great role in geological discontinuities for determining the global rock stress condition of the longwall mining. If the roof and floor strata are divided by them into several independently movable plates, the abutment pressure will increase with decreasing of the area of unmined coal seam supporting the over-lying roof plate by itself. This is considered to be an explanation compatible with the concentrations of the N-type events in the regions I, III and V.

4.2 Evaluation of energy release density and rate of de-stress

In order to detect and locate the potential regions of instability in front of the face, a practical rating scheme of de-stressing in the coal seam has been presented, and its excellent applicability has been verified as will be seen later. In this scheme, the energy release density is required to be calculated from the seismic energy of AE. This is defined as the released energy per unit area on the coal seam plane, and is computed as follows. Firstly, the coal seam plane is divided into 1m meshes and the seismic energy, E, of each AE event is distributed evenly to the meshes within a circle of radius R computed according to eq.(5),

$$E = \xi S = \xi \pi R^2, \tag{5}$$

where ξ is the energy release per unit area by one event, and $S = \pi R^2$ represents the de-stressed area by one event. After the iteration of such a procedure for all the AE observed, the energy release density of each mesh is given by the summation of distributed energy. Therefore, the energy release density is a multiple of ξ. For the computation of the energy release density in the No.60 district, ξ has been assumed to be a constant, and its value has been back-analyzed from the seismic energy radiated by a coal burst and the area of the fractured coal seam in front of the damaged face (Sugawara and co-worker, 1987b).

If the energy release in the coal seam is always associated with observable AE events and its magnitude is proportional to the seismic energy radiated, the de-stressed region will then be expected to be found out as a region of high energy release density. However, for such a direct decision according to the distribution of the energy release density, the assumption of eq.(5) is much too rough and the accuracy of the present source location is considered to be insufficient. To avoid the risk of misunderstanding, a mean value of the energy release density over some area is considered to be more applicable for locating the de-stressed area and the potential regions of instability. Therefore,

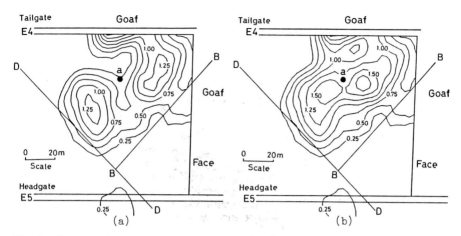

Fig.9 Change of the rate of de-stress (RDS) in the main coal seam
       resulted by a large scale AE event located at the point a:
       (a) contour map of RDS just before the event;
       (b) contour map of RDS after the event.

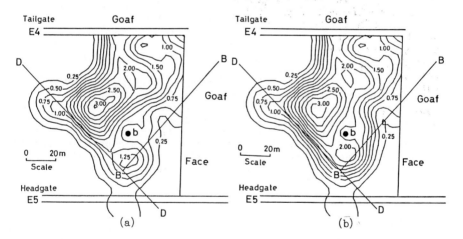

Fig.10 Change of the rate of de-stress (RDS) in the main coal seam
        resulted by a large scale AE event located at the point b:
        (a) contour map of RDS just before the event;
        (b) contour map of RDS after the event.

the rate of de-stress (RDS) in the present rating scheme has been
defined as the mean value of the energy release density within a
circle of 20m in diameter divided by ξ. Such a dimensionless RDS is
given to the center point of the circle, and the contour map of RDS
has been drawn as shown in Figs. 9 and 10.
   A remarkable change in the contour of RDS has been caused by the
intermittent occurrence of larger scale AE events, as demonstrated in
Figs. 9 and 10. In Fig.9, the epicenter of the event is indicated by
point a and there is a distance of 47m between the face and the

epicenter. The de-stressing is clearly progressive in region I. In Fig.10, the epicenter of the event is point b, 28m apart from the face. It is noticeable that the large scale AE events have a clear tendency to occur on the circumference of the de-stressed area of RDS>0.75.

In region I, no coal burst occurs. This result can be explained by the completion of de-stressing in front of the face. It can be empirically noted that the vacant area of de-stressing within regions I, III and V is most dangerous when it comes close to the face in the tailgate side. Since the limitation of the present rating scheme is conditioned by the accuracy of the source location and seismic energy evaluation, it must be emphasized that the extent of the failure zone and the potential region of instability should be re-examined prior to the final decision by means of in-seam test-drilling and so on. Additionally, a de-stress operation, such as de-stress blasting and de-stress drilling, needs to be reinforced while the face goes through the forecasted hazardous area (Sugawara and co-workers, 1987b).

5.CONCLUSION

It has been shown that acoustic emission in the potential field of coal burst is characterized by a periodical fluctuation of the intensity of AE activity having a high amplitude and an intermittent occurrence of large scale AE events in front of the longwall face. Additionally it is noted that the regional differences of AE activity result from the ununiformity of global stress conditions, and dominant discontinuities like faults play an important role in determining the global stress conditions. Classification of AE events according to the geometrical relationship between the epicenter and the working face, and a practical rating scheme of de-stressing to detect and locate the potential regions of instability near the working face has been developed successfully for the prediction of coal bursts. From this successful application and subsequent AE monitoring, it is concluded that AE monitoring is one of the most useful means presently available for evaluating global rock stress conditions in multi-layered coal mining fields.

REFERENCES

Hardy, H.R.Jr. 1973. Microseismic techniques - Basic and applied research, Rock Mechanics, Suppl.2, pp.93-114.
Kimura, O., K.Sugawara and K.Kaneko 1982. Study on the controlling of coal burst in Miike mine, Proc. of 7th Int. Strata Control Conf., Liege (Belgium), pp.431-448.
Sato, K., T.Isobe, N.Mori and T.Goto 1986. Microseismic activity associated with hydraulic mining, Int. J. Rock Mech. Min. Sci. & Geomech. Abstr., 23, 1, pp.85-94.
Sugawara, K and K.Kaneko 1986. Rock pressure and Acoustic Emission in the longwall coal mining, J. of MMIJ, 102, 1177, pp.561-566
Sugawara, K., H.Okamura, Y.Obara and O.Kimura 1987a. Transversal coal outburst in Miike coal mine, Int. Sympo. on Coal Mining and Safety, Seoul (Korea).
Sugawara, K., K.Kaneko, Y.Obara and T.Aoki 1987b.   Prediction of coal outburst, Proc. of 6th Int. Congr. of ISRM, Montreal (Canada).

2nd International Symposium on Field Measurements in Geomechanics, Sakurai (ed.)
© 1988 Balkema, Rotterdam. ISBN 90 6191 778 6

Evaluation of the stress state of cast-in-place underground structures by a FEM analysis and 'in situ' measurements

V.S.Ivanov & R.D.Parashkevov
Rock Mechanics Laboratory, Higher Institute of Mining and Geology, Sofia, Bulgaria
V.H.Vassilev
Institute of Water Problems, Bulgarian Academy of Sciences, Sofia

1 INTRODUCTION

A description is given of a geomechanical investigation for evaluating the stability of a group of cast-in-place support structures in a problem section of a highway tunnel resulting from its passage through a loosened tectonic zone.

An analysis of the stress-strain state based on the Finite Element Method has been applied and 'in situ' measurements of the active stresses in the concrete structures have been performed according to the Partial Stress Relief Method.

Geomechanical conclusions regarding the construction of tunnels in the problem zone are drawn.

2 GEOLOGICAL STRUCTURES AND TECHNICAL MINING CONDITIONS

The rock strata in which the tunnels are built, are composed of breccio-conglomerates, argillites, quartzites and phillites. The rocks have anticlinal and synclinal mode of occurrence, They are slightly weathered, irregularly alternating in type and degree of variability, with a number of fault zones. The track of tunnels in the abovementioned section crosses the plane of the tectonic fault.

The maximum depth at which the tunnels are driven is 80 m. The cross section of the tunnels is shaped like a horse shoe with an area of $76m^2$, each tunnel having roads consisting of three lanes.

The support lining of the tunnels is two-layered (the first layer played the role of a temporary support), built up of pre-cast concrete with total thickness of 90 cm.

3 PROBLEM

3.1 Conditions

The right tunnel is completely built and the construction of the left one has begun, its face being still far from the fault zone.

3.2 Problem for investigation

What is the distribution of the stress fields in the problem section and

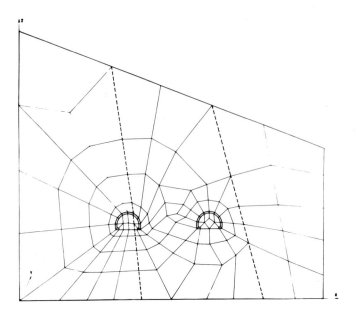

Figure 1. Finite element mesh

how will they be redistributed when the second tunnel reaches the fault?
Will the stability of the two tunnels be preserved under the conditions
of the dynamic stage (i.e. during construction works) at comparatively
small depth of drivage and relatively large cross section of the tunnels
whose track is not a straight line and changes character near the problem zone.

4 INVESTIGATIONS

4.1 FEM Analysis

A problem package according to the Finite Element Method is used for
studying the stress and strain state of the tunnel linings Vassilev(1986).
The package includes volumetric isoparametric elements with 8-20 nodals,
each nodal having three degrees of freedom.
   The finite element mesh in the first vertical section from the area of
rock strata under investigation is shown in (Figure 1). Depending on
whether the tunnels are investigated in a plane strain state or considered as a three-dimensional system, the third dimension of the area(along
axis Y) obtains a different value - 3 m. or 120 m., respectively. The
threedimensional topological description of the area studied is made automatically on the basis of the data about the plane finite elements in
the first vertical section.
   Three main problems had to be solved in the course of the investigation:
   1. Modelling the natural stress state of the rock strata for an inclined terrain line and in the presence of zones with various physico-mechanical characteristica (Figure 2:.

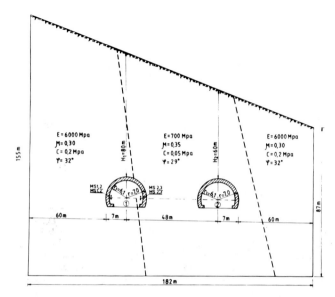

Figure 2. Cross section with geometrical dimensions, geological conditions and location of measuring stations given.

2. Determining the interaction of the two tunnels.
3. Taking into account the effect of the sequence of construction of each tunnel.

As the tunnel overburden is not thick, it is assumed that the rock strata act in an elastic stage and rheological processes are not taken into consideration in the calculations.

In determining the natural stresses $\{\sigma\}_H$ in the rock strata (from dead load) according to FEM, the relative strains $\{\varepsilon\}_H$ corresponding to these stresses, are obtained. If, at this calculation stage, the construction of the tunnel is modelled, then wrong values will be obtained for the stresses and strains in the tunnel lining. This applies particularly to the case under study where the rock strata around the tunnels are heterogeneous. That is why here we have used the so-called 'Initial Stress Method', Hristov (1977) by which, during the first calculation stage, the natural stress state of the rock strata is determined for the particular terrain line assuming that the strata are homogeneous. During the second calculation stage, the drivage of each tunnel and the lining are modelled, and the initial stresses $\{\sigma_o\} = \{\sigma\}_H$ are introduced in each finite element of the rock strata, the latter being divided in zones determined by geological investigations. In this way the lining is loaded only as a result of rock strata deformations caused by the tunnel drivage.

In order to determine the interaction between the two tunnel tubes, the problem is solved in three caculation stages (plane strain state):
1. Determination of the natural stress state.
2. First tube built.
3. Second tube built.

The numerical results show that for the existing interaxial distance between the tunnels (48 m), the stresses in the tunnel lining are not

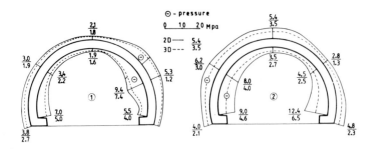

Figure 3. Diagrams of stresses obtained by 2D and 3D FEM analysis.

affected by the construction of the neighbouring tube. The stresses
obtained along the outer and inner contour of the linings of both tunnels under conditions of plane strain state are shown (Figure 3).

The construction of either tunnel is in fact a sequence of two main
processes – digging the rock strata and supporting the exavation. At
the time of constructing the tunnel lining, part of the strata deformations are realized and the lining is loaded as a result of the strata
deformations obtained as the face advances.

The determination of the stresses in the lining caused by the face
advance is carried out by means of 3D FEM analysis, assuming that the
length of the advance is equal to 0,25D (D – tunnel diameter), and the
lining is set up after the exavation has been finished. The increase
of stresses in the lining cross section studied practically stops after
four drivings, i.e. at a distance from the section to the face equal
to 1D. We should note here that during the first driving the stresses
obtained in the lining are 75% of the maximum stresses, while the remaining 25% are formed during the other three drivings. There are shown
the stresses along the inner and outer contour of the lining obtained
by means of a 3D FEM analysis. It is seen (Figure 3) that the stresses
in the lining obtained by taking into account the spatial bearing reaction of the rock strata in the face area are lower as compared to the
stresses obtained when the tunnel is assumed to be in a plane strain
state. The stresses in the most heavily loaded part of the lining (the
area where it is crossed by the horizontal diameter) considerably decrease.

The analysis shows that in this area the stresses are two times lower.

4.2 'In Situ' measurements of the stresses in the tunnel linings

The partial stress relief method is applied for measuring the stresses
in characteristic points of the tunnel linings, Ivanov et al (1983). An
improved modification of the method is used by multi-base measurements
of the displacements providing higher reliability and accurate determination of the active stresses, Ivanov et al (1985). Four mesuring stations have been built, the elevation location of which, for both support
parts in the fault zone, is shown (Figure 2). Two cycles of preliminary
control measurements of the stresses for each mesuring station have
been conducted, the second measurements being made during a definite
interval of time after the first ones. The results from the measurements are given in table 1.

Table 1

	I cycle of measurements		
Measuring station No	σ_1 MPa	σ_2 MPa	θ degrees
1	$-(3,3 \div 8,1)$	$+(0,2 \div 0,7)$	$1 \div 8$
2	$-(4,2 \div 6,0)$	$+(1,4 \div 2,2)$	$40 \div 42$
3	$-(1,0 \div 1,5)$	$-(2,7 \div 4,0)$	$41 \div 47$
4	$-(2,4 \div 8,2)$	$-(4,2 \div 1C,4)$	$25 \div 27$
	II cycle of measurements		
1'	$-(2,0 \div 4,9)$	$+(0,2 \div 0,5)$	$10 \div 18$
2'	$-(2,1 \div 3,1)$	$+(1,5 \div 2,2)$	$-(80 \div 84)$
3'	$-(1,0 \div 1,4)$	$-(2,2 \div 4,1)$	$40 \div 44$
4'	$-(0,3 \div 0,9)$	$-(1,7 \div 4,2)$	$80 \div 84$

Here σ_1, σ_2 are the magnitudes of the quasi-main stresses, and the angle θ (taken between the positive direction of axis Y and the direction of σ_2) which determines their orientation. The symbols $\div$ designate, respectively, the caracter of the stress - pressure stress or tensile stress.

From the measurements it was found that the stress in two support structures seated in the fault zone are irregularly distributed. A quantitative similarity was also found in the character of loading for the measuring points corresponding in disposition in the left and right walls of the support structures controlled, as well as presence of tensile stresses in separate points.

The second cycle of measurements showed changes in the magnitudes and orientation of the stresses σ_1 and σ_2 which, in our opinion, is due to attenuating rheological processes occurring in the rock strata and the tunnel lining.

5 CONCLUSIONS

The analysis of the changes of the stress fields according to FEMA shows that the interaction between the two tunnels during dimensioning of the tunnel linings can be ignored.

A qualitative and quatitative similarity of the results obtained by

FEMA and 'in situ' stress measurements in the linings is established. The combination of an experimental and analytical method of investigation allows for making a more reliable assessment of the stress and strain state of the support structures and the rock strata.

The stresses in the linings determined by both methods do not exceed the allowable strength of the concrete and this gives us good ground to evaluate positively the safety of the structures.

In the investigation program we envisage an increase in the number of measuring stations and the use of Back Analysis for evaluating the stress state of the rock strata and the lining during the construction of the second tunnel.

REFERENCES

Hristov, T. 1977. Considering the natural stress of the rock strata in calculating underground hydrotechnical equipment by FEM. Water Problems, 6, Sofia.
Vassilev, V. A program pacage for spatial investigation of underground equipment by FEM. Technical Mind, 5, 1986, Sofia.
Ivanov, V., R. Parashkevov & S. Popov. 1983. Deformations measurement with the method of partial stress relief and geomechanical processing of the results. FMGM'83, vol.2.
Ivanov, V & S. Popov. 1985. Planing the experiment in measuring the stresses by the partial stress relief method. Studies of stresses in rock strata , Novosibirsk.

The phenomena, prediction and control of rockburst in some Chinese underground engineering

Mei Jianyun
Institute of Geophysics, Academia Sinica, Beijing, People's Republic of China
Lu Jiayou
Institute of Water Conservancy and Hydroelectric Power Research, Beijing, People's Republic of China

1. INTRODUCTION

Rockburst is one of the main disasters in mines and underground works especially in coal mines. In China, the
earliest record of rockburst can be traced back to 1933
and an incomplete statistic shows that about 2,000 rockbursts have occurred in 32 coal mines since 1949. It
can be expected that the occurrence of rockburst will
tend to increase in number with increasing coal production
and mining depth. Rockbursts of various intensity have
also been encountered in metal mines, hydropower stations
and railway tunnels. In the past decade, numerous studies
on the prediction and control of rockburst were carried
out in China and encouragingtresults in eliminating and
reducing the catestrophic consequence of disasters were
achieved. In this paper, the term "rockburst" refers
to coal burst and largearea roof falling as well.

2. PHENOMENA OF ROCKBURST

Rockburst mostly occurs in underground works built in hard
rocks. Some typical case histories are given below.
1. The railway tunnel at Guanchunba. This tunnel is
situated in a mountain and-canyon region in Southwestern
China. Its buried depth is 600-1500 m. The surrounding
rock is a limestone hard and intact with a uniaxial compressive strength of 95-128 MPa. During its construction,
7 larger rockbursts occurred in the period from January
to March, 1976. They have two types of manifestation.
in one type, smaller rock fragments were projected out
with a rather loud pop. The area of fragments is generally
a few square centimeters and the largest volume is 0.5X
0.37X0.08 m^3. In the other type, the lumpiness of the
rock block thrown out is larger and is generally 2.5-
4.2 m long, 1.0-3.0 m wide and 0.1-0.3 m thick. The largest volume amounts to 3.5X1.5X0.3 m^3. In contrast, the
accompanying pop is less loud. Rockbursts mostly occurred
within 2-3 hours after shoot. Some of them even delayed

for several days and thus led to some troubles to the
construction.

2. The headrace tunnel of Tianshengqiao Hydropower Station.
The tunnel is 10 m in diameter and 10 km in length, of
which 7 km pass through the limestone formation. Its average
buried depth is about 400 m and the maximum buried depth
is about 800 m. Rockburst occurred in tunnel No. 2, which
is a construction branch tunnel also in the limestone,
nearly perpendicular to the main tunnel. This branch tunnel
is 1.3 km in length, 10 m in diameter and also circular
in cross section. Both the main tunnel and this branch
tunnel were excavated by TBM. The rockburst occurred in
sections where the buried depth ranges from 200 m to 250 m.
Five major rockbursts occurred, at places where the limestone
is rather hard with a uniaxial compressive strength of
60-80 MPa, and no fissures detectable by visual observation.
The rock in this region is very dry since the groudwater
level is below the tunnel. The rockburst took place mostly
in the top part of the tunnel. The distance from the excavated face to the location where rockburst first occurred
is the same as the diameter of this tunnel. The rockburst
was very active in 24 hours and it weakened and ceased
after two months. The mechanism of rockburst is cleavage
fracture and the fracturing sound is audible. The thickness of the rockburst zone is found to be dependent on
the overbunden of the tunnel.

3. Ertan Hydropower Station. The host rock there is
a hard and intact syenite with a uniaxial compressive
strength of 200 MPa, elastic modulus of 70 GPa and Poisson's
ratio of 0.22. Among the 203 boreholes in the dam area,
disc-shaped cores occurred in 84 of them. In 40 of the
48 boreholes in the river bed zone with high stress concentration, core discing was observed. The disc thickness
is 1-3 cm and the major principal stress measured in the
borehole is 65 MPa. Besides, when preparing rock specimens
40×40 cm^2 in area for shear tests in an adit with a 400 m
thick overburden, the specimen threw rock fragment out or fratured
along the interface of specimen and host rock with a pop whenever the
cutting depth around it attained to 10 cm or so. When continued to cut,
the same phenomenon occurred again and thus obstructed the preparation
of specimen. It is suspected that rockburst is likely to occur in the
future when excavating the underground power house.

4. Pangushan Tungsten Mine. The rock surrounding the
gallery is a quartzite which is hard and intact with a
uniaxial compressive strength of 200 MPa, elastic modulus
of 60-70 GPa, Poisson's ratio of 0.18 and a longitudinal
wave velocity of 5 km/s. Disc-shaped cores with a disc
thickness of 0.4-1.5 cm were found in boreholes. The measured geostress is 20-35 MPa and the stress concentration
factor of the gallery is 3-6. 13 rockbursts happened when
mining at the depth of 400-600 m. When bursting, the rock
projected powder or fragments out or outburst by a length
of up to 10 m. Consequently, pillars were crushed.

5. Jinchuan Nickel Mine. The uniaxial compressive strength
of the ore body is 75 MPa. Disced cores were found in
boreholes. At the mining depth of 400-500 m, rockburst

occurred in the gallery. The principal stresses are horizontal in direction with a major one of 30 MPa and increase
with depth.

6. The coal burst in some coal mines. According to
imcomplete statistics, coal burst have occurred in about
32 coal mines up to 1985. Among them, the destructive
ones amount to 2,000 in times. The maximum seismic magnitude
ranges from 2.5 to 3.8 in Richter scale. The total length
of destroyed gallery attained to 17 km. Nowadays, the
depth of coal mining in China extends 20-30 m downward
every year and thus leads to an increasing potential
menace of coal burst.

The characteristics of rockbursts in China can be
summarized as follows.

1. Rockbursts mostly occur in rocks which are hard
and intact and of high geostress. Disc-shaped cores
are often found in boreholes. According to the statistics
of several hydraulic tunnels in China where rockbursts
have taken place, the maximum peripheral stress at the
entrance of the tunnel is less than or close to one
half of the ultimate comprhessive strength of the rock.
It indicate that the condition of rockburst occurrence
is the fracture under low stress of a brittle material
with high strength.

2. The coal seams in which coal bursts occur are usually
0.7-10.0 m in thickness with a variety of dip angles.
The roof rock is hard and intact with a uniaxial compressive
strength of 100-200 MPa. The coal has a water content
less than 30%, uniaxial compressive strength 10-50 MPa,
elastic modulus 2-9 GPa, Poisson's ratio 0.2-0.3, bursting
liability index 3-7 and Richter magnitude 1.5-3.8.

3. The occurrence of rockburst in generally at the
depth of 200-700 m and increases in number with increasing
depth. Rockbursts mostly occur in coal pillars and advancing faces in highly stressed zones. Some of them may
also occur in zones having subjected to intense tectonic
movements or showing local geological anomalies.

4. For a hard sandstone roof overlying the coal seam,
large area integral falling often happens to result
in disaster.

3. PREDICTION OF ROCKBURST

The methods commonly used for rockburst prediction in
China are as follows.

3.1 Identification of rockburst liability

Strain energy storage index W_{ET} (A. Kidybinski, 1981)
is rather widely used for this purpose. It is defined as:

$$W_{ET} = \phi_{sp}/\phi_{st} =$$ elastic strain energy/dissipated strain
energy where ϕ_{st} and ϕ_{st} are determined from the areas
under the load-deformation curves for loading up to
and unloading from 80-90% of the ultimate strength resp-

Table 1. Class of rockburst and testing standard

Case of Rockburst	Maxium drilling yield (kg/m)			Constituent percentage of drill chips larger than 3 mm (%)	Dynamic Phenomena	Estimated supporting pressure (MPa)
	L<4 m	L=5-6 m	L>6 m			
I	>5.0	>5.5		>30	drill-jamming, impulsive sound, jerking of drill rod	>26
II	3.5-5.0	4.0-5.5	>6.0	>30	pop drill-jamming	15-26
III	<3.5	<4.0	<5.5	<30	none	about 15

ectively. It is specified that

$$W_{ET} \geqslant 5.0, \text{ high bursting liability,}$$

$$2.0 \leqslant W_{ET} < 5.0, \text{ low bursting liability,}$$

$$W_{ET} < 2.0, \text{ no bursting liability.}$$

This specification has been verified by experiments and agrees well with the field situations in China.

3.2 Drilling yield testing

This method is one of those which are ripe and very widely used for rockburst prediction in the world. When this method is used, the distance from the working face to the location of peak stress, the depth of possible rockburst and the dynamic effect as indicated by drill-jamming are also measured along with the drilling yield.
   The classification of rockburst and testing standard of Longfeng Mine are listed in Table 1.
In this table, class III indicates no bursting liability, class II indicates bursting liability which requires careful observation, strengthened supporting and necessary depressurizing treatment and Class I indicates serious bursting liability which requires depress-urizing measures to be adopted immediately and mining activities to be stopped until the bursting liability has already been eliminated as confirmed by drilling yield testing. The similar indices are also adopted in other coal mines of China.

3.3 Accoustic emission and microseismic monitoring

Since the drilling yield testing is laborous and time-consuming and can hardly be carried out continuously, accoustic and microseismic methods are used as important complementary means.
   The rock releases strain energy in the form of elastic

pulse and thus radiates stress wave when it fractures.
This phenomenon is called acoustic emission (AE). To
monitor the AE signal caused by rock or coal fracture
enables the prediction of rock or coal burst to be made.
In addition, the rock or coal releases elastic energy
of low frequency and high intensity in the form of microseismic activity. To monitor the microseismic activities
in rock or coal makes the prediction of rockburst in
a larger scale possible. The frequency of microseismic
activity is lower than that of acoustic emission by 4 orders
of magnitude. The former is a manifestation of macrofractur-
ing, while the latter is that of microfracturing. The following experiences in applying AE and microseismic techniques
to rockburst monitoring have been accumulated in China.

1. The Research Institute of Safety Technology of the
Ministry of Metallurgical Industry and Pangushan Tungsten
Mine have systematically monitored the failure of hard
rockmasses using AE technique and established the following
criterion (Li, 1980):

$N<10$, $R<10\%$, the rockmass is in a relatively stable
condition;

$10< N<20$, $R<10\%$, the rockmass is in the stage of slow
fracturing;

$20< N<30$, $R<60\%$, the rockmass is in the stage of accelerate
fracturing;

$N>30$, $R>60\%$, the rockmass is in the stage of violent
fracturing;

where N is the accumulated count of sounding and R is
the percentage ratio of the count of loud and ultra-loud
sounding in the total count.

This standard has been used to monitor the rockburst and
rockmass failure in Pangushan Tungsten Mine, Kuimeishan
Tungsten Mine, Bali Tin Mine and Daye Iron Mine and successful forecasts were made. However, it should be pointed
out that the values of N and R may be different for various
rocks and should be modified in practice depending on the
experience of the user. For this reason, long-term monitoring
and data accumulation should be upheld in order to determine
the location and extent of rockmass filure accurately.

2. AE technique has been used in Langfeng Mine to study
the relationship between the energy rate of AE and rockburst
occurence (see Fig. 1), to determine the regions of stress
concentration and to select the locations for drilling
yield testing. By adopting such a compreheensive means
in rockburst liability prediction, the labor and time
spent on testing were largely reduced and the density
of testing increased.

3. Microseismie monitoring has been successfully used
in Taozhuang Mine to determine the amount of energy released
by coal burst (in Richter scale), as shown in Fig. 2,
to circle the region of stress concentration and the extent
and time of coalburst occurrence approximately, and to
indicate the precursory sign of coal burst.

4. Large-scale microseismic monitoring together with AE

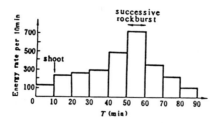

Fig. 1. Variation of energy rate after shoot

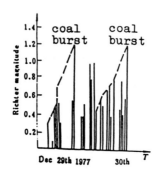

Fig. 2. Gradual increase of microseismic intensity leads to coal burst

sourse location were used successfully in Datong Mine
to predict the regularity of roof movement. By doing so,
the harm to production brought about by large-area roof
caving has been largely reduced.

3.4 Indices of core discing

This method has been used to evaluate the bursting liability
in metal mines in many countries because core discing
is caused by high stress concentration. However, it does
not apply to coal seams since drilling in highly stress
concentrated coal seams will make the coal powdered. Therefore, it is the drilling yield of coal that has been used
for evaluating coalburst liability.
  Disc-shaped cores were found in boreholes in all the
above-mentioned places where rockbursts occurred, viz.,
the Ertan Hydropower Station, Pangushan Tungsten Mine
and Jinchuan Nickel Mine.

3.5 Preliminary estimation of the bursting location by
calculation according to the theory of rock dilatation.

By this method, the region of microfracturing can be
determined from the criterion of $\sigma_{oct} > f'_{3}$ and the potential

energy and volume of this region can be evaluated. Therefore,
the amount of work for predictive monitoring can be reduced
(Tan, 1986). In addition, there are also the methods
of measuring the geostress, crustal deformation, surrounding
rock deformation, etc., which will not be discussed here
owing to the limited space.

It should be pointed out that rockburst prediction
is an extremely complicated problem which requires the
coordination of various methods and comprehensive analysis
in order to achieve satisfactory results. The goal of
rockburst prediction can hardly be attained by any individual
method.

4. CONTROL OF ROCKBURST

There are two types of measures for rockburst control.
One is the measure which has an overall and fundamental
feature. It is to make efforts to eliminate the condition
of rockburst occurrence: This type of measures includes
controlling the stress within an acceptable level, using
reasonable stope arrangement and mining method, and giving
none-bursting treatment to the rock or coal body. The
other type is the measure which has a local and temporary
feature, i.e., to give the burst prone region some liability
releasing treatment. It includes high-pressure water-
injection, unloading blast, borehole unloading, liability
releasing treatment on pillars, etc.. Owing to the complexity of geological conditions and the difficulties encountered in judging the condition of rockburst occurrence,
it seems that both types of measures are necessary.

Measures now adopted for rockburst control in China
are as follows.

4.1 Stress control

The aim of stress control is to avoid the formation of
highly stress-concentrated region or to change the shape
of stress concentration zone by improving the stope arrangement and mining procedure. It should be noted that coalbursts
are mostly related to the choice of mining procedure.
In other words, they mostly occur when excavation is
carried out in highly stress-concentrated zone or partly
through a region under high stress. Therefore, once the
cause, location and conditions for transfer of stress
concentration are found out and reasonable mining procedure
is adopted, it is quite possible to reduce the frequency
of rockburst occurrence and the bursting intensity.

The measures for stress control commonly adopted in
coal mines of China can be stated as follows:

1. Mining in faces towards each other and excavating
in isolated coal body should be avoided.

2. Excavation in highly stressed region can be avoided
by extracting and excavating altermately in different
sections.

3. In regard to the mining procedure, to cut multiple
extraction galleries in coal seams in a same section
is advised. The spacing between parallel galleries should
generally be greater than 10-20 m and two galleries should
intersect, when necessary, at right angle.
    4. The arrangement of pillars should be improved.

4.2 High-pressure water-injection

Water injection method has been successfully used in
many countries and is also widely used in China. Its
role is to improve the texture of coal body so as to
cause a varition of the existing stress concentration
region, and to change the physico-mechanical properties
of coal so that high stress concentration can be avoided.
    In Mentougou Mine, it is found that water-injection
can soften the coal and reduce its elastic energy as
well as the value of W_{ET}, as shown in Fig. 3. By doing
so, rockbursts of intensity higher than 2.3 have been
reduced by 79% in number.
    In Langfeng Mine, before injecting water to the coal
body, the width of stress concentration zone is 15-20 m,
the width of peak stress zone is 2-3 m the distance from
peak stress zone to the working face is 3-7 m and the
peak stress value is 30-50 MPa, respectively. After water
injection, the peak stress zone extends by 2 m towards
the depth, the peak stress reduces by 20%, the value
of W_{ET} reduces by 40-60% on average, and the maximum
drilling yield reduces by 20%. It is shown by experiments
that the bursting liability can substantially be eliminated
when the water content in coal exceeds 3.8%.
    Testing in Taozhuang Mine shows that water injection
can reduce the energy released by a coal body when it
fails and lengthen the time of dynamic failure. As a
result, the energy release decreases significantly and
tends to be stabilized. Fig. 4 gives the curves of time

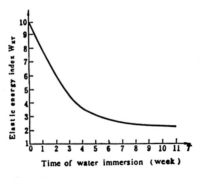

Fig. 3. Elastic energy storage index as a
function of the time of water immersion

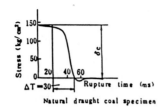

Natural draught coal specimen

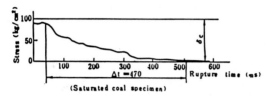

(Saturated coal specimen)

Fig. 4. Curves showing the process of
dynamic failure of coal specimen.

of dynamic failure versus stress for coal specimens
under natural and water saturated conditions.

   The purpose of injecting water to roof is to propagate
the cracks in roof rock, to increase the water content,
to reduce the strength and to increase the possibility
of caving.

   The sudden large-area caving of thick-layered graywacke
roof in Datong Mine has once caused serious harm to
coal production. After studying the microscopic feature
of rock and the mechanism of water-injection softening,
special measures of water-injection have been adopted
and good results achieved. The strength of rock reduces
by 26-49% and the number of cracks increases after water-
injection. Accordingly, measures have been taken to
control the roof caving manually. By making the roof
to cave part by part and layer by layer instead of to
cave as a whole, the cost of roof ripping has been reduced
by 85.7% and the cost of equipment exhaustion by 73.7%
(Pan et al., 1983).

   In Longfeng Mine, the roof subsidence rate before
water-injection is 0.167 mm/day whilst it increases
to 4.16 mm/day during the period of water-injection
which is 25 times of the former. This fact shows that
water-injection can reduce the roof rigidity and thus
increase its deformability.

   It must be noted, however, that the coal seam for
water-injection should have a water-weakening feature
and its porosity should generally be no less than 4%.
Furthermore, the higher the rock stress, the higher
the resistance against water-injection and the lower
the efficiency of water-injection will be. In view of
this, it is inadvisable to drill holes for water-injection
in highly stressed zones.

4.3 Unloading blast Method

This method is to drill holes and shoot in stress concentration zone and to release elastic strain energy by blast
in order to change the shape of stress concentration zone
and the value of stress, so that the bursting liability
can be eliminated by unloading. The position of shoot
should be as close as possible to the location of park
stress. In using this method, hard roof is generally desired.
  In Mentougou Mine, once a maximum drilling yield was
detected at a distance of 5.54 m from the coal face. After
blast unloading, the maximum energy rate monitored by
Geophone reduced to one third of the critical energy rate
and the bursting liability was eliminated.
  In Longfeng Mine, the location of peak stress concentration
is generally 3 m apart from the coal face. By arranging
boreholes for blasting 4-6 m in depth and 3-5 m in spacing
along vertical working face, the drilling yield is reduced
after blast, the stress concentration is mitigated and
hence unloading effect is achieved.
  In Tangshan Mine, the drilling yield is reduced by 7.5-
17.3% and the location of peak stress concentration shifts
towards the depth after blast unloading.
  In many mines, bursts often occur in coal pillars. Therefore, bursting liability mitigation by lossening blast
on coal pillars has also been used. For instance, the
coal extraction in Tianchi Mine has been increased up
to 84% using this method. Nevertheless, drilling large
boreholes in pillars is inadvisable since this will menace
the safety. This method has also been used in metal mines
limitedly.
  Using the method of blast unloading requires safety,
reliability and practical economic efficiency.

4.4 Borehole unloading method

This method is to drill holes of large diameter (ϕ100 mm)
so as to unload the ore body or lower the peak stress
and to increase its distance to the working face. The
higher the stress concentration, the larger the range
affected by borehole unloading and the better the unloading
effect will be. The selection of diameter, spacing and
depth of boreholes should be made in accordance with actua-
lities and complemented by drilling yield and microseismic
testings.

4.5 Excavation of protective seams

By doing so, the surrounding rock and coal body are unloaded
beforehand so that stress concentration can be avoided and
the bursting liability eliminated. This method applies
to multiseam working and requires suitable protective
seams to exist for excavation. Generally, the protective
seam of the roof, coal seams of lowest bursting liability
and those of smallest thickness are excavated first.

4.6 Strengthening supporting

This method is mostly used in metal mines and underground
excavations in hard rocks. In the railway tunnel of Guancunba,
the time for the rock surface to expose freely in air
was shortened as for as possible by lining immediately
after excavation. Safe operation can be assured by using
anchor bolts together with steel wire gauze. In addition,
some effects were also achieved by mist spray and water
sprinkling in the tunnel, controlling the amount of explosive and excavating parallel leading adit, etc..
    The inital rockburst during tunnelling cannot be easily
controlled. The purpose of strengthening by rockbolt is
mainly to prevent the secondary rockburst, which occurs
when fissures and microfisures are caused by rheological
fracture, dynamic effect, etc. The loading in the rock
bolt cannot be correctly calculated, but the effect of
rock strengthening by rock bolts is to lock the fissures
and microfissures from extension. Thus, the earlier the
rock bolts are provided, the better will be the effect
of the strengthening. Because the dangerous zone is near
to the boundary surface of the hole, short rock bolt will
be satisfactory.
    In the variety of engineering works mentioned above,
especially in coal mines, significant safety and economic
effects have been achieved by using the measures for rockburst
control described above.
    In recent years, no casualty due to rockburst happened
in Mentougou Mine and Langfeng Mine. 1.225 million tons
of coal were extracted safety from burst prone regions
in 1981-1985. The average increase in annual output value
in the four years amounts to 1.6 million of RMB yuan.
    In Tianchi Mine, no rockburst casualty has happened
since 1976. Merely in the year of 1984, output value of
0.82 million RMB yuan was created by extracting coal from
burt prone regions.
    In Tao zhuang Mine and Tangshan Mine, casualty has been
reduced and production promoted as well by taking measures
to control rockburst.

5. PRELIMINARY PROBE INTO THE MECHANISM OF ROCK BURST

Generally, the failure mechanism of rock specimen in laboratory test may be classified into three kinds, namely,
brittle cleavage failure, shearing failure accompanied
by tensile failure, and shearing failure (Fig. 5). Especially
in low confining σ_3 or in uniaxial stress (σ_3=0), the
failure of the specimen is in a disorderly pattern (Fig. 5a).
    If the confining stress σ_3 is rather high, the failure
is of shearing type (Fig. 5c).
    In the fracturing process of brittle rock, a series of
acoustic emission signals can be recorded. Every acoustic
emission signal is a response of the growth of a microfracture point. But during the process of loading the specimens
do not exhibit the cleavage fracture as the rockburst
and the failure surface consisting of a lot of micro frac-

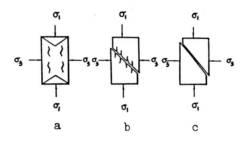

Fig. 5 Failure mechanism of rock specimens

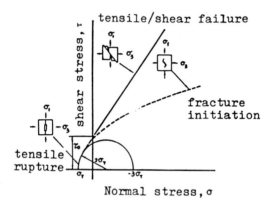

Fig. 6. Failure criterion of brittle rock

tures appears only after the macro failure has been formed. It is reported in Scholz's work (Scholz, C.H., 1968) that the fractured points concentrate near the macro failure surface of the specimens.

Under the condition of compression, the Griffith criterion can be used to reflect the initiation of fracture. Under tension condition, the failure of specimen happens at once when microfracture is initiated. A lot of rock specimen test results show that the Coulomb-Navier criterion can fairly well describe the macro failure of rock. In certain cases, the Modified Griffith criterion may be better because it takes into consideration the close up of the cracks in rock. Thus, in brittle rock the Griffith criterion is a criterion for initial fracture and the Coulomb-Navier or Modified Griffith criterion is the criterion for ultimate failure, as illustrated in Fig. 6.

According to the limit equilibrium theory, the boundary surface after failure of circular tunnel is a logarithmic spiral line as illustrated in Fig. 7b. Some model tests also give the same results, because the models are made of elastoplastic material, which obeys the Coulomb-Navier criterion in ultimate state and the brittle fracture

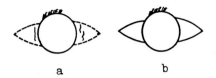

Fig. 7. Possible failure mechanism of adjoining rock
of circular tunnel

is not obvious. Because the radial stress σ_r at the surrounding of the tunnel is rather small and σ_r equals zero
at the boundary of the tunnel, so the major principal
stress σ_r will be σ_θ. The stress state near the boundary
is like the uniaxial stress state. If the behavior of
the rock is brittle, fracture will be initiated near the
boundary under low stress condition, leading to cleavage
spall. Failure might be formed by one crack or by the
connection of a series of cracks and the fractured surface
must be parallel tothe boundary (Fig. 7a) (Lu, etc., 1984).
The mechanism of rockburst in Tianshengqiao tunnel is
typical of brittle fracture.

By comparing the failure mechanism of rockburst with
that of specimen of brittle rock, it can be seen that:

1. The cracks at the adjoining rock of tunnel are much
more in quantity and in size than the microcracks in the
specimen, so the adjoining rock has smaller fracture tough-
nesses, K_{IC} and K_{IIC}, and larger stress-intensity factors
K_I and K_{II} than the rock specimen.

2. The boundary conditions for the specimen and the
tunnel are different. The specimen is laterally constrained
at the top end while the tunnel is free to move in the
radial direction.

3. The stress σ_1 in the specimen has a uniform distribution but the σ_θ in the rock surrounding the tunnel shows
variation with great gradient.

Hence, the microfracture and fracture at the boundary
of the tunnel are liable to spall. Conclusion may be drawn
that the initiation of rockburst and the fracture in the
specimenare similar in characteristics, and they both
obey Griffith criterion. The differences are that the
magnitudes of cracks are not of the same order and the
strength parameters in the Griffith criterion are different.
The macro failure of rock surrounding the tunnel and the
ultimate failure of the specimen are the same and the same
and they obey the Coulomb-Navier criterion or Modified
Griffith criterion. In the prototype the rock mass surrounding
the tunnel is cut by joints and fissures and the boundary
surface after failure is very irregular, no longer being
logarithmic spiral line.

6. CONCLUSION

Rockburst is an extremely complex dynamic phenomenon. In China, some effective experiences in rockburst control have been accumulated through laboratory experiments and in-situ tests: Studies on the mechanism of rockburst are being strengthened. However, further improvement based on practice, communication and cooperation between scholars and engineers from different countries are still needed. Though the problem is very complicated, these still exists the possibility of eliminating the menace of rockburst according to the regularities of its occurrence.

ACKNOWLEDGEMENTS

The author wishes to express his thanks to the Academy of Coalmining Sciences of the Ministry of Coal Industry for offering him the collected paper, "Research on the mechanism of shock ground pressures and experiences in their prevention", from which many data concerning coal mines have been cited. He also achnowledges the help from Mr. Chen Chengzong and Mr. Ding Jiayu who have sent him the data of Guancunba and Pangushan.

REFERENCES

Ganzhou Research Institut of Non-ferrous Metallurgy, Jiangxi Research Institute of Metallurgy and Pangushan Tungsten mine, 1985, A research report of the study on the ground pressure activity in the lower-middle section of Pangushan Tunsten Mine and methods for its control.
Kidybinski, A., 1981. Bursting liability indices of coal. Int. J. Rock Min. Sci. No. 4, V.18, 295-304.
Hou Faliang, Jia Yuru 1986. The relations between rockburst and surrounding rock stress in underground chambers- with a tentative gradation of rockburst intensity Proc. Int. Symp. on Engineering in Complex Rock Formation. Beijing, 297-505.
Li Dianwen, 1980. Study and application of the technique of acoustic emission detection in rockmasses, Proc. 1st Annual meeting on mining Sciences Chinese Society of Metal Sciences.
Lu Jiayou, Ye Jinhan, Chen Fengxiang, 1984. Review and prospect of rock mechanics in underground engineering. Chinese Journal of Rock Mechanics and Engineering. Vol. 3, No. 1.
Pan Qinglian, Xing Yumei, Wang Shukun, Niu Xizhuo, 1983. An application to research on rock behavior and its microstructure for coal mining engineering, Chinese Journal of Rock Mechanics and Engineering, V. 2, No.1, 77-83.
Scholz, C.H., 1968. Experimental of fracturing process in brittle. J. Geophy.Res, Vol. 73, No. 4.
Scientific Documentation Center for Mining Pressure of the Ministery of Coal and shoock Ground Pressure Station,

1985. Research on the mechanism of shock ground pressures
and experiences in their preventions.
Tan Tjong Kie, 1986. Rockburst, case record, theory and
control. Special lecture in Proc. Int. Symp. ECRF'86,
Beijing.
Wang Shukun, 1985. An experimental study on the bursting
liability of coal.
Zhou Simeng, Wu Yushan, Lin Zhuoying, 1986. Post-failure
behavior of rocks under uniaxial compression, Proc.
Int. Symp. on Engineering in Complex Rock Formation,
Beijing, 253-261.

Study on smooth blasting in hard granite tunnels using an automatic drilling machine based on field measurements

Ken-Ichi Yoshimi & Masayuki Suzuki
Civil Engineering Division, Hazama-Gumi Ltd, Tokyo, Japan
Kohei Furukawa & Koji Nakagawa
Department of Civil Engineering, Yamaguchi University, Ube, Japan

1 INTRODUCTION

In recent years, a tunneling method using rockbolts and shotcrete has become common in Japan. Smooth blasting (SB) is regarded as an important technique, not only because of quality control, but also from an economical standpoint. In spite of this fact, we find very few test results of SB which treat it systematically. In many cases, the tests have been done through trial and error on each construction site.

In the study of SB under restricted conditions at construction sites, the following disadvantages have been found :

(1) It is difficult to obtain the precise input/output data for studying the effects of SB.

(2) Rock type and joint conditions may differ in each excavation cycle at the actual construction site. The profiles of excavation openings are not the same, therefore, even when the same blast pattern is used.

(3) There are many factors which have some effect on the SB results. It is, therefore, very difficult to evaluate mutual relations between them.

Recently, with the advance of micro-computers and control techniques, an automatic drilling machine has been developed. The automatic drilling machine, which is numerically controlled, can automatically drill bore-holes following the designed drilling pattern. Using the machine, we may settle the first of the above problems in some form or other. The drilling data, such as drilling position, drilling length, and drilling time, are recorded in the memory bank. They can be used as input data for studying the SB results. Another advantage of using the automatic drilling machine is its ability to drill holes without the skill of workers.

In order to evaluate the results of SB, the profile of the excavated opening must be measured. For this measurement, a range finder (tunnel profile measuring equipment) which utilizes a laser beam may be used.

By using the automatic drilling machine (AD jumbo, Mazda) and tunnel profile measurement equipment (Mikasa type), we may, therefore, get the input and output data of SB without disturbing the progress of the tunnel heading at the construction site. This report deals with a study of SB using the automatic drilling machine to progressively excavate the actual tunnel heading.

2 BASIC CONCEPTS OF SB

SB techniques are used to minimize both overbreak beyond the designed boundary and damage to the remaining rock. As for the minimization of overbreak, the SB technique may be divided into the following three stages :

2.1 1st stage : Blasting technique

The results of SB in the 1st stage are said to be successful when the blast holes are linked by blast cracks and the drill holes remain in shape. Figure 1(a) and Figure 1(b) show an example where the SB in this stage is successful. The result of SB in the 1st stage is evaluated mainly by the drill hole contour

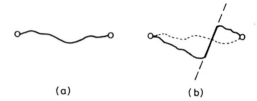

(a)   (b)

Figure 1   Linkage of holes by cracks

ratio. The success of the technique in this stage depends primarily on the geology of the rock formation to be blasted. In order to get improved SB results in this stage, we can choose hole spacing, burden, charge weight, and delay time which are suitable for the rock (Langefors and Kihlstrom (1979) and Blasters' Handbook (1980)).

2.2 2nd stage : Realization of designed tunnel profile

The SB results in this stage are evaluated by the difference between the excavated tunnel profile and the designed tunnel profile. When the difference is small, it is said that SB in the 2nd stage is successful. In addition to the success of SB in the 1st stage, it is necessary to drill holes precisely, according to the drilling design, to get a good SB result in this stage.

In Figure 2, an example is shown where the SB results in the 1st stage are successful but in the 2nd stage are not. The holes are linked by cracks, but the holes are not drilled along the designed contour line.

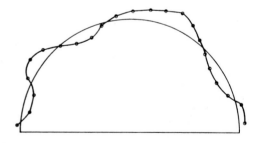

Figure 2 Results of blasting with improperly drilled holes

2.3 3rd stage : Optimization of SB

In actual tunneling, the profile of excavated openings is different from the designed one to a greater or lesser degree. The difference is called an overbreak when the former is bigger than the latter, and an underbreak when the former is smaller than the latter.

Thus, when excavating large openings, large overbreaks and small underbreaks will occur along the designed tunnel profile, and vice versa. SB in the 3rd stage is evaluated by the attainment of cost minimization which includes the cost of revising the overbreaks and underbreaks.

In addition to the factors mentioned in the 1st and 2nd stages, it is necessary in this stage to grasp the overbreak and underbreak distribution from the statistical standpoint.

3 EXPERIMENTAL CONDITIONS

The experimental work was executed in the west side of the Shiwa Tunnel project, San-yo Expressway, ordered by the Hiroshima Construction Bureau, Japan Public Road Corporation. The tunneling method mainly employed in this project used rockbolts and shotcrete.

The rock at the tunnelling site was granodiorite, and the seismic wave velocity was 4.0-4.5 km/sec. The uniaxial strength was 1210-1470 kgf/cm^2, and the Brazilian tensile strength was 68-83 kgf/cm^2.

When our study started, excavation of the top heading had already advanced to more than half the planned length of excavation. The drilling and blasting designs are shown in Figure 3 and Table 1. These drilling and blasting designs were obtained through the method of trial and error. The drill hole contour ratio obtained from the pattern was considerably high. The results of SB in the

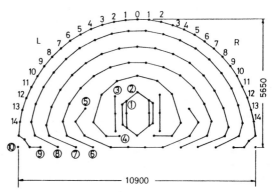

Figure 3 Blasting pattern used in this study

1st stage was, therefore, estimated to be successful. The drill hole contour ratio is defined as the visible length of drill holes remaining after blasting, divided by the total length of perimeter holes. The ratio was calculated using the data obtained from the sketch of drill hole contour as shown in Figure 4.

4 EXPERIMENTS AT CONSTRUCTION SITE, THE RESULTS AND DISCUSSIONS

4.1 Experiment 1 (Drilling by automatic drilling mode)

The drilling mode where all holes are drilled automatically following the designed drilling pattern is called drilling by a fully-automatic drilling mode. The experiment conducted with this mode is called Experiment 1.

The drilling pattern and blast design used in this test are shown in

Table 1 Blasting design

1. Rock material   : Granite
   Seismic wave velocity   : 4.3 km/s

2. Sectional area   : 49.3 m^2
3. Length of excavation   : 2.5 m
4. Length of drilling   : 2.7 m
5. Bit gauge   : 41 mmφ
6. Explosives : No.2 Enoki dynamite (30 mmφ, 100 g)
                Slurry explosive for SB (20 mmφ, 200 g)

7. Volume of blasting   : 123.25 m^3

8. Unit amount of explosive : 1.04 kg/m^3

9. Unit number of holes   : 2.74 holes/m^2

DSD	Holes	Explosive /hole(kg)	Total amount of explosive(kg)
1	6	1.1	6.6
2	6	1.0	6.0
3	6	1.0	6.0
4	11	1.0	11.0
5	6	1.0	6.0
	4	1.1	4.4
6	14	1.0	14.0
	2	1.1	2.2
7	17	1.0	17.0
	2	1.1	2.2
8	20	1.0	20.0
	2	1.1	2.2
9	31	0.7	21.7
	6	1.1	6.6
10	2	1.1	2.2
Total	135		128.1

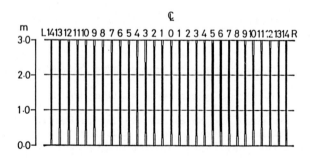

Figure 4   Sketch of drill hole contour

Figure 3 and Table 1. The look-out angle was fixed at 7 degrees. As mentioned above, SB results were evaluated by the profile of excavated tunnel openings. This study was only concerned with the upper semi-circle part of the excavation opening. The distance from the center of the circle to the rock face was measured every 5 degrees. The measured distance was compared to the designed radius of the circle. The distribution of the difference between the measured value and the designed radius (hereafter called "difference" or simply "D") was analyzed in order to discuss the SB results. It is shown in a histogram in Figure 5.

Figure 5 shows that the $\bar{D}$ (mean value of difference D) in Experiment 1 is 6.8 cm. This value is small enough for the average amount of overbreak. The standard deviation σ of D is, however, is as large as 13.4 cm.

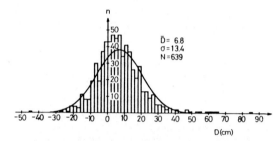

Figure 5 Histogram of the results of measurements of actual blast faces in Experiment 1

In this experiment, drilling was performed in the fully-automatic drilling mode. In spite of the moderate distribution of underbreak expected in the design stage, the actual value was extremely large. This may be explained by the following fact. The drilling machine had been operating for more than 1500 hours, and was not capable of securing the original drilling accuracy of ±5 cm due to a loosening in some mechanical parts. It was presumed that the many underbreaks were caused by this. (The data in Experiment 1 compensated for the machine's error analysis with the measured tunnel profile data.)

4.2 Experiment 2 (Drilling by semi-automatic drilling mode)

When drilling by the semi-automatic drilling mode, the collaring was performed manually at the circle drawn on the tunnel face. After moving the boom to the designed position, the machine was changed to the automatic mode, and drilling was done automatically. The experiment conducted in this mode is called Experiment 2.

The distribution of the difference between the measured distance and the designed radius is shown in Figure 6. Figure 6 shows that $\bar{D}$ is 20.8 cm and σ is 10.6 cm. $\bar{D}$ is larger and

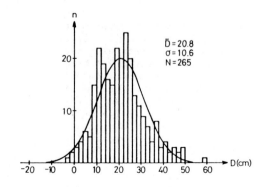

Figure 6 Histogram of the results of measurements of actual blast faces in Experiment 2

Underground openings

σ is smaller than those in Experiment 1. Consequently, the number of underbreaks (D < 0) is almost zero. The mean overbreak ratio is 6.6 %. In Figure 6, the distribution of the difference between the drilling position and the designed contour circle is shown. $\bar{D}$ and σ values in Figure 6 are almost the same as those in Figure 7. This means that in this study, the blast results are mainly controlled by the drilling positions.

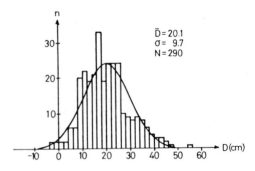

Figure 7 Histogram of the results of drilling in Experiment 2

In Figure 8, the drill hole contour ratio of each hole is shown. Averaging is done in all 13 cycles in Experiment 2. In most holes, the ratio exceeds 70 %.

From the facts mentioned above, SB in Experiment 2 is successful as far as SB in the 1st stage is concerned. As for SB in the 2nd stage the results are not, however, successful because of a relatively large $\bar{D}$ and a somewhat large σ .

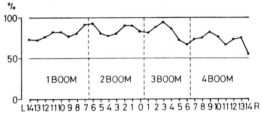

Figure 8 Mean drill hole contour ratio in Experiment 2

4.3 Experiment 3 (Drilling by semi-automatic drilling mode)

As an improvement on Experiment 2, a circle on the tunnel face was drawn more precisely than that in Experiment 2. The center of the circle was obtained more accurately and the line of the circle was drawn as finely as possible. Such fineness aimed at a psychological effect to make workers recognize the importance of exact collaring of holes.

Experimental results are shown in Figure 9, namely, that $\bar{D}$ is 8.2 cm and σ is 8.0 cm. These values are lower than those in Experiment 2. The

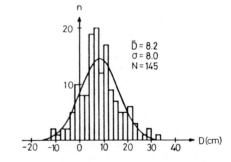

Figure 9 Histogram of the results of measurements of actual blasted faces in Experiment 3

average overbreak ratio is only 1.6 %.
Figure 10 shows the drill hole contour ratio in Experiment 3. The ratio is as high as that in Experiment 2.

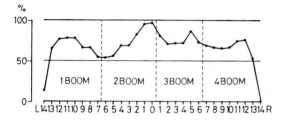

Figure 10   Mean drill hole contour ratio in Experiment 3

4.4 Characteristics of drilling machine by booms

The heading was drilled with two automatic drilling jumbos each having two booms. Figure 11 shows the drilling zone which each boom is in charge of. Table 2 shows the D̄ and σ of each boom in Experiments 1-3. From Figures 8 and 10 SB in the 1st stage shows good results regardless of booms. As shown in Table 2, D̄ and σ are larger for side booms than those for central booms. This means that the machine makes larger errors in drilling side holes than in drilling upper holes.

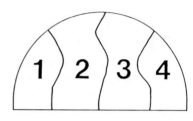

Figure 11   Drilling zones covered by each boom

Table 2   D̄ and σ for each boom

Experiment No.	Boom No.	D̄(cm)	σ(cm)
1	1	11.8	14.2
	2	4.0	9.2
	3	3.7	10.1
	4	7.4	16.6
2	1	23.5	11.1
	2	16.2	8.3
	3	17.1	8.7
	4	24.6	10.5
3	1	11.2	7.2
	2	6.9	6.8
	3	8.3	7.1
	4	6.4	9.2

4.5 Distribution of the difference between the measured distance and the designed radius

The distribution of the difference D between the measured distance and the designed radius is considered to be approximately the normal Gaussian distribution. This is justified by statistical treatment (χ^2 inspection and significant level 5 %).

4.6 Evaluation of SB in the 3rd stage

In discussing SB in the 3rd stage, it is necessary to assess the costs for treating underbreaks and overbreaks exactly to a certain degree.

While it is relatively easy to estimate the cost of revising overbreaks, it is hard to estimate the cost of revising underbreaks. This is due to the facts that the revision of underbreaks is a time consuming task and it is difficult to convert the consumed time into cost. Therefore, it is cumbersome to evaluate Experiments 1, 2, and 3 from an economical standpoint. Here, we try to evaluate the experiments from another standpoint: a method to determine a drilling pattern by fixing the permissible underbreak ratio.

If 16 % of the underbreak ratio is allowed, one can equalize the mean value $\bar{D}$ to the standard deviation σ. In Experiments 1, 2, and 3, the mean values are 13.4, 10.6, and 8.0 cm, respectively.

Actually $\bar{D}$ was 6.8, 20.8, and 8.2 cm in each experiment. This indicates that $\bar{D}$ was too small in Experiment 1, too large in Experiment 2, and moderate in Experiment 3. Though this approach is simple, the answer obtained from the experiments seems to be reasonable.

5 CONCLUSION

Results obtained from the experiments can be summarized as follows :

(1) The basic concept of SB was divided into three stages. In the 1st stage, SB is mainly related to the blasting technique. In the 2nd stage SB is mainly related to the accuracy of drilling, and in the 3rd stage optimum design of SB is important.

(2) In hard and massive rock, SB in the 1st stage is usually successful. It was clarified in this study that the drilling precision governs the SB results in the 2nd stage. In this study, it can be said that each experiment was successful as far as SB in the 1st stage is concerned.

(3) Distribution of difference D between the actually measured blast face and the designed contour follows the normal Gaussian distribution.

(4) In order to drill holes precisely, it is necessary to move the boom to the indicated position manually and to change the machine to the automatic drilling mode. An accurate and fine circle is advisable for precise hole drilling. It was possible to reduce the standard deviation of D by the above mentioned techniques.

(5) SB in the 3rd stage, namely, the minimum cost blasting design, can be done by using the mean value $\bar{D}$ and the standard deviation σ. They are obtained corresponding to a given blasting design. A simple discussion on SB in the 3rd stage showed a reasonable answer in the decision of $\bar{D}$.

REFERENCES

Blasters' Handbook. 1980. E.I. du Pont de Nemours & Co.
Langefors, U. and B, Kihlstrom. 1979. The modern technique of rock blasting : John Wiley & Sons.

2nd International Symposium on Field Measurements in Geomechanics, Sakurai (ed.)
© 1988 Balkema, Rotterdam. ISBN 90 6191 778 6

In-situ loading test of coal seams with respect to the problem of rock burst hazard

Jacob A.Bich & Ilya A.Feldman
All-Union Research Institute of Rock Mechanics and Mine Surveying, Leningrad, USSR
Alexander B.Fadeev
Civil-Constructional College, Leningrad, USSR

1 INTRODUCTION

The mechanical properties of rocks surrounding underground opening are among the principal natural factors stipulating the possibility of onset of the dynamic forms of the rock mass failure. Prediction of such fenomena must be based on determination of the characteristics defining the rock tendency to both elastic deformation and brittle failure. The results of laboratory tests are satisfactory only in case of the monolithic rock masses, homogeneous in both composition and structure. With non-homogeneity of structure and jointing, the laboratory test results are suffice to decide only on separate structural blocks (seams/benches) the specimen is made of but not on the rock mass as a whole. The known techniques of laboratory tests fail to allow for the effects of jointing, stratification, intercalation and different strength of benches in the rock mass. Besides, the rock mass natural humidity, gas content, air temperature, relaxation processes, etc., have no little effect on the mechanical properties of rock. The monolith extraction from the rock mass, its transportation and preparation of specimens lead to damaging the material considerably. Therefore, the objective information on the rock mass mechanical properties can be obtained mainly under the full-scale conditions with a wide range of the most characteristic actual rates of loading.

The full-scale tests must first of all furnish the representative data, i.e., the data reflecting the rock mass mechanical properties from the point of view of the problems to be solved. Therefore, the requirements for the full-scale test methods can be neither general nor universal.

When predicting and developing measures to prevent the dynamic phenomena, the rocks stripped by the working are of a special interest. Thus, the jointing density and orientation, availability of rock strata having their own individual thickness, strength and deformability that should be considered, are determined by the size of the working. Hence, the requirement for choosing necessary and sufficient volume and area of a full-scale specimen to be tested.

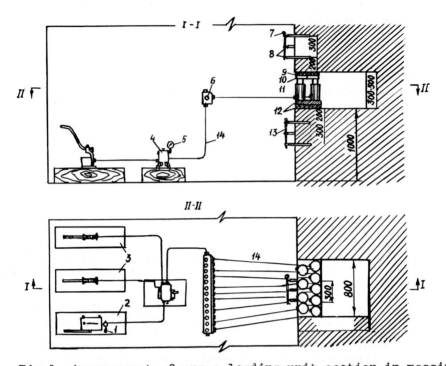

Fig.1   Arrangement of press-loading unit section in massif
at static loading.   Schematic diagram.

1 – low pressure gauge (1000 kgf/cm^2); 2 – low-pres-
sure pump (1000 kgf/cm^2); 3 – high-pressure pumps
(3000 kgf/cm^2); 4 – distributer; 5 – high pressure
gauge (4000 kgf/cm^2); 6 – collector; 7 – clock-type
indicators; 8 – bench marks; 9 – rubber; 10 – steel
plates; 11 – hydraulic jacks; 12 – steel bars; 13 –
dish-shaped stops; 14 – pipelines.

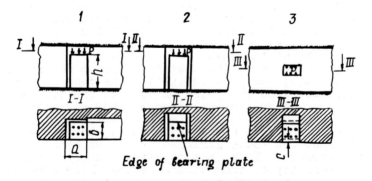

Fig.2   Coal massif loading diagrams

1 – full mapping of prismatic specimen; 2 – partial
mapping; 3 – without lateral mapping (one horizontal
slit).

as well as regime of loading. Of course, these requirements will not be analogous for solving other mining art
problems.

During the in-situ test it is necessary to take into consideration the factors affecting the results, such as the
rock mass loading diagram, type of the loading equipment,
location of the full-scale specimen to be tested, minimum
size of the loaded block, means used for its mapping and
the initial stressed state.

2 STATIC LOADING

Compactness, simplicity and reliability of equipment used
to obtain information on the rock mechanical properties and
stress-strain state under uniaxial and three-dimentional
compression at static, dynamic and continuous loading is an
important requirement the technique of the in-situ testing
should meet. The sectional hydraulic press-loading unit
developed in the All-Union Research Institute of Rock Mechanics and Mine Surveying (VNIMI) is the best to answer
the purpose.

The habitual sectional hydraulic testing unit includes
steel plates, hydraulic jacks, hand-operated or power-driv-
en high-pressure pumps (up to 300 MPa), the distributor
(header), low- (100 MPa) and high- (400 MPa) pressure
gauges, strain measuring appliances (Fig.1). The plant
sectionalization makes it possible to create the non-uni-
form loading due to the distance from the face.

In the beginning of the test cycle, one of the prisms is
crushed to establish approximate strength of the material.
All subsequent prisms are used to define the coal (rock)
deformation properties in the course of the cyclic test under the "load-unload" conditions. The load maximums in the
cycles make up correspondingly 20%, 50%, 60%, 70%, 80%,
90%, 100% of the compression strength σ_e established earlier. No less than three prisms are tested usually and no
less than five prisms in case of a disturbed or non-homoge-
neous rock mass.

The coal massif properties have been defined during the
test carried out following one of the diagrams shown in
Fig.2.

For predicting the rock tendency to bursts any loading
diagram is suitable. However, considering the lower labour input required to prepare the prisms as per diagram 2,
its application is admittedly more preferable. The maximum
discordance in the uniaxial compression strength values obtained for both diagrams is not over 20%. With loading
realized in accordance with diagram 3, the results obtained
depend considerably on the relation of side dimensions of
bearing plates and on the depth the latter are driven to
from the rock mass edge. Dimensions of monoliths to be
tested were chosen with due allowance for the scale factor.
It was ascertained that with the monolith base side of 40-
-50 cm, it reflects sufficiently closely the properties of
the actual rock mass. The strains were measured between

the bench marks installed in holes, with the help of the
clock-type indicators. The indicator gauge length equal to
0.2-0.3 m was adopted proceeding from the most complete determination of strains in separate benches and in the bed
as a whole.

The deformation properties have been determined as a rule
during cycling test under the "load-unload" conditions
which made it possible to obtain the modulus of elasticity,
Poisson's ratio and burst hazard factor K_ε.

The value of stress (Pa) was calculated from the formula

$$(1) \qquad \sigma = \frac{0.785 \, d \, n \, P\beta}{S} \, ,$$

where d – diameter of the jack ram, m;
n – number of jacks;
P – pressure in the hydraulic system, Pa;
β – friction loss factor (0.94-0.97);
S – area to be loaded, m^2.

The ultimate elastic relative strain (ε_{el}) – to total
relative strain (ε_{full}) ratio at loads no less than 80%
of the breaking load was assumed as the generalized criterion of the rock brittle failure:

$$(2) \qquad K_\varepsilon = \frac{\varepsilon_{el}}{\varepsilon_{full}} \cdot 100\% \, .$$

The burst hazard factor K_ε was defined either from the
formula (2) or by the graphical method from the "stress-
-strain" diagrams. The rock mass tends to brittle failure
if $K_\varepsilon \geqslant 70\%$.

If coal seam thickness exceeds the height of prism (diagrams 1-2) or the size of the plate along a wall of the
opening (diagram 3), the trend to burst hazard was determined at loads of 80% of the breaking ones using the formula

$$(3) \qquad K_\varepsilon = \frac{\Sigma \, \varepsilon_{i \, el} \cdot m_i}{\Sigma \, \varepsilon_{i \, full} \cdot m_i} \cdot 100\% \, ,$$

where $\varepsilon_{i \, el}$ – ultimate elastic strain of a separate stratum within the whole seam;
$\varepsilon_{i \, full}$ – ultimate total strain of a stratum;
m_i – thickness of a stratum, m.

The rock mass is burst hazardous if $K_\varepsilon \geqslant 70\%$.

3 LONG-TERM LOADING

The method of direct loading applied for determination of
the rock (coal) massif rheological properties (creep, long-
-term strength) consists essentially of the following.

First, the place is selected and specimens prepared as
has been shown for static test. The static strength is determined on the basis of 2-3 specimens or no less than 4-5
specimens – in case of considerable non-homogeneity or

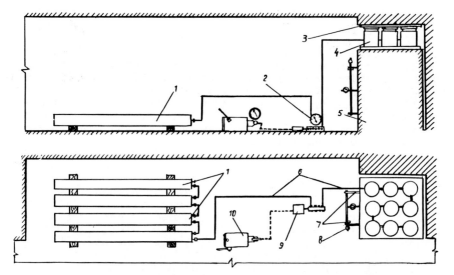

Fig.3 Plant for long-term test. Schematic diagram.

jointing of the massif. The load application rate at static test as well as at creep or long-term strength tests must be constant (2-5 kgf/cm^2 per second).

The bearing steel plates 3 and hydraulic jacks 4 are installed on the specimen - prism 5 (Fig.3) in 1.5-2 hours after it had been prepared (the time needed for mapping the specimen and mounting the test unit). The pump 10 is used to create pressure in the hydraulic jacks which is maintained constant in time with the help of the bank of hydraulic receivers 1. Some elements of the test unit are mounted with the help of the high-head pipe-lines 6. The pressure in the hydraulic system and hence the stress in the specimen are checked by pressure gauges 2. The longitudinal and the lateral strains are measured by the clock-type indicators 8 mounted on bench marks 7 with the dish-shaped stops. The bench marks are secured by wedge-type metal locks in the holes drilled in the specimen specially for the purpose before it has been separated from the massif. It should be born in mind that wooden plugs will be crushed under conditions of the continuously (months) acting load and their application for securing the bench marks is therefore restricted by the press-loading rates.

With the test carried out for several prisms simultaneously, the different load levels are obtained through the use of a different number of jacks with the constant total pressure in the hydraulic system of through the individual loading. The letter consists in that after creating the required load on the prism, the hydraulic system is shut-off from the pump 10 by the valve 9 and the pump can be used then for loading the next prism. During the long-run test, the pump 10 is also used for raising pressure in hydraulic

systems of several specimens being under load simultaneously. During the long-run test, the valve 9 prevents pressure drop in the system, i.e., eliminates one of the main
disadvantages of the majority of the high-head pumps. As a
result, the decrease in pressure in the system is observed
during the first 1-2 days (not over 1.5-2%), the only cause
being the massif deformation and not leaks.

With long-run tests, the attained rate of strain ranges
from 10^{-11} to 10^{-7} I/s and below, the rate of strain during
the test being variable, it is assumed as the average for
the sustained creep area.

During the static test preceding the long-run test, not
only the strength but also the strain characteristics of
the massif up to sustained compression are determined.

The creep test or the long-term strength test commences
from static application of a permanent sustained calculated
load σ_t the value of which is assumed in per cent of the
instantaneous strength σ_e. The values of σ_t as recommended for the massif are 90%, 80%, 70%, 60% and 50% of σ_e.
The frequency of taking the readings depends on the rate of
strain. The test is carried to failure of the specimen or
to stopping the deformation process during 5-7 days (creep
damping). If the sustained load did not cause failure at
of the specimen (due to creep), the latter is unloaded. The
elastic strain at unloading and the elastic aftereffect are
determined. Then the test is carried to failure of the
specimen under the "load-unload" cyclic conditions at constant rate.

The rock mass moisture content before and after the sustained loading (both on the surface of the specimen and inside it) is measured in the course of the long-term mine
test. The temperature in both the massif and the working
is measured when taking readings.

The test results are presented in the form of the "strain-
-time" diagram with different stress levels, these diagrams
being actually the basic characteristics of creep of the
massif. The "stress-strain" diagrams before and after the
sustained compression are also plotted and the burst hazard
criterion K_ε and the long-term strength σ_l are determined.
The latter characteristic can be obtained only when the sus-
tained-to-progressing creep area inflection points are revealed on the "$\varepsilon-t$" diagrams for no less than three specimens with different values of σ_t. If, in all the cases,
the creep is damped, then the half-sum of the maximum load
leading to hardening of the material and the minimum load
leading to loss of strength may be assumed as the long-term
strength.

The massif is considered burst hazardous if one of the
following conditions is satisfied: (a) $\sigma_l \geqslant 0.9 \, \sigma_t$; (b) with
calculated $\sigma_t \approx (0.8-0.9) \, \sigma_e$, strains are damped in the
sustained creep area within 7-10 days; (c) $K_\varepsilon \geqslant 70\%$ after a
sustained (7-10 days) compression with $\sigma_t \approx (0.8-0.9) \sigma_e$.

The creep diagrams for both burst hazardous and burst non-
-hazardous coal seams are shown in Fig.4.

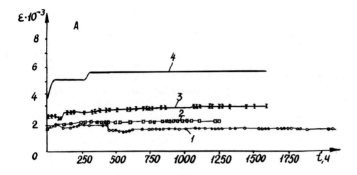

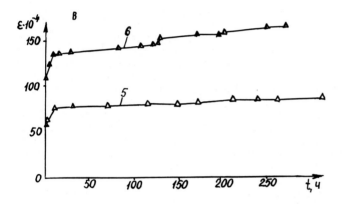

Fig.4 Creep diagrams for burst hazardous (A) and non-hazar-
dous (B) coal massifs with the following values of
sustained load σ_t in fractions of σ_e : 0.2 (curve 1);
0.5 (2); 0.7 (3); 0.6 (4); 0.8 (5 and 6).

4 DYNAMIC LOADING

The rock properties in the massif at dynamic loading have
been determined with the use of the press-loading unit described earlier which was additionally fitted with receivers, the quick-acting valve and recording equipment. The
high rate of massif loading and deformation is attained due
to elastic expansion of the compressed liquid in the receivers, the sudden increase of the volume being ensured by
opening the valve. The application time varied from several tens of minutes to 0.02 s. The results of the experiments have also been presented in the "strain-stress" diagrams. Since there was no unloading, the elastic characteristics have been estimated but roughly.

5 SOME RESULTS OBTAINED IN THE VORKUTA COAL BASIN

Being extremely labour-intensive, the full-scale tests of
coal or rock representative volumes are not always realiz-
able. It is desirable therefore that a method of passing
over from specimens test results to the indices characterizing the entire massif or approaching such characterization should be available. Besides, an express-method of
determining the mechanical properties of a stratum under
conditions of its natural occurence would have been very
useful also. To this end, more than 60 experiments have
been carried out in mines of the Vorkuta occurence with the
help of the press-loading unit (with different coal massif
loading diagrams); large number of coal samples taken following the special technique ensuring the maximum possible
retention of the initial properties of the material were
used to prepare more than 200 coal specimens subjected to
laboratory test. In addition to that, the express-method
as suggested by the Institute n.a. A.A.Skotchinsky for determining the coal strength under the mine conditions have
been comprehensively tried-out. According to that method,
the coal strength is determined by the depth h (in millimeters) of indentation left by the steel punch dynamically
forced into the coal surface after receiving a metered impact energy from the spring-loaded mechanism of the device
Π -1 (strength meter). The empirical relation between the
value of h obtained in the massif and the value of σ_e obtained in the specimens may be expressed as

$$(4) \qquad \sigma_e = \frac{80}{h - 5} .$$

The total number of determinations made with the help of
the Π -1 device was about 700. Disagreement between the
coal compression strength values obtained by recalculating
the device readings in accordance with formula (4) and
those obtained by the direct testing of coal specimens was
not over 18 %. This however is true on one condition: that
the seam tested is partly overmined by the adjacent seam situated at a distance of no more than 20-30 m, the time
elapsed from the partial overmining not exceeding 1 year.
As regards the relationship between the full-scale and laboratory test results obtained for one and the same coal
(the "Thick" and the "Fourth" seams of the Low-Vorkuta strata), it may be stated that quantitative relationships between the uniaxial compression strength σ_e values, the
seam material internal friction angle ρ , cohesion C
in the seam and modulus of elasticity E obtained in the
coal specimens and similar values obtained in the massif
are 4; 1; 2.7; 2.5 respectively. The result is important
and is explained mainly by the effect of the scale factor.
The tests carried out in the Vorkuta mines proved that
the coal full-scale compression strength with different
loading diagrams is characterized by the values given in
table 1 below.

Table 1

Characteristic	Compression strength of prism with		Cube strength σ_{cube}	Massif strength with one horizontal slit
	4 lateral faces	3 lateral faces		
Absolute value, MPa	2.8	4.0–4.5	6.0	10.0–11.0
Absolute value-to 4-faces prism strength ratio	1.0	1.5	2.0	3.5–3.7

6 CONCLUSIONS

6.1 The full-scale in-situ test method makes it possible to objectively predict the coal seams tendency to bursts in the wide range of mining engineering conditions. Such a prediction enables the design bodies to duly specify measures preventing bursts in the course of further use of coal fields.

6.2 There is a certain function between some ultimate mechanical parameters of coal obtained in-situ (massif), on one hand, and obtained in laboratory (specimens), - on another one (see above).

REFERENCES

Bich, J.A., Minin, J.J. & A.I.Bazhenov. 1977. Handbook for determining mechanical properties of Rock Massif with respect to the problem of bursting in coal mines. Leningrad, VNIMI-Institute, 35p.
Feldman, I.A. 1983. Mechanical properties of coal seams subjected to sudden outbursts of gas and coal. Proceedings of the Institute n.a. Skotchinsky, issue 217:41–47, Moscow.

2nd International Symposium on Field Measurements in Geomechanics, Sakurai (ed.)
© 1988 Balkema, Rotterdam. ISBN 90 6191 778 6

Observation of ground movements caused by tunnelling

E.J.Cording
Department of Civil Engineering, University of Illinois, Urbana-Champaign, USA

1 INTRODUCTION

Measurement of ground movement has proven to be the most reliable
and useful means of monitoring the performance of excavations and
underground openings in both rock and soil.

In this paper, examples of ground displacement measurements from
the writer's experience are described. In most of the cases, the
measurements of ground movement were of direct use on the project,
either to confirm the adequacy of the excavation and support procedures, or to serve as a means of identifying conditions in which
corrective measures were needed to prevent instability or damage.

The cases are broadly divided into three categories:

1. Measurement of ground movements in large chambers in rock, used
to monitor the stability of the chamber as the various stages are
excavated. In Nevada, two chambers were excavated at large depth in
a weak tuff subject to stress-induced fracturing. On the Washington
Metro eight station chambers were excavated at shallow depth in a
blocky and seamy, foliated schistose gneiss with a ratio of rock
cover to chamber span that was only on the order of 1/3 to 1/2.

2. Measurement of ground movements in tunnels in squeezing rock,
used to assist in determining the need for additional support or the
ultimate load that will develop on the support. The case described
is a U.S. Bureau of Reclamation water supply tunnel, the Stillwater
Tunnel, in Utah, excavated by TBM at a depth of 600 m in a shale that
was often closely jointed and contained some shear zones subject to
squeezing.

3. Measurement of ground movements around tunnels and excavations
in soil that could distort and damage nearby structures. Results
from a series of tunnel projects are described. (Ground movements
aroung braced excavations are not included in this paper.) Ground
movements were observed from their source at the tunnel perimeter to
the structures that were to be protected.

Described in this third category is a recently developed application of the compaction grouting technique in which ground movements
were controlled during active tunneling on several U.S. projects.
The case history meets the criteria of the observational method, as
described by Peck (1969), which requires that a course of action be
established so that the construction or the design can be modified in
a timely fashion, if so indicated by the observations.

2 STABILITY OF LARGE CHAMBERS

Displacement measurements have proven to be a useful means of
assessing the stability of large chambers during and subsequent to
their construction. Because large chambers are excavated incremen-
tally, it is possible to install and read instruments that will
record the displacements occurring as subsequent excavation stages
are carried out. Based on the results of the measurements, adjustment can be made in the support placed or in the excavation sequence
in subsequent stages.
    Several conditions need to be considered in assessing stability
from displacement measurements. They include:
    1. Magnitude of displacement.
        The magnitude of rock displacement in a chamber can be compared
with the displacements that would be expected for an elastic continuum, or with the displacements previously measured in well-
supported sections of the project. The rock movements measured in
large chambers, such as those constructed for powerhouses, are typically in the range of 2 to 8 mm, on the order of 1 to 3 times the
movements that would be estimated from an elastic analysis, using an
appropriate in-situ modulus that accounts for the stiffness of the
joints in the rock mass. Movements that occurred where shear zones
or other major discontinuities were not adequately supported were
typically in the range of 12 to 80 mm, approximately 5 to 10 times
the elastic modulus (Cording, et al, 1971).
    2. Rate of displacement.
        Displacements should be plotted with time and compared with the
construction events. High rates of movement that are unrelated to
excavation or that continue after the face has advanced well beyond
the extensometer location may be indicative of an unstable condition.
    3. Volume of displacing rock mass.
        The depth and lateral extent of the displacing rock mass should
be determined in order to properly assess the severity of the problem
and the corrective measures to be applied. As illustrated in the
following two case histories, multiple position borehole extensometers are quite useful in determining the depth of the zone of
movement. If it is possible that significant movements (particularly
those related to opening and shearing along discontinuities) will
take place beyond the longest extensometer anchor, then the total
displacement at the head of the extensometer should be determined
independently, using precise surveys or convergence gages extending
between chamber walls.
    4. Displacement capacity of the rock mass and the support.
        Acceptable displacements in a rock chamber will be determined
by the displacement capacity of the support system and the rock mass.
Displacements should not exceed the displacements that would cause
unacceptable distress or failure of the support system; furthermore,
displacements should not exceed the capacity of the rock mass to
maintain its strength and coherence, unless the support system is
capable of supporting the increased rock loads.
    5. Supplemental observations
        Observations that accompany the displacement measurements may
include: (a) visual evidence of opening of joints or movement of
rock blocks, deterioration of the rock surface; (b) mapping of
joints, shear zones, other structural features of the rock, observation of groundwater flows and pressures; (c) evidence of distress

or movement in the support system: dishing of rock bolt bearing
plates, breaking of bolts, cracking or spalling of shotcrete, distortion of steel ribs, crushing or loosening of timber blocking,
measurement of strains or loads in the support system.

Two case histories are summarized. The first is a deep chamber in
a weak tuff subject to stress-induced fracturing.

Nevada Chambers

Two large hemispherical chambers, each 36 m in diameter, were excavated in a tuff at a depth of 400 m (Figs 1 and 2). The strength of
the tuff was 20 MPa. The ratio of rock strength to overburden stress
was 1.5. Thus, the walls of the chamber were subject to stress-
slabbing, resulting in formation of new fractures to depths of 1 to
2.5 m and formation of a loosened, slabby zone typically 1 m deep.
Borehole extensometers revealed the loosening in the outer 1 m and
showed that the movements stabilized upon application of gunite
(shotcrete without coarse aggregate) to protect the tuff against
drying and cracking between the rock bolts, which were spaced on 1 to
2 m centers (Cording, et al, 1971).

The planar face of the hemispherical chambers was recognized to be
the most critical area of the chamber in terms of stability.
However, it was a project requirement that rock bolt support be minimized on the planar surface. For this reason, the displacements on
the plane face were closely monitored with extensometers. The rock
mass surrounding the first chamber excavated had a high RQD and contained few joints or bedding plane weaknesses. It was supported with
a bolt spacing of 2 m on the plane face. In the second chamber excavated, a continuous, flat-lying bedding plane weakness and vertical
jointing intersected the face and combined with newly-formed stress
relief fractures behind the face to cause movement of a large mass of
rock over a 20 m by 20 m area of the face. Movements were observed
in all four borehole extensometers on the face, and revealed that the
large rock displacements were developing at a depth of 3 to 9 m
behind the face. Rock bolts, spaced 2 m on center on the face began
to break as rock displacement reached 50 mm. To stabilize the face,
additional bolts were placed on the face, reducing the spacing to 1 m
and increasing the total bolt capacity by a factor of four. The rate
of movement was dramatically reduced. The records revealed that the
displacement measurements provided several warnings of increased
movement before the rock bolts began to break (Fig 1, Chamber 2).

Washington Metro:

Displacement monitoring conducted in the 18- to 24-m-wide station
chambers excavated in schistose gneiss on the Washington, D.C. Metro
provided a means for evaluating the adequacy of the support and excavation sequences. One of the first stations excavated was Dupont
Circle. The rock was very blocky and seamy, containing shear zones
parallel to foliation as well as planar joints and shears in other
orientations (Cording, et al, 1977).

Although it was recognized that support on the sidewalls of the
initial excavation stages of the Dupont Circle Station was low and
that movements might occur on the wall, the full extent of the movements was only revealed by the borehole extensometer in the roof. It
showed 12 mm of downward movement of a 6-m-thick mass of rock above
the roof (Fig 3). The movement developed in response to the inward

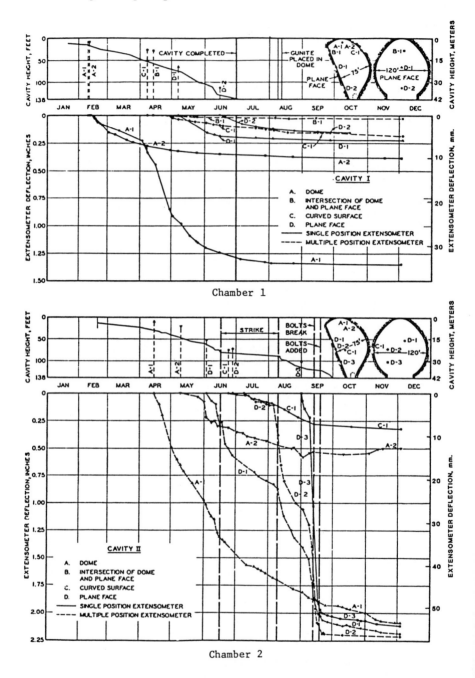

Chamber 1

Chamber 2

Fig. 1. Displacement of Surface with respect to
Deep (30- and 50-ft) Extensometer Anchors,
Nevada Chambers in Tuff.

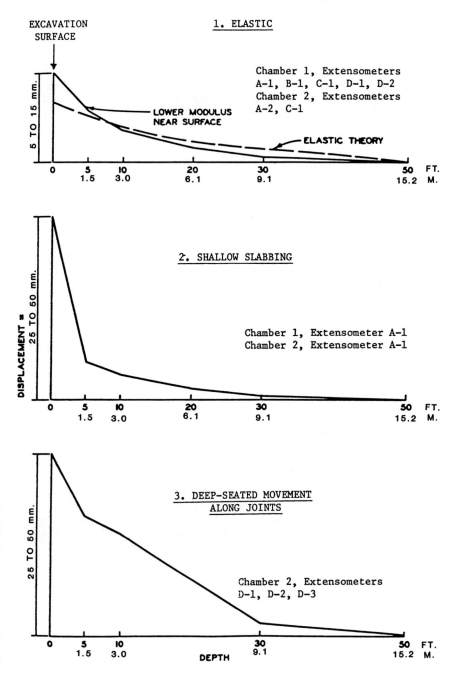

Fig. 2.   Typical Axial Extensometer Displacements
in Chambers 1 and 2, Nevada

lateral displacement of 16 mm on the side walls, which allowed
settlement of the overlying rock mass. The rock mass above the
opening moved as a block, with the movement concentrated beyond the
rock bolts in the crown, 6 m above the roof.
  Without the extensometer results, the extent of the movements would
not have been immediately recognized. The measurement results provided the facts needed to modify the construction contract to permit:
(1) reduction in the height of the Stage 1 sidewall and (2) increase
in the length and number of bolts on the wall.
  Displacement measurements in the other seven station chambers in
rock provided further illustration of the large extent of the movements that could develop along the very planar and continuous shears
that were present in the more heavily sheared rock. Most of the
chambers were aligned with their long axes parallel to the strike of
the steeply dipping foliation shears. Thus, sidewall movements and
failures would tend to occur, unless the walls were heavily supported. The chamber arches also required heavy support. Rock bolt
lengths of 6 m in the arch of 20-m-wide openings that would normally
be adequate in jointed rock that is well keyed, were not capable
alone of preventing settlement of the ground surface. Early placement of a shotcrete-encased steel rib provided the continuous lining
necessary to prevent the large displacements.

3   TIME-DEPENDENT DISPLACEMENT IN TUNNELS

Displacement monitoring is a useful tool for evaluating the adequacy
of the support in tunnels in rock subject to time-dependent displacements.
  In order to evaluate the potential effect of the rock movement on
the tunnel support requirements, it is imperative that the stiffness
of the displacing rock be properly assessed. For example, large
inward displacements may develop if a rock block loosens from the
wall, but the support required to restrain the block may be very
small. In effect, the gap developing behind the block results in a
very low stiffness for the rock block. In other cases, in which the
rock mass displaces inward as a continuum, the support pressures
required to restrain the movement may be very high. Tape or rod
extensometers extended across the tunnel are often used to measure
convergence but do not provide information on the depth of movement.
Multiple position borehole extensometers are useful for determining
the distribution of movements and differentiating between movements
caused by loosening or squeezing.
  In tunnels in squeezing ground, it is difficult to obtain measurements of the initial movement of the ground. Displacements begin
ahead of the tunnel face, and access to the face for installation of
instruments is difficult, particularly in tunnels driven with tunnel
boring machines.
  In the Completion Contract for the Stillwater Tunnel, a
3-m-diameter tunnel excavated by Tunnel Boring Machines at a depth of
600 m in a shale, the principal ground behavior affecting support and
tunnel progress consisted of both loosening and squeezing. Radial
displacement of the rock ahead of the face was measured at one test
section using horizontal inclinometers extended ahead of the face,
just outside the tunnel perimeter. In other test sections borehole
extensometers and tape extensometers were used to monitor radial

CROSS-SECTION
VIEW NORTH

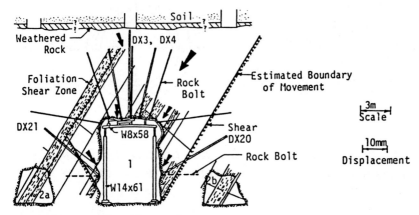

a. Geology, construction, and
zone of rock movement

Note: Stage 1 Sidewall
Bolts Placed
Locally, 6 to 25m
Behind the Face

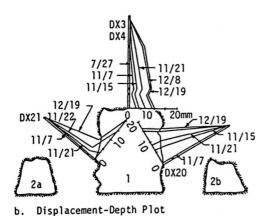

b. Displacement-Depth Plot

Fig. 3. Displacements during initial stages,
Dupont Circle Station, Washington Metro

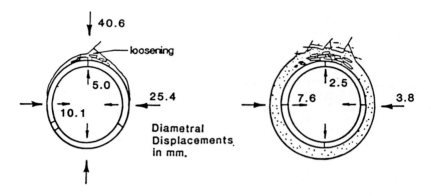

4 – INCH STEEL RIBS

TEST SECTION 1, OUTLET

CLASS III

5 – INCH CONCRETE SEGMENTS

TEST SECTION 4, OUTLET

CLASS II

Fig. 4.   Outward Deflection of the Lining due to
Loosening in the Crown, Stillwater Tunnel

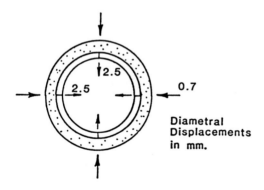

5 – INCH CONCRETE SEGMENTS

TEST SECTION 2, OUTLET

CLASS II

Fig. 5.   Inward Movements of the Lining,
Stillwater Tunnel

displacements. Because of the presence of the TBM, the extensometers
often could be placed no closer than 4.5 m behind the heading. Tape
extensometers were used to monitor the displacement of the support,
which was installed 5 m behind the face.

The progress in the tunnel was excellent. Most of the 8-km tunnel
was excavated by a TBM that averaged 45 m/day over a period of 5
months, with steel ribs installed on .2-m centers. Precast concrete
segments were used as initial and final support in a portion of the
tunnel, however, steel ribs were used as initial support, followed by
a cast in place concrete lining throughout most of the tunnel.

Squeezing occurred in closely jointed zones and in shear zones.
Upon excavation, inward radial displacements took place in all directions. Where the rock was fractured, the largest inward displacements would take place in the tunnel crown, as a result of loosening.
(Cording, 1984).

The pattern of displacement of linings installed in the tunnel was
often significantly different than the displacement of the rock
surrounding the lining. For both the steel ribs and the concrete
segments, inward lining movements developed in the side wall, but, in
many cases, the lining would move upward into the crown where
loosening had taken place. The sidewalls of the lining were being
loaded by the inward squeezing, and, in response, the crown of the
lining displaced upward into the low stiffness zone in the crown. As
a result, locally high bending moments would develop in the upper
arch. In the case of the concrete segments, a longitudinal bending
crack would develop in the middle of the upper sections of the
4-piece segments (Fig 4).

In cases in which loosening in the arch was small, the lining would
be in good contact with the surrounding rock and the lining would
displace inward in the crown and invert as well as on the sides
(Fig 5).

Light steel ribs (W4x13) spaced 1.2 m on center were used
throughout most of the tunnel to support the ravelling and squeezing
ground. In the ground which was subject to light to moderate
squeezing, measured strains in the ribs revealed that stresses
reached the full yield capacity of the ribs quite rapidly. Because
the ribs were expanded directly against the smooth rock surface
created by the tunnel boring machine, they had good contact over the
full perimeter and were able to provide good support even though
yielded. With further inward displacement, the ribs would tend to
deflect out of the cross section, longitudinally along the tunnel.
Placement of additional collar braces (longitudinal support extending
between ribs) provided restraint against the longitudinal deflection
and allowed the ribs to continue to support the rock. The presence
of overbreak, loosened rock, or timber blocking behind the ribs would
create a soft zone into which the rib could displace in some cases,
resulting in distortion and buckling of the rib. Thus, the performance of the support in the squeezing ground was strongly influenced
by the local loosening that took place.

The light steel ribs were able to support the squeezing ground
until the 90-m-long trailing gear passed the rib location. At this
point, it was possible to reinforce and resupport the ground.
Average advance rates were 45 m/day in the squeezing ground, almost
identical to the rates in ground not subject to squeezing.

Analyses of rock movement and lining loads have been performed
using visco-elastic and plastic, visco-elastic models. The creep

rates predicted from these models, using the results of constant
load laboratory creep tests on samples of the shale and the clay
gouge in the shear zones has correlated well with the rates observed
in the tunnels (Phienweja, N., 1987).

4   GROUND MOVEMENTS ASSOCIATED WITH TUNNELING IN SOIL

In the past 20 years significant advances have been made in the procedures for evaluating and controlling ground movements around tunnels in soil.

Emphasis on monitoring (both within the tunnel and at the ground
surface), improved shield design, and control procedures such as compaction grouting have resulted in significant improvement in the ability of the owner/designer to be able to prepare specifications that
allow control of ground movements. Measurements made with deep
settlement points after each shove of the shield have helped pinpoint
the cause of ground movements and permit corrections to be made to
the tunneling process. Improvements in shield design and in
construction procedures in the tunnel have resulted in reduction of
the percentage of ground loss normally encountered along the tunnel.

Measures can also be applied outside the tunnel, before or during
tunneling, to protect structures. No longer is it necessary to
consider underpinning as the only, or even the preferred, method for
protecting the structure.

Several key developments in tunneling practice have contributed to
these improvements. Examples from the writer's experience are used
as illustration.

4.1 Collection of Ground Movement Data and Development of Relationships
Between Construction Procedures and Movements

Ground displacement data have provided the engineer and contractor
with the ability to recognize factors influencing ground loss and to
specify or use methods that will minimize loss.

In 1969, Peck summarized, in his state of the art paper on deep
excavations and tunnels in soft ground, available data on ground
movement measurements. He summarized surface settlement data and
proposed the use of the normal distribution curve for quantifying the
shape of the surface settlement trough and relating trough width to
tunnel depth. At that time, very little data was available on the
distribution of ground movements within the soil mass and the inter-
relation of tunneling procedure and ground movements.

These conditions were investigated in the early 1970's in one of
the first tunnels excavated on the Washington, D.C. Metro.
Inclinometers using servo-accelerometers had just become available,
improving the precision of lateral deflection measurements by an
order of magnitude. An array of inclinometers and extensometers were
placed in the soil mass through which tunneling was to take place.
Lateral, longitudinal, and vertical displacements were then monitored
as the two tunnels were mined through the array. The results clearly
pinpointed the source of the ground movements (in this case, most of
the ground loss occurred around the shield, not in the face of the
shield). From these measurements the distribution of volume changes

and shear strains were also obtained (Hansmire and Cording, 1972, 1985).

Using the relationships between trough width and tunnel depth, and between volume loss volume changes in the soil mass, it is possible to relate the ground loss around the tunnel to the surface settlement (Fig 6). The magnitude of the volume changes in the soil mass, for a given volume of lost ground are controlled by the maximum volume change the soil mass is able to accept. Thus, the deeper the tunnel and the denser the soil, the more volume change that can be accommodated. Figure 7 illustrates the results of the field data collected by Hansmire in the interbedded dense sands and stiff clays in Washington and the additional results obtained from model testing in sands by Hong (1984). Hong showed that the relation (Dr x Z/2R) could be used to quantify the potential volume change capacity of the soil mass, where Dr is relative density, Z is depth to tunnel centerline and R is the tunnel radius.

4.2 Relationships Between Ground Movements and Damage to Structures

As tunneling and excavation projects continued in Washington, D.C., our attention was focussed on the influence of the ground movements on nearby structures. A large number of brick-bearing wall structures with shallow foundations built 75 to 100 years ago were located along the subway routes. Many were not underpinned. Thus, the opportunity was afforded to study the relationship between ground movement and the distortion and damage to the structures, resulting from both excavation of deep cuts as well as tunneling. At one location, prior to tunneling, an unoccupied structure was instrumented with tape extensometers extending horizontally and diagonally between walls inside the structure (Fig 8), pendulums and crack gauges on the exterior, and survey measurements on the ground and on the structure (Boscardin, 1980).

From such observations, it was apparent that the damage to the structures was caused by both vertical and lateral displacement. Thus, the criteria used for buildings settling under their own weight with relatively little lateral movement (Skempton and McDonald, 1956), could not be used without modification.

A criterion was developed that considered the effects of both settlement and lateral displacement on the maximum tensile strain developed in the structure. The magnitude of this strain is then related to the various damage levels (Boscardin, 1980). The resulting relationship, shown in Fig 9, is compatible with the angular distortion relationship developed for buildings setting under their own weight (abscissa of Fig 9) and with the lateral strain relationship developed for damage resulting from subsidence over deep mines, where angular distortions over the length of a structure are small (ordinate of Fig 9). Because the brick bearing wall structures studied are very flexible, the angular distortions and lateral strains imposed on the structures are close to the values of the distortion and strain that would occur in the ground without the structure being present. For stiffer structures, the angular distortions and lateral strains are reduced. Grade beams will have a particularly pronounced effect in reducing the lateral strains in the structure to very low values (Boscardin, 1980).

From the studies described above, it was possible to relate tunnel

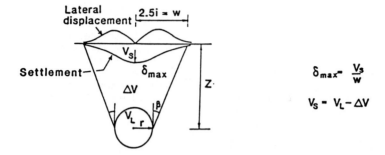

Fig. 6.   Volume Changes and Distribution of Ground
          Movements for Tunnels

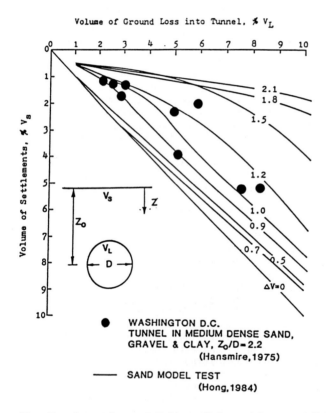

Fig. 7.   Comparison of Volume of Ground Loss and Volume and
          Surface Settlement Trough for Tunnels in Sand

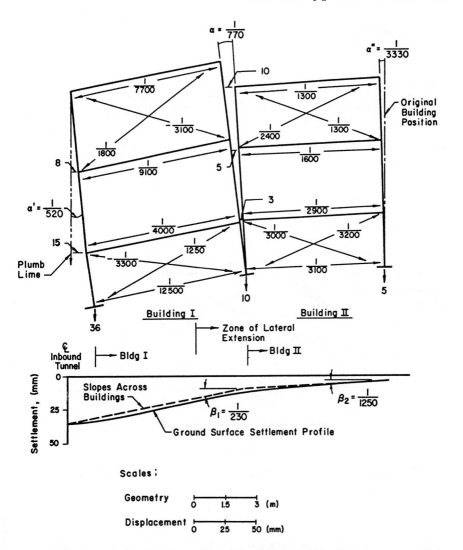

Fig. 8.   Measured Distortion of Brick Bearing Wall
Structures Adjacent to Tunnel, Washington, Metro

ANGULAR DISTORTION $\beta_z = \dfrac{\delta v}{L}$ LATERAL STRAIN $\in = \dfrac{\delta h}{L}$

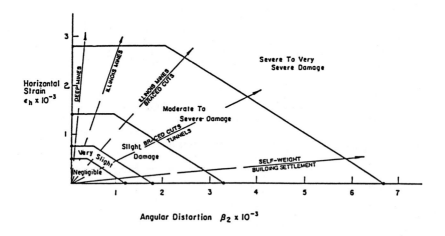

Fig. 9. Damage Criteria for Masonry Bearing Wall Structures,
         Boscardin 1980

construction procedures, shield characteristics and soil conditions
to ground loss, then ground loss to volume change and movements
throughout the soil mass, and finally the surface settlement to the
distortion and damage to structures.

Subsequent work has focussed on measures to control ground movements and minimize damage to structures, as described in the
following section.

4.3 Measures to Control Ground Movements Around Tunnels

Control measures used in the tunnel should be selected with due consideration to the standup time of the soil. Some of the measures
used in the tunnel which are effective in reducing ground loss are

the following:
1. Maintain stable face, with groundwater under control, yet maintain rapid progress. If the face is not controlled, ground movements can be very large and sudden. A variety of shield designs can be used to control the face such as high capacity digger shields that can maintain a muck-filled face, mechanical doors or shelves that can be used to immediately breast the face, use of compressed air or dewatering, shields with rotating heads (wheels) whose doors can be closed, and earth balance or slurry shields that maintain closed, pressurized faces.
2. Minimize external appendages that can result in changes in shield diameter. Use a shield whose dimensiona are properly scaled; articulate long shields for ease of steering.
3. Maintain small tailskin and tight lining to minimize size of tail void and provide for rapid expansion of lining or for grouting behind lining during the shove. Provide effective grout seals at the tail of the shield.

Protection measures can also be carried out outside the tunnel, using procedures such as underpinning, leveling of the structure, chemical grouting, or compaction grouting.

Compaction grouting was recently applied for the first time to control movements during the advance of a tunnel, on the Bolton Hill project for the Baltimore Rapid Transit System (Fig 10). A program of instrumentation was conducted and interpreted in order to evaluate the method (Baker, et al, 1981, 1983). The procedure has since been successfully used on several other projects, and has been required in the contract specifications on other current projects.

Compaction grouting consists of placing a low slump-soil cement at high pressure through holes drilled from the ground surface to replace the volume of ground lost around the tunnel during its advance.

One of the primary advantages of the method is that it can be placed during the tunnel advance, as needed in order to keep the movements of structures within acceptable levels. The amount of compaction grouting carried out can be controlled based on settlement measurements during construction, and on the volume and pressure of grout injected.

The specifications on several recent tunnel projects have been written in such a way that compaction grouting need only be placed if settlement or differential settlement of the structure exceeds a certain value, such as 6 or 12 mm. In a recent project in which the writer participated, this specification produced a strong economic incentive for the contractor to control ground loss from within the tunnel so that he did not have to utilize compaction grouting and other protection methods on certain structures adjacent to the tunnel.

Figures 11 through 14 summarize the results of control measures used in Minneapolis by the contractor, the W. & J. Lewis Corporation, to advance a tunnel beneath the base of an old stone arch culvert. The vertical clearance between the two tunnels was 4.5 m. Ground movements were controlled at their source by using a shield with a thin (12-mm) tailskin and by placing a lining tight within the tailskin in order to minimize the tail void so that very little ground loss could occur prior to lining expansion. The shield was also designed with enough thrust and a large enough pan that the sandy muck could be maintained at its natural angle of repose in the face.

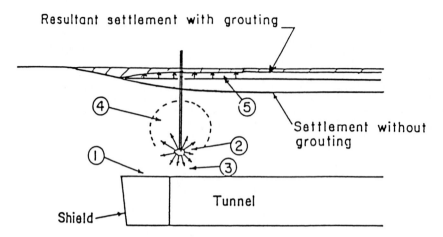

Resultant settlement with grouting

Settlement without grouting

Shield

Tunnel

① Ground loss over front of shield.

② Insertion of low slump compaction grout behind tail of shield.

③ Densification of soil around grout bulb, recovery of volume loss in vicinity of tunnel before it can reach the surface.

④ Heave of soil above tunnel.

⑤ Recovery of previous surface settlement.

Fig. 10.   Effect of Compaction Grouting on Surface Settlements

Prior to construction, settlements caused by tunneling with this shield were predicted to be 25 mm at a distance of 4.5 mm) above the tunnel crown (the location of the invert of the culvert). Two sets of deep and shallow settlement points were installed approximately 100 and 200 m prior to reaching the stone arch culvert in order to verify that movements were under control, or to allow changes to be made in the tunneling procedure if movements were larger than expected. The observed movements (Fig. 11) were almost precisely the same as those predicted.

To further reduce settlements, compaction grout was placed as the tunnel passed beneath the culvert. The compaction grout was proposed by the contractor as an alternative to the chemical grout specified in the contract documents, which would have required approximately ten times as many holes as the compaction grouting and a more detailed and time-consuming sequence of grout placement. The compaction grout was placed through ten holes, five drilled from the ground surface and five through the invert of the culvert (Fig 12).

The compaction grout was placed just behind the tail of the shield, midway between the tunnel and the culvert, as the shield advanced beneath the culvert; a total advance of 24 m in a 32-hour period. An average of .9 cu m of grout was placed per hole in order to replace the volume of soil lost around the advancing tunnel. Grouting was usually carried out after each shove of the shield. The procedure tends to densify the soils surrounding the grout bulb, heave the overlying soils, and push down on the tunnel lining. Monitoring of the lining of the tunnel being excavated showed that it deflected downward approximately 12 mm beneath the grout bulb as it was placed.

Careful monitoring was essential. Although it was estimated, from inspection of the stone arch culvert, that settlements in excess of 25 mm could be accommodated without significant damage, it might have been disastrous if the invert slab had been heaved and cracked due to the compaction grouting. The culvert was flowing with 0.7 m of water during the advance of the tunnel beneath the culvert, and cracking of the culvert invert could have flooded the tunnel.

The compaction grouting operation was directed from a trailer over the tunnel. Loudspeaker and TV communication was maintained between the surveyors in the culvert, an observer in the tunnel, and grout crew at the surface (Fig. 12). Settlement readings on the invert slab and the walls of the culvert were taken at 10-minute intervals.

Figures 13 and 14 show a typical sequence of shield advance and grouting. Settlements of 3 to 6 mm took place after each shove of the shield, then grouting produced heaves that restored the wall and invert slab of the culvert to their original position. Total settlements of the culvert walls after the shield passed were less than 9 mm, and there was no evidence of water inflows into the tunnel.

5 CONCLUSIONS

Ground movement measurements have proven to be a reliable means for evaluating the stability of chambers and tunnels in rock and in determining the potential for damage to structures and utilities adjacent to tunnels and excavations in soil. The measurements can be integrated in an observational program that permits planned modifications and adjustments to be made in the construction so that excessive damage or instability will not result.

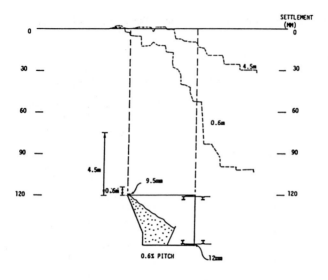

Fig. 11.   Monitoring of Deep and Shallow Settlements
Prior to Passing Beneath Culvert, Minneapolis

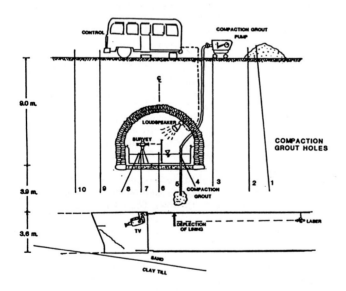

Fig. 12.   Control of Compaction Grouting During
Tunneling, Minneapolis

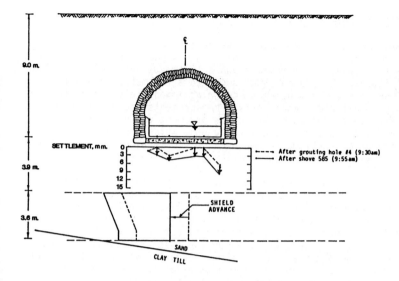

Fig. 13.   Settlement of Culvert Due to One
Shove of the Shield

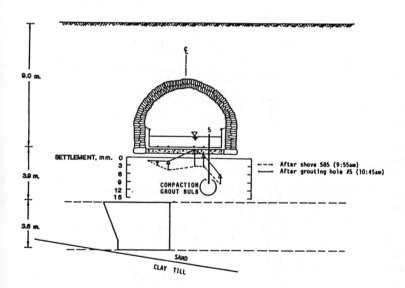

Fig. 14.   Heave of Culvert Due to Compaction
Grouting Through Hole 5

5.1   Tunnels in Soil

Instrumentation to measure ground movements should be easy to read, reliable, rugged. For tunnels in soil, surveying accuracy (+- 1 mm) is generally adequate, and most ground settlements are monitored with survey equipment. Settlement rods anchored a few meters above the tunnel crown permit the movements occurring immediately around the tunnel to be monitored. Near-surface measurements are obtained with shallow settlement rods extending below pavement and utilities. Other techniques, such as fluid level gages, which provide a continuous readout of settlement, can be used. Lateral deflections can be measured using borehole inclinometers, which are repeatible to slopes of 1/10000.

In tunnels and excavations in soil, the ground movement measurements provide a means of evaluating the adequacy of the tunneling and excavation procedures, and are also used to determine the level of damage that may develop at an adjacent structure.

The ability to not only observe movements but to make adjustments in the construction procedures based on the observations permits the application of the observational method. Ground displacements should be measured close to the wall of the excavation or the side of the tunnel in order to correlate the movement with its cause. Ground movements should be measured at the surface and at the level of structure foundations, before the structures are reached, in order to assess the potential magnitude of movements that will be imposed on the structure. If the movements at this point are observed to be excessive, adjustments in the tunneling procedures should be made to reduce the movements, or additional protection measures should be carried out in the vicinity of the structures. Finally, the displacements of the structure and the surrounding soil should be monitored as the tunnel passes. If movements are observed to occur at this point, it may still be possible to prevent excessive distortion of the structure. For example, ground movements near the tunnel can be arrested by placing a replacement or compaction grout into the soil near the movement zone, replacing the volume of ground lost and preventing the movements from progressing to the surface and the structure.

5.2   Chambers and Tunnels in Rock

More accurate measurements are required in excavations in hard rock than for excavations in soil. Displacements of borehole extensometers are commonly measured to 0.02 mm.

Previously, difficulties associated with drift and failure of the electrical systems used to measure displacements and load led to recommendations to use mechanical systems or at least a system that could be checked with a mechanical readout. Difficulties in installing and protecting electrical readouts and conduit as excavation and support took place also made it difficult to use the electrical remote reading systems. Instrumentation is now commercially available that has a proven record of performance, is precise, can be calibrated, and can be checked to ensure that it is functioning properly and providing accurate information. Some of the available borehole extensometers can accept a mechanical readout, a portable probe that can be inserted in the hole and removed after the reading, or an integral readout that permits continuous and automatic recording of the displacements.

In rock chambers, the principal use of the displacement measurements is to evaluate the stability of the chamber and the need for modification of excavation and support. In the cases described, modifications to the design and construction of the chambers were made based on the results of the field observations. To evaluate the displacement reading, the folllowing must be considered: magnitude, depth and lateral extent of moving mass, rate of movement with respect to excavation and time, effect of movement on the strength of the rock mass and the stability of the support.

REFERENCES

Baker, W. H., H. H. MacPherson and E. J. Cording, "Compaction Grouting to Limit Ground Movements, Instrumented Case History Evaluation of the Bolton Hills Tunnels," U. S. Department Of Transportation, Urban Mass Transportation Administration, Office of Technology, Development and Deployment, Office of Rail and Construction Technology, Report No. UMTA-MD-06-0036-B1-1, Washington, D. C., 1981.

Baker, W. H., E. J. Cording and H. H. MacPherson, "Compaction Grouting to Control Ground Movements During Tunneling," Underground Space, v. 7, N3, p.205-213, Pergamon Press, Jan. 1983.

Boscardin, M. D., "Building Response to Excavation-Induced Ground Movements," Ph.D. Thesis, University of Illinois, 1980.

Cording, E. J., D. U. Deere, A. J. Hendron Jr., "Rock Engineering for Underground Rock Chambers, Underground Rock Chambers, ASCE, p.567-600, 1971.

Cording, E. J., "State of the Art: Rock Tunneling", Geotech III Specialty Conf., ASCE, Atlanta, May, 1984.

Cording, E. J., J. W. Mahar and G . S. Brierley, "Observations for Shallow Chambers in Rock," Int. Symp. Field Measurements in Rock Mechanics, Zurich, p.485-508, 1977.

Hansmire, W. H. and E. J. Cording, "Performance of a Soft Ground Tunnel on the Washington Metro," Proceedings, 1st North American Rapid Excavation and Tunneling Conference, AIME, Vol. 2, 1972, p.371-389.

Hansmire, W. H. and E. J. Cording, "Soil Tunnel Test Section: Case History Summary," Jour. Geotec. Eng., ASCE, Vol. III, No. 11, Nov, 1985.

Hong, S. W., "Ground Movements Around Model Tunnels in Sand," Ph.D. Thesis, University of Illinois, 1984.

Peck, R. B., "Advantages and Limitations of the Observational Method in Applied Soil Mechanics," Geotechnique, Vol. 19, June, 1969, p.171-187.

Peck, R. B., "Deep Excavations and Tunneling in Soft Ground,"
Proc. 7th Int'l Conf. on Soil Mech. and Found. Engr., Mexico City,
State-of-the-Art Vol., 1969, p.225-290.

Phienweja, N., "Ground Response and Support Performance, in a Deep
Faulted Shale, Stillwater Tunnel, Utah," PhD Thesis, University of
Illinois, 1987.

Skempton, A. W. and D. H. MacDonald, "The Allowable Settlement of
Buildings," Proc., Institute of Civil Engineers, Vol. 5, Part III,
p.727-784, 1956.

2nd International Symposium on Field Measurements in Geomechanics, Sakurai (ed.)   821
© 1988 Balkema, Rotterdam. ISBN 90 6191 778 6

Discontinuous deformational behaviour of soft rocks due to tunnel excavation – Field measurements of Shioyadanigawa Flood Control Tunnel

T.Sugano, A.Shigeno & K.Suenaga
Public Works Bureau, Kobe, Japan
Y.Okabe & O.Sugita
OYO Corporation, Osaka, Japan
S.Sakurai
Kobe University, Japan

1.Introduction

The Shioyadanigawa flood control tunnel extends from the Western Shimobata District through Mt.Hachibuse in Kobe City to the estuary of Sakai River.   The total tunnel measures about 1,500m in length.   As presented in Figs.1 and 2, the tunnel is geologically situated in the Kobe group of formations (alternate layers of mudstone and sandstone) along the upper reachs of the tunnel and in Rokko granite along its downstream, with the Yokooyama fault dividing the two stretches.   Both zones in this tunnel were constructed (partially as box culverts) by NATM.

Type A measurements (including convergence displacement measurements and tunnel face observations) were implemented at 20m intervals. In locations demanding specific consideration, Type B measurements (including instrumentation performed from the outside of the tunnel), guided the construction(See Fig.2).

This paper specifically reports the results of Type B measurements conducted within the Kobe Group of formations(ST.No. 69+10) from the ground surface. Additionally, results of back analysis are shown, and displacements measured by horizontal inclinometers placed on the Yokooyama Fault are also presented.

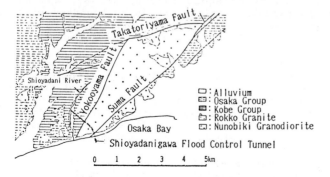

Fig.1 Geolgical Map around the Tunnel

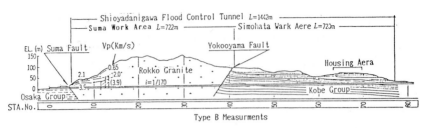

Fig.2 Geological Profile

2. Measured displacements and back analysis

2.1 Measurements

Fig.3 illustrates the geological structure of a measured section (ST.No.69+10), and the layout of instruments (inclinometers and sliding micrometers). The tunnel has been excavated in the mudstone layer at a position of about 45m below ground level (See Fig.4).

(1) Measurement by inclinometer
The inclinometers were installed at locations 2m away from the wall on each side of the tunnel. Fig.3 shows the line of measurements, and sign convention for the readings. Figs.5 through 8 present some of the horizontal displacement readings from inclinometers. As a results of these measurements, horizontal displacement was found to occur primarily in the mudstone layer and not significantly in the other layers. This is probably due to the following factors; that is, a considerable variance in the modulus of elasticity between mudstone and upper/ lower conglomerates and sandstone layers (9,170kg/cm2 for conglomerate, 3,120kg/cm2 for mudstone and 6,100kg/cm2 for sandstone as determined by borehole loading tests), and the presence of permeable formations along the layer boundary which causes the mudstone layer to turn into clay and its surface to become slippery. The results of the inclinometer measurements are summarized as follows;

Figs.6 and 8 demonstrate that the mudstone layer tends to be squeezed out toward the inside of the tunnel. The displacement starts when the tunnel face approaches the measuring section at a distance equal to the tunnel's diameter, and it practically converges when the tunnel face has passed through at a distance of double the tunnel's diameter. A detailed observation reveals that the mudstone gives particularly large displacements at the lower half of the tunnel.

The results for displacements along the tunnel axis (Figs.5 and 7) show a rather complex behaviour. The inclinometer (SS 3) readings indicate that the mudstone layer is extruded parallel to

STA. No. 69+10

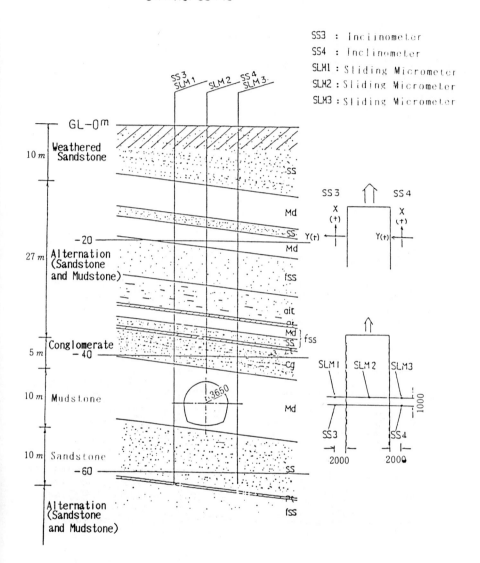

Fig.3 Instrumentation Layout and Geological Section

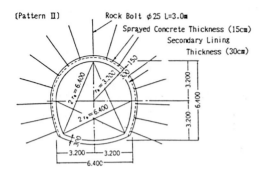

Fig.4 Standard Cross Section of the Tunnel

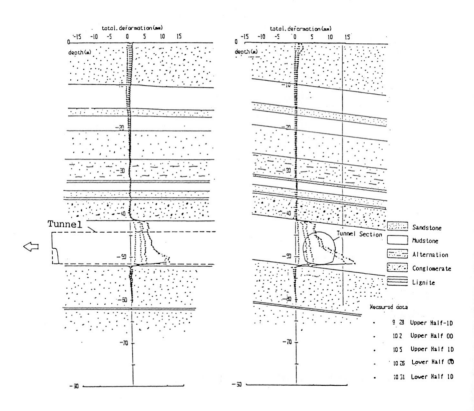

Fig.5 Measurements Obtained
by Inclinometer (SS3: X axis)

Fig.6 Measurements Obtained
by Inclinometer (SS3: Y axis)

the tunnel axis in opposition to the direction of the excavation, and undergoes a particularly large displacement at the lower half. The reading of the inclinometer (SS4) shows that a small displacement occurs in the same direction as SS 3, when the upper half of the tunnel face approaches the measuring section. However, after the tunnel face has passed through the measuring section, the direction of the displacement begins to reverse.

(2) Measurement by sliding micrometer

Three sliding micrometers were installed; two at locations 2 meters apart from the wall on each side of the tunnel and one in the center of the tunnel, as shown in Fig.3.

Figs.9 through 11 present examples of measurements of vertical displacements using sliding micrometers. It is seen from the results that vertical strain is compressive in the mudstone layer where the tunnel is excavated, but slightly tensile as far as the boundary between the mudstone and conglomerate layers from the ground level goes.

However, unlike the horizontal displacement shown in Figs.5 through 8, no discontinuity of displacement distribution occurs along the upper and lower boundaries of the mudstone layer.

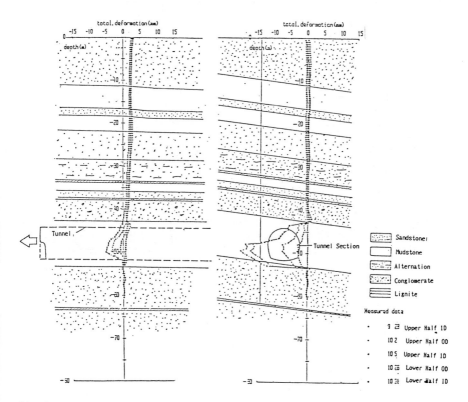

Fig.7 Measurements Obtained   Fig.8 Measurements Obtained
by Inclinometer (SS4: X axis)   by Inclinometer (SS4: Y axis)

2.2 Back analysis of measurements

One of the authors has developed a back analysis method for interpreting the results of measured displacements in a tunnel.
The method is formulated by FEM, and its computer program is named DBAP (Sakurai and Takeuchi,1983; Sakurai and Shinji,1984).
This program is designed to determine the macroscopic modulus of elasticity and the initial stress of the ground based on displacements due to the excavation of tunnels. Once these values have been determined, they are used to analyze the deformational behaviour of tunnel, and optimal supervision of the construction could be possible.
The measured displacements and the initial stress are related as follows;

$$\{u\} = \frac{1}{E} \ (A)\{\sigma_\theta\} = (A) \ \{\sigma_\theta^*\} \qquad \cdots\cdots\cdots\cdots (1)$$

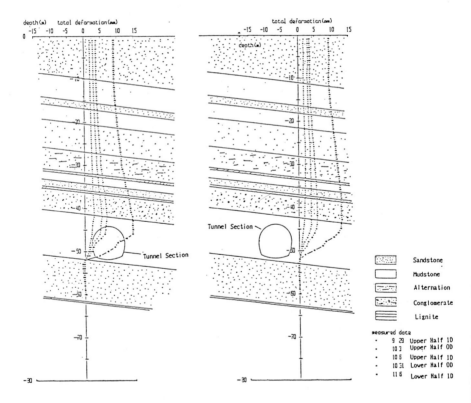

Fig. 9 Measurements Obtained   Fig. 10 Measurements Obtained
    by Sliding Micrometer (SLM1)   by Sliding Micrometer (SLM3)

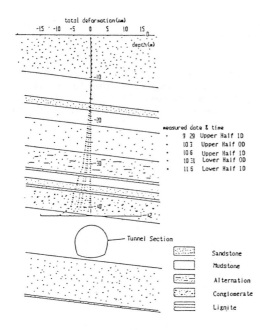

Fig. 11 Measurements Obtained
        by Sliding Micrometer (SLM2)

Where, $\{\sigma_0{}^{\cdot}\} = \{\ \sigma_{x0}/E,\ \sigma_{u0}/E,\ \tau_{xu0}/E\ \}^T,$ and
it is called "normalized initial stress." (A) is a flexibility
matrix determined as a function of measuring positions, Poison's
ratio and the ratio of $E_1/E_0, E_2/E_0, \cdots, E_0/E_0$ ($E_0, E_1, \cdots, E_0$:
modulus of elasticity for each layer of formations). Then,
$\{\sigma_0{}^{\cdot}\}$is determined by the least squares method from measured
displacements, as follows:

$$\{\sigma_0{}^{\cdot}\} = [[A]^T[A]]^{-1}[A]^T\{u_m\} \qquad \cdots\cdots\cdots\cdots (2)$$

Assuming that the vertical components initial stress,σ_{u0},equals
rH (r: average unit volume weight of bedrock, H :overburden
height), the modulus of elasticity and other components of
initial stress are all determined.
In order to express the discontinuous deformational behaviours
of boundaries between mudstone and conglomerate/sandstone, a
joint element is used in FEM for this back analysis.
Back analysis, coupled with the measurement of displacements by
inclinometers and sliding micrometers, is conducted after the
upper and lower halves of the face excavation have been completed
and their effects cease to exist. The modulus of elasticity
determined by the back analysis is 8,480kg/cm2 after completion
of the upper half, and 5,470kg/cm2 after completion of the lower
half section, (A loading test performed in a borehole show
results of 3,120kg/cm2).

Figs.12 and 13 compare measured displacements with theoretical displacements derived from an ordinary FEM analysis using input data obtained from the back analysis. These figures demonstrate a good coincidence between the two.

Figs.14 and 15 present estimated maximum shear strain distributions due to excavation of the tunnel, which is obtained on the basis of the results of back analysis.

Considering laboratory results, critical strain ε_θ of mudstone is about 1 percent.

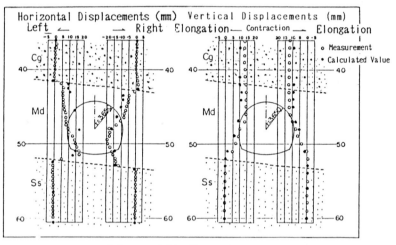

Fig.12 Comparison of Measured Displacements with Those
Derived from Back Analysis (Upper Half Excavation)

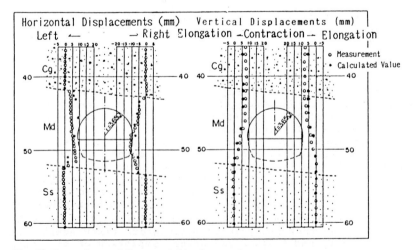

Fig.13 Comparison of Measured Displacements with Those
Derived from Back Analysis (Lower Half Excavation)

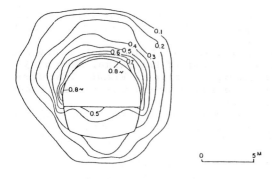

Fig.14 Distribution of Shear Strain (at the final stage
       of the upper half excavation)

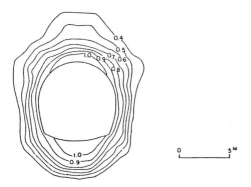

Fig.15 Distribution of Shear Strain (at the final stage
       of the lower half excavation)

Therefore, critical shear strain becomes γ_{max} $(=(1+\nu)\varepsilon_0)=1.45$ %
assuming $\nu=0.45$. It is understood from Fig.18 that the maximum
shear strain occurring around the tunnel arch crown is smaller
than the critical shear strain. This leads us to assume that a
satisfactory excavation has been made.

3. Measurements of Yokooyama Fault by inclinometer

The Yokooyama Fault (strike/inclination:N34° E/63° SE) intersects
the tunnel's center line at an approximate angle of 86°. The
tunnel was excavated toward the fault, difficulty in construction
work was expected. Therefore, in order to acquire fundamental
data for judging the geological conditions, and for determining

auxiliary construction methods, an inclinometer was laid near the
tunnel crown to measure displacements which occurred as face
excavation progressed.

The inclinometer was installed at the ST.No.36+7.7, along the
tunnel axis from 50cm below the crown of the tunnel face at 5°
elevation angle with length L=60m, as shown in Fig.16. Therefore,
the zone covering 48m from the place where the inclinometer
reached the crown was investigated.

Fig.17 shows the measurement results. In the course of the
measurements, an initial value was taken when the upper and lower
half tunnel faces reached steel rib No.643 R and 621 R,
respectively.

These measurements provide the following information.

As the upper half tunnel face was excavated at steel rib Nos.
647-662, a large displacement occurred at the zone between steel
ribs Nos.645-652. The displacement peaked at 72mm at No.648.5,
when the upper half tunnel face reached No.662. The largest
strain (relative displacement/measuring length) in this zone
reached 1.7%. However, as the upper half tunnel face passed
through No.662, the displacement in this zone almost converged.

The displacement at the other places was not as large as the
zone between No.645-652, and it converged to approximately 30mm.

After the upper half tunnel face excavation had reached the
point of No.683, lower half tunnel face excavation proceeded
through the zone where the inclinometer had been installed.

However, as evidenced by the measurements, little or no
increase in displacements due to the progression of the lower
half was observed.

Fig.18 shows the strain distribution along the axis of the
inclinometer.

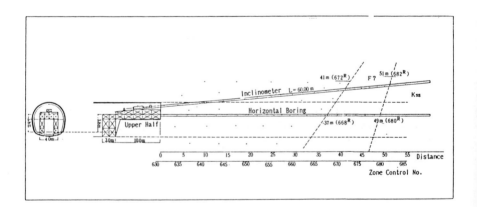

Fig.16 Location of Horizontal Boring

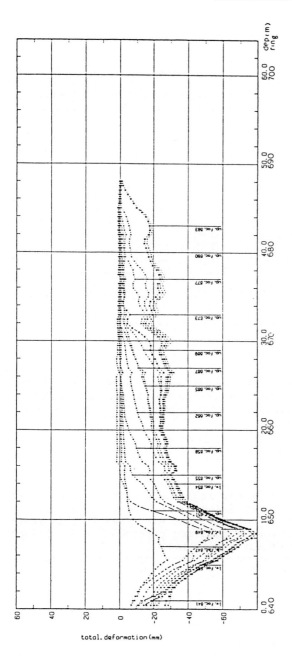

Fig.17 Displacement measured by horizontal inclinometer

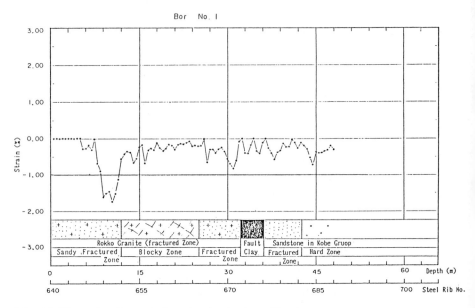

Fig. 18 Distribution of Strain (relative displacement measuring length)

Figs.17 and 18 indicate that an extremely large displacement and strain occurred in the zone of Nos.648-653 where the sandy layer (fractured zone) begins to transform into the conglomerate layer.

4. Concluding Remarks

This paper presents a summary of particularly significant measurements obtained from the instrumentational/analytical project on the Shioyadanigawa Flood Control Tunnel. The authors believe that the results of measurements disclosed in this paper will provide useful reference data and information for future tunnel construction and measuring schemes.

The authors would like to conclude this report by thanking all the personnel who joined this project for their valuable contributions.

References

1) Sakurai, S. and K. Takeuchi, 1983.
   Back analysis of measured displacements of tunnels, Rock Mechanics and Rock Engineering, vol.16, PP.173-180.
2) Sakurai, S. and M. Shinji, 1984.
   A monitoring system for the excavation of underground openings based on microcomputers, Proc. ISRM Symposium, Design and Performance of Underground Excavations, Cambridge, U.K. ,PP471-476.

2nd International Symposium on Field Measurements in Geomechanics, Sakurai (ed.)
© 1988 Balkema, Rotterdam. ISBN 90 6191 778 6

Measurement and monitoring in a tunnel excavated in detritus soil ground with thin overburden, behavior of ground, and effect of supporting members – Tomari Tunnel of Hokuriku Expressway

T.Kawata
Shimizu Construction Co. Ltd, Tokyo, Japan
K.Tanii
Japan Highway Public Corporation, Toyama

1 INTRODUCTION

The Tomari Tunnel, 713 m long, is located at the southernmost end of the Hokuriku Expressway which is under construction between Joetsu and Asahi. (Fig. 1) The bedrock is silty mudstone, and detritus soil is deposited between about a 200 m section from the tunnel mouth on the Toyama side with over 5 to 15 m of overburden on the tunnel. (Hirayama & Nanasawa 1985) As a result of carrying out a detailed geological investigation on the detritus soil section prior to excavation of the tunnel, the following problematical points were predicted. (Tanii & Kawata 1986)

(1) Deformation of tunnel due to fracture of ground
(2) Settlement of tunnel due to insufficient bearing capacity of ground
(3) Collapse of face due to unconsolidated ground and spring water
(4) Landslide due to fracture of ground

In order to solve the above-mentioned problematical points, the tunnel was excavated by adopting optimum excavation and auxiliary execution methods while conducting careful measurement and monitoring inside and outside the tunnel. Consequently, excavation of the detritus soil stratum section and that of the bed-rock section were completed without causing any major deformation or any serious collapse of face which might have caused problems in terms of the structure.

Discussed in this report are the excavation records, measurement and monitoring procedures, as well as consideration to the behavior of the ground and effect of a supporting system based upon the results of measurement and monitoring with respect to the Tomari Tunnel project.

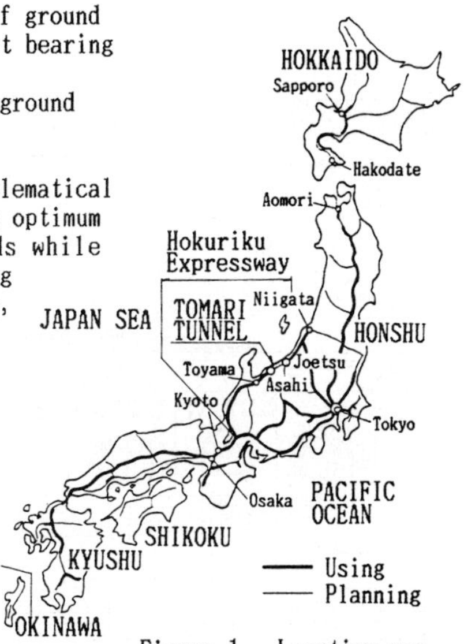

Figure 1. Location map for Tomari Tunnel

2 OUTLINE OF TOPOGRAPHY AND GEOLOGY

This tunnel is located on the eastern foot of the fan of the Kurobe River from the Northern Alps. The bedrock of the main mountain hill is comprised of the Yatsuo Formation of the Tertiary period, while the detritus soil on the Toyama side is composed of clayey soil containing gravel which has been sedimented through many years of collapse of this main mountain hill. Geologically speaking, the detritus soil consists of alluvial soil stratum and diluvial soil stratum. Between a 70 m section from the tunnel mouth, an alluvial soil stratum is distributed at the top heading of the tunnel, while a diluvial soil stratum is distributed over the entire deeper section therefrom.

Fig. 2 is the geological profile of the detritus soil stratum section. The results of soil tests are indicated in Table 1. As shown in Fig. 1, the tunnel adjacent to STA. 666+80 passes under the Yamada River bed with an overburden of 6 to 10 m.

Table 1.   Results of soil tests

Test items			Alluvial clay	Diluvial clay		Mudstone	
			Boring	Boring	Block sample	Boring	Block sample
Geological character-ristics	Grain size classification (%)	Gravel content	16	6	9	—	0
		Sand Content	28	15	16	—	28
		Silt content	35	47	56	—	62
		Clay content	21	32	19	—	10
	Natural water content (%)		41	38	38	—	26
	Consistency (%)	Liquid limit	53	49	60	—	36
		Plastic limit	35	30	27	—	24
	Void ratio		1.11	1.07	1.21	—	0.75
	Degree of saturation (%)		96	96	96	—	—
	Specific gravity		2.67	2.64	2.69	—	—
	Unit weight γ (t/m³)		1.77	1.79	1.75	—	1.77
Mechanical properties	Unconfined compression test	qu (kgf/cm²)	0.31	1.72	1.68	79.0	35.6
		E (kgf/cm²)	13	67	57	—	8600
	Triaxial compression test	C (kgf/cm²)	0.23	0.56	1.12	—	25.6
		ϕ (deg)	7.1	13	0	—	12.1
	Borehole jack test	D (kgf/cm²)	60	159	—	—	—
	Plate load test	Ultimate bearing capacity (t/m²)	—	—	93	—	—
	N value		5~30	5~30	—	—	—
	Ground strength ratio qu/γh		0.26	0.54	0.90	15.4	4.7

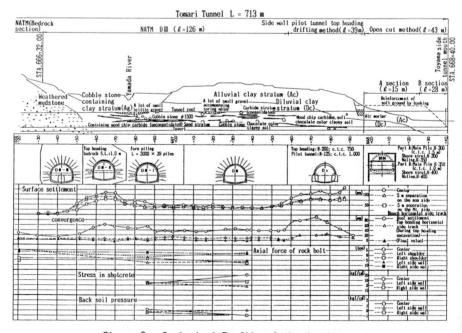

Figure 2.   Geological Profile of the detritus

3 DESIGN OF TUNNEL

Abnormal phenomena expected prior to the start of the tunnel's excavation were examined, and studies were carried out on specifications for the support members and tunnel excavation methods. As a result, tunnel excavation methods for the respective sections were determined as presented in Table 2.

3.1 Selection of excavation method

a) Section between STA. 668 + 40 (tunnel mouth on the Toyama side)
   and 667 + 97 (ℓ = 43 m)
The topography around the tunnel mouth on the Toyama side assumed a slope inclined at about 8°. Since a landslide block was observed, which is assumed to have caused a landslide in the past, the safety factor against landslides was calculated to be 1.0 or less. As a result, the necessity to take some countermeasures against landslides was confirmed. Moreover, an alluvial soil stratum is distributed over the entire section of this span, which is loose and contains water. Thereby, it was considered difficult to expect self-standing of the face. Also, the leg section might possibly be subjected to settlement due to an insufficient allowable bearing capacity of the ground as well. Therefore, in order to cope with such geological conditions, the landslide protective open excavation method according to a master pile horizontal sheet pile system was adopted for excavating this section.

b) Section between STA. 667 + 97 and 667 + 58 (ℓ = 39 m)
Geologically speaking, the alluvial soil stratum is distributed at the top heading part of this secton. The ground water level is so high and the degree of consolidation is so low, that at the time of excavation there was a problem of self-standing of the face. In light of the possibility of settlement expected at the support leg part due to the allowable bearing capacity being as small as 13 t/ m^2, a side wall pilot tunnel top heading drifting method (conventional method with steel support and laggings) was used to make it possible to obtain a drainage effect and support by means of the diluvial soil stratum with a comparatively large bearing capacity for the ground. In addition, since the allowable bearing capacity of the diluvial stratum appearing in the pilot tunnel section equivalent to 28 t/ m^2 was not necessarily large enough, the bottom bed width of the side wall was re-examined according to the loosening load calculation formula for shallow stratum sand ground of Terzaghi. As a result, the width was widened from 1.4 m to 2.0 m.

c) Section between STA. 667 + 58 and 666 + 32 (ℓ = 126 m)
In terms of geology, the diluvial soil stratum with a comparatively high degree of consolidation, is distributed in this section. Thus, it was expected possible to attain a self-standing face and to execute excavation by the NATM. In addition, it was necessary to minimize loosening of the ground as far as practicable, so as to prevent the occurrence of ground surface settlement and landslides due to the excavation at the hill part (which has a topography causing landslide) and the swamp area around the Yamada River. Judging that the NATM would be more advantageous than the conventional method under the above-mentioned situations, the NATM, a standard method of the Japan Highway Public Corporation was adopted for excavation of the tunnel along this section. In view of the possibility that the insufficient allowable bearing capacity of the diluvial clayey soil could cause settlement of the upper and lower half leg parts during excavation, the allowable bearing capacity was examined. As a result, the leg width was extended to 1.1 m by shotcrete as shown in Fig. 3.

Table 2.   Predicted abnormalities and preventive countermeasures taken

Section	Predicted abnormal phenomena	Causes	Investigation and tests	Method of study	Results of study	Counter-measure and methods	Key control items	Measuring points
Detritus soil section	(1) Deformation of tunnel	Distortion due to excavation Insufficient stiffness of support members	Investigation of physical and dynamic characteristics of landslide clay by boring survey	Prediction of the deformation of tunnel based on FEM analysis	Since the roof was settled by 15.5cm as, a result of analysis it is necessary to heighten the stiffness and restrict loosening in front of the face.	Reinforcement of the stiffness of tunnel.	Observation in tunnel	
							Convergence	5 m pitch
							Roof settlement	5 m pitch
							Extensometer	2 sections
							Stress in support members	2 sections
	(2) Settlement of tunnel	Loosening due to excavation Insufficient bearing capacity of leg support ground Insufficient ground contact width of support members	Allowable bearing capacity of supporting ground Investigation of and physical dynamic characteristics of upper ground and loosening ground by boring survey	Required support width was calculated on the basis of allowable bearing capacity of ground and loosening load of upper ground.	ℓ=2.0m is required as the support width for side wall pilot tunnel top heading drifting method. ℓ=1.1m is required for the NATM.	Widening of the support width of tunnel leg part.	Observation in tunnel	
							Settlement of leg	5 m pitch
	(3) Collapse of face	Consolidation degree of ground Spring water Loosening due to excavation	Observation of excavation slope on exposed ground work Investigation of physical and dynamic characteristics of ground by boring survey	Study of face stability according to Murayama method and Broms method	Soil retaining capacity is required to stabilize the face in case the spring line is adopted and upper half bedrock.	Ring cut method and use of fore-piling	Observation in tunnel	
							Observation outside tunnel	
							Settlement of ground surface	5 m pitch
	(4) Landslide	Loosening due to excavation	Topographical survey on landslide by surface reconnaissance Investigation of the continuity of landslide clay by boring survey Investigation of physical and dynamic characteristics of landslide clay survey	Calculation of landslide stability	The safety factor became less than 1.2 at a part of the tunnel mouth and the hill part.	Adoption of soil retaining open excavation method and reforcement of dynamic observation at the tunnel mouth part. Adopting of the NATM at the hill part.	Observation outside tunnel	
							Expansion and contraction of landslide	16 side tracks
							In-ground inclination	3 positions
Bedrock section	(1) Collapse of face	Abnormal spring water	Horizontal boring	Prediction of the amount of spring water	Spring water could possibly cause unstable face and make shotcrete method not practicable.	Drainage pilot tunnel	Observation in tunnel	
							Convergence	30 m pitch
							Roof settlement	30 m pitch

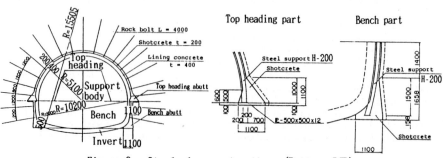

Figure 3.   Standard support pattern (Pattern D III)

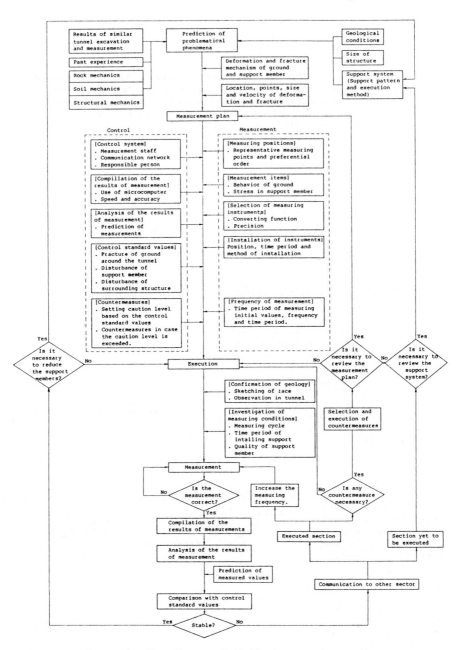

Figure 4. Flow diagram of field measurement execution

4 MEASUREMENT AND MONITORING PLAN

4.1 Monitoring system

As trouble causing abnormal phenomena were predicated due to excavation of the Tomari Tunnel between the detritus soil stratum, the monitoring items were established in consideration of the expected abnormal phenomena, and measurements were carried out on the basis of a detailed monitoring plan. Presented in Fig. 4 and Table 2 are a monitoring flow chart, and the expected abnormal phenomena and monitoring/control items, respectively.

4.2 Monitoring standards

The strain in semi-infinitely elastic ground is expressed according to the following formula with reference to Fig. 5.

(1) $\quad \varepsilon\,\theta = \dfrac{Ur}{r} + \dfrac{1}{r}\dfrac{\partial U\theta}{\partial\theta}$

(2) $\quad \varepsilon\,r = \dfrac{\partial Ur}{\partial r}$

where
$\varepsilon\,\theta$: Strain in tangential direction
$\varepsilon\,r$: Strain in radius direction
$U\theta$: Deformation in tangential direction
Ur: Deformation in radius direction
ro: Tunnel excavation diameter (Circular form)
Po: Initial stress in ground
$\sigma\,r$: Stress in radius direction
$\sigma\,\theta$: Stress in tangential direction
When the second term in Formula (1) is disregarded (since it is smaller than the first term), Formula (1) can be expressed according to the following formula:

(3) $\quad \varepsilon\,\theta = \dfrac{Ur}{r}$, $Ur = \varepsilon\,\theta \cdot r$

As a result of investigation, the unconfined compressive strength, fracture strain and ratio of assumed preceding deformation were obtained as indecated in Table 3.
The monitoring standards have been orked out as shown in Table 4, by obtaining the amount of allowable wall deformation on the basis of this fracture strain, and establishing the ratio of the preceding deformation.

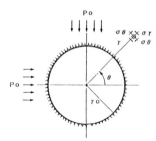

Figure 5.   Stress strain in ground

Table 3.   Fracture strain in ground

Geology	Unconfined compressive strength qu (kgf/cm²)	Fracture strain ε_r (%)	Ratio of preceding deformation
Detritus soil	1.71	2.5	0.2
Mudstone	79	1.0	0.2

Table 4. Control standards for tunnel wall deformation

		Roof settlement (mm)	Gradient of deformation (mm/m)			horizontal line (mm)	Gradient of deformation (mm/m)			Cumulating trend
			Face distance				Face distance			
			0~0.5D	0.5D~1.5D	1.5D~		0~0.5D	0.5D~1.5D	1.5D~	
Detritus soil section	Standard value	40	8.0	4.0	2.0	40	8.0	4.0	2.0	Slightly remarkable
	Caution value	80	16.0	8.0	4.0	80	16.0	8.0	4.0	Slightly remarkable
	Limit value	120	24.0	12.0	6.0	120	24.0	12.0	6.0	Slightly remarkable
Bedrock section	Standard value	15	3.0	1.5	1.0	30	6.0	3.0	2.0	Slightly remarkable
	Caution value	30	6.0	3.0	2.0	60	12.0	6.0	4.0	Remarkable
	Limit value	44	9.0	4.5	3.0	88	18.0	9.0	6.0	Very remarkable

5 RESULTS OF MEASUREMENTS AND CONSIDERATIONS

5.1 Section of landslide protection open excavation method

The excavation section was begun from the first stage after driving master piles. With the depth of excavation being 1.0 m below the strut, excavation of the second and third stages was carried out after piling and the installation of the strut. Along with the installation of the strut, the axial force was measured by attaching an axial force meter to the strut. Although a maximun of as much as 46 tons of axial force took place at the second stage strut as a result of measurements, this value was less than the design load, and was not considered to be a load resulting from a landslide.

5.2 Sections between tunneling method

The support pattern and auxiliary excavation methods are as indicated in Fig. 2. Excavation of the detritus soil stratum section was carried out according to ring cut method, that of the ring section by manual excavation using hand breakers, and that of the support body by mechanical excavation using backhoes. Meanwhile, excavation of the bedrock section was carried out according to a blasting method.

a) Observations in tunnel
The ultimate values of the results of measurements are shown in Fig. 2. Roughly 80 mm of roof settlement occurred, and resulted in a breakdown phenomenon of the laggings in the pattern Ds. The broken condition of the laggings was investigated and examined. As a result, since a breakdown of the laggings was deemed to have progressed along with a drifting of the top heading, the ground was reinforced with shotcrete (t =10cm) from above the laggings. Subsequent to the reinforcement, deformation was converged.

The pattern DⅢ did not cause any phenomenon which might cause any particular problem.

b) Convergence and roof settlement
Fig. 6 indicates the deformation characteristics of convergence with regard to the respective support patterns.

From Fig. 6, the deformation pattrns of the top heading horizontal line can be classified into the following three patterns:

1. A pattern wherein deformation takes place toward the inside simultaneously upon starting measurement, and is almost free from the effects of bench excavation (Pattern Ds).

2. A pattern wherein deformation toward the outside occurring simultaneously after starting

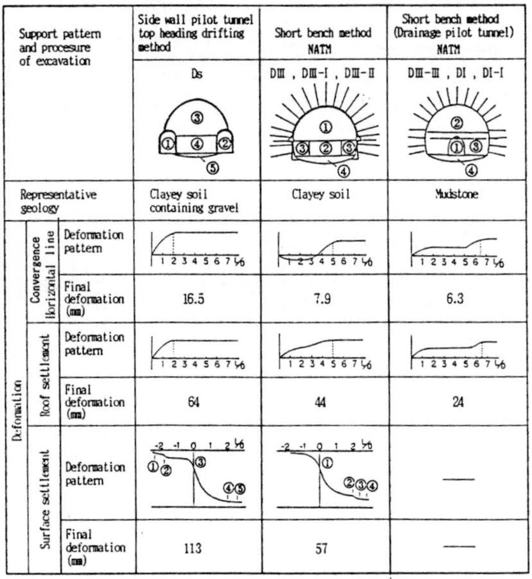

Support pattern and procedure of excavation		Side wall pilot tunnel top heading drifting method	Short bench method NATM	Short bench method (Drainage pilot tunnel) NATM
		Ds	DⅢ , DⅢ-I , DⅢ-Ⅱ	DⅢ-Ⅲ , DI , DI-I
Representative geology		Clayey soil containing gravel	Clayey soil	Mudstone
Convergence Horizontal line	Deformation pattern			
	Final deformation (mm)	16.5	7.9	6.3
Roof settlement	Deformation pattern			
	Final deformation (mm)	64	44	24
Surface settlement	Deformation pattern			—
	Final deformation (mm)	113	57	—

Figure 6. Deformation characteristics by support patterns

excavation. It is converged once and deformation toward the inside takes place as the bench approaches (Patterns DⅢ, DⅢ-I, and DⅢ-Ⅱ).

3. A pattern wherein deformation toward the inside arising simultaneously upon commencement of measurement is converged once and the same deformation arises again as the bench approaches (Patterns DⅢ-Ⅲ, DI, and DI-I).

The settlement of the roof and that of the leg part indicate a roughly equivalent trend. The settlement which has taken place simultaneously upon the start of measurement is converged once, but the settlement increases again as the bench approaches, and is converged after excavation of the invert.

One of the deformation characteristics mentioned above, namely, the fact that the top heading horizontal line indicates deformation extending outwardly at the time of top heading excavation (particularly in the detritus soil excavation section) is considered to result from the fact that the tunnel leg part was settled and opened since the load of overburden was predominant, while the bearing capacity of the leg supporting ground was insufficient in the case of the clayey soil ground with thin overburden.

c) Surface settlement

Fig. 7 indicates the change of surface settlement with elapse of time in typical sections in the side wall pilot tunnel top heading drifting section and the NATM section.

In the side wall pilot tunnel top heading drifting method section, roughly 17% of settlement progressed when both of the pilot tunnels were excavated, and settlement began to start again when the top heading approached about 1D (D:tunnel diameter) in front of the face. When of passing through the top heading face, 31% of the settlement took place. 90% of the settlement had occurred by the time the face had passed through 2D, and little settlement associated with excavation of the bench and the invert occurred. This is considered to be due to the effect of the side wall concrete.

In the case of the NATM section, moreover, settlement was also initiated by the time the top heading face approached nearly 1D in front; followed by 31% of the settlement by the time it passed through the top heading face; 80% by the time the face advanced to 2D; and 84% by the time the face passed through the bench. Then, settlement advanced again along with the excavation of the bench and reached its ultimate settlement.

The relationships between the overburden and the amount of settlement according to the side wall pilot tunnel drifting method and NATM are given in Fig. 8.

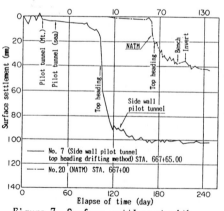

Figure 7. Surface settlement with elapse of time

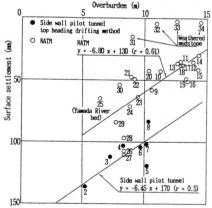

Figure 8. Relationship surface settlement and overburden

Namely, according to both of the methods, the deeper the overburden, the smaller the amount of surface settlement. However, the amount of settlement in the case of the NATM was smaller by about 40 mm than that in the case of the side wall pilot tunnel drifting method, provided that the thickness of the over-burden was the same. This is considered to show the significance of the NATM in clayly soil ground with a thin overburden.

d) Effect of support members

Fig. 9 indicates changes due to the elapse of time of the stress in shotcrete, and the axial force of rock bolts (Depth: 1.5 m) in the detritus soil ground.

Along with the progress of the excavation of the top heading, stress in the shotcrete increased, and nearly reached a state of equilibrium when the face advanced to 1D. The maximum stress at this time was 30 kgf/ cm² at the leg on the mountain side, and the stress in the roof increased along with the excavation of the bench. However, the stress decreased rapidly at the leg part, and was converged respectively after excavation of the invert.

With regard to the axial force of rock bolts, a compressive force of 2.5 tons took place respectively on the roof during excavation of the top heading. Along with the excavation of the bench, the compressive force increased to 5 tons, which eventually converged. This tendency is considered to be a phenomenon inherent to ground with thin overburden, having resulted from a loosening of the entire ground extending to the ground surface.

Although the compressive force acted at the top heading side leg part during excavation of the top heading, a maximum tensile force of 5 tons acted during excavation of the bench and decreased to 0 - 2 tons after excavation invert. The reason that the axial force at this leg part was converted from the compressive force to tensile force is deemed to be due to the fact that the stress in shotcrete at the leg part rapidly decreased due to excavation of the bench and the deformation (settlement of the leg) increased.

In order to feed back the results of this measurement of actual excavation work, the portion of 52 m from STA. 667 + 5 was changed to a support pattern from which five rock bolts adjacent to the roof were eliminated.

When a large stress in shotcrete has taken place at the leg part of the tunnel due to a loosening load at the time of the top heading excavation, and has

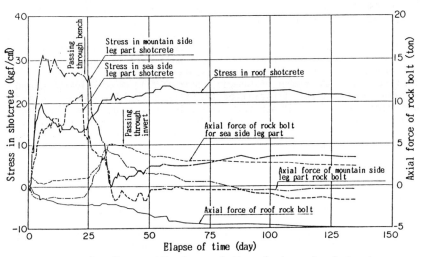

Figure 9. Change with elapse of time of stress in shotcrete and axial force of rock bolt (Depth: 1.5 m point)

exceeded the bearing capacity of the ground, the tunnel will be subjected to major settlement. At this time, it is considered possible to restrict the settlement with rock bolts. In the case of a sandy ground where such effects cannot be expected, a phenomenon of settlement will arise. Therefore, when the loosening load is predominant, the bearing capacity of the supporting ground is insufficient against the load, and it is impossible to expect good results from the use of rock bolts. Therefore, the construction for closing a ring is considered necessary.

6 CONCLUSION

In this paper, the excavation records of the Tomari tunnel, as well as the measurement and monitoring procedures have been discussed. At the same time, the behavior of ground in detritus soil stratum and mudstone, and the effect of support members have also been presented. The results are concluded to be as follows;

1. In the detritus soil ground with shallow overburden, the loosening load is predominant and the deformation of the tunnel is dominated by vertical deformation. In such ground, it is considered effective to widen the tunnel leg part by shotcrete. Depending upon the ground conditions, however, it is required to close a ring.

2. The deformational characteristics of convergence and surface settlement vary depending on the differences in geology and support systems. In cases where deformation characteristics are predicted in advance, based upon the existing data, and such abnormal deformation characteristics are indicated, it is considered necessary to investigate the causes, and take appropriate counter-measures as required.

3. The amount of surface settlement in the case of the NATM is smaller by about 40 mm than that in the case of the side wall pilot tunnel drifting method, whose supporting members are steel supports and laggings, and where the overburden in the case of the NATM is the same. This indicates the significance of the NATM in excavation of a tunnel in clayey soil ground with thin overburden.

4. The major support member during the excavation of the top heading in the detritus soil stratum is the shotcrete. Since a comparatively large stress arises at the top heading leg part, a support system is required in which the bearing capacity of the supporting ground is taken into account. Along with the excavation of the bench, the stress in the shotcrete at the top heading leg part decreases rapidly, and is borne by the rock bolts driven from the shoulder to leg parts. The rock bolt for the side wall part is considered to be effective even in clayey soil ground with shallow overburden.

In recent years, the number of tunnels constructed in unconsolidated ground have increased. Since deformation and stress rapidly change in such unconsolidated ground, it is necessary to sufficiently predict possible abnormal phenomena prior to the start of excavation, and conduct measurements and monitoring so as to ensure preparation of countermeasures for the prevention of such abnormal phenomena.

REFERENCE

Hirayama, K. & K. Nanasawa 1985. Excavating Clayey Soil Ground with Shallow Overburden. Tunnels and Underground Vol. 16, No.12. 13-22
Tanii,K. & T. Kawata 1986. Consideration of Measurement Result of the Tunnel Constructed in Detritus. Procedings of the 18th Symposium on Rock Mechanics, Japan. 46-50

Deformation monitoring and analyses on a shallow railway tunnel in a loess stratum

Weishen Zhu, Jiaoqiao Zhu & Caizhao Zhan
Institute of Rock and Soil Mechanics, Academia Sinica, Wuhan, People's Republic of China

1 INTRODUCTION

A new railway line being in construction, the Datong Qinhu-
angdao, will pass through a 9 km long shallow tunnel which
is located in the northern suburbs of Beijing, 65 km away
from the city. The tunnel named Junduoshan Tunnel has a cross
section of 11.86×11.32 (m×m). At the entrance part of the
tunnel, the overburden with a thickness of about 12 m consists of loess. In order to understand the stability of this
shallow tunnel with such a large cross section when passing
loess, a 200-300 m long section of the tunnel eastbound from
the entrance was chosen to be a safe monitoring testing segment. A number of boreholes were drilled downwards from the
ground surface above this future tunnel, in which extensometers installed ahead of excavation to monitor the deform-
bility of the tunnel roof during the construction of it.
And the levelling measurement was carried out simutaneously
to monitor the ground settlement. Special attention is paid
to the description of measuring results with extensometer,
back analysis on the in-situ measured data, time effect caused by loess deformability and special effects caused by
working face advance.

2 GENERALIZATION

The media or strata surrounding the testing segment consist
of neoloess, the upper part of them is a 7-9 m thick sandy
clay with rudaceous lenses and the lower part is a fine sand
layer with a thickness of 4-6 m as the direct roof of the
tunnel. There exists in the middle of this loess a pure and
noncohensive thick sand layer containing a very small amount
of clay, this sand layer has a poor self-stable capability
because of a X-shaped joint set in it. The physical and mechanical indexes are: $\gamma=1.97t/cm^3$, cohension C=0.05-0.08 Mpa
and inner frictional angle $\phi=25°$. The NATM was employed to
excavate the testing tunnel, the part by part excavating
operation by stages was introduced for the loess segment,
the kernal part of the soil was kept undisturbed all along
and the construction sequences are shown in Fig.1.

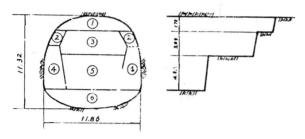

Fig. 1 Sequence of construction

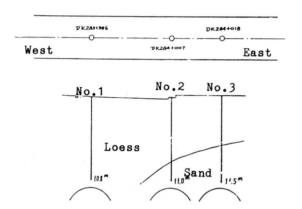

Fig. 2 Layout of boreholes for extensometer

Blasting method was not used to reduce vibration and the kernal part was excavated by the heavy machinary.

3 MONITORING THE ROOF DEFORMATION DURING CONSTRUCTION

Three testing sections were arranged in the testing tunnel, on each of them, three mechanical extensometers were installed in the pre-drilled boreholes. The arrangement layout of the boreholes is shown in Fig. 2. Because of the irregularity of borehole diometers in loess, the reasonable anchoring measure is quite essential to tests. The authors used small compressive wooden cylinders as anchoring ends, the volumatrical expansion of them after being immerged into water, can amount to 10%. The measuring points were installed individually, in the case of over-sized boreholes, a small amount of cement mortar was grounded into the hole. There was six anchoring or measuring points in all for each borehole (Fig. 3). WRM-3 mechanical extensometers developed by our institute were used for measuring, which characterized by reliable high accuracy for repeating measurements, high sensitivity of 0.01 mm, generally unlimited measuring range. For the measuring invar steel wires were not strained tightly untill reading out a constant tensile force which can be al-

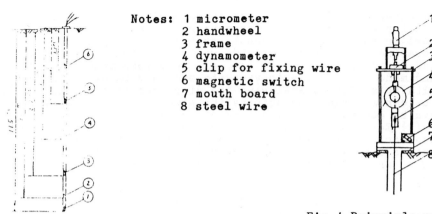

Notes: 1 micrometer
2 handwheel
3 frame
4 dynamometer
5 clip for fixing wire
6 magnetic switch
7 mouth board
8 steel wire

Fig.4 Principle of extensometer

Fig. 3 Layout of measuring points

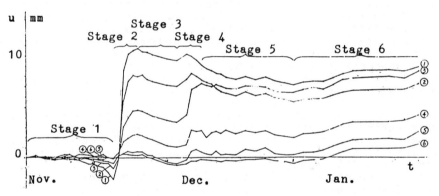

Fig. 5 Relation between displacements and time

ways kept while measuring, and therefore creep of the wires resulting from being under long term tension condition, this may cause reading errors, can be usefully avoided. The schematic diagram of this instrument principle is shown in Fig.4.

Observation to borehole 1 lasted 161 days, from Nov. 1 to the end of 1985. when the tunnel was excavated beyond this borehole by 10 m, roof collapse occurred through the overburden to the ground surface. This forced the excavation operation to be held up for engineering treatment. Consequently, the later deformation values measured when the tunnel passed through this area were small. These irregular data will not be discussed in this paper.

In Fig. 5, are shown the relation curves between displacements of anchoring points and time. By using these curves along with the large deformation trace of major cracks occurring on the ground surface, a potential danger situation of disastrous roof collapse was successfully predicated.

After that, a series of measurements was introduced to support or reinforce the tunnel, such as grouting in roof,increasing bolt density at side walls, temporary arch invert and local concrete line, ect, all this took about 12 days and made it possible to excavate smoothly below the borehole with no accident. Analysing the displacement curve in Fig.5, one can see that the deformation procedure of the roof loess can be generally divided into six stages.

(1) Early Deformation. This stage started far away from monitoring boreholes and ended just below the hole. Most relative displacements were negative, with a maximun value of -2.13 mm, and lasted about 19 days, the deforming rate was e=-0.12 mm,per day.

(2) Rapid Deformation. The maximun displacement occurred when the excavation reached at the monitoring borehole. When the excavation went beyond the hole, the displacement became positive and rapidly increased. Five days later, the deformation got its peak value. This stage with an average deformation rate of e=2.57 mm/day, is characterized by that the max. deformation took place at the measuring point near the roof, the farther away from the roof the measuring point, the smaller the displacement, exhibiting an obvious rule.

(3) Controlled Deformation After Supporting. Out of worries about collapse probably taking place, the site engineers took some supporting measure, afterwards, the deformation was markedly controlled, shown by regular reduction of displacement curve slope reduced to e =0.15 mm/day.

(4) Second Peak Deformation After Recovery of Excavation. The normal reexcavation started after a 12 days holding up work period for supporting the tunnel, the displacement increased again. The displacement of each point reached early or late to its max. one in 5-day. The deforming rate of this stage is e=0.6 mm/day.

(5) Deformation Tending to Stability. After the excavation had advanced beyond the monitoring hole by 7 m, the deformation became gentle, lasting about 20 days, with a rate of e=0.1 mm/day.

(6) Slightly Varying Deformation Caused by Temperature. With the tunnel advancing, the displacements which had been stable increased again. Such a uniform synchronous increasement of displacements during later period might be caused by temperature variation. The excavation of this stage started after the local nadir air temperature had gone and the working face was advanced to the place 1.7 times the tunnel span away from the monitoring hole. The deforming rate of this stage, e, is 0.06 mm/day.

From the above stated deformation stage, it can be known that the roof deformation is closely related to excavation footages and supporting measure and to stradum properties as well. The fact of displacement reduction at point 1 and 2 and no obvious displacement variation for others in those 4 stages (Fig.5) suggests that the sandy strata where point 1 and 2 are located in, are easily reconsolidated after undergoing deformation, while for other points the deformation of clayey strata difficultly recovered and the consolidation phenomenon is not serious. The excavation of the tunnel's

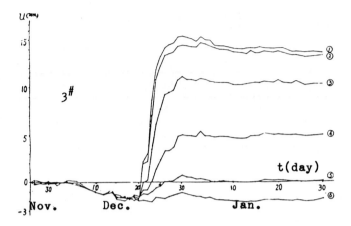

Fig. 6 Relation between displacement and time

lower part lasted one and a half months, the displacements of the roof were in general 3-5 mm, the deforming rate, e, equals to 0.09 mm/day. The displacement curves for each monitoring points indicate that the displacements at 2 or 3 points near the roof exhibit an abrupt variation reflecting the property of sandy strata.

For borehole 3, one can see the displacement curve with time elapsing (Fig.6). The maximun deformation of the deepest point (point 1) is U_{max}= 15.29 mm, the deformation procedure can be divided into 3 stages.

(a) Early Deformation. The maximun compressive deformation was -1.91 mm, lasting 24 days with a rate of 0.08 mm/day.

(b) Abrupt Deformation. The maximun displacement was as high as 15.29 mm, lasting 10 days, with a rate of e=1.73 mm per day.

(c) Stable and Gentle Deformation. The displacement did not increase any more with a little fluctuation, however, when the excavation activity was advanced to the place 14 m beyond the monitoring borehole. The deformation rate, e, was 0.06 mm/day. The stage lasted for about one month, with the only exception of two deepest hole points (1 and 2), where the maximun displacement, at the earlist several days, somewhat decreased. This abnormal phenomenon might be caused by the consolidation occurring locally. Later on, when the excavation of lower part was carried out, the displacement of this segment of the tunnel measured by 1-5 mm. The deformation rate, e, is 0.14 mm/day.

According to the data of levelling measurement, the comparison between the relative displacements measured by extensometers in the drilled hole and the settlement of the ground surface is given in table 1. It is shown that the later is 2-3 times as many as the former.

Table 1. Comparison between the relative displacement and the settlement of ground surface

Max U (mm)	No 1	2	3
Vertical relative displacement	2.14	10.7	15.29
Settlement of ground surface	27.71	27.3	31.22
%	8	39	49

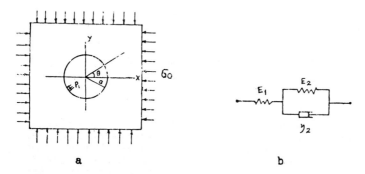

a   b

Fig. 7. Visco-elastic analytic model

4 VISCO-ELASTIC BACK ANALYSES

(A) Symmetric model for analyses
The numerical analysis of the tunnel should belong to the three dimensional problem, in order to avoid too large amount of calculation work, however, the axial symmetric analysis was firstly carried out for approximate extimation. At the beginning of excavation, the excavation height and width of the upper part of the tunnel are 4 m and 10 m respectively. For the sake of approximate simulation analysis, the equivalent radius of the tunnel was chosen as the average of its height and span, i.e., a=7 m.
   (a) The model for approximate analysis is shown in Fig.7.a.
   (b) The rheology model of the medium is assumed as in Fig. 7b.
(B) Imaginary supporting force calculation
   Considering spacial effects, the authors choose the equivalent supporting force as $p=(1-\lambda)\sigma_0$ to simulate the real one, where $\lambda=0.5 +\frac{1}{\pi}arctg(t-t_0-2)$, t_0 is the time necessary to excavate to the measuring points (Gesta, 1986).
(C) Displacement determination
   Assuming that the volume strain is elastic and the poisson's ratio is constant, the physical equation of the deviator parts follows the rheology model of 3-element (Fig.7.b), from visco-elastic correspondence law, one can obtain.

$$U=\frac{1+\mu}{E_1}\sigma\bullet\frac{R^2}{r}\{ \ 0.5+\frac{1}{\pi}arctg \ (t-t_\bullet-2)\}\{ \ 1+\frac{E_1}{E_2}- \ \frac{E_1}{E_2}exp(-\frac{E_2}{\eta_2}(t-t_\bullet))\}$$

According to the in-situ measured curves and the equation above, E_1 and E_2, η_2 can be calculated by back analysis method as follows:

$E_1= 64$ Mpa $E_2=192.7$ Mpa $\eta_2=5504.3$ Mpa

Substituting E_1, E_2 and η_2 into equation of displacement, a curve (Fig.8) can be obtained through numerical calculation which is fairly coincidant with the in-situ measured results.

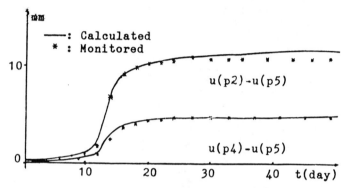

Fig. 8 Comparison between measured and calculated

Note that it is in the tenth day (Fig.8) that the working face was advanced to the place just below the measuring points.

5 DISCUSSION AND CONCLUSION

(1) The extensometer, developed by author's institution which was designed originally for the use in rock engineering, is also suitable for monitoring of deformability of a tunnel in loess, and the monitoring curves exhibit a satisfactory rule. By using this monitoring method, potential danger of collapse can be successfully predicated.

(2) The upper strata above the shallow tunnel in loess deform compressively at first when the working face is approaching to them, and this part of deformation makes up 10-20% of the total one. The deformation abruptly turns to tensile when the working face is advanced just beneath the measuring borehole. The settlement law of ground surface is very similar to that caused by excavation of the working face in coal mining.

(3) The deformation of the superstrata above the tunnel exhibit obvious variation with stages. The deformation law is maily related to working face proceeding, supporting measure and strata properties. The maximun relative displacements measured in the monitoring points installed ahead of

excavation are about 1/3-1/2 of ground surface settlement.
The obvious secondary consolidation phenomenon is found in
sand strata.
   (4) Commenly used rhological model of 3-element can be introduced to carry out numerical back analysis on this kind
of stratum and the corresponding rhological parameters of a
rock mass can also be derived.

REFERENCES

Gesta, P. 1986. Tunnel support and lining. Tunnel. 73:
Sakurai, S. 1983. Displacement measurements assiciated with
   the design of underground openings. Proc. of Intern. Symp.
   on Field Measurements in Geom., Zurich.

2nd International Symposium on Field Measurements in Geomechanics, Sakurai (ed.)
© 1988 Balkema, Rotterdam. ISBN 90 6191 778 6

The monitoring of a tunnel excavated in shallow depth

H.Noami & S.Nagano
Arai-Gumi Ltd, Hyogo-ken, Japan
S.Sakurai
Kobe University, Japan

1 INTRODUCTION

Recently the demand for field measurements in Geotechnical Engineering
has increased tremendously. Nevertheless, it is still questionable how
to evaluate these measurement results, how to use them for verifying
the stability of structures, and if necessary, how to use them for modifying the scheme of design and construction.

This paper deals with the interpretation of field measurement results
obtained during the excavation of a two-lane road tunnel. The stability
of the tunnel was monitored through use of the Direct Strain Evaluation
Technique (DSET) proposed by Sakurai (1981). This technique is based on
the comparison of the strain occurring in the ground around the tunnel
to the critical strain of the ground materials. If the occurring strain
is smaller than the critical strain, stability of the tunnel is guaranteed. The computer program DBAP is used for determining the strain distribution from measured displacements.

2 BRIEF DESCRIPTION OF THE TUNNEL

The total length of the tunnel is 126 m. The height of the maximum
overburden is 25 m, while some sections have an overburden of only
1.5 - 10 m. Thus, the tunnel can be classified as a very shallow one.

The ground consists of weathered granite which becomes a nearly soil-
like material. The velocity of elastic wave (Vp) travelling through
the ground is approximately 0.8 - 1.0 km/sec. Little underground water
comes out at the tunnel surface, and its magnitude is about 20 - 30
l/min at some portions of the tunnel.

Excavation of the tunnel was conducted by NATM, adopting a top heading method. Excavation of the lower half followed approximately 30 m
behind the face of the top heading. A permanent arch lining was installed 30 m behind the face of the lower half, and invert concrete was
laid after completion of the arch lining.

Support measures consist of the combination of shotcrete, rock bolts
and steel ribs, depending on the geological conditions. The standard
tunnel cross section with support measures is shown in Fig. 1.

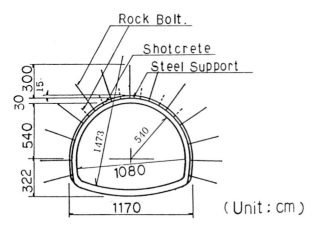

Fig.1 Standard Tunnel Cross Section

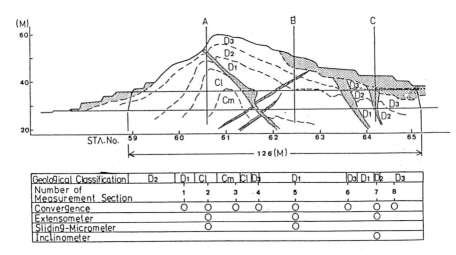

Fig.2 Location of Measurement Section

3   FIELD MEASUREMENTS

Field measurements were taken at eight different tunnel sections, as
shown in Fig. 2, where displacements of the ground, axial force of rock
bolts, and stress acting in shotcrete were measured. A convergence
meter, multi-rod extensometer and inclinometer were used to measure the
displacements around the tunnel, and a sliding micrometer developed at
ETH, Zurich was installed from the ground surface. The settlement at
the tunnel crown was also measured. The places where the instruments
were installed for taking the displacement measurements are shown in
Fig. 3.
   The results of the displacement measurements are shown in the following.

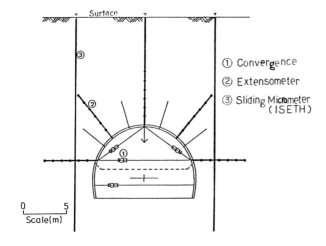

Fig. 3 Installation of Instruments (Section A and B)

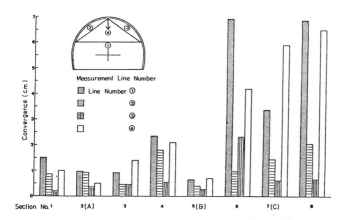

Fig. 4 Final Value of Convergence at Different Measurement Sections

(1) Convergence measurements

The final values of convergence at each measurement section are illustrated in Fig. 4. It should be noted that the largest value obtained
is either for measuring line No. 1, or for the settlement at the tunnel
crown. The figure also indicates that large displacements appear at
Section Nos. 6, 7, and 8, where the height of the overburden is as
small as less than 10 m.
    In order to secure the stability of the tunnel excavated in ground
with a small overburden, additional rock bolts and shotcrete were installed. The invert at the top heading of Section No. 8 was temporarily covered with shotcrete to increase the tunnel's stability.

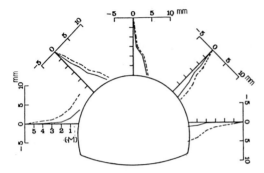

Fig. 5   Results of Extensometer Measurements   (Section A)

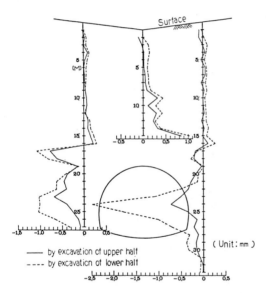

Fig. 6   Results of Sliding Micrometer Measurements   (Section A)

(2)   Multi-rod extensometer and sliding micrometer measurements

The multi-rod extensometer and sliding micrometer were used to measure the displacements caused by excavation at Sections A and B.   The results are shown in part, in Figs. 5 and 6, indicating the displacements measured by use of multi-rod extensometers and the sliding micrometer, respectively.   In these figures, the displacements measured before and after excavation of the lower half are compared.

4 BACK ANALYSIS OF MEASURED DISPLACEMENTS

Back analysis was used for obtaining strain distributions from measured displacements. The computer program used for this purpose was DBAP, developed by Sakurai, et al. Since details of this program have been described elsewhere (Sakurai and Takeuchi 1983, Sakurai and Shinji 1984), only a brief summary is presented below.

The formulation of back analysis is based on the finite element method. Assuming the ground in which a tunnel is bored consists of homogeneous isotropic elastic materials, the following equation can be derived,

$$[A] \{ \sigma_0 \} = \{ u \} \tag{1}$$

where [A] is a matrix of only a function of the location of measuring points and Poisson's ratio. $\{ u \}$ is a vector of measured displacements, and $\{ \sigma_0 \}$ is a normalized initial stress vector defined in a two-dimensional state as,

$$\{ \sigma_0 \} = \{ \sigma_{x0}/E \quad \sigma_{y0}/E \quad \tau_{xy0}/E \}^T \tag{2}$$

where σ_{x0} , σ_{y0} , τ_{xy0} are the components of initial stress existing in the ground before excavation, and E denotes Young's modulus of ground materials.

The finite element mesh used here is shown in Fig. 7. The 6 and 8-mode isoparametric elements are used. Poisson's ratio of the ground materials is assumed to be 0.3, while Young's modulus and Poisson's ratio of shotcrete are 4×10^4 kg/cm^2 and 0.25, respectively.

In back analysis, four cases are considered which all depend on the use of measured displacements for input data, that is, 1) convergence measurements only, 2) both convergence and extensometer measurements, 3) extensometer measurements only, and 4) both extensometer and sliding micrometer measurements.

It should be noted that the sliding micrometer can measure total displacements of the ground due to tunnel excavation. For an easy com-

Total Number of Elements = 46

Total Number of Points = 148

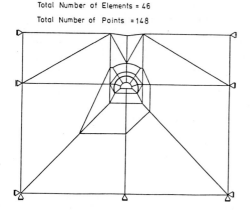

Fig. 7   Finite Element Mesh

Table 1   Results of Back Analysis

	Case 1	Case 2	Case 3	Case 4
	Convergence	Convergence Extensometer	Extensometer	Extensometer Sliding-Micrometer
σ_{xo}/E	-0.977×10^{-3}	-0.123×10^{-2}	-0.221×10^{-2}	-0.233×10^{-2}
σ_{yo}/E	-0.669×10^{-3}	-0.978×10^{-3}	-0.178×10^{-2}	-0.233×10^{-2}
τ_{xyo}/E	-0.243×10^{-3}	-0.957×10^{-4}	0.245×10^{-3}	0.206×10^{-3}
$\sigma_{xo}(kg/cm^2)$	-7.00	-6.04	-5.97	-4.80
$\sigma_{yo}(kg/cm^2)$	-4.80	-4.80	-4.80	-4.80
$\tau_{xyo}(kg/cm^2)$	-1.74	-0.47	0.66	0.42
$E(kg/cm^2)$	7175.	4908.	2703.	2057.
ν(assumed)	0.3	0.3	0.3	0.3

parison of the results, however, only the incremental displacements measured after the tunnel face passed through the measuring section are used for the back analysis in Case 4.

The results of the back analysis conducted for Section B are summarized in Table 1.

Assuming that the vertical component of initial stress is equal to the overburden pressure, the other components of initial stress and Young's modulus of the ground material can be separated from the value of the normalized initial stress.

Once the initial stress and Young's modulus are determined, they are used as input data for an ordinary finite element analysis to determine stress, strain and displacement distributions throughout the ground. The comparison between calculated and measured displacements proves the accuracy of back analysis, as shown in Figs. 8 and 9. Fig. 8 is for a case in which only the data of convergence measurements are used as input data. It is no wonder that we obtain a good agreement, while a large discrepancy can be seen in the extensometer measurements, because only convergence measurement results have been taken into consideration.

On the other hand, the results shown in Fig. 9 indicate that a good agreement exists between the computed and measured values of extensometer measurements, because the extensometer measurement results are used for this back analysis.

It is understood from Fig. 9 that the measured values of the convergence measurements are smaller than the values expected for ground assumed to be of a homogeneous isotropic elastic material. This means that a loosened zone near the tunnel surface, occurring because of stress relief due to excavation, tends to be compacted as the excavation progresses. Thus, although relatively large displacements occur in the surrounding ground, small convergence values are measured. This may be due to the fact that convergence is restricted by the high rigidity of support structures.

The displacements back analyzed through use of both the extensometer and sliding micrometer (Case 4) data are also compared to the measured displacements shown in Fig. 10. It is seen from this figure that there

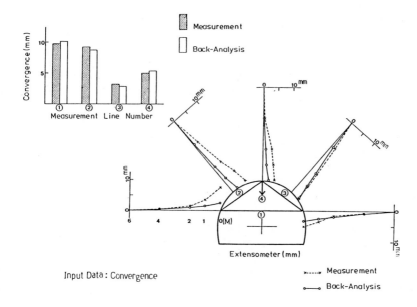

Fig. 8 Comparison Between Measured and Back Analyzed Displacements
(Input Data: Convergence)

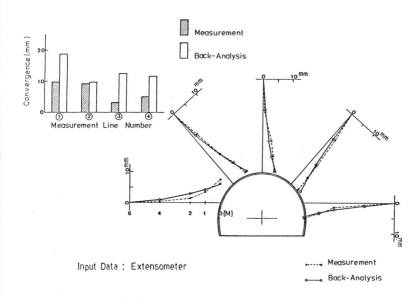

Fig. 9 Comparison Between Measured and Back Analyzed Displacements
(Input Data: Extensometer)

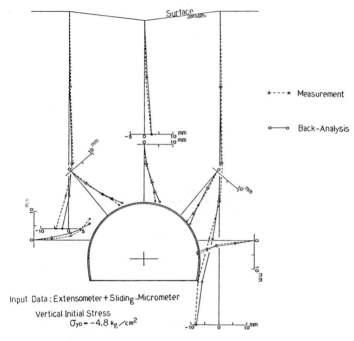

Fig. 10   Comparison Between Measured and Back Analyzed Displacements
(Input Data: Both Extensometer and Sliding Micrometer)

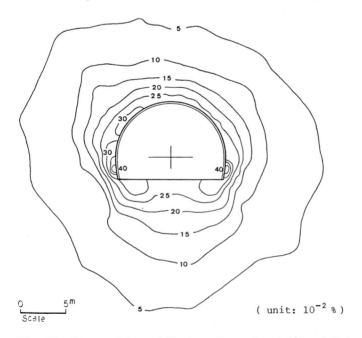

Fig. 11   Contour Line of Maximum Shear Strain (Input Data: Extensometer)

is a good agreement between the two, and it demonstrates that a large extent of the ground deforms continuously, just like a homogeneous isotropic elastic material. In this case, Young's modulus becomes the smallest among all the cases shown in Table 1. And, it is also of interest to know that if in back analysis we use the data of the measured displacements for a large extent of the ground, the back analyzed Young's modulus of the ground becomes a smaller value.

Based on the direct strain evaluation technique, the stability of the tunnel was monitored by evaluating the strain of the ground. The strain distribution was calculated for each case of the back analysis. One of the results for Case 3 is shown in Fig. 11, as contour lines of maximum shear strain. This figure shows that the largest value of maximum shear strain reaches about 0.4%, and ranges from 0.2 - 0.3% in the ground at a distance of 5 m from the tunnel surface where the rock bolts are installed.

On the other hand, in laboratory tests the critical strain proposed by Sakurai (1981) is evaluated at approximately 1.2% under a uniaxial condition. Therefore, as the strain occurring in the tunnel's surrounding ground is still smaller than the critical strain, stability of the tunnel is guaranteed.

5 CONCLUSION

Back analysis of measured displacements can demonstrate a deformational mechanism of the ground due to tunnel excavation. In the particular tunnel described here, convergence values and the displacements in the surrounding ground cannot be explained by a single mechanical model, such as a homogeneous isotropic elastic body. It seems that a loosened zone exists near the tunnel surface, and tends to be compacted as the excavation progresses, so that a stiffened zone develops in the vicinity of the tunnel surface.

It is worthwhile to note that back analyzed Young's modulus becomes a smaller value if the measured displacement data is taken from various points in a large extent of the ground.

The stability of the tunnel is well monitored by comparing the strain occurring in the ground with the critical strain of the ground materials.

REFERENCES

Sakurai, S. 1981. Direct strain evaluation technique in construction of underground opening. Proc. 22nd US symposium on rock mechanics, MIT. pp. 298-302.
Sakurai, S. and K. Takeuchi 1983. Back analysis of measured displacements of tunnels. Rock mechanics and rock engineering, 16, pp. 173-180.
Sakurai, S. and M. Shinji 1984. A monitoring system for the excavation of underground openings based on microcomputers, Proc. ISRM symposium, Design and performance of underground excavations, Cambridge, pp. 471-476.

2nd International Symposium on Field Measurements in Geomechanics, Sakurai (ed.)
© 1988 Balkema, Rotterdam. ISBN 90 6191 778 6

Field measurements of the Kobe Municipal Subway Tunnel excavated in soil ground by NATM

S.Sakurai
Kobe University, Japan
R.Izunami
Transportation Bureau, Kobe City Office, Japan

1 INTRODUCTION

The Kobe Municipal Subway System started operation of a newly completed 5.9 km extension line connecting Gakuen-Toshi with Seishin-Chuo, in March of 1987. The total length of the system has now become 22.7 km. The Omoteyama Tunnel is located at the middle of the extension. The tunnel was excavated by NATM in soil ground consisting of tertiary and/or quaternary deposits. Various types of field measurements were performed for monitoring the stability of the tunnel during and after its excavation.

In this paper, the results of the field measurements together with the results of back analysis are introduced. These provide useful, technical information for tunnels being bored into soil ground.

2 BRIEF OUTLINE OF THE TUNNEL

The Omoteyama Tunnel is a double track tunnel, whose length is 770 m. The vertical alignment rises constantly with a gradient of 30/1000 from the south entrance, while the horizontal alignment is a circular, which has a radius 2200 m over the distance up to 300 m from the entrance, and then is a straight line. The overburden is rather small, that is, the smallest overburden height is 13 m, and the largest is 41 m. The ground surface is covered with forests and cultivated fields. Both ends of the tunnel near the portal were constructed by the cut-and-cover method, thus, the total length of NATM's application is 577.5 m. The tunnel support measures consist of 15 and 20 cm thick shotcrete, and 10 rock bolts per one meter along the tunnel axis. H-shaped steel ribs were also installed in some sections of particularly weak ground.

3 GEOLOGY

The geology of the ground in which the tunnel was bored, consists mainly of the Osaka strata, which is tertiary and has quaternary deposits. It is covered with a terrace deposit and talus cone along the valley.

3.1 Osaka strata

The Osaka strata located near the tunnel construction site is often

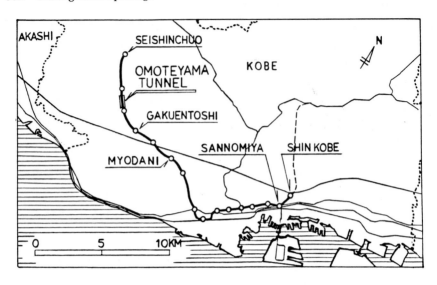

Fig. 1   Kobe Municipal Rapid Transit System

called the Akashi Strata, which mainly consists of sand and clay layers
with gravel deposits.
  (a) Gravel deposits
Gravel deposits dominate the Osaka strata, and it is well compacted but
unsolidified. Its colors are yellow brown to brown gray. The diameter
of the gravel is between 20 and 100 mm. The mixture ratio of gravel is
between 25 and 80%. N-value is more than 50, mostly between 60 and 140.
The modulus of elasticity (E) is 1100 kg/cm^2, and the coefficient of
permeability (k) is approximately 1 x 10^{-5}cm/sec.

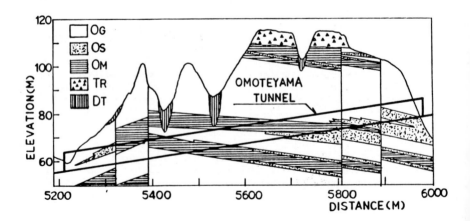

Fig. 2   Geological Profile

(b) Sand deposits

Sand deposits constitute thin layers, distributed on and beneath the clay deposits. The uniformity coefficient (Uc) is approximately 100 to 500. N-value is more than 60, k is 1×10^{-5} cm/sec, and E is 1100 kg/cm^2.

(c) Clay deposits

Two consecutive layers of clay deposits are located at elevations of 65 m and 75 m, respectively. Their colors are blue green or green gray. The mixture content of silt accounts for 50%, and that of clay and sand is about 20 to 30% for each. The fresh clay deposit is fairly solidified, and its N-value is more than 50. Its unconfined compressive strength is 13 to 16 kg/cm^2. The upper part of the clay layer is often weathered, which causes browning and softening. The modulus of elasticity (E) is more than 3000 kg/cm^2 for the fresh portion, while 900 kg/cm^2 is E for the weathered one.

3.2 Terrace deposit

The terrace deposit is 7 to 8 m thick, and mainly consists of gravel. The size of the gravel is 5 to 10 cm in diameter. Its mixture ratio is approximately 30%. The matrix consists of sand and clay, with a mixture ratio of 30 and 40%, respectively. N-value is 40, and the modulus of elasticity is approximately 200 kg/cm^2.

3.3 Talus cone deposit

The talus deposits are located in the valley at a distance of 5420 m and 5540 m as shown in Fig. 2. The mixture ratio of gravel, sand, and clay is approximately 40-50%, 30%, and 20%, respectively. Compared to the gravel of the Osaka strata, there is less clay content. The coefficient of permeability (k) is 5.0×10^{-4} cm/sec, and the N-value is 5 to 8; an extremely small value. The modulus of elasticity is approximately 80 kg/cm^2.

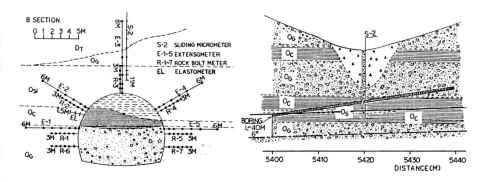

Fig. 3 Position of Instruments   Fig. 4 Location of Horizontal Inclinometer

4   MEASUREMENTS

The following field measurements were performed;
   (a)   Displacement of the ground (Extensometer, Sliding Micrometer
ISETH, and Horizontal Inclinometer)
   (b)   Axial force of rock bolts
   (c)   Stress of shotcrete lining (Curvometer and Deformeter ISETH)
   (d)   Modulus of elasticity (Elastometer)
   The measurements were carried out at a distance of 5380 m (Section A)
and 5420 m (Section B).   The arrangement of the instruments is shown in
Figs. 3 and 4.

5   MEASUREMENT RESULTS

5.1   Displacement (Extensometer)

The relative displacements were measured by multi-rod extensometers.   The
results obtained at Section A are shown in Fig. 5.   It is seen from this
figure that horizontal displacements E-1 and E-5 are slightly larger than
the others.   This tendency is also found in Section B.
   Comparing the settlement at the crown with the relative displacement
of E-3, the following relationship is obtained, that is,
   Crown settlement (10.0 mm) > E-3 (5.1 mm), at Section A
   Crown settlement (11.0 mm) > E-3 (2.6 mm), at Section B
   The difference between the crown settlement and the relative displacement indicates a downward movement of the deepest fixed point of extensometer E-3.   In other words, the deepest fixed point located 8 m from the
tunnel crown, shows some displacement, and its magnitude is approximately
5 to 8 mm.
   On the other hand, when the horizontal convergence measurement is compared with the horizontal extensometer measurement results, the following
relationship is obtained.
   Convergence (13.0 mm) ≑ E-1 + E-5 (11.4 mm) at Section A
   Convergence (10.2 mm) ≑ E-1 + E-5 (9.2 mm) at Section B
This means that the deepest fixed points of extensometer E-1 and E-5
(6 m from the tunnel surface) indicate almost no displacement.

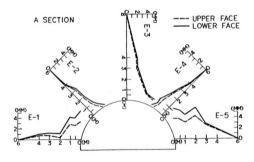
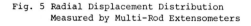
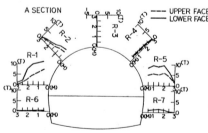

Fig. 5 Radial Displacement Distribution
        Measured by Multi-Rod Extensometers

Fig. 6   Axial Force of Rock Bolts

5.2 Axial force of rock bolts

The results obtained at Section A are shown in Fig. 6. Similar to the displacement measurements, large axial forces occur at the horizontally installed rock bolts, R-1 and R-5. The axial forces of R-6 and R-7 are small. This is due to the fact that the horizontal displacements had already taken place when rock bolts R-6 and R-7 were installed after the excavation of the lower half.

The maximum values of axial force in both Sections A and B are as follows;

	Section A	Section B
After excavation of upper half	5.9 ton (R-5)	11.4 ton (R-5)
After excavation of lower half	11.5 ton (R-1)	14.0 ton (R-5)

5.3 Stress acting on shotcrete

Fig. 7 shows the distribution of axial stress acting on shotcrete at Section B. This was obtained from the relative displacements measured by the Curvometer and Deformeter developed by Kovari et al (Kovari, Amstad, and Fritz, 1977). Although the stress largely changes from place to place because of the irregularity in thickness of shotcrete, the average axial stress is approximately 20 kg/cm^2.

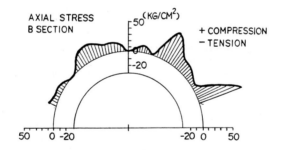

Fig. 7 Axial Stress of Shotcrete Observed
by Curvometer and Deformeter-ISETH

5.4 Displacement (Sliding Micrometer, ISETH)

The displacements around the tunnel due to excavation were measured from the ground surface using an instrument called a Sliding Micrometer, developed by Kovari et al (Kovari, Amstad, and Koppel, 1979). The results obtained at Section B are shown in Fig. 8. This is the vertical displacement distribution along a measuring line right above the tunnel crown. In this figure, the displacement at the ground surface is taken to be zero. It is seen from this figure that the displacement increases largely at the depth of 8 m, where the boundary between the Osaka strata and the talus deposits exists.

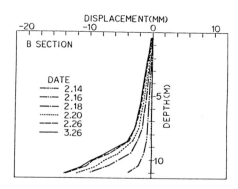

Fig. 8 Displacements Measured by a Sliding Micrometer Along the Vertical
Axis Above the Tunnel Crown

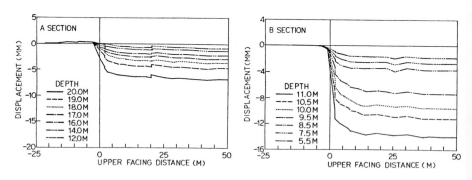

Fig. 9 Displacements Measured by a Sliding Micrometer in Terms of the
Distance of Tunnel Face

Fig. 10 Displacements Measured by a Sliding Micrometer in Terms of the
Distance of Tunnel Face

The maximum values of the measurements are as follows;

	Section A	Section B
After excavation of upper half	6.3 mm	13.7 mm
After excavation of lower half	6.8 mm	14.0 mm

These values were recorded at 1.5 m above the tunnel crown.
   It is noted from these results that there is no difference between
the results obtained after excavation of the upper half and after that
of the lower half. This means that the excavation of the lower half
has no influence on the ground around the tunnel crown.
   In order to investigate how much displacement has already taken place
when the tunnel face arrived at a measuring section, the displacements
are plotted versus the distance of tunnel face of the upper half from
the measuring section. The results are shown in Figs. 9 and 10. It
is seen from these figures that at Section A about 40% of the total
displacements had already taken place when the tunnel face arrived,
while at Section B only 25% of the total displacements had taken place.

5.5 Displacement (Horizontal Inclinometer)

The horizontal inclinometer was installed from the crown of the tunnel face into the forward ground as shown in Fig. 4. The results are shown in Fig. 11 in terms of the continuous location of the upper half tunnel face. Special care must be paid, however, when interpreting the results because the inclinometer is not parallel to the tunnel axis, and measured displacements tend to become small as the tunnel face progresses.

The maximum displacement amounts to 12 mm. This is the absolute displacement due to tunnel excavation. In order to give a clear picture of the relationship between the displacements and the location of the tunnel face, Fig. 12 is shown. It is understood that the displacement tends to converge to a certain value when the tunnel face passes through the measuring section at a distance of the tunnel diameter.

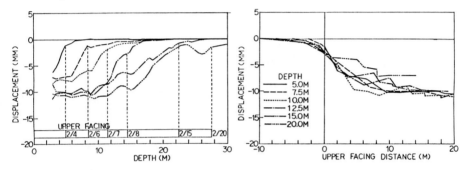

Fig. 11 Displacements Measured by a
Horizontal Inclinometer

Fig. 12 Displacements Measured by a
Horizontal Inclinometer in
Terms of the Distance of Tunnel
Face

5.6 Modulus of elasticity

The modulus of elasticity was measured by loading tests in a borehole drilled from the tunnel surface. Elastometer 200 was used for this purpose.

The tests were performed at positions of 0.5 m, 1.0 m, and 1.5 m from the tunnel surface, where the ground mainly consists of clay. The measurements were taken twice, namely, when the tunnel face arrived at the measuring sections, and after the tunnel face had passed through the sections. The results are shown in Fig. 13 for both Sections A and B.

It is obvious from the figure that the modulus of elasticity is 800-1100 kg/cm^2 near the tunnel surface, and 1800-2500 kg/cm^2 at 1.5 m from the surface. These results indicate that there is a loosened zone caused by excavation, in which a small value of the modulus of elasticity was obtained. It is of interest to note that the modulus of elasticity in the loosened zone increases with the progression of the tunnel face from 810 kg/cm^2 to 1210 kg/cm^2, and from 1140 kg/cm^2 to 1520 kg/cm^2 near the tunnel surface at Sections A and B, respectively.

Although the amount of data is small, it is supposed that a loosened zone appears right after excavation within a depth of 1 m from the tunnel surface, and then tends to gradually contract with the progress of the tunnel face. This contraction may be due to the fact that the lining is stiff enough to support the deformation of the surrounding ground.

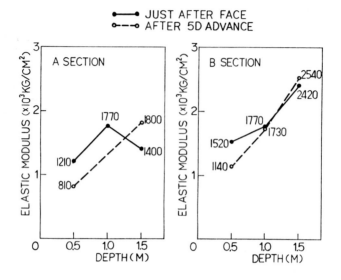

Fig. 13   Modulus of Elasticity Measured by Elastometer - 200

6   BACK ANALYSIS

Back analysis was performed for measured displacements to obtain the mechanical constants and initial stress of the ground. The computer program named DBAP, which was developed by Sakurai and Shinji (Sakurai and Shinji, 1984) was used. Using the results obtained by the sliding micrometers installed from the ground surface, all the displacements measured from the inside of the tunnel were calibrated to the total displacements. Then they were used as input data for the back analysis. Once the mechanical constants and initial stress are obtained, strain distributions can be calculated.

The back analysis was performed for every excavation phase of the upper and lower halves of tunnel section. The modulus of elasticity determined by the back analysis is as follows;

	Section A	Section B
Average	1500 kg/cm^2	900 kg/cm^2
Sand Deposit	1000	600
Clay Deposit	2000	1200

Some examples of the results of strain distributions calculated using the back analyzed modulus of elasticity are shown in Figs. 14 and 15 for principal strain and maximum shear strain distributions, respectively.

The strain occurring around the tunnel is compared with the "critical strain" (Sakurai 1981) of ground materials to verify the stability of the tunnel.

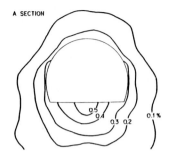
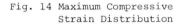

Fig. 14 Maximum Compressive
Strain Distribution

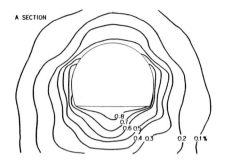

Fig. 15 Maximum Shear Strain
Distribution

The critical strain is approximately 0.3-0.4%, and 0.8% for compressive and shear strains, respectively. It is seen from these figures that since the strain occurring in the ground near the tunnel arch is still smaller than the critical strain, the tunnel must be sufficiently stable.

7   CONCLUSION

Construction of the Omoteyama Tunnel, bored in tertiary and quaternary deposits, was controlled by field measurements. The results of the measurements revealed the deformational behaviour of the surrounding ground, that is, there first occurred a loosened zone near the tunnel surface, however, it tends to gradually contract with the progression of the tunnel face.
    Comparing the strain occurring in the ground due to excavation with the critical strain of ground materials, the stability of the tunnel was monitored, and rational support measures were chosen. The tunnel was constructed safely without any problems.
    This paper has been written on the basis of "The Report of Measurements and Analyses of Omoteyama Tunnel in Seishin-Enshin Line of Kobe Rapid Transit", prepared by the Construction Engineering Research Institute, Dec. 1985.

REFERENCES

Kovari, K., Ch. Amstad & P. Fritz 1977. Integrated measuring technique for rock pressure determination. Proc. of Int. Sympo. field measurements in rock mechanics, Zurich, vol. 1, pp. 289-316.
Kovari, K., Ch. Amstad & J. Koppel 1979. New developments in the instrumentation of underground openings. Proc. 4th rapid excavation and tunnelling conference, Atlanta, U.S.A.
Sakurai, S. & M. Shinji 1984. A monitoring system for the excavation of underground openings based on microcomputers. Proc. ISRM Sympo., design and performance of underground excavations, Cambridge, U.K., pp. 471-476.
Sakurai, S. 1981. Direct strain evaluation technique in construction of underground openings. Proc. 22nd US symposium on rock mechanics, MIT, pp. 298-302.

An example of measurements on the behavior of oblate and large section tunnel in unconsolidated ground

Yukio Yamashita & Toshio Fujiwara
Technical Research Institute, Ohbayashi Corp., Tokyo, Japan

1 INTRODUCTION

The tunneling method, of which the main support materials are shotcrete and rock bolts, has been widely used since being introduced in Japan, owing to its ability to the guarantee of an economical and safe tunnel consturction by utilizing, at its maximum, the loading capacity of the ground itself. In the meantimes, many studies have been carried out based on on-site measured results and numerical analyses as well as on model tests and the deformational behaviors pertaining to a tunnel and its peripheral ground have been clarified. However, we now consider this clarification to be insufficient.

This paper discusses the measured results of behaviors of a tunnel with an extremely oblate sectional form constructed in unconsolidated ground, based on the construction of the Asahikawa Tunnel, as an example.

The tunnel concerned has a large section (approximately 100 m²), and an oblate sectional form with a large width compared with the tunnel height. The geological conditions consist of gravel, welded tuffs, volcanic cohesive soils, and so forth, which belong to the Tokachi Welded Tuffs of the Diluvium, but the ground is unconsolidated, without a welded structure. Moreover, since the overburden is as thin as 3.5 m at the thinnest portion, we were apprehensive about the safety of the tunnel during construction. For this reason, various monitorings were executed to ensure the behavior of the tunnel lining and peripheral ground. We discuss here the results of these monitorings.

2 OUTLINE OF THE ASAHIKAWA TUNNEL

The Asahikawa Tunnel is a 860 m long road tunnel constructed on National Highway No. 12, named the Asahikawa New Way. The area concerned is located at the eastern edge of the Kamuikotan Metamorphic Rock Belt, and a hill belt with a 100 − 150 m elevation is formed here. The geological conditions of the tunnel protion consist of gravel, welded tuffs and volcanic cohesive soils belonging to the Tokachi Welded Tuffs (Figure 1), but ground is unconsolidated with little welded structure. The tunnel passes though with 3.5 − 35.5 m of overburden, under a hill belt on which private houses are dotted near the summit portion.

The cross section of the tunnel is shown in Figure 2. The short bench method (by ring cutting of upper half section) was adopted as the excavation method. Shotcrete, steel supports and rock bolts were prepared just after the excavation. The roof bolt-to-crown part and the shotcrete-to-cutting face were executed for the protection of the cutting face. The monitoring parameters measured during the tunnel excavation are listed in Table 1.

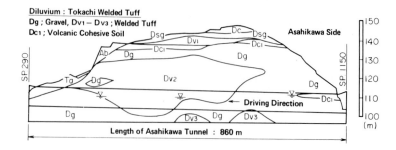

Fig. 1   Geological Profile

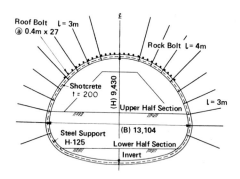

Fig. 2   Tunnel Typical Cross Section

Table 1   Measuring Parameters

1)	Relative Displacement of Tunnel Inner Section
2)	Tunnel Crown Settlement
3)	Ground Surface Settlement
4)	Subsurface Displacement in the Peripheral Ground
5)	Axial Force of Rock Bolt
6)	Stress of Shotcrete
7)	Stress of Steel Support

3 OBSERVATIONS AND THE MEASURED RESULTS OF DEFORMATIONAL BEHAVIOR

In tunnel excavation, a deformation occurs at the tunnel wall surface and peripheral ground due to the release of in-site stress at the cutting face. Such a deformation is not only affected by the difference in geological conditions, strength characteristics of the ground, tunnel sectional form, and construction method, but also is due to differences in overburden. In this report, the deformational properties of an unconsolidated ground due to differences in overburden are mainly described, based upon the deformational monitoring results measured in the Asahikawa Tunnel.

3.1 Displacement at Tunnel Wall Surface

Figure 3 shows the relation between the relative displacement of the inner section of the horizontal measuring line (C measuring line) set at the side wall of the upper half section and the overburden. Both the overburden height, z, and the relative displacement, δc became dimensionless numbers divided by the inside diameter of the tunnel, D (=B). From this figure, we understand that the occurrence of displacement is different from the boundary where the ratio, Z/D of the overburden height to the tunnel inside diameter is 1.2. In short, the inner section displacement increases in proportion to the overburden in the case where Z/D is smaller than 1.2, while it becomes stable showing a roughly constant value if Z/D is greater than 1.2. These behaviors are closely linked to the formation of the ground arch in the peripheral ground of the tunnel. In the case where Z/D is less than 0.8, we recognize that the C measuring line extends to the upper half section stage of excavation. This means that the foot portion of the upper half section lining was pushed to the ground side, as a ground arch is not formed in the ground and the ground at the upper portion of the crown is rapidly settled if the overburden is thin.

The elapsed changes of the tunnel crown settlement are shown in Figure 4. The abscissa represents the ratio of the distance, L, between the cutting face of the upper half section and the measured point, and the tunnel inside diameter. Here, the measured sections, of which Z/D are (1) 0.33, (2) 0.60, (3) 1.12 and (4) 1.63, were described as representative examples of different overburdens. In cases (1) and (2), for which the overburden is thin, we understand that the initial displacement is large and sudden settlement occurs in the area of the cutting face apart from 0.5D, after the upper half section excavation. The final convergent values have, however, roughly similar values, not depending on the difference of the overburden.

Figure 5 shows all the measured sections by indicating the relation between the convergent values of crown settlement and the overburden. Although there are dispersions by each section, the final settlement shows, on the whole, that the values within the constant

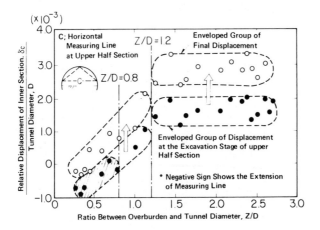

Fig. 3   Relation between Inner Section Displacement
          (C Measuring Line) and Overburden

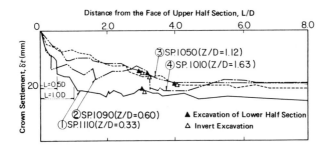

Fig. 4   Elapsed Changes of Tunnel Crown Settlement

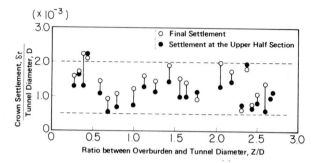

Fig. 5   Relation between Tunnel Crown Settlement and Overburden

range are not affected by the overburden. It is for this reason, we believe, that the settlement is restrained by the resistance of the lining exetcuted at the tunnel's inside face. Crown settlement is finally dominated by the scale of the back load and the rigidity of the lining, as well as the ground bearing capacity near the lining foot portion.

3.2 Ground Surface Settlement

The final settlement on the ground surface measured on the tunnel center line indicates a gradual decreasing tendency in accordance with an increase of the overburden as shown in Figure 6. So far as the elapsed changes indicate, the relation with the overburden does not always show a continuous change. Figure 7 shows the elapsed changes of the representative measured sections. At (1) and (2), where the overburden is thin, being the same as the crown settlement mentioned above, a sudden settlement occurs at the early stage and the time to converge on the final values is short. This means that the ground surface goes down together with the tunnel crown settlement. On the other hand, there is no sudden settlement near the cutting face at (3) and (4) where the overburden is relatively thick, and the settlement of the ground surface changes gradually between the upper half section excavation and the final excavation stage.

The relation between the ground surface settlement and the tunnel crown settlement is shown in Figure 8 by comparison of each common measured point. The final convergent

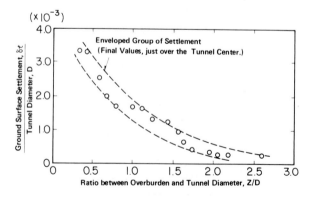

Fig. 6   Relation between Ground Surface Settlement and Overburden

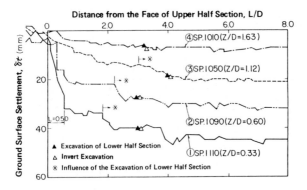

Fig. 7   Elapsed Changes of Ground Surface Settlement

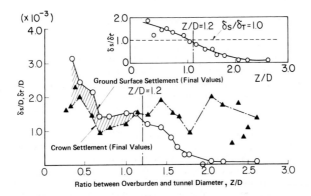

Fig. 8   Relation between Ground Surface Settlement after the Cutting Face
of Upper Half Section is Reached and Tunnel Crown Settlement

values, after the full-face excavation, were adopted at each settlement. Grund surface settlement excludes the pre-displacement occurring before the upper half section face is reached. Accordingly, both displacement amounts shown in the figure represent the movement after the face is reached. As clarified in the figure, the ground surface settlement is larger than the crown settlement in the case where Z/D is approximately smaller than 1.2, so the so-called inverse phenomenon of settlement occurs. The ratio between the two settlements has an approximate maximum value of 1.9 times. The reason why the ground surface settlement is larger seems to be due to the ground arch formation in the ground and the confined displacement of the lining. This means that a sufficient ground arch can not be formed in the ground and the sudden settlement occurs on the ground surface together with the tunnel crown settlement in the case where the overburden is thin. However, the displacement is confined, as the effect of the internal pressure on the lining is displayed at the tunnel crown. Since the confinement of the ground displacement by the lining is not exerted on the ground surface it appears to occur that the ground surface settlement is larger than the crown settlement. Conversely, it indicates that a sufficient ground arch can not be formed in the area where a reverse phenomenon of settlement occurs. The value of Z/D = 1.2, which represents a boundary, coincides with the measured results of the inner section displacement mentioned above.

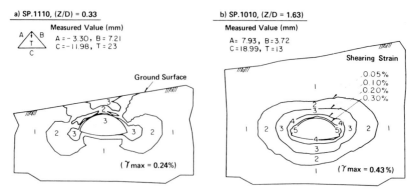

Fig. 9   Analytic Results by Back Analysis – Shear Strain Distribution
at Peripheral Ground of Tunnel

3.3 Strain Distribution on Peripheral Ground of Tunnel

It has been clarified that the deformational behavior of the tunnel wall surface and the peripheral ground were greatly affected according to differences in the overburden. In Figure 9. the strain distribution on the peripheral ground of the tunnel is obtained at the stage by the upper half section excavation, by carrying out a back analysis by means of the method of Sakurai (1983). We analyzed two cases where Z/D were larger than 1.2, and smaller than 1.2. Results of the displacement measured in the tunnel inner section were used as the input data.

Among these, (a) is an example where the overburden is thin, which corresponds to section (1) shown in Figure 4 and Figure 7. We understand that the shear strain occurs to a large extent on the lower portion of the side wall due to a pushing of the lining foot portion toward the ground side. Furthermore, we can estimate that the contour of the shear strain extends to a ground surface just over the tunnel, so a ground arch can not be formed and the ground has fallen to the cavity side. This is a very dangerous situation for tunnel excavation. In such an area, it is important to support the ground of the tunnel's upper portion at as early a stage as possible, and we judge that the roof bolt method applied in this tunnel work was especially effective. In the roof bolt method, a ground can be supported further forward than the cutting face. On the other hand, (b) corresponds to section (4) where the overburden is a little thicker. Since the shear strain contour is distributed with the wall surface of the tunnel as a center in the shape of a concentric circle, the above mentioned phenomenon is not recognized. The maximum shear strain is, however, larger than that of the former.

3.4 Characteristic Curve of Ground Displacement
(The measured results of a subsurface displacement)

It is known that the ground displacement accompanied by the tunnel excavation occurs further forward than when the cutting face is reached. In this tunnel, the subsurface displacement before and after the cutting face advances, were measured by executing the boring from the ground surface and by burying the in-situ strain meter in the ground over the center of the tunnel. Overburden at the measured position is 1.77 of Z/D. Accordingly, we judge from the above mentioned ground behavior that the measurement was made in the ground in a relative stable state.

The relation between the advancing cutting face and the subsurface displacement is generally expressed as a characteristic curve of ground displacement. Figure 10 shows the

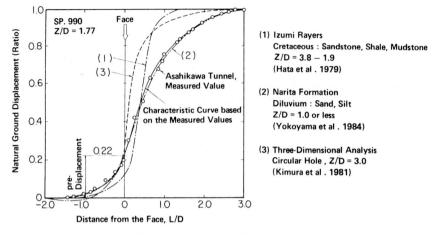

Fig. 10   Characteristic Curve of Ground Displacement

characteristic curve measured in the Asahikawa Tunnel. The subsurface displacement which measured 3.2 m over the tunnel crown was normalized, as a reference, by the measured value at the time apart from 3.0 D after the passage of the upper half section cutting face. From this figure, we understand that 22% of the pre-displacement has occurred in the ground beyond the cutting face. Hata et al (1979) and Yokoyama et al (1984) made the same measurements for other tunnels. Moreover, Kimura et al (1981) obtained the characteristic curve by performing the three-dimensional analysis of a circular tunnel.   These curves are described in the figure.   The measured values of the Asahikawa Tunnel show a similar curve to that of the Narita Formation which is mainly composed of unconsolidated ground of the same Diluvium.

4   OBSERVATIONS AND THE MEASURED RESULTS OF ROCK BOLT AXIAL FORCE

A rock bolt is one of the main support materials in the tunnel excation method. Here, the axial force occurring in the rock bolt (axial stress) was measured by using a strain gauge. The measured results are shown in Figure 11.   Among these, (a) is the measured example in the case where the overburden is shallow and $Z/D = 0.79$.   The crown rock bolt was not

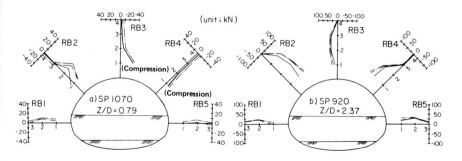

Fig. 11   Axial Force Distribution of Rock Bolts

disposed at the initial design but the measurement was made to ensure the behavior of the entire ground and the effect of the rock bolt here. Since the compressive force occurred in the crown rock bolt, we find that the ground on the tunnel's upper portion is in the compressed state. This shows that the nearer to the tunnel wall surface, the greater this axial force is, and the crown displacement is confined by a lining. However, the maximum axial force does not indicate as large a value as 32KN. The compressive force which activates the rock bolt on the right shoulder portion according to the measured examples, is caused by a slight action of unsymmetrical pressure due to the declining of the ground surface to the left side. The two bolts on the side wall portion are installed upon excavating the lower half section. Attention must be paid because they are different from the starting time of the monitoring from the three bolts disposed on the upper portion. It has been actually observed that the side wall portion is greatly displaced to the inner section side prior to the excavation of the lower half section. However, only partial measurements are made here because the greater part of the displacement occurs before the installation of the bolts. For the rock bolt of the lowest portion placed on the upper half section, greater axial force than the measured value may occur. On the other hand, (b) shows an example where the overburden is thick and Z/D = 2.37. The tensile force acts on all the rock bolts including the crown. Furthermore, the maximum axial force reaches 100 KN. Throughout these monitorings, since it was clarified that the axial force of the rock bolt has a tendency to increase as the overburden increases, the design change was carried out to place the rock bolt at the crown portion, too.

5 OBSERVATIONS AND THE MEASURED RESULTS OF PRIMARY LINING STRESS

In the Asahikawa Tunnel, the steel supports and shotcrete were installed as a primary lining. Here, the stress of the lining was measured by the method shown in Figure 12, covering four different sections of overburden. Further, two sections were added for the measurement of the shotcrete.

Figure 13 shows the measured results of the axial stress of the steel supports and shotcrete, with the relation of the overburden. The final convergent values, after the full-face excavation, were adopted as an axial stress. It comes to light that the axial stress occurring on the lining increases as the overburden increases. Maximum values of 20.9 for MPa (when Z/D = 2.60) at the shotcrete and 241 MPa (when Z/D = 2.67) at the steel supports were obtained. The occurrence of a maximum axial stress is ensured near the positions where the overburden becomes the deepest in this tunnel. As mentioned in Chapter 3, a ground arch begins to form in the ground and the deformational behavior has a stable state when Z/D is greater than 1.2. It may be considered, however, that the external force

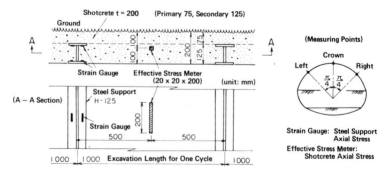

Fig. 12   Measuring Points of Lining Stress and Installing Method of Instruments

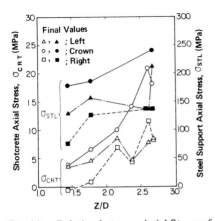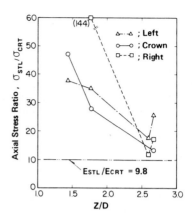

Fig. 13   Relation between Axial Stress of Lining and Overburden

Fig. 14   Relation between Axial Stress Ratio and Overburden

acting on the lining increases gradually as the overburden becomes deeper so that the arch itself is not so strong in unconsolidated ground, as different from a rock.

The ratio of the axial stress of the steel supports to that of shotcrete $(\sigma_{STL}/\sigma_{CRT})$ is shown in Figure 14. Contrary to the results of Figure 13, it comes to light that the ratio of axial stress become large as Z/D decreases. $\sigma_{STL}/\sigma_{CRT}$ represents the load distribution proportion based on the axial stress. The greater ratio means that the load of the steel support increases compared with that of the shotcrete. As clarified by the measured results of displacement, the ground displacement proceeds rapidly at the tunnel's upper portion as the overburden becomes thin. Since shotcrete has the tendency to increase the strength with age, it cannot support if the load increases at an early stage after the concrete placing, so that a concentration of stress occurs in the steel support. Therefore, the steel support plays an important role in unconsolidated ground, especially in the area where the overburden is thin.

The ratio of the elastic modulus of the steel supports to that of shotcrete (E_{STL}/E_{CRT}) is also shown in the figure. Tsuchiya (1986) reported in detail the elastic modulus of shotcrete. When applying the value for 28-days, which may represent a long term mechanical property, the ratio of both is as follows:

$$\frac{E_{STL}}{E_{CRT\,(28)}} = \frac{2.08 \times 10^5 \text{ MPa } (2.10 \times 10^6 \text{ kgf/cm}^2)}{2.10 \times 10^4 \text{ MPa } (2.14 \times 10^5 \text{ kgf/cm}^2)} = 9.8 \quad \ldots\ldots\ldots\ldots (1)$$

Formula (1) expresses the lower limit value of $\sigma_{STL}/\sigma_{CRT}$ representing the value when the steel supports and shotcrete may be charged with the same loading, just after execution. As a matter of course, as the young aged shotcrete has a small elastic modulus, the lower limit value observed in the actual works has a slightly greater value than this. We understand that $\sigma_{STL}/\sigma_{CRT}$ approaches the value of $E_{STL}/E_{CRT} = 9.8$ given in Formula (1), when the overburden becomes deep.

6 SUMMARY

We have mainly stated here the behavioral properties of tunnel linings and peripheral ground according to the differences in overburden in unconsolidated ground based on the measured results of the Asahikawa Tunnel. The results are as follows:

1) In the case where the ratio of the overburden height to the tunnel inside diameter, Z/D,

is roughly less than 1.2, a ground arch cannot be formed in the peripheral ground of the tunnel, and thus, unstable behaviors are shown. The formation of a ground arch for the overburden changes is clarified based on the relative displacement of the tunnel inner section and the measured results of the ground surface settlement, etc. (Fig. 3, 8 and 9).

2) In the case where Z/D is smaller than 1.2, the ground surface settlement is larger than the crown settlement, so the reverse phenomenon of displacement occurs. (Fig. 8) This seems to be caused by the ground arch formation in the ground and the confined displacement due to the lining. We could confirm that the ground over the crown's upper portion was in the compressed state from the results of the axial force measurement of the rock bolts (Fig. 11).

3) The ground displacement occurs rapidly as the overburden becomes thinner, and the time to converge to the final value is short. Especially, when Z/D is not more than 0.6, a greater part of all the displacements in the upper half section excavation occur during the distance from the cutting face of the upper half section, $L = 0.5$ (Fig. 4 and 7).

4) Although the settlement at the tunnel crown is affected by overburden in elapsed changes, the final convergent value is almost within a constant range (Fig. 4 and 5). Because we believe that displacement is confined, due to the registance of the lining. The crown settlement is finally dominated by the scale of the back loading, the rigidity of the lining and the ground bearing capacity.

5) The axial force of the rock bolt and the axial stress of the tunnel lining have a tendency to gradually increase as the overburden increases, even if Z/D becomes larger than 1.2 (Fig. 11 and 13). We believe that this is different from the case of rock, at the strength of a ground arch itself is lowered in an unconsolidated ground and the external loading increases.

6) As a result of the examaination of the ratio of the axial stress of the steel supports to that of shotcrete $(\sigma_{STL}/\sigma_{CRT})$, this ratio increases as the overburden becomes thinner (Fig. 14). The $\sigma_{STL}/\sigma_{CRT}$ ratio represents the loading charges of both, in the case where the axial stress is set as reference. It shows that the charge of the steel support becomes, in particular, large for the external loading when the overburden is thin. As the overburden becomes deeper, $\sigma_{STL}/\sigma_{CRT}$ approaches $E_{STL}/E_{CRT} = 9.8$ which seems to represent the lower limit value.

It has becomes clear that the deformation and scale of the external loading acting on the lining occurring in the tunnel excavation depends upon the difference of the tunnel sectional form, ground strength properties and construction method. However, in many other tunnel works it is reported that the compressive axial forces of the crown's rock bolts occur if the overburden is shallow. It also indicates that the ground of the crown's upper portion reaches a compressed state. Therefore, the above mentioned behavior must be qualitatively confirmed in other tunnel works, though the absolute values of the displacement and others are different according to the geological conditions. We want to add further examination of other ground conditions in the future.

REFERENCES

Sakurai, S. & Takeuchi, K. 1983. Back analysis of displacement measurements in tunneling Proceedings of JSCE. No. 337: 137 – 145
Hata, S., Tanimoto, C. & Kimura, K. 1979. Field measurement and consideration on deformability of the Izumi layers. Rock Mechanics, Suppl. 8: 349 – 367
Yokoyama, A. & Takase, A. 1984. Study of ground behavior during excavation of thin overburden tunnel in unconsolidated ground. Proceedings of JSCE, No. 352, III-2: 79 – 88.
Kimura, H., Kamemura, K. Harada, H. & Sato, M. 1981. A study for tunnel analysis considering face progress. 21th JNCSMFE: 1565 – 1568
Tsuchiya, T. 1986. Study on the design of the tunnel method using rock bolt and shotcrete. Doctor's Thesis: 44 – 47.

2nd International Symposium on Field Measurements in Geomechanics, Sakurai (ed.)
© 1988 Balkema, Rotterdam. ISBN 90 6191 778 6

Field measurements at a multiple tunnel interaction site

K.W.Lo, L.K.Chang, C.F.Leung & S.L.Lee
Department of Civil Engineering, National University of Singapore
H.Makino & T.Mihara
Kajima-Keppel J V, Singapore

SYNOPSIS

Field measurements were carried out at the site of an intricate
pattern of four interweaving tunnels driven in stiff to hard Old
Alluvium partly overlain by a pre-grouted buried channel of
waterlogged sediments. The instruments adopted were monitored over a
continuous period of some 18 months. After multiple tunnel
interaction, combined axial thrusts and bending moments were well
within safe design capacities of precast concrete lining segments
determined by standard empirical methods. Bending moments deduced
either by empirical or finite element prediction, or measurement,
tended to increase with greater tunnel interaction. In the case of
axial thrusts, however, corresponding increases in measured values
exceeded those by prediction to the extent that a far greater margin
of safety in lining load capacity that anticipated was actually
available. Interaction factors for bending moments generally varied
within 25% of empirically-determined values. For single tunnels,
ground surface settlement measurements could be reasonably fitted
with normal distribution functions laterally and error functions
longitudinally, as usually assumed in practice. Also, following
known behaviour, lateral profiles from field measurements after
interaction were biased towards an initial tunnel drive, and
corresponding cumulative settlements exceeded those determined by
practical empirical procedure. Furthermore, incremental longitudinal
surface settlements due to tunnel interactions were also found to
follow error function dustributions.

1 INTRODUCTION

Between February 1985 and September 1986, an intricate pattern of
four interweaving tunnels of 5.85m diameter each was constructed
between Raffles Place and City Hall Stations of the Singapore mass
rapid transit (MRT) system, thereby providing a unique opportunity to
study an unusual case of multiple tunnel-ground interaction. Fig 1
shows a 90m stretch of the alignment in the Connaught Drive area,
within which four lateral lines of field instrumentation were
installed prior to tunnelling in the vicinity. Fig 2 shows a typical
instrumentation section and Fig 3 indicates the longitudinal soil
profile along one of the tunnels. Generally at the site, a surface

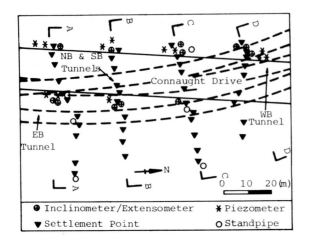

Figure 1.   Instrumentation Layout

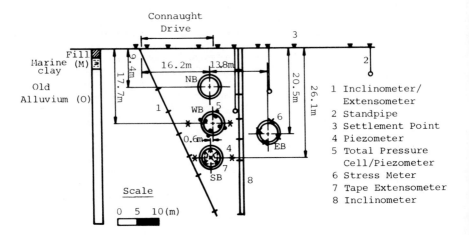

Figure 2.   Section A-A

cover of fill followed by waterlogged intercalations of very soft to
soft clay or loose to medium dense sand deposits of the Kallang
formation is underlain by the stiff to hard clayey sand of the Old
Alluvium formation.

   The initial deeper southbound (SB), followed by eastbound (EB)
headings were entirely advanced within the Old Alluvium formation by
conventional semi-mechanical shield tunnelling, face excavations
being carried out by a backactor. Usually, five standard segments,
each of approximately 3.5m extrados length and 1m width, and one key
segment of approximately 1m extrados length and similar width were
bolted together to form a lining ring. To minimise the peripheral
take, primary grouting of the annular gap of about 75mm between

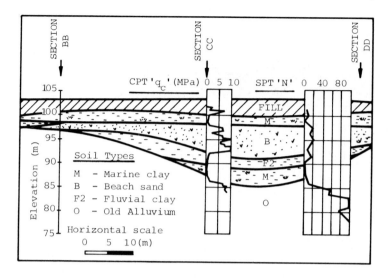

Figure 3. Longitudinal soil profile

excavated ground surface and extrados was carried out using a
bentonite-cement mix, via pre-formed holes in the lining. Some two
weeks after primary grouting, high pressure secondary grouting took
place via the same holes to fill in any persistent voids in the
annulus. The shallower westbound (WB), followed by northbound (NB)
tunnels, on the other hand, were partially and totally constructed,
respectively, through the buried channel of Kallang deposits. To
minimise ground movements during tunnelling, these waterlogged
sediments were initially stabilised by cement followed by chemical
jet grouting. Compressed air support of up to 1.5 bars was applied
in conjunction with face timber breasting within headings advanced
through the buried channel, excavations being carried out with
pneumatic spades.

2 MULTIPLE TUNNEL INTERACTION

2.1 Ground Surface Settlements

Fig 4 shows the surface settlement profile for cumulative ground
losses due to excavations at a comparable site on the stretch of
alignment - from Dhoby Ghaut to Somerset Stations - contiguous with
that of Connaught Drive. The settlement profile has been determined
via the empirical practice of superposing individual tunnelling
responses. Field settlements superimposed on the same plot generally
exceed empirical values - in excess of 100% at maximum settlement,
thereby substantiating the view that ground in the vicinity of an
initial tunnel drive (SB) is effectively weakened, resulting in
relatively greater ground losses during excavations for a subsequent
tunnel arrival (NB). Furthermore, in accordance with known field
observation (Hansmire, 1975) and as also justified by ground

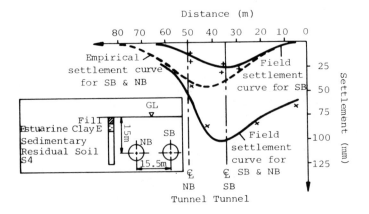

+ Measurement for SB alone
X Measurement for SB & NB

Figure 4. Surface settlements between Somerset and Dhoby Ghaut
Stations

weakening around an initial tunnel, the measurement profile was
significantly biased towards the first arrival tunnel.

Figs 5(a) and 5(b) show respective longitudinal profile fits to
incremental surface settlement measurements for WB and NB tunnel
excavations respectively. Since each tunnel was preceded by others,
the profiles reflect only ground losses due to individual tunnel
excavations and effects of interactions with existing tunnels.
Accordingly, an error function may also be applied to incremental
field settlement data, as hitherto only assumed in practice for the
single tunnel case (Yoshikoshi et al, 1978). Furthermore, from
settlement measurements at both sites, it was found that the
practical assumptions of a normally-distributed lateral ground
surface profile and an error function approximation to the
longitudinal profile (Peck, 1969a) were indeed applicable to the
initial tunnels.

2.2 Pore Water Pressures

Fig 6 shows a typical pneumatic piezometer response to multiple
tunnel passbys. With the approach of the initial SB tunnel, an
excess pore water pressure developed which increased to a maximum of
some 6m head when the tunnel face was about half diameter preceding
the instrument location. During this initial phase of tunnelling,
the compressive effects of forward shoving of the shield by hydraulic
jacks reacting against installed lining obviously outweighed those of
pressure relief due to face excavations. However, as the heading
receded from the instrument location, the pore pressure fell rapidly,
attaining a minimum head of about 5m below groundwater table level
when the separation distance between heading face and instrument
location was some two tunnel diameters. In this latter phase of

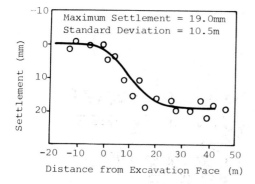

Figure 5(a). Longitudinal profile for WB tunnel

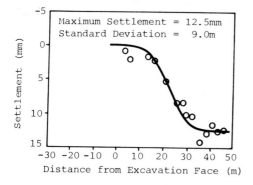

Figure 5(b). Longitudinal profile for NB tunnel

pore pressure development, the effects of pressure relief due to tunnel excavations and ground stretching as the shield pulled away were clearly dominant. Nevertheless, due to a surrounding water table, the pore pressure recovered partially with time, reflecting a permanent drawdown of about 1.5m, probably as a result of seepage into the tunnel. With the arrival of subsequent EB, WB and hence NB tunnels, similar trends in pore pressure development occurred, the degree of pore pressure response to a given tunnel arrival depending on orientation and proximity or otherwise of tunnel to piezometer, as well as the extent of its interaction with existing tunnels.

2.3 Lining Loads

Fig 7 provides a comparison between bending moment interaction factors based on Peck's (1969b) empirical approach, and as determined from stress-meter readings at various locations of the Connaught Drive tunnels, as functions of tunnel separation. Accordingly, the factors deduced from measurements generally varied within 25% of

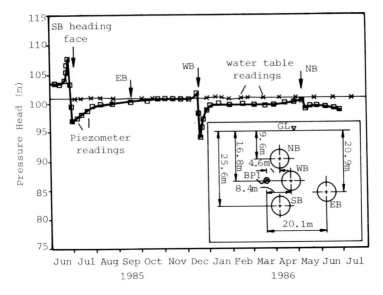

Figure 6. Variation in pressure head of piezometer BP1

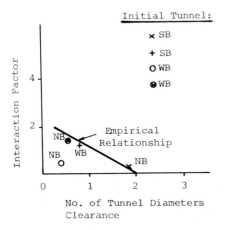

Figure 7. Interaction factors for moments

those obtained empirically. A few extreme discrepancies were also found which may be partly attributed to the presence of nearby joints, staggered in both longitudinal and transverse directions, at which significant local re-distribution of bending moments could have taken place. The other source of disparity could have been relative rotation between contact faces of adjacent lining segments resulting in eccentricity of thrust loading. Nevertheless, as shown in Fig 8, the combined effects of thrusts and bending moments at a typical

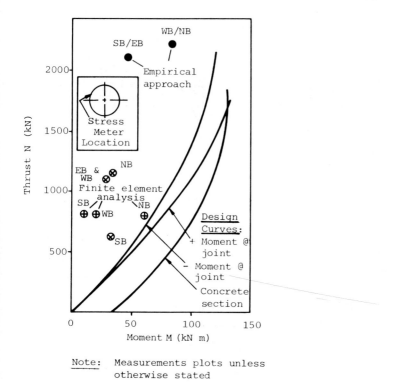

Note: Measurements plots unless otherwise stated

Figure 8. Combined thrust and moment loadings at Section A-A

stress-meter location in SB tunnel remained well within the design envelopes for safe load capacity of the precast concrete lining, even after interaction. These envelopes constitute upper bounds on plots of empirically-determined bending moments and axial thrusts based on assumed interaction factors (Kajima-Keppel J.V., 1985) applied to Curtis' (1976) single tunnel design loading.

For the purpose of comparison, those empirical plots relevant to the measurement results, upon which the design envelopes are based, as well as corresponding results from finite element analyses adopting Kajima-Keppel J.V. (1984) parameters have also been included in the figure. It is noteworthy that although, as also the case for the analytical and empirical plots, measurement results indicate increasing bending moments with greater tunnel interaction, the latter results show a clear tendency for relatively greater thrusts to develop in these circumstances than either method of prediction, so that the actual margin of safety will be far better than designed.

2.4 Lining Distortion

Fig 9 shows the development of incremental lining displacements at various tape extensometer reference stud positions in initial SB

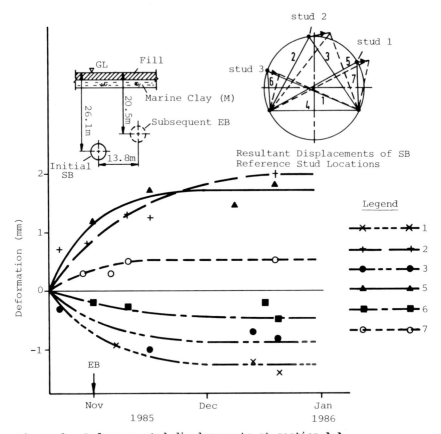

Figure 9.   Reference stud displacements at section A-A

tunnel due to interaction with second arrival EB tunnel. Accordingly, reference stud 1 registered about 1.5mm movement at approximately 15° to the horizontal towards EB tunnel.   The displacement of stud 2 of some 3mm at a shallower angle of about 10° - as to be expected - was also directed towards EB tunnel.   Also reflecting the general tendency of SB tunnel to bulge towards EB tunnel and hence squash in the perpendicular direction, stud 3 indicates some 1.5mm movement at about -15° to the horizontal.   These small order of magnitude movements are consistent with tunnel drives in very stiff to hard Old Alluvium followed by the installation of relatively stiff lining.

ACKNOWLEDGEMENTS

The research upon which this paper is based was funded in part by the Science Council of Singapore under RDAS Grant No. C/81/04-06 which is gratefully acknowledged.   The authors also wish to express their thanks to the Mass Rapid Transit Corporation of Singapore for their co-operation in the research project.

REFERENCES

Curtis, D.J. 1976.   Discussion on the circular tunnel in elastic ground. Geotechnique 26:231-237.
Hansmire, W.H. 1975. Field measurement of ground displacement about a tunnel in soil. PhD thesis. University of Illinois at Urbana-Champaign.
Kajima-Keppel J.V. 1984. Finite element analysis of bored tunnels. Report for Singapore MRT Contract 107.
Kajima-Keppel J.V. 1985. Design calculation for bored tunnels. Report No. C/MC/107/CB for Singapore MRT Contract 107.
Peck, R.B. 1969a. Deep excavation and tunnelling in soft ground. State-of-the-Art Report, 7th Int. Conf. on Soil Mech. and Fdn. Engg. Mexico City: 255-290.
Peck, R.B. 1969b. Design of tunnel liners and support systems. Final Report for Office of High Speed Ground Transportation, Washington D.C. 20591.
Yoshikoshi, W., O. Watanabe & N. Takagi 1978. Prediction of ground settlements associated with shield tunnelling. Soils and Foundations 18:47-59.

2nd International Symposium on Field Measurements in Geomechanics, Sakurai (ed.)
© 1988 Balkema, Rotterdam. ISBN 90 6191 778 6

The results of the tunnel crossing intersection measurements under different conditions

Nobuhiro Fukao & Tsuguo Takebayashi
Shimizu Construction Co. Ltd, Japan

1 PREFACE

Tunnels excavated in recent years have tended to become increasingly longer and more complicated due to improvements in technology, social needs, and topographical restrictions.

Since the excavation of a tunnel's intersection has been avoided, due to the technical difficulty of such work, only a few examples of excavating intersections in tunnels exist.

Reported in this paper are the results of monitoring during execution of work for the three intersections in a tunnel with different geological conditions, namely, the Aobayama Tunnel along the Sendai-Nishi Road in Sendai City, Miyagi Pref.

2 OUTLINE OF THE WORK

The Aobayama Tunnel consists of twin two-way tunnels, an east-bound tunnel with a total length of 2,233 m. These tunnels are interconnected by three connect tunnels, and three shelter adits have been provided. Presented in this paper are the measurement results of the east-bound tunnel. Meanwhile, the west-bound tunnel has already been completed and is in service.

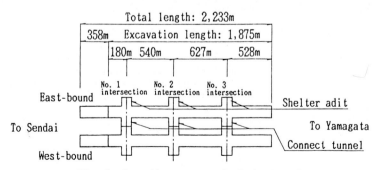

Fig. 1 Overall arrangement of the tunnel

In conducting the excavation work of the east-bound tunnel, three inter-sections were arranged at intervals as shown in Fig.1, and of a system wherein these intersections are bifurcated at a right angle against the main tunnel so that each corner is cut off by 2 m. Therefore, their fit portions are a construction extending in the form of a bell. The interconnecting parts of the main tunnel, connect tunnels, and shelter adits are as indicated in Fig.2. The intersections are called No.1, No.2, and No.3 intersections from the Sendai side with different individual geological conditions, respectively. Excavation of the main tunnel was carried out according to a bottom drifting method, while the No.1 and No.2 intersections were excavated by the New Austrian Tunnel Method (NATM), and the No.3 intersection by upper half drifting according to a breast boards method.

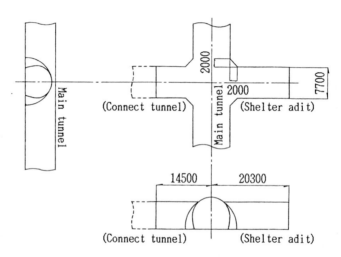

Fig. 2   Shape of intersections

3 OUTLINE OF GEOLOGY

Geologically speaking, the ground around the Aobayama Tunnel roughly consists of andesite and tuffaceous rock.
   The andesite is classified into (1) very hard type with few joints; and (2) andesite susceptible to spalling due to the development of joints, while the tuffaceous rock is classified into (1) very hard type containing a large amount of gravel; (2) reddish brown rock with high water absorption; and (3) green type containing a large amount of sandy materials.
   For the Nos. 1 and 2 intersections, moreover, a boring survey of $\ell = 15$ m was conducted from the bottom pilot tunnel of the main tunnel toward the upper half of the pilot tunnel. The results of boring, physical properties and lithology of the ground are presented in Fig.3.

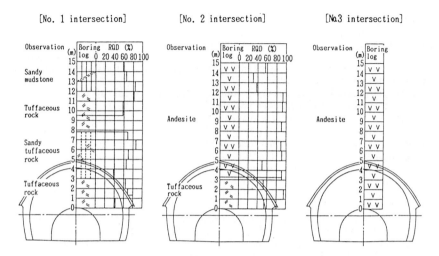

Fig. 3 Boring log of Nos. 1, 2 and 3 intersections

Table 1 Physical properties and lithology of the ground at
No. 1, 2 and 3 intersections

	No.1 Intersection	No. 2 Intersection	No. 3 Intersection
Unit weight	1.87 t/m^3	2.70 t/m^3	2.78 t/m^3
Unconfined compressive strength	44.4 kg/cm^2	556.3 kg/cm^2	1,017.2 kg/cm^2
Destructed strain	3.0%	0.5%	-
Young's modulus	1,013 kg/cm^2	130,000 kg/cm^2	753,000 kg/cm^2
Lithology	Both the upper and lower half parts consist of soft sandy tuffaceous rock	The upper half part is comprised of andesite highly susceptible to falling due to development of joints, while the lower half consists of reddish brown tuffaceous rock.	Both the upper and lower half parts are comprised of very hard andesite.
Ground overburden	57 m	67 m	78 m

4 DESIGN OF INTERSECTIONS

Since a part of the arching concrete lining came from a cut form in the intersection, the concrete lining of this section acted as a three-dimensional shell
structure rather than as an arching, and thus, it was inevitable that the
stiffness be lowered as a structure.

Due to duplicated work, namely, simultaneous excavation of both the main
tunnel and the branch tunnel, the fractured zone around the tunnels was greater
than that for a single tunnel excavation.

In the design of the intersection, therefore, rock reinforcement is generally
studied as shown in Table 2, in order to improve the rigidity of the supporting
structure and to minimize the fracture zone during excavation of the main tunnel
and branch tunnel, as far as is practicable.

In order to minimize fracture of the ground around the No.1 and No.2 intersections, the NATM was adopted for excavation of both the main tunnel and the
branch tunnel. As the ground around No.1 intersection was comprised of soft
tuffaceous rock, a high stiffness was required of the shell structure of the
tunnel lining for the branch part during excavation of the branch tunnel. It
was determined that the branch tunnel be excavated after completing the concrete
lining of the main tunnel by arranging reinforcing bars.

Since comparatively desirable andesite was distributed around the roof rock
on the Nos.2 and 3 intersections, reinforcement mainly on the basis of a
hanging effect and rock reinforcement effect of rock bolts was adopted as the
support members. Therefore, it was determined that the concrete lining of the
intersection should be laid after excavation of the branch tunnel.

Table 2   Design of support members

| | Concept of support | Major support members | | |
		Main tunnel	Approach section of intersection	Connect tunnel and shelter adit
No. 1 intersection (NATM)	(1) Because of small modulus of deformation of the ground, stiff support was obtained by reinforcement with rock bolt, shotcrete and steel supports (2) Excavation of branch tunnel was carried out after placing lining concrete for main tunnel.	Steel support: H-200 ctc: 1.0 m Shotcrete t: 150 mm Rock bolt L: 3 m 5 m 7 m	Steel support: H-200 ctc: 0.9 m Shotcrete t: 100 mm Rock bolt L: 7 m	Steel support: H-200 ctc: 1.00 m Shotcrete t: 100 mm Rock bolt L: 3 m
No. 2 intersection (NATM)	(1) Suspension effect of rock bolt (2) Lining concrete for main tunnel was placed after excavation of branch tunnel	Steel support: H-200 ctc: 1.0 m Shotcrete t: 150 mm Rock bolt L: 3 m 5 m	Steel support: H-200 ctc: 0.90 m Shotcrete t: 100 mm Rock bolt L: 5 m	Steel support: H-200 ctc: 1.0 m Shotcrete t: 100 mm Rock bolt L: 5 m
No. 3 intersection (Breast boards method)	(1) Suspension effect of expansion rock bolt	Steel support: H-175 ctc: 1.1 m	Steel support: H-175 ctc: 1.1 m Expansion rock bolt L: 2.5 m	Steel support: H-175 ctc: 1.1 m

5 EXCAVATION OF INTERSECTIONS

In order to minimize fractures during excavation, due to the distribution of soft tuffaceous rock, No.1 intersection was excavated mechanically, using arm type excavators, together with the main tunnel and branch tunnel. In the case of Nos. 2 and 3 intersections, the adoption of a drill and blast method was inevitable, because of the distribution of andeside. Moreover, to minimize fractures during excavation, smooth blasting was conducted, taking into account the limit of explosive volume, improvement of drilling precision, etc. The excavation methods adopted are as presented in Table 3.

Table 3   Excavation method

	Geological conditions	Excavation system	Method
No. 1 intersection	Upper and lower half parts: Sandy tuffaceous rock	Mechanical excavation (CL 1010B)	NATM (Upper half drifting)
No. 2 intersection	Andesite highly susceptible to falling due to development of joints	Drill and blast method	NATM (Upper half drifting)
No. 3 intersection	Upper and lower half parts: Hard andesite	Drill and blast method	Conventional (Upper half drifting)

Excavation of the intersections was conducted in the following order:

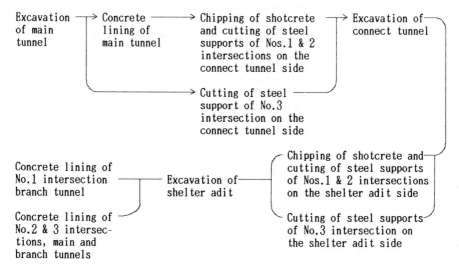

6 ITEM OF MONITORING

Prior to excavation of the intersections, a measurement plan was prepared in order to obtain the data for studying the justifiability of the design policy and countermeasures to be taken during excavation of branch tunnels.
The items of monitoring are shown in Table 4.

Table 4   Monitoring items

Item Positions	Convergence	Roof settlement	Axial force meter of rock bolt	Extensometer
No. 1 inter- section	6 sections 14 traverse lines	7 sections	L = 3 m B-1,2,3 L = 5 m B-5 L = 7 m B-4,6	L = 6 m E-1,2,3 L = 10 m E-5 L = 14 m E-4,6
No. 2 inter- section	6 sections 14 traverse line	7 sections	L = 3 m B-1,2,3,5 L = 5 m E-4,6	L = 6 m E-1,2,3 L = 8 m E-4,5,6
No. 3 inter- section	---	Center of intersection	---	---

Since the ground around the No.1 intersection was composed of soft tuffaceous rock, long extensometers and rock bolt axial force meters were arranged from the main tunnel prior to excavation of the branch tunnel.
As it was judged from geological observations during main tunnel's excavation that joints of andesite adjacent to the roof of the No.2 intersection had developed, and as a result of investigating the bearing capacity of the tuffaceous rock distributed around the lower half portion, monitoring of the roof settlement and of the convergence were carried out, and extensometers and rock bolt axial force meters were arranged.
Since excellent andesite was clarified to be distributed around the No.3 intersection as a result of geological observation during the main tunnel's excavation, it was determined that measurement of the roof settlement and an in-tunnel observation should be carried out.
The major monitored sections are indicated in Fig.4 .

7 RESULTS OF MONITORING

The results of monitoring the No.3 section of the intersection between the main tunnel, connect tunnel and shelter adit are shown in Fig.5, and the results of monitoring the No.2 section of the corner cut part are in Fig.6 .
According to the axial force meter of rock bolts of the No.1 intersection, the majority of measurements indicated tensile stress. This in turn indicated that the reinforcement of soft ground with rock bolts would be effective.

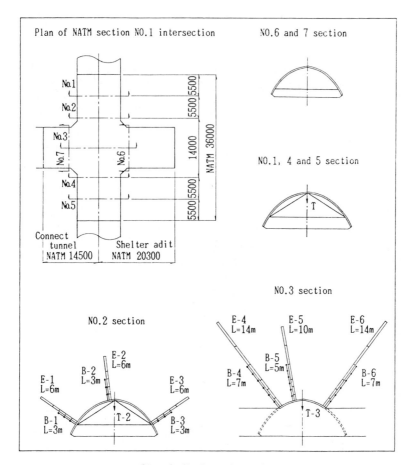

Fig. 4 Monitored sections

During excavation of the shelter adit at No.2 intersection, the axial force meter for B-4 rock bolt indicated a rapid increase of the axial force to as much as 14 tons. Therefore, excavation was suspended immediately and a long (7 m) rock bolt was additionally provided. Moreover, re-shotcreting was carried out to increase the width of the bearing of the shotcrete spring. The excavation was restarted. This rapid increase of the axial force is considered to have been caused by an insufficient suspension effect resulting from the development of andesite joints on the roof. At the approach portion of the Nos.1 and 2 intersections, the steel supports and shotcrete arching are considered to have been mainly predominant, while the effect of rock bolts was not remarkable.

Presented in Fig.7 is the roof settlement at the center of the inter-sections in each stage of excavation of Nos.1, 2 and 3 intersections.

According to this table, the amount of roof settlement was smallest at No.3 intersection where excellent andesite was distributed even though the breast boards method was adopted for excavation.

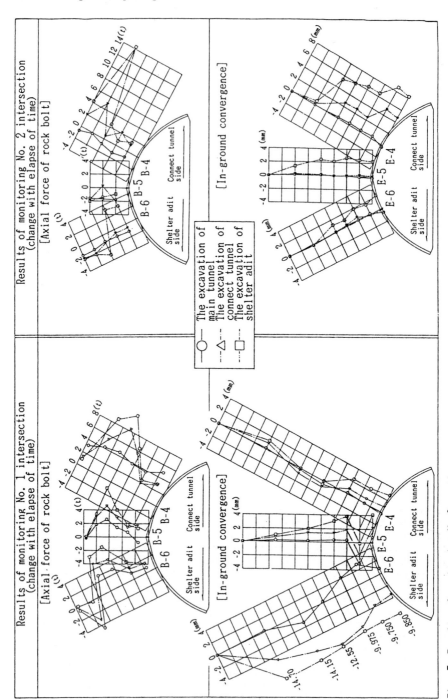

Fig. 5   Results of monitoring No.3 section

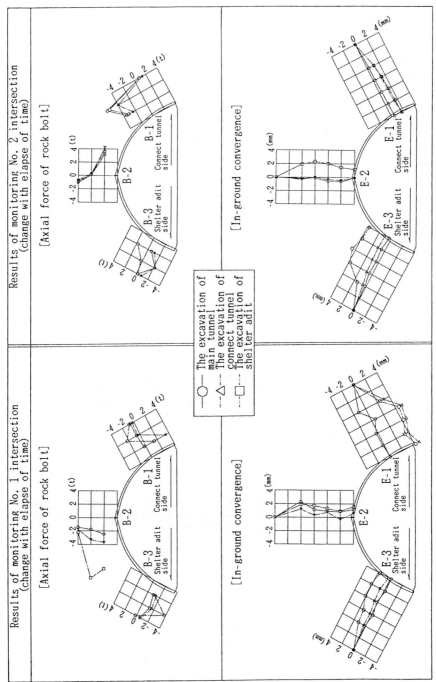

Fig. 6 Results of monitoring No.2 section

Although andesite was distributed around the roof rock at No.2 intersection, the amount of roof settlement was unexpectedly large due to the development of joints.

In the case of the No.1 intersection, the branch tunnel was excavated after providing a concrete lining to the main tunnel and forming a stiff shell structure. However, roof settlement took place also during excavation of the branch tunnel. Judging from the above facts, the geological conditions are considered to constitute the most important factor in executing excavation of the intersections.

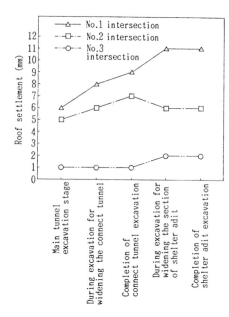

Fig. 7 Roof settlement

8 SUMMARY

Discussed in this paper have been the results of monitoring the cross form intersections of the Aobayama Tunnel under different conditions (whether the rock is hard or soft). The results can be summarized as follows:

(1) As the springs of steel supports and shotcrete were cut off in the case of intersecting in a cross form during excavation of intersections, it was inevitable to mainly use rock bolts as support members, and possible to expect not only the suspension effect of hard rock, but also the ground improvement effect of soft ground to a certain extent.

(2) The support members of the approach section of the main tunnel and branch tunnel are considered to be supported by steel arching supports, shotcrete and concrete linings rather than by rock bolts.

(3) In conducting the excavation of the intersection, the geological conditions constitute the most siginificant factor, that is, the worse the geological conditions are the larger the stiffness is required for the support structure.

This report has presented the results of monitoring the intersection during excavation of the upper half. However, monitoring of the lower half has not been discussed in this paper since the excavation work of this portion was carried out smoothly without specifying any particular provisions.

The authors would be only too happy if evaluation of these monitoring results should be of any help for the design and excavation of the intersections of tunnels.

2nd International Symposium on Field Measurements in Geomechanics, Sakurai (ed.)
© 1988 Balkema, Rotterdam. ISBN 90 6191 778 6

Breakage face measurement and ways of improving coal seam stability

Ju.A.Veksler
Polytechnical Institute, Karaganda, USSR
E.P.Bragin
Scientific-research Coal Institute, Karaganda, USSR
G.M.Present & S.A.Chesnokov
Kostenko Mine, Karaganda, USSR

Miners' work in breakage faces in medium to thick seams
equipped with powered supports is endangered by coal
squeeze and the resulting caving of the near-to-face
part of the roof. Decreases in the technical-and-econo-
mic indices of the breakage faces are also connected with
the squeeze.

Special experimental and theoretical studies were being
carried out for a number of years by Karaganda scientists
in order to develop technical measures aimed at decreasing the volume and neutralizing the negative consequences
of the squeeze.

Engineers of Kostenko mine together with scientists of
the Polytechnic institute and the scientific-research
coal institute carried out instrumental measurement of
the face displacement, wall rock convergence, influence
of the power characteristic of supports and technological
processes in the long face of the roof and face displacements in the process of working out thick sloping seams.

The displacements were measured by repeater test racks
(struts) of CYU-II-type equipped with dial indicators,
their accuracy of reading being 0.0I mm.

For horizontal displacement measurements the rack with
indicators is set horizontally between the face and the
front hydraulic props of the support's sections, vertical
displacement being measured by racks with indicators set
up vertically between the roof and the floor. Measurements of the powered support's behaviour during the operation cycle are recorded by self-recording pressure
gauges mounted on the hydraulic props.

Experiments were carried out in seams K2 "Lower Middle"
KIO "Felix" and KI2 "Upper Maryanna" of the upper and
lower layers with powered supports of 2M8I and MI30-
types.

An experimental check was carried out of the opinion
that one of the most effective ways of preventing or reducing coal squeeze is firstly, a great outward thrust
of the powered support and secondly, the effectiveness
of the bench mining as a measure of reducing dangerous
consequences of the coal squeeze.

The experiments showed that with off-loaded supports the indicators registered O.I-3.9 mm of coal squeeze opposite the section. This was obviously caused by the increase of pressure on the seam edge. Subsequent loading on the section till the initial thrust level also resulted in O.I-I.26 mm displacement of the face opposite the section in the direction of outcrop. It is likely to be the result of the seam's wall resistance in conditions of reduced roof pressure on the edge of the face.

Analysis of the experimental data showed that the outward thrust of powered supports and its increase do not lead to the decrease of the face displacement in the direction of outcrop and, consequently, do not lead to the decrease of coal squeeze. This is proved in practice: not withstanding the application of powered supports "Pioma", 4KMT-I30 with high operational resistance and initial thrust, the coal squeeze remained intensive.

Thus, substitution of powered supports of 2M8I -and MI30-types (at Kostenko mine) with initial thrust of 440 and 200 KH correspondingly for "Pioma" supports with the 3000 KH thrust the coal squeeze didn't reduce and in a number of cases it considerably increased. On the whole, judging from the readings the thrust of the support doesn't noticeably influence the coal squeeze.

Fig.I and 2 present the diagrams of the influence of technological processes on the magnitude and speed of rock displacements in long faces of K2 seam during shuttle and bench mining. In shuttle mining coal is won during each run of the coal cutter operating at full capacity. Conveyor and supports follow the coal-cutter. In bench mining the coal-cutter moves in the direction from the belt-to air-heading: at first the upper part of the face is won, a bench being left unmined with dimentions: 0.8-I.2 m high and the cutter's web size-deep. The conveyor and the powered support do not follow the coal cutter. When the coal-cutter moves in the opposite direction the unmined bench is won, the conveyor and the support following the coal-cutter. Thus , the downward run of the coal cutter makes the face straight (vertical) again..

Fig.I shows the down operation of the coal-cutter in a section 75 meters long. The repeater test rack (strut) was set at a distance of I22 metres from the belt heading. The overall rock dosplacement was I.65 mm, the magnitude of displacements increasing with the coal-cutter's approaching the line of observation.

The backward operation of the coal-cutter (from bottom upwards "a" in fig.2) was observed along 26 metres of long face's section. Coal was won from the upper part of the face, rock displacement amounting to 0.2I mm. The top-down operation of the coal-cutter ("b") was observed along 40 metres of long face's section, rock displacement amounting to 0.3I mm.

The diagrams show that the influence of technological processes and support's outward thrust on the face contour is limited. The magnitude and speed of rock displacements are insignificant. However, the comparison of the results of shuttle and bench mining shows that the operation even with a temporary bench in similar sections results in a significant reduction in the magnitude of rock displacement. According to the experimental data the general average rock displacement in bench mining decreases compared with shuttle mining by I.6-I.7 times.

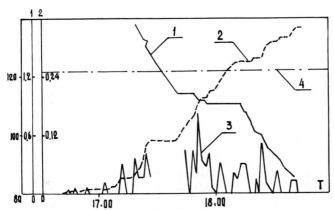

Fig.I The influence of technological processes on rock displacement in long face 3IK2-3 of Kostenko mine in bench mining.
I. coal-cutter's operation, 2. rock displacement
3. displacement rate, 4. observation line.

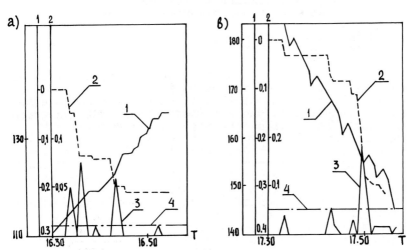

Fig.2 The influence of technological processes on rock displacement in long face 3IK2-3 in winning the upper part of "a" face and in winning bench "b":
I. coal-cutter's operation, 2. rock displacement,
3. displacement rate, 4. observation line.

Practical experience of mines operation and special observations carried out in mines show that after winning the lower bench as a result of exposing the seam and straightening up the heading the near-to-face section of the seam fails, this leading to occupational injuries, sometimes rather serious ones. One of the measures preventing the coal squeeze may be constant bench contouring of the seam. In constant bench mining the face is divided into 2 partssimilar to 2 faces of lessened thickness, the distorted edge of each part being confined to the bench or to the bottom.

Theoretical studies of the stressed-strained state of the seam and country rock around the breakage face were carried out to substantiate the way of preventing coal squeeze by means of changing the contour of the face. Stability analysis of different forms of breakage faces were carried out with the help of the numerical method of finite elements for the elastic-viscous model of rock with regard to failure caused by factors of damage accumulation and maximal tensile stress. In solving these problems the algorithm and program developed in the polytechnic institute of Karaganda were made use of.Mining specifitations of certain long faces with straight and bench faces of Kostenko mine were simulated by the ECIO33 -computer.

Theoretical studies confirmed the expediency of bench mining, which enhances the contour stability of the breakage face.

Technology of constant bench mining by coal-cutters with augers of various widths was experimentally implemented at Kostenko and Shakhstinskaya mines of "Karagandaygol" production amalgamation in order to put into practice the results of experimental and theoretical studies.

This method was used in cyclic coal mining of upper and lower benches. (Fig.3)

The depth of the upper bench mining exceeds that of the lower one. This was made possible due to the application of the auger 0.8 m wide on the belt-heading side and of the auger 0.5 m on the air-heading side. The coal-cutter operates according to the one-side scheme. The operation of the coal-cutter along the main cut is carried out during the upward run along the long face, the auger being lifted to the roof. When the coal-cutter moves downward the coal is cut by a long auger lowered by 30 cm, the shortened auger lowered to the ground cutting the lower bench and cleaning the floor.

The bench mining technology was tested in faces 43KI2-I-3 and 32K2-B of Kostenko mine.

Bench mining was watched during the whole period of operation with augers of enlarged width. It was found that with the old technology (when the face after bench mining was straight in contour) the number of downtime cases caused either by coal cutter's failure or by supporting the roof or by eliminating the consequences of coal squeeze was considerably greater compared with the

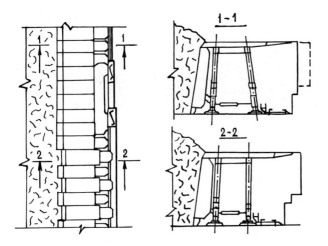

Fig.3 Coal winning according to the constant bench
mining scheme.

bench mining technology.
The drop in coal-cutter's failure was obviously connected with uniform and reduced loading on the excavating and
feeding parts owing to a considerable reduction of every
auger's web. During the upward operation of the coal-
cutter the magnitude of the front web is 0.2 m, that of
the rear one being 0.3 m.
Watching over the conditions of the face and roof in
the workings with powered supports showed that despite
the presence of unstable rocks in the roof and lengthening the distance between the baffle (lip) and the face
up to 0.5 metres (with the existing structure of support
-MI30 for bench mining) instead of 0.2 m for staright-
line face, the condition of the face or near-to-face section of the roof did not deteriorate. The idle time of
long faces caused by stowing the domes reduced by 2-3
times. However, a certain expansion of the near-to-face
section of the roof caused by the undersliding of the
baffle (lip) of support MI30 sections should be considered
a drawback. In future this drawback will be overcome by
applying telescopic baffle (shield).
The use of OKII70 supports some sections of which are
shifted for 7I0 mm following the coal-cutter will obviously prevent expansion of the exposed part of the near-
to-face roof section.
Down-time caused by coal squeeze with the old method of
mining amounted to I5 hours I0 min. a month,with the bench
method the outward thrust of the seam edge considerably
reduced and there were practically no downtime periods
caused by coal squeeze.
Thus, bench mining by coal-cutters with augers of various width is reasonable; it reduces coal squeeze, roof
caving, loading the excavating parts of the coal cutter;

it excludes down-time caused by coal squeeze, considerably increases safety.

The technology of constant bench shape of the coal face may be used in long faces with KMI30 coal-cutters provided the strength and stability of lower layers of the immediate roof allow the one-way coal mining. Bench mining can be used in all cases in long faces with shield supports. The influence of bench mining is expected to be maximal in longwall advancing to the rize.

Adaptability of convergence forecasting method for controlling tunnel construction

Jun'ichi Seki, Masayuki Okada, Hiroyuki Inoue & Toshihiko Miwa
Technical Research Institute, Maeda Construction Co. Ltd, Tokyo, Japan

SUMMARY

In tunnelling, it is important to ascertain the stability of the ground
and the safety of support members by daily observation and measurement.
And when danger is foreseen, countermeasures, such as the reinforcement
and alteration of support or excavation methods have to be taken
immediately. For this purpose, it is necessary to comprehend the rock
behavior at an early stage of excavation and carry out the works, with
forecasting the process and ultimate of displacement.
    In general, displacement caused by tunnel excavation is expressed as
a function of time elapsed, or the distance to the tunnel face. The
authors have taken note of the correlation between deformation of a
tunnel and distance to the face, and made an approximation thereof with
an exponential equation. A method has been investigated, by which the
ultimate and convergent process of displacement is forecasted in the
early construction stage using the least squares method with measurement data (Takahashi 1986, Seki 1986).
    In this paper, by introducing the outline of this method, adaptation
to actual tunnelling is considered, with data obtained during the upper
half excavation stage using the bench cutting method. Moreover, the
results ate indicated herein, expanding this method to forecast the
convergence of the lower half excavation stage.

1. INTRODUCTION

It is necessary to know the value of displacement and to observe the
convergent process of displacement when monitoring construction by
measurement. The former is established by the monitoring criterion of
displacement. The latter is also established by the monitoring
criterion of the rate of displacement, or the region of displacement
control in the majority of cases. As mentioned before, it is important
to judge as soon as possible whether the tunnel becomes stable or not
from the forecasted ultimate displacement.
    Various techniques and applications about the convergence forecasting
method have been reported. These are divided into two major groups
depending on the parameters used. The primary factors of displacement
are used on one side, such as the characteristics of the ground like

the compressive strength, geological conditions like the crack distribution and over burden height. And on the other hand, measurement results, such as displacement or the rate of displacement, are used. In comparing the two groups, in the case of the methods using primary factors of displacement, each factor is difficult to establish simply because of the many conceivable factors being related to each other complicatedly, and much measurement data is required. On the other hand, the method which uses measurement results, can be used with only a few kinds of parameters and highly accurate data, and its technique is very simple. In general, since the data can be collected one after another in actual tunnelling, the latter is more suitable for forecasting the displacement.

2. CASE STUDY ON FORECASTING DISPLACEMENT IN THE ACTUAL TUNNELLING

As an example of forecasting displacement in controlling the construction by measurement, the method adopted in actual tunnelling is explained hereinafter. Displacement is expressed as a function of time elapsed or the distance to the tunnel face in many cases. In the case of this tunnel, displacement has been observed depending on distance rather than time elapsed. As shown in Fig. 1, therefore, convergence has been expressed as the distance to the tunnel face, and, based on forecasting results, the controlling construction has been carried out by measurement.

The monitoring criterion of the upper half displacement is determined by equation (1) with a distance to face at 1D (10m) and upper ultimate displacement data obtained before the decision of monitoring criterion (Fig. 2).

$$U_C = 1.66 \ U_B + 7.47 \qquad (1)$$

where U_C : Upper half ultimate displacement (mm)
           U_B : Displacement of distance to upper half tunnel face at
           1D (mm)

And for the ultimate displacement after construction of the lower half, the monitoring criterion is determined by equation (2) with the upper half ultimate displacement.

$$U_L = 1.89 \ U_C \qquad (2)$$

where U_L : Ultimate displacement caused by excavation of lower
           half (mm)

And with regard to comprehension of the convergent process of displacement, the relation of displacement and distance to the face is expressed by the following equation (3).

$$U_i = U_C \ (1 - e^{-\beta X_i}) \qquad (3)$$

where U_i : Displacement (mm)
           X_i : Distance to upper half tunnel face (m)
           β : Ultimate coefficient
     The ultimate coefficient β in equation (3) means that if its value becomes larger, the initial incline gradually increases and

displacement converges in the early stages. In order to facilitate the comparison of each measurement section, displacement U_i is described by a ratio of displacement α (= U_i/U_C x 100) against forecasted ultimate displacement U_C and the scale of displacement curves is standardized. Also, based on the measurement results obtained in advance, the allowable limit of ultimate coefficient β is calculated and the region of displacement control is determined (the extent of the oblique line in Fig. 4). By plotting each measurement data on the graph, the behavior of the displacement is judged to be extraordinary or not. Fig. 4 shows one of the operative examples of the region of displacement control. Each measurement data has been stated in the control region and the state of convergence at almost all measurement sections has been stable as the example figure shows. This method, in which the region of displacement control is used with the rate of displacement, is recognized as one of the controlling construction methods by measurement in actual tunnelling. As • marks in several sections of Fig. 4 show, the measured displacement is larger than the forecasted ultimate displacement, and the rate of displacement exceeds the allowable amount in the control region. Namely, since the convergent process of displacement is affected by the accuracy of the forecasted ultimate displacement obtained from equation (1), proper construction control is unable to be carried out depending on whether or not the accuracy is good or bad. Thus, in order to make up for this defect, forecasting the ultimate displacement and the convergent process simultaneously is required as soon as possible.

3. FORECASTING DISPLACEMENT IN UPPER HALF EXCAVATION STAGE

3.1 FORECASTING METHOD

Ultimate displacement U_C and ultimate coefficient β are calculated simultaneously by using the least squares method directly on forecasting equation (3). Non-linear equation (3) is replaced as follows :

$$f_i(U_C, \beta) = U_C(1 - e^{-\beta X_i}) - U_i \qquad (4)$$

At first, calculating the initial presumed values of U_C and β , and performing Taylor expansion around (U_{CO}, β_0) on equation (4), the following linear equation is obtained, where the terms of higher orders are disregarded.

$$f_i(U_C, \beta) = f_i(U_{CO}, \beta_0) + \left.\frac{\partial f}{\partial U_C}\right|_{U_{CO}, \beta_0} (U_C - U_{CO})$$

$$+ \left.\frac{\partial f}{\partial \beta}\right|_{U_{CO}, \beta} (\beta - \beta_0) + \cdots\cdots \quad (5)$$

Unknown U_C and β are obtained by using the linear least squares method on equation (5). This operation is iterated and desired values U_C and β are obtained. In this method, therefore, it is possible to forecast U_C and β with more than three measurement data.

3.2 EXAMINATION OF ADAPTABILITY

An example is shown in Fig. 5, of forecasting displacement in the upper half excavation stage using actual measurement data obtained by the above method.

This method has a characteristic that the precision of forecasting ultimate displacement is improved with an increase in data; in other words, with the advance of tunnel excavation.

The variation of error is shown in Fig. 6, in forecasting the ulti-mate displacement and the ultimate coefficient with the advance of distance to face. In the case of Fig. 6, the error is noticeable in the forecasting displacement when it is approximately 1D distance to the face. The error is less with the advance of tunnel excavation and when it is approximately 2.5D distance to the face, the error is 30% per measured ultimate displacement, and more when it is 4D distance to the face, forecasting displacement agrees with the measured result. As shown in the above case, this method has a characteristic that the precision of forecasting displacement obtained is improved with an increase in distance to the face.

The authors have examined the adaptability of this method. The data used belongs to four kinds of rock which correspond to A~D of a rock classification in the Japan Highway Public Corporation. The seismic wave velocity of the ground in the above rock classification is approximately 5(km/sec) for A, $3 \sim 5$(km/sec) for B, $2.5 \sim 4.5$(km/sec) for C, and $1 \sim 3$(km/sec) for D.

The tendency of the variation in average error is shown in Fig. 7, in forecasting ultimate displacement with the advance of distance to the face. In all the rock classifications, with the advance of distance to the face, the error is less and the precision of forecasting is improved. From Fig. 7 it is proved that displacement can be forecasted within 30% error when distance to the face is less than about 2.5D. Now, required precision of forecasting is different depending on many factors such as the value of deformation or the importance of the structure and so on. Therefore, in the case of the actual forecasting, the reliability must be judged by taking into consideration the required precision of forecasting.

A histogram of the ultimate coefficient β at the time of ultimate displacement is shown in Fig. 8. The distribution of β has a different tendency depending on the rock classification and the better the ground is, the larger β is. This shows that time leading to the ultimate displacement is different depending on the rock classification.

The relation is shown in Fig. 9, between the ultimate coefficient and the measured ultimate displacement in the stage of deformation leading to the ultimate displacement. The ultimate coefficient and the measur-ed ultimate displacement are shown on the average of each rock clas-sification. It is shown that the displacement is large in the case of weaker ground. The ultimate coefficient is inclined to be small as the displacement is large. That is to say, in class A which is the best condition, the displacement is at the smallest value and the convergent at the earliest time. On the other hand, when the ground is weaker, the displacement is large and the convergent late.

Thus, the process of deformation in this method can be confirmed quantatively by using the ultimate coefficient and the ultimate dis-placement. And the characteristics of the process of deformation can be clarified.

4. FORECASTING DISPLACEMENT IN LOWER HALF EXCAVATION STAGE

4.1 FORECASTING METHOD

By extending the forecasting method in the upper half excavation stage, the ultimate displacement and coefficient are forecasted for the displacement caused by the lower half excavation. The total displacement Ui is expressed as follows :

$$U_i = U_{1i} + U_d + U_{2i} \qquad (6)$$

where   Ui   : Total displacement (mm)
         U1i : Displacement caused by the upper half excavation (mm)
         Ud   : Prior displacement before the lower half excavation
                reaches (mm)
         U2i : Displacement caused by the lower half excavation (mm)

The relation between the lower half displacement U2i and distance to lower half tunnel face Zi is expressed as follows :

$$U_{2i} = U_e \left(1 - e^{-\gamma Z_i}\right) \qquad (7)$$

where   Ue   : Ultimate displacement caused by lower half excavation
                (mm)
         γ : Ultimate coefficient of the lower half displacement
         Zi   : Distance to lower half tunnel face (mm)

where U2i is obtained from equation (6) :

$$U_{2i} = U_i - U_{1i} - U_d \qquad (8)$$

By using Uc and β which are forecasted with several data less than distance Xm, Ui is calculated from the following equation.

$$U_{1i} = U_c(1 - e^{-\beta X_i}) \qquad (9)$$

where   UC   : Upper half ultimate displacement (mm)
         β : Ultimate coefficient of upper half displacement
         Xi   : Distance to upper half tunnel face (m)

By introducing equations (8) and (9) in equation (7), and using the least squares method in the same way as the upper half stage, it is possible to forecast the ultimate displacement Ue and the ultimate coefficient γ for the lower half excavation.

4.2 EXAMINATION OF ADAPTABILITY

An example of forecasting displacement in the lower half excavation stage is shown in Fig. 11. It is the same section data as shown in Fig. 5. Simularly in the case of the upper half, the forecasting ultimate displacement comes close to the measured one with the increase of data, in other words, with the advance of tunnel excavation. The variation in error is shown in Fig. 12, in forecasting the ultimate displacement and the ultimate coefficient with the advance of distance to the lower half tunnel face. The error is less than about 10% and similar to the upper half tunnel face, thus, the precision of forecasting displacement is improved.

The authors have examined the adaptability of this method in the lower half excavation stage. The data used belongs to rock classifications C and D which have been examined in chapter 3. The tendency is shown in Fig. 13, of the variation in average error in forecasting the ultimate displacement with the advance of the distance to the face. Frome Fig. 13, it is proved that when it is 4D distance to the face, the error is 40%, and after 4D, it is 10 - 15%. In this tunnel, the excavation speed of the lower half is faster than that of the upper half. Therefore, at less than 3D distance to face, as the number of data is scarce, the error is large.

When using this method, it seems that the measurement has to be conducted after due consideration of the distance to the face. The histogram of the ultimate coefficient is shown in Fig. 14, at the time of the ultimate displacement. A distribution of γ has characteristics depending on the rock classification. In class C, the value is larger than that in class D, and this means that the ultimate is faster. The average of γ value is C - 0.14, D - 0.06 and similar to β of the upper face (C - 0.11, D - 0.08). Therefore, in the same rock classification, the deformation process of the upper half is very similar to that of the lower half.

5. CONCLUSION

The authors have examined the method by which we can forecast both the ultimate displacement and the ultimate coefficient (the convergent process) directly at the same time. As a result, the following has been clarified.

As a result of applying this method to the forecasting of displacement in the upper half excavation stage by the bench cutting method, it is found that we can forecast displacement with improved precision as the tunnel excavation advances. Further, extending the method to forecast displacement in the lower half excavation stage, it is proved that the method is adaptable, similarly to the upper half. It is possible to confirm the behavior of deformation with an amendment to the forecasting displacement one after another as excavation advances.

The authors have forecasted displacement by this method using many kinds of data which are equivalent to rock classifications A ~ D, and found characteristics of tendency for the deformation process to some extent. With class C and D data, forecasting the lower half displacement, it is found that the tendency of the deformation process in the lower half excavation stage is similar to that of the upper half. By considering the practicability on actual tunnelling, the method has a system which can be operated with interactive type. By using more data, we intend to further improve the usefulness of this method.

REFERENCE

Takahashi, H., K.Sanagi, J.Seki & M.Okada. 1986. Case study on controlling construction by measuring large deformation in tunnel. Japanese Society of Soil Mechanics and Foundation Engineering. Vol.34, No.2. 53-58.

Seki,J., T.Nakamura, M.Okada & T.Miwa. 1986. Adaptability of method
for forecasting displacement under NATM. 18th symposium on Rock
Mechanics. Committee on Rock Mech. J.S.C.E. Proc. 21-25.

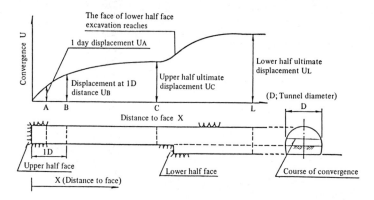

Fig. 1   Schematic diagram of the behavior of deformation in tunnelling

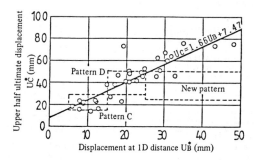

Fig. 2   Relation between displacement at 1D distance and upper half
ultimate displacement

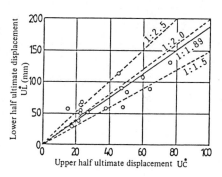

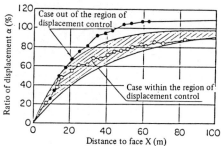

Fig. 3   Relation between the upper
half ultimate displacement
and the lower half ultimate
displacement

Fig. 4   Example of region of
displacement control

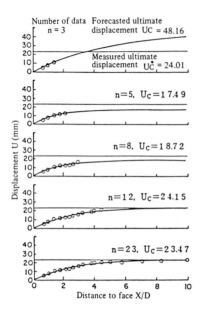

Fig. 5   Example of forecasting as distance to face (the number of data) increases

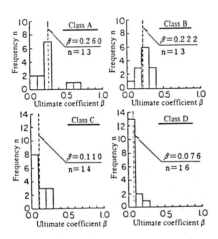

Fig. 8   Histogram of ultimate coefficient

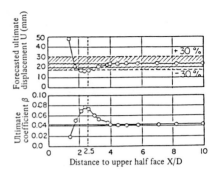

Fig. 6   Behavior of the error of forecasted displacement and ultimate coefficient as distance to face increases

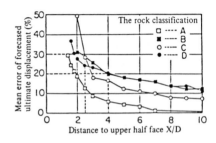

Fig. 7   Variation of the mean error of forecasted displacement at each distance to upper face depending on the rock classification

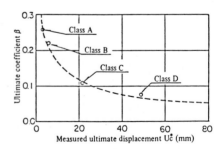

Fig. 9   Relation between the mean measured ultimate displacement and the mean ultimate coefficient depending on the rock classification

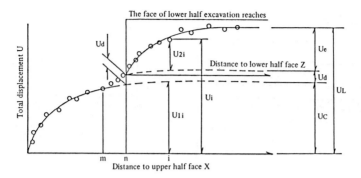

Fig. 10 Schematic diagram of deformation in each excavation stage

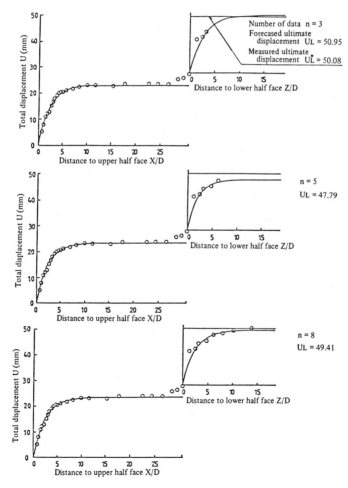

Fig. 11 Example of forecasting displacement as distance to face (the number of data) increases after excavation of lower half face

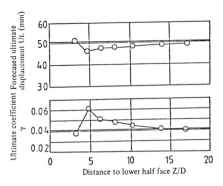

Fig. 12 Behavior of the error of
forecasted displacement and
ultimate coefficient as
distance to lower face
increases

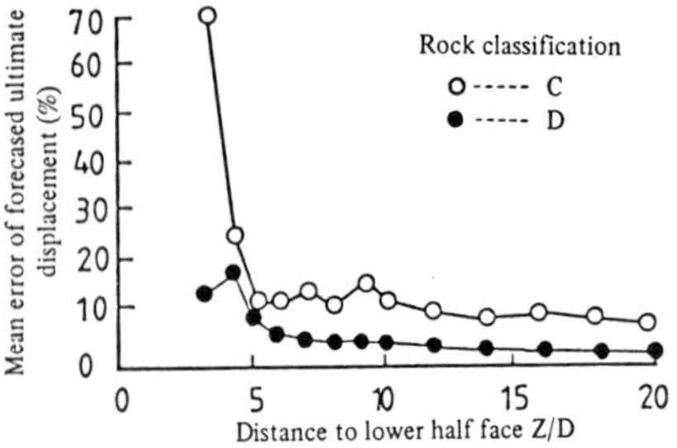

Fig. 13 Variation of the mean
error of forecasted
displacement at each
distance to lower half
face depending on the rock
classification

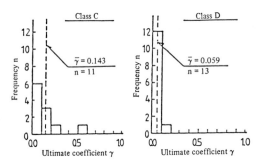

Fig. 14 Histogram of ultimate coefficient for lower half displacement
depending on the rock classification

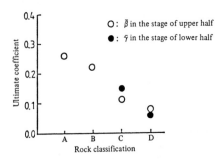

Fig. 15 Mean ultimate coefficient of upper and lower stage depending on
the rock classification

2nd International Symposium on Field Measurements in Geomechanics, Sakurai (ed.)
© 1988 Balkema, Rotterdam. ISBN 90 6191 778 6

Countermeasures for ground displacement in earth pressure balance shield tunnelling

W.Miyazaki
Transportation Bureau of Tokyo Metropolitan Government, Japan
T.Hatakeyama & M.Komori
Okumura Corporation, Japan

1 PREFACE

Since tunnel work in an urban area is frequently executed near existing structures, strict conditions are imposed on site acquisition,
traffic movement and the construction environment. Furthermore, in
several cases the ground in an urban area coincides with soft alluvial deposits. Thus, under these construction conditions, the shield
tunnelling method is beginning to be adopted because it has minimal
influence on adjacent structures, is good for preserving the environment, and does not require a large work area.

As tunnelling techniques advance, the adoption of the slurry shield
method and earth pressure balance shield method is growing, because
these shield tunnel methods have less influence on peripheral ground
and adjacent structures as compared to the conventional pneumatic
shield method.

However it is difficult for the current technique to completely
avoid displacement in peripheral ground, since settlement due to consolidation caused by the disturbance of the soft clayey ground occurs
and the ground deformation is therefore inevitable.

This paper presents measurement results on operational control of
an earth pressure balance shield in soft clayey ground in the City
of Tokyo and deformation of the peripheral ground.

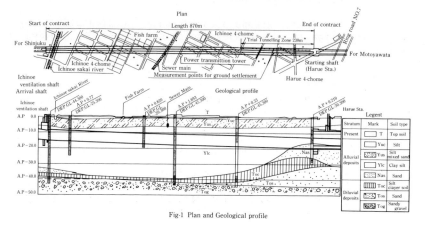

Fig-1 Plan and Geological profile

2 OUTLINE OF WORKS

Name of project   : Tokyo subway Shinjuku line (No.10 line)
Construction period   : From May 1982 to December 1984
Total length of tunnels : 870m x 2 lines (Two single track subway)
Distance between tunnel center lines   : 11.0m to 14.6m
Shield machine   : 2 earth pressure balance shields
                             Outer diameter : 7450mm
                             Machine length : 6900mm

In this work, two single track tunnels were constructed entirely under private land. Overburden of the tunnels was settled at a depth of between 9m and 12m. Houses were densely located on the ground under which the shield passed (Fig-1).

3 GEOLOGICAL PROFILE

The construction site is located in the delta zone of Tokyo lowland which spreads between the river-mouths of the Arakawa River and the Edogawa River. The ground in this section consists of a gravel layer and silt layer of diluvial deposits, situated at a depth of 40m or more, and clayey silt layer, sand layer and top soil of alluvial deposits. The majority of ground through which the shield passed was a clayey silt layer in alluvial deposits.
The properties of this clayey silt layer are as follows; Clay silt content is 96%, N-value is very low (0 to 1), moisture content is similar to the liquid limit value (60% to 80%), and sensitivity ratio is very high (Fig-1, Fig-2).

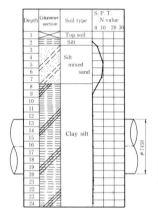

Fig-2 Geologic columner section

4 THE CAUSE AND PRE-COUNTERMEASURES OF GROUND DEFORMATION

4.1 The cause of ground deformation

The causes of peripheral ground deformation associated with shield tunnel construction in soft clayey ground can roughly be divided into the following three points.
    1. Ground deformation in the case where the taking in of soil at the shield facing is not properly executed.
    In general, the shield facing can be held by balancing pressure in the chamber (earth pressure or slurry pressure) against the earth's pressure at the cutter face (shown in Fig-3). If this balance is lost, it means that too much or too little soil is taken in, and an upheaval or settlement occurs when the shield face passes.
    2. Ground deformation when backfill grouting is not properly executed.
    Tail void, which is defined as the difference between the outer diameter of the shield machine and that of the segment, should be immediately filled with backfilling material, since ground deformation occurs especially easily in soft ground.

If materials of backfilling, grouting pressure and volume are not adequate, ground settlement or ground upheaval occurs when the shield machine tail passes.

3. Settlement due to consolidation associated with the disturbance of the ground.

It is supposed to be diffucult for the current technique to completely eliminate peripheral ground disturbance associated with the shield tunnelling. Therefore, after the shield machine passes in the soft clayer ground, settlement due to consolidation occurs because of the disturbance of soil.

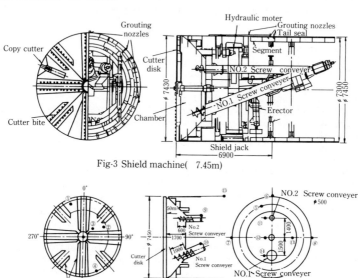

Fig-3 Shield machine(7.45m)

Earth pressure gauge NO	①~④	⑤	⑪~⑭	⑥~⑧	⑨~⑩	⑮
Earth pressure gauge position	Cutter disk	Chamber (rotating)	Chamber (fixity)	Circum ference	Screw conveyer	tail

Fig-4 Pressure gauge arrangement

4.2 Adoption of earth pressure balance shield and pre-countermeasures against ground deformation

In Fig-3, the earth pressure balance shield machine used in this section is shown. As a rotating cutter disk is mounted at the front of the shield machine, and has slit openings, the muck excavated by cutter bits can be taken through the openings of this cutter disk. There is a chamber to store the muck behind the cutter disk, and two screw conveyors by which the proper quantity of muck can be removed. The stability of the earth pressure at the cutter face is held by adjusting the rotation speed of the screw conveyors which control the earth pressure within the chamber.

1. Pre-countermeasures against ground deformation over the shield face.

In this section, the following countermeasures were done to minimize ground deformation when the shield cutter face passed.

i . Two screw conveyors are mounted on the upper and lower parts of the shield machine to control the earth pressure within the chamber and to hold the muck discharge balance better within the chamber.

ii . The opening ratio at the periphery of the cutter face is similar to that at the central part to take in a uniform soil quantity from the total cutter face (opening ratio = 27%).

iii . Earth pressure, earth pressure distribution and the earth pressure ratio between that at the cutter face and that inside the chamber must be constantly measured by earth pressure gauges mounted on the cutter face and inside the chamber.

2. Pre-countermeasures against the ground deformation when the tail passed.

i . Formerly backfill grouting has been usually carried out through the grout hole of the segment. Instead of this, however, backfill grouting will be done through 5 grouting nozzles installed on a tail skin plate of the shield machine at the same time as tunnelling.

ii . Quick hardening mortar of the two-liquid mixture (gell time: 30 seconds) was used as the grouting material.

iii . Backfill grouting is controlled by grouting pressure, which is monitored by the earth pressure gauge near the grouting pipe at the tunnel crown.

5 MEASUREMENT AND TRIAL TUNNELLING

It is difficult for the current technique to determine the tunnelling control values (facing pressure, backfill grouting pressure) beforehand, that minimize ground deformation in shield tunnelling. This is caused by the variety of ground conditions and the uncertainty of the behaviour of the peripheral ground associated with earth pressure balance shield tunnelling with a large diameter, which has been conducted a few times.

Furthermore, each ground displacements in tunnelling progresses in the soft clayey ground (displacement when the cutter face passes, displacement by backfill grouting, settlement due to consolidation) and is difficult to quantitatively grasp.

From the above-mentioned viewpoints, trial tunnellings of 230m in length were carried out. Based on measurement results of trial tunnelling, the optimum control values for shield tunnelling were determined.

The measurement system consists of a micro-computer which collects and processes on shield machine operation and ground displacement simultaneously, as shown in Fig-5.

Based upon the soil characteristics, the vertical displacement was supposed to be larger than the horizontal displacement. Then, only vertical ground displacement was measured at the 16 measurement cross sections, as shown in Fig-6. The main instruments on the shield machine are shown in Fig-4.

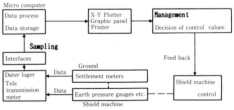

Fig-5 Measurement system

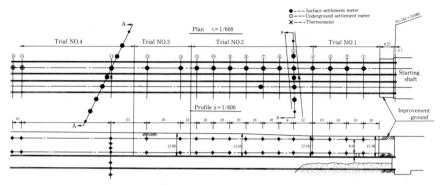

Fig-6 Settlement meter arrangement

Table-1 Shieid tunneling data

					Trial	NO.1							Trial		NO.2			Trial NO.3	Trial NO.4	General part
Case NO.	1	2	3	4	5	6	7	8	9	10	11	1	2--1	2--2	3--1	3--2	4	1	2	
Earth pressure at cutter disk (kg/cm²)	3.0	3.3	2.8	3.6	3.0	2.5	3.0	3.3	2.8	2.8	3.3	2.8			2.6			2.6	2.6	2.6
Earth pressure in chamber (kg/cm²)	2.0	2.3	1.8	2.6	2.0	1.5	2.0	2.3	1.8	1.8	2.3	1.8			1.6			1.6	1.6	1.6
Muck discharge ratio of screw conveyer (NO.1 : NO.2)		1 : 0			1:0.5	1:0		1 : 1	1:0.5	1:1	1:0.5	1:0.5		1 : 0			1:0.5	1 : 0.5	1 : 0.5	1 : 0.5
Cutter disk rotation speed (kg/cm²)						0.5								0.5				0.3 0.7	0.5	0.5
Backfill grout	Type (1-liquid or 2-liquid)				2							1	2	1		2	2	2	2	2
	Percentage (%)				1 5 0							1 5 0	220	200	180	180		150	1 5 0	1 8 0
	Pressure (kg/cm²)				1 . 7							1 . 4	2.2	2.0	1.7	2.0		1 . 7	1 . 7	2 . 0

6 MEASUREMENT RESULTS FROM TRIAL TUNNELLINGS

Trial tunnellings were done in 4 successive sections to determine the tunnelling control values.

Results of the shield machine operation for each trial and the general part are shown in Table-1.

6.1 Results from the first trial

Ground displacement from the first trial is shown in Fig-7 and Fig-8. Regarding ground deformation caused by the shield facing, pre-settlement occurs when the cutter pressure is 2.5kgf/cm² while pre-upheaval occurs when the cutter pressure is 3.0kgf/cm². According to there results, an appropriate earth pressure exists between 2.5kgf/cm² and 3.0kgf/cm².

Fig-9 shows the earth pressure distribution corresponding to several cutter pressures. In the case of the earth pressure

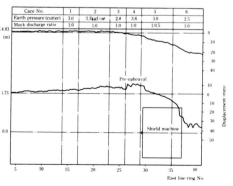

Fig-7 Ground displacement (measurement section No.2)

distribution of the cutter face when the cutter pressure is more than 3.0 kgf/cm², the earth pressure is larger in the central part and smaller at the periphery. On the cutter face when the cutter pressure is less than 2.8kgf-/cm², the earth pressure almost shows a distribution of earth pressure at rest (coefficient of earth pressure at rest 1.0). From these results, an earth pressure of less than 3.0kgf/cm² is favorable as a control value.

The relationship between thrust and the earth pressure, earth pressure at the cutter face and within the chamber was verified that there exists correlation by thrust; in addition to this, there is approximately 1.0kgf/cm² difference between the earth pressure at the cutter face and the earth pressure in the chamber.

There is a variant in the cutter earth pressure distribution caused by the difference in the muck discharge ratio by the screw conveyor under the same set earth pressure. As for the earth pressure distribution corresponding to the muck discharge ratio of 1:1, a large dispersion exists when compared to other muck discharge ratios, and it is judged to be inadeguate.

6.2 Results of the second trial

Only two cutter earth pressures of 2.6kgf/cm² and 2.8kgf/cm², were taken into account, based on the results obtained from the first trial tunnelling. As shown in Fig-10 and Fig-11, there is not a great difference in ground displacement above the cutter face. As pre-upheaval at 2.8kgf/cm² is slightly larger, the tunnelling should be done with a

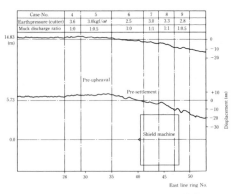

Fig-8 Ground displacement (measurement section No.3)

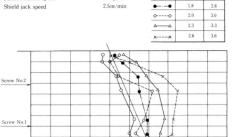

Fig-9 Distribution of earth pressure by decided earth pressure

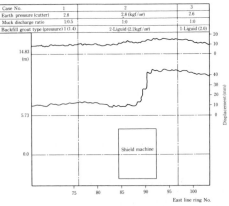

Fig-10 Ground displacement (measurement section No.7)

cutter pressure of 2.6 kgf/cm² and an earth pressure in the chamber of 1.6kgf/cm² hereafter, in order to minimize ground disturbance.

When backfill grouting is executed with one kind of liquid, as shown in Fig-11, immediate settlement is observed after the tail has passed. Comparing the grouting material of the single mix variety (mortar only) to that of the two-liquid (quick hardening mortar), the initial strength of the former was less and the grouting effi ciency of the former was inferior to the latter.

When the backfill grouting pressure was set to be greater than 2.0kgf/cm², upheaval occurred at the ground surface, and when it was set to be less than 1.4kgf/cm², ground settlement occurred. Judging from these results, there exists a proper grouting pressure of between 1.4-kgf/cm² and 2.0kgf/cm².

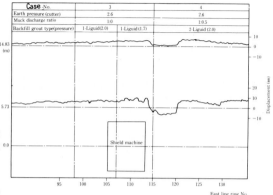

Fig-11 Ground displacement (measurement section No.9)

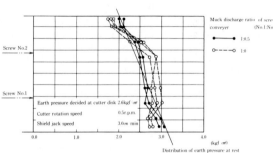

Fig-12 Distribution of earth pressure at cutter disk by muck discharge ratio

In Fig-12, earth pressure distribution, when the cutter earth pres sure was set constantly at 2.6kgf/cm² and the muck discharge ratio of two screw conveyors was 1:0 or 1:0.5, is shown. When the muck dis charge ratio 1:0 was set, earth pressure at the cutter face was larger in the central part and smaller at the periphery. On the otherhand, when the ratio was 1:0.5, the uniform distribution was observed and was similar to the earth pres sure at rest. From these results, the muck ratio between the lower and upper screws was set at 1:0.5.

6.3 Results of the third trial

Fig-13 shows ground dis placement in the third trial tunnelling. As little ground displace ment was observed until the cutter face passed, validity of the cutter pressure control value was verified.

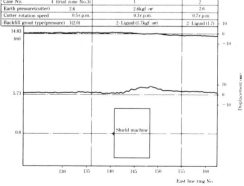

Fig-13 Ground displacement (measurement section No.12)

Only the backfill grouting
material of the two-liquid mix-
ture(quick hardening mortar)
were used. The pressure of
1.7kgf/cm² which is a medium
value of 1.4kgf/cm² and 2.0
kgf/cm² from results of the
second trial were judged to
be adequate and consequently
adopted as the backfill grout
ing pressure. As shown in
Fig-13, although upheaval of
approximately 10mm occurred
2m above the shield machine,
only a little displacement was
observed on the ground surface.

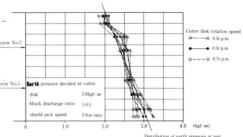

Fig-14 Distribution of earth pressure at cutter disk by cutter rotation speed

With a cutter pressure of 2.6kgf/cm² and a muck discharge ratio of
1:0.5, Fig-14 shows corresponding variation due to the cutter rotation speed (0.3, 0.5, 0.7 r.p.m.). Generally a large difference was
not observed in the earth pressure distributions which showed the
earth pressure nearly at rest, and the most uniform distribution was
observed for 0.5 r.p.m. Ground displacement related to cutter rotation speed was not observed. Therefore, the optimal cutter rotation
speed was determined to be 0.5 r.p.m.

6.4 Results of the fourth trial

The tunnelling control values judged to be optimal until the third
trial tunnelling were verified. The maximum ground upheaval above the
shield machine about four days after the shield machine had passed
was approximately 2 or 3mm.

When the backfill grouting pressure was set at 1.7kgf/cm², only a
little ground displacement was observed when the shield machine tail
passed. As shown in Fig-15, when the ground had been temporarily heaved due to backfilling with a pressure of 2.2kgf/cm², the observed settlement four months after tunnelling showed the minimum value. This
is due to the fact that absolute settlement from the original ground
decreases if the ground is heaved and the relative settlement from the
upheaved ground increases. When the ground is heaved, the influence on

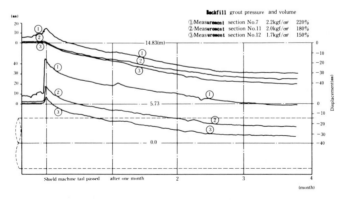

Fig-15 Ground settlement after shield machine passed

the peripheral structure caused by the upheaval is larger in compari
son with the settlement; a backfill grouting pressure of 2.0kgf/cm²
which caused little upheaval, is used as the control value hereafter.

7 ON THE MEASUREMENT RESULTS

In earth pressure balance shield tunnelling through soft clayey
ground, influence of the shield facing during excavation, and ground
displacement caused by the tailvoid were managed by determining the
tunnelling control values from trial tunnellings based on the ground
displacement measurements. However, it was difficult to completely
avoid post-settlement after the shield machine passed. After the
shield machine had passed, displacement at point F, shown in Fig-16
were as follows; +3mm, when it passed -15mm after one month, -30mm
after two months, -36mm after three and half months.

   Post-settlement is considered to be settlement due to consolidation
caused by the disturbance in the peripheral ground associated with the
shield machine passing. The following items are considered to be
causes of the disturbance in the peripheral ground.

   1. Pulsation of pressure does exist when the backfill grouting is
done evenly at a constant pressure. It is difficult to control the
small pressure variation.

   2. As earth pressure distribution on the cutter is complex, it's difficult to make it completely balance with the earth pressure at rest.

   3. There is a small variation in the shield tunnelling speed. This
speed variation causes a pressure variation at the cutter face.

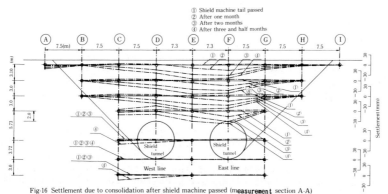

Fig-16 Settlement due to consolidation after shield machine passed (measurement section A-A)

7.1 Comparison with calculated value

According to Mori et alii, variations in the disturbed soil structure are due to the disturbance of the ground caused by the shearing
strain under the undrained shearing condition generated by the force
applied to the ground. Thus, effective stress in the ground decreases
and pore water pressure increases. The rise in pore water pressure
diffuses and settlement due to consolidation occurs, and the effective
stress then recovers. The settlement due to consolidation, which is
given in the following formula is in proportion to the shear strain.

$$\alpha = \frac{0.3\ C_c}{1+e_o}\ \log\{(0.33\ I_p{}^{-0.37}\cdot\log S_t)\gamma + 1\}$$

α : coefficient of volume compressivity C_c: compression index
I_p : plasticity index S_t: sensitivity ratio
e_0 : initial void ratio γ : shear strain

Incorporating the above, the calculation was executed according to the following FEM model.
Calculated settlements in comparison to the measured values are shown in Fig-17. Calculated with respect to the ground surface, settlements due to consolidation were 24 to 33mm while there were 2 to 8mm upheavals by the backfill grouting.

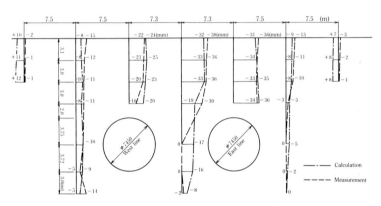

Fig-17 Displacement comparison calculation and measurement

8 CONCLUSION

In trial tunnellings using the earth pressure balance shield, the optimum control values based on measurement results from the shield machine operation and ground displacement are reported. Though it was impossible to completely avoid post-settlement after the shield machine had passed, shield tunnelling control values seem to have been found to minimize ground settlement and influence on the peripheral structures by restricting soil disturbance. Furthermore, the behaviour of the ground caused by earth pressure balance shield tunnelling is also clarified to some degree. In the future, there will be further advancement in the study of ground behaviour asscociated with shield tunnel works by collecting additional measurement data.

REFERENCES

Mori,A., Akagi,H.; The consolidation phenomena of soil due to disturb-nce caused by shearing deformations, Proc. The 16th Japan conference on soil mechanics and foundation engineering, 1981, pp. 225-228.
Miyazaki,W., Abe,K. and Hatakeyama,T.; The excavation control of earth pressure balance shield with a large diameter, Proc. Tunnels and Underground, Vol.15, NO.5, May, 1984, pp. 33-41.(Japan)
Mori,A., Akagi,H.; Relations between the injection volume of backfill materials and its sinkage prevention effects at shield work in soft cohesive soil, Proc. The 29th Japan symposium on soil mechanics and foundation engineering, 1984, pp. 35-40.

2nd International Symposium on Field Measurements in Geomechanics, Sakurai (ed.)
© 1988 Balkema, Rotterdam. ISBN 90 6191 778 6

Development of tunnel data bank system

Toshihiro Asakura
Railway Technical Research Institute, Tokyo, Japan

1 INTRODUCTION

It is said that tunnel engineering is the most typical kind of empirical
engineering. The reason is as follows. A tunnel is a peculiar structure
that is constructed in deep ground along a line. Therefore, it is difficult to know either the exact external force (that is, the initial
stress caused by the overburden and its release by excavation) at the
designing stage, or the properties of materials (that is, strength,
deformation characteristics, crack distribution etc. of the rock mass in
which the tunnel is constructed). And it is equally difficult to estimate the effect of each support member. It is necessary, therefore, to
take an empirical designing method for deciding the actual design, by
observing the geological condition at the faces and the support conditions at the construction stage. This above-mentioned situation will not
change, although the analytical designing method has come to be positively adopted for tunnel design.
  Since the concept of NATM was introduced from Europe to Japan about 10
years ago, results of research on rock mechanics have been widely
incorporated into tunnel design and execution. Furthermore, an increased
importance is being attached to construction control by measurement.
Consequently, measured results which are objective information can now
be applied to the actual design, which has been dependent on empirical
judgement in the past. But documents about old tunnel constructions
which should provide important information for new tunnel construction
works are liable to be scattered and lost. They are rarely preserved as
construction records etc., but usually the information they offer is not
sufficient to be practically useful for new constructions. For the
above-mentioned reason, a Tunnel Data Bank system (TDB for short) has
been developed for practical application in the future construction of
tunnels and in their maintenance. An outline of TDB is given as follows.

2 OUTLINE OF SYSTEM

2.1 Composition of the system

The computer employed to formulate the system is FACOM M360, which is
available at RTRI and is compatible with computers installed in each
construction division of JNR. FORTRAN77 is used as the program language
because calculation should be carried out in a statistical analysis
function. A conversational system using a TSS work station is adopted

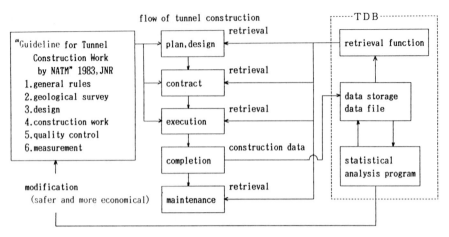

Fig. 1 Outline of TDB

as a reference function.   The user can obtain the results of retrieval
easily in order to feed the necessary information in to the computer
according to the instructions from the monitor at the work station.
    The composition of the system is shown in Fig. 2.   The data which can
be presented as digital information and character information are compiled by the computer, and diagrams are compiled as micro film.

2.2 Combination of programs

In the TDB system, all of the functions are modulized and can be maintained and improved easily.

2.3 In/Out-put data and retrieval items

(1) Input items
The data of tunnel construction is divided broadly into four categories
as follows(Fig. 3).   The system should be formulated such that this data
which has varied characters can be retrieved multilaterally.
    ① General data common to a tunnel
    ② Data common to a sector (omitted in the case of 1 sector for
       a tunnel)
    ③ Data on each sub-sector
    ④ Data on each cross section of tunnel
① and ② are fixed length data that require constant number cards.
③ and ④ are variable length data that require variable number cards
according to the construction work conditions.
    Data compiled by the computer as digital information and character
information is shown in Fig. 4.   The information is presented in 69 kinds
of cards.   Each card contains various attributes for each item.   For
example, the card concerning rock bolts contains subdivided information
such as rock bolt number, length, materials, arrangement, etc.   The number of all input items is about 600.
    TDB contains 74 sectors, 1713 cross sectional data of 65 railway
tunnels constructed in Japan.

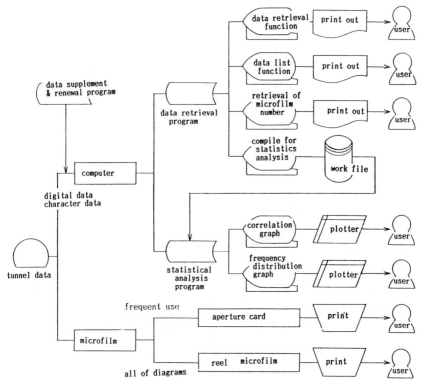

Fig. 2 Composition of TDB

(2) Retrieval items
41 items are selected as key words to retrieve required data. These
items are considered important for retrieving tunnel construction information. A user can retrieve the required data by logical sum (or) and
logical product (and).

(3) Output items

Output items are recompiled to 143 items from about 600 items of input
data. When a user gets output data, he can select any item as his option from among the 143 items.

3 FUNCTION OF THE SYSTEM

3.1 Retrieval functions

Retrieval functions of TDB are as follows:
① function to retrieve the data
② function to draw up the work file for statistical analysis
③ function to print out the data list
④ function to retrieve the micro film number
Above functions are carried out by selecting from the menu displayed on
the monitor at the work station.

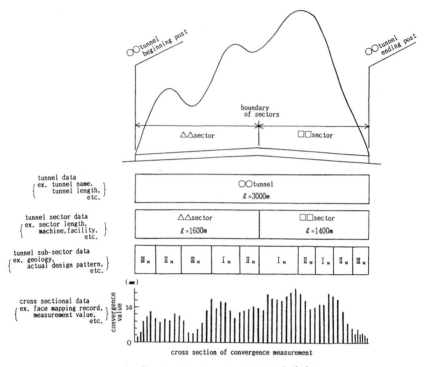

tunnel data
{ ex. tunnel name,
    tunnel length,
    etc. }

tunnel sector data
{ ex. sector length,
    machine, facility,
    etc. }

tunnel sub-sector data
{ ex. geology,
    actual design pattern,
    etc. }

cross sectional data
{ ex. face mapping record,
    measurement value,
    etc. }

Fig. 3 Structure of tunnel data

(1) Data retrieval function
A user selects a logical sum or a logical product and sector data or
cross sectional data to be employed before retrieving the data. When
he selects the retrieval by sector data, the required data is retrieved
by representative and typical data in the sector, consequently representative values in the sector are printed out as results of retrieval.
Data which is represented by numerical values is yielded as the maximum
value or the average value of the section or the cross section.
     An example of the monitor frame is illustrated in Fig. 5.   Retrieval
data represented by numerical values is selected by a combination of
their maximum and minimum values.
     When retrieval is finished for each item, the number of selected data
can be confirmed.   And if a user happens to make a mistake in his key
operation, he can turn to the initial frame of the retrieval function in
a moment.
     An example of a result of retrieval is shown in Fig. 6.

(2) Function to draw up the work file for statistical analysis
The function is to provide the data necessary for formulation of statistical diagrams.   In the same operation as the above-mentioned data
retrieval, data is selected to formulate a statistical diagram.

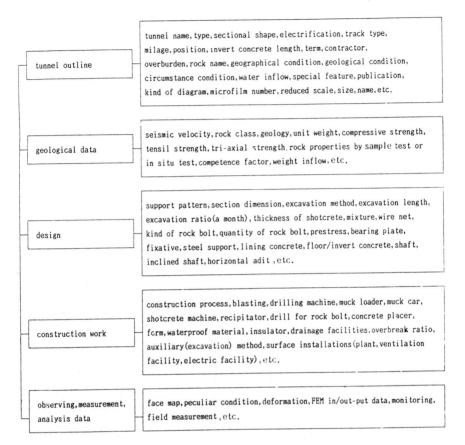

tunnel outline — tunnel name, type, sectional shape, electrification, track type, milage, position, invert concrete length, term, contractor, overburden, rock name, geographical condition, geological condition, circumstance condition, water inflow, special feature, publication, kind of diagram, microfilm number, reduced scale, size, name, etc.

geological data — seismic velocity, rock class, geology, unit weight, compressive strength, tensil strength, tri-axial strength, rock properties by sample test or in situ test, competence factor, weight inflow, etc.

design — support pattern, section dimension, excavation method, excavation length, excavation ratio(a month), thickness of shotcrete, mixture, wire net, kind of rock bolt, quantity of rock bolt, prestress, bearing plate, fixative, steel support, lining concrete, floor/invert concrete, shaft, inclined shaft, horizontal adit, etc.

construction work — construction process, blasting, drilling machine, muck loader, muck car, shotcrete machine, recipitator, drill for rock bolt, concrete placer, form, waterproof material, insulator, drainage facilities, overbreak ratio, auxiliary(excavation) method, surface installations(plant, ventilation facility, electric facility), etc.

observing, measurement, analysis data — face map, peculiar condition, deformation, FEM in/out-put data, monitoring, field measurement, etc.

Fig. 4 Contents of in/out-put data

(3) Function to print out the data list
This is a function to print out all of the data on every sector or every item. It is used to check the data continuity of the section, the data number, etc.

(4) Function to find out the micro film number
This is a function to find out the film number to demand the print outs of necessary diagrams. Four reference methods are available as follows:
① reference by tunnel sector
② reference by types of diagrams
③ reference by both tunnel sector and diagram types
④ reference by micro film number
Since reference key words for micro film numbers are tunnel sector name, types of diagram, and micro film number only, detailed reference such as data retrieval is impossible. In this case, a primary retrieval by the data retrieval function should come first.

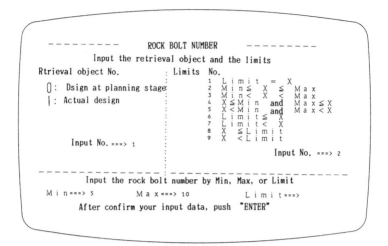

Fig. 5 An example of work station monitor

Tunnel name			Dishaka(south)	Narita(8)	Chokubetsu
Rout			Ohu main line	Narita Shinkansen	Nemuro main line
JNR/JRCC			JNR	JNR	JNR
Construction division			Morioka const. div.	Tokyo I const. div.	Kushiro rail. ope. div.
Type of cross section			Single track	Shinkansen double track	Single track
Shape of cross section			Standard	Special	Standard
Purpose of construction			Track addition	New line	Disaster prevention
Excavation direction			From exit to entrance	From exit to entrance	From entrance to exit
Tunnel locat.	Entrance (km)		468.590	62.990	261.985
	Exit (km)		470.830	65.000	262.285
Sector locat.	Starting point		468.590	63.145	261.985
	Ending point		469.700	63.400	262.285
Length by NATM (m)			810.00	216.00	240.00
Construction term	Contract		Oct. 1982	Nov. 1978	Jul. 1979
	Start		Nov. 1982	Jul. 1979	Aug. 1979
	Completion		Mar. 1984	Oct. 1981	Aug. 1981
Max. overburden (m)			28.90	8.00	25.00
Average overburden (m)			16.10	7.00	17.00
Geology	1	Rock type	E	F	F
		Rock name	Tuff	Clay Silt	Silt Clay
	2	Rock type	F	F	F
		Rock name	Sand Gravel	Sand	Silt Sand
	3	Rock type	F		F
		Rock name	Sand stone		Gravel Sand

Fig. 6 An example of printed out put

3.2 Function to draw up the statistical graph

A correlation graph and a frequency distribution graph are drawn up by the static induction plotter using the work file for the statistical analysis described in 3.1 (2). Only this function is carried out in the conversational remote batch process. This function has the required minimum function to search for data which can draw the graph of the data subdivided by the tunnel section type, rock type, and excavation method. The number of items to be analyzed statistically is 55. The scale of the graph can be either a log scale or a linear scale.

(1) Correlation graph
Two items are selected for two coordinate axes and it is the user's option to consider the correlation of them. Each datum is plotted by a code number of support pattern shown in "Guideline for Tunnel Construction Work by NATM" (1983, JNR). A regression line and confidential lines of 70% and 90% are given on the graph. An example of the correlation graph is shown in Fig. 7.

(2) Frequency distribution graph
A frequency distribution graph for each support pattern is drawn up together with the average value and the standard deviation. An example of the frequency distribution graph is given in Fig. 8.

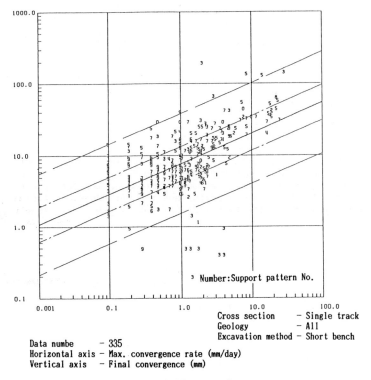

³ Number:Support pattern No.

Cross section   – Single track
Geology   – All
Excavation method – Short bench

Data numbe   – 335
Horizontal axis – Max. convergence rate (mm/day)
Vertical axis   – Final convergence (mm)

Fig. 7 Correlation graph

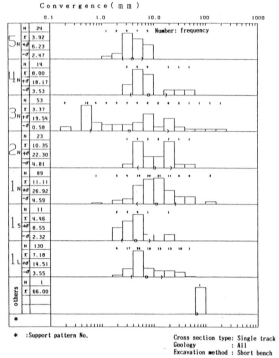

Fig. 8 Frequency distribution

4 SUBJECT FOR FUTURE STUDY

TDB has been developed for the internal service of JNR, forming a link in the chain of IR(Information Retrieval) system.   However, the system should be improved to meet the changing user demands.   For example, utilization of disks to preserve diagrams, and utilization of personal computers to apply the system at the site must be promoted to complete the functions.   Furthermore, a support system for the interpretation of reference results by tunnel experts, and the introduction of knowledge engineering (expert system by artificial intelligence) is also being considered.

REFERENCES

Asakura, T. & Onoda, S. 1987. Development of tunnel data bank system, Railway Tecnical Research Report, No. 1335 (in Japanese)
Kawata, H. & Asakura, T. 1986. Data base and expert system (Engineering system development in the National Railways), Tunnels and Underground, Vol. 17, No. 12 (in Japanse)

5 Waste disposal problems

2nd International Symposium on Field Measurements in Geomechanics, Sakurai (ed.)
© 1988 Balkema, Rotterdam. ISBN 90 6191 778 6

Progress of the Swedish radioactive waste management program

Hans S.Carlsson
Swedish Nuclear Fuel and Waste Management Co., SKB, Stockholm

1 INTRODUCTION

The nuclear power program in Sweden consists of 9 BWRs and 3 PWRs with a total capacity of 9650 MW electric power. These nuclear power plants generates approximately 50 % of the total Swedish electric power, the remainder being hydropower.

Based on an intense debate on the nuclear issue during the 70's and a referendum in 1980 the Parliament decided that no more reactors are to be built in Sweden and that the existing ones shall not be operated beyond the year 2010.

The primary responsibility for the management of the nuclear waste produced by the reactors lies with the nuclear power utilities. They have, in concurrence with the Swedish government, delegated this duty to the Swedish Nuclear Fuel and Waste Management Co., SKB, jointly owned by the four utilities that operate nuclear reactors in Sweden. Hence, SKB is responsible for all handling, transportation, temporary storage and final disposal of the spent fuel and radioactive waste from the power plants.

Furthermore, SKB is responsible for the planning, construction and operation of all facilities required for the management of spent fuel and radioactive waste and for the comprehensive research and development work necessary to provide such facilities.

The Swedish waste management program is described below. Special emphasis is given to instrumentation used for investigation of potential sites for final disposal of spent nuclear fuel.

2 THE SWEDISH WASTE MANAGEMENT SYSTEM

The basic strategy of the Swedish radioactive waste management system (Figure 1) is that short-lived waste (<500 years) should be finally disposed of as soon as feasible without interim storage, whereas for spent fuel and other long-lived wastes, an interim storage period of 30-40 years is foreseen prior to final direct disposal in crystalline rock formations.

Essential parts of the waste management system are already in operation or under construction. The central interim storage facility for spent nuclear fuel, CLAB, and the transportation system are both in operation. The final repository for reactor waste, SFR, is under construction and scheduled for operation in 1988. Extensive research

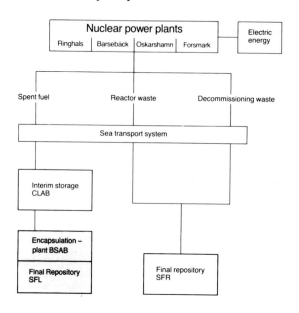

Figure 1.   The Swedish waste management
system

is being conducted for the design and location of the treatment plant
for spent fuel, BSAB, and the final repository for spent nuclear fuel,
SFL. According to present plans, these facilities will not be built
until after 2010.

2.1 Final repository for reactor waste

The waste to be disposed of in SFR is low- and intermediate-level and
short-lived, which means that it will have decayed to a harmless level
within a few hundred years.
    The entire repository is located in crystalline rock 50 metres below
the seabed and about 1 km from the shore at the Forsmark power plant
(Hedman et al 1986). The repository consists of rock caverns of
various designs depending on the type of waste to be disposed of
(Figure 2). The intermediate level waste, which contains most of the
activity, will be deposited in a silo-like concrete structure cast
inside a cylindrical rock cavern and the waste will be isolated from
the surrounding rock by concrete walls and a layer of clay backfill
(bentonite) between the silo and the rock. The low level waste will be
deposited without extra barriers in rock caverns designed for the
particular type of containers being used for such waste.

2.2 Interim storage for spent nuclear fuel

The spent fuel will be stored for 30-40 years in the central interim
storage facility, CLAB, (figure 3), located adjacent to the Oskarshamn

Figure 2. Final repository for reactor waste (SFR)

Figure 3. Transport flask in the fuel reception
building at CLAB

Figure 4. Transport vehicle with fuel transport
cask on its way out of M/S Sigyn's cargo hold

power station on the south-east coast of Sweden. During this period,
the fuel's activity content and residual heat will decline considerably. CLAB was taken into operation in 1985, thereby relieving
the pressure on the storage capacity for spent fuel in the power
plants.

CLAB consists of an above-ground receiving building and an underground storage complex in rock (Gustafsson et al 1986). The fuel is
handled and stored under water. The capacity of the facility is now
about 3000 tonnes of spent fuel in four pools but one or possibly two
expansions of the capacity are planned.

The CLAB facility constitutes a fundamental strategic element in the
Swedish spent fuel management scheme. It will ensure uninterrupted
nuclear power production and it will provide ample time for R&D work,
site selection, system design and optimization for the development of
a final repository for spent nuclear fuel.

2.3 Transport system for spent fuel and radioactive waste

All Swedish nuclear power stations, as well as the storage facilities,
are located along the coast. A sea transportation system has therefore
been developed. It consists of a ship, M/S Sigyn, transport containers, see figure 4, and terminal equipment. The transport containers meet the stringent requirements on radiation shielding and
ability to withstand external stresses that have been issued by the
International Atomic Energy Agency, IAEA.

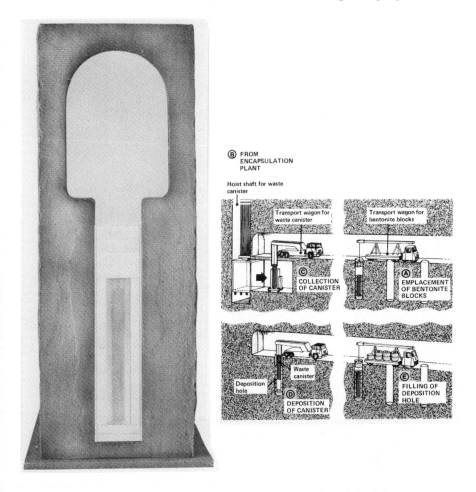

Figure 5. According to KBS-3, canisters and blocks of highly compacted bentonite are deposited in holes drilled in the floor of the
tunnel. The tunnel is then filled with a bentonite/sand mixture

2.4 Final repository for spent fuel

In Sweden, it is currently planned to store the waste without repro-
cessing. The so called KBS-3 concept for disposal in crystalline rock,
prepared by SKB, has been evaluated by the Government after a comprehensive international peer review. It was concluded that the KBS-3
concept was a method acceptable with regard to safety and radiation
protection.

The KBS-3 method involves depositing copper canisters containing
spent fuel in holes drilled in the floor of tunnels at a depth of
about 500 meters in a selected suitable rock formation. In the
boreholes, the canisters are embedded in compacted bentonite, see
figure 5. When deposition in a tunnel is concluded, the tunnel is
sealed with a sand/bentonite backfill.

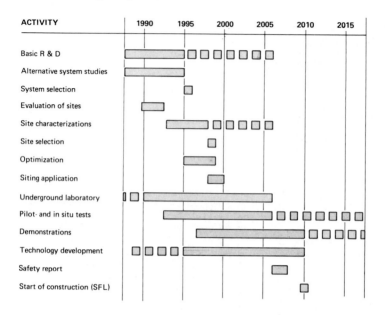

Figure 6. General timetable for realization of a final
repository for spent nuclear fuel

3 RESEARCH AND DEVELOPMENT

In the KBS-3 report, <u>one</u> feasible method has been demonstrated.
However, considerable work remains to be done in order to develop the
optimal method and to select the site for the final repository. The
future work comprises:

- Continued research and development work in order to further deepen
  the scientific knowledge that constitutes the base for the performance and safety assessment

- Studies and evaluation of alternatives to the methods and concepts
  investigated so far

- Opimization of systems in terms of technology, economy and resource
  utilization in view of the improved scientific knowledge base

- Investigation for site selection

   In order to further study the geological conditions at repository
depth and to develop and test investigation methods, an underground
rock research laboratory will be built. The possibilities to do this
close to the Oskarshamn power plant and CLAB, is currently being
investigated.
   In September 1986 a comprehensive plan for the future R&D work was
presented to the Swedish Government. The basis for the planning of the
R&D work is the overall timetable shown in figure 6. This timetable is

based on forty years interim storage of spent fuel in CLAB. An
application for licensing of a repository is foreseen around the year
2000.

Cooperation and exchange of information on an international or
bilateral basis is an integrated part of the R&D-activities of SKB.

As an example, an international research project is being executed
at Stripa, an abandoned iron mine in central Sweden, with the par-
ticipation of nine OECD countries. Different aspects of the geological
and engineered barrier systems are being investigated on a large scale
and in a realistic underground environment at Stripa.

4 SITE INVESTIGATIONS AND SELECTION

The long-term safety of a geological repository for high-level nuclear
waste depends on the isolating capacity of the barriers. The properties of the natural geological barrier are important not only for its
own function. Its chemical and hydrogeological characteristics also
affect the long-term stability of the canisters and backfill. The
availability of reliable methods and instruments for characterization
of geological formations down to great depths are therefore of crucial
importance.

The world's most extensive site investigations in crystalline rock
have been carried out in Sweden during the last ten years. Fourteen
sites have been investigated since 1977. On 8 of these a full investigation program has been performed from the surface and in boreholes
down to a maximum depth of 1000 m. In total, about 60000 m of cored
boreholes have been drilled and measured. The investigations follow a
standard program and are divided in four different phases as outlined
in figure 7. This work has resulted in a comprehensive data base and
broad experience.

The investigations have confirmed that the Swedish bedrock, dominated by granite and gneissic formations of old age, provides acceptable conditions for safe disposal at many locations. In the beginning
of the 1990's a couple of sites will be chosen for more detailed
studies, including a pilot shaft down to the foreseen repository
level. Final site selection will be made at the end of the 1990's.

Special downhole instruments have been developed for geophysical
logging and measurements in slim boreholes of rock stress, hydraulic
head and conductivity and chemical properties of the groundwater such
as pH, Eh and sulphur and oxygen content.

The instrumentation used and developed in the Swedish site
characterization program together with an experienced operating staff
and specialists for evaluation of results are available also to other
nations on a commercial basis.

5 INSTRUMENTATION USED FOR SITE CHARACTERIZATION

A complete set of methods and instrumentation for geological, geophysical, hydrological, hydrogeochemical and rock mechanical investigations are used within the site characterization program (Almén et al
1986). Many of the instruments used are of a standard type. However,
some of the information needed requires specially developed instrumentation. A short description of this unique instrumentation for
borehole investigations is given below.

STUDY-SITE INVESTIGATIONS ACCORDING
TO STANDARD PROGRAMME ARE DIVIDED
INTO FOUR PHASES.

PHASE 1 RECONNAISSANCE

A   Map and literature studies
   * Rock blocks bordered by fracture zones
     in area with flat topography

B   Field reconnaissance
   * Overall geological assessment, possibly
     supplemented by geophysical measure-
     ments

C   Reconnaissance drilling
   * Geological, geophysical and geohydro-
     logical investigations.

PHASE 2 SURFACE INVESTIGATIONS

A   Geological surface mapping

B   Geophysical measurements
   * Magnetic, electromagnetic, electrical,
     gravimetric and seismic methods

C   Evaluation

PHASE 3 SUBSURFACE INVESTIGATIONS

A   Percussion drilling
   * determination of fracture zones
   * registration of groundwater tables

B   Core drilling of deep boreholes
   * core mapping
   * geophysical logging by means of electrical,
     radiometric and acoustic methods
   * hydraulic investigations for calculation of
     hydraulic conductivity and groundwater head
   * hydrochemical investigations for analysis of
     chemical/physical composition of water

PHASE 4 EVALUATION

A   Descriptive models
   * geology, geohydrology and geochemistry

B   Numeric models
   * water transport and nuclide transport

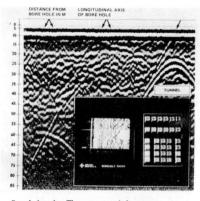

Bore hole radar. The arrows mark fractures.

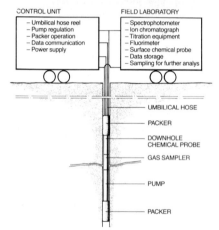

Water chemistry field laboratory.

Figure 7.   Standard program for study-site investigations
illustrated with examples of investigation methods

   Drilling of boreholes is an expensive activity. However, the costs
can be drastically lowered if the borehole diameter is reduced
resulting in more boreholes within the same budget. The experience
gained from the Swedish site selection studies have shown the importance of many boreholes within the investigated site. Therefore, all
instruments used within the Swedish site characterization program can
be fitted into holes with a diameter as small as 56 mm.

5.1 Borehole radar

A prototype radar system has been developed within the international
Stripa Project. It uses radio waves to determine the location, extent,
thickness and physical properties of fracture zones at large distances
from the borehole. Fracture zones with a width down to about 10 cm can

be detected at distances up to 150 m. This makes the radar a unique instrument for mapping structures in the rock.

A radar system further developed by SKB, called RAMAC, see figure 7, is now available for commercial measurements. This system is suitable for routine field measurements in minimum 56 mm boreholes and at maximum depths of 1000 m.

The RAMAC borehole radar system consists of six different parts:

- Microcomputer with two 5 1/4 inch floppy disc units for control of measurement, data storage, data presentation and signal analysis

- Winchester hard disc unit with 20 MByte data storage capacity

- Control unit for timing control, storage and stacking of a single radar measurement

- Borehole transmitter for sending out radar pulses in the rock

- Borehole unit for receiving and digitizing radar pulses

- Fiber optical borehole cable and winch

The borehole radar measurement procedure is simple. After selection of the variable parameters, the radar probes are inserted into the borehole and the measurement can start.

Measurements are normally carried out with a separation of measurement points of 1 m. The measurement at each position takes about 20-40 seconds depending on the number of samples and stacks. This corresponds to about 100 m of measured borehole per hour in one meter steps.

5.2 Hydraulic conductivity measurement system

Several methods are available to determine hydraulic conductivity in single boreholes. A comprehensive study in Sweden showed that a transient injection test gives the most relevant results and covers the largest conductivity range.

In the Swedish site selection program about 8000 m of cored boreholes are measured at each site. The great number of test sections calls for efficient and reliable equipment.

Specially developed equipment called the "Umbilical hose system" (the Umbilical) has been used in the Swedish program for several years and has proven to be a reliable system. The system is based on three mobile units; the measurement trailer, the recording trailer and the power supply. The interior of the measurement trailer is shown in figure 8.

The Umbilical is operated by a computer program. The operator starts the test sequence when the packers are lowered into position and inflated. The computer program starts the injection test by giving 200 kPa pressure and putting a downhole valve in an open position. One of two flowmeters for different ranges is selected automatically. After a chosen period of time the test section is closed and the pressure fall-off in the section is monitored. No manipulations are needed during the entire test. If the section has a conductivity below the

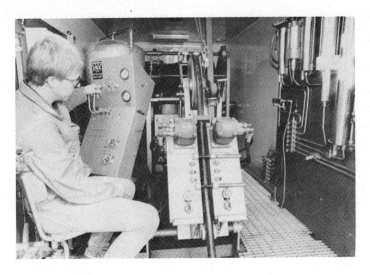

Figure 8. The feed-down device in the measuring
trailer of the Umbilical hose system

measurement limit, or if the fall-off is very rapid, the operator
interrupts the sequence in order to save time. Experience has shown
that the test sequence takes about five hours. This includes two hours
injection and two hours fall-off registration. After completion of a
test the packers are moved to the next position and a new test is
started. During this test, the previous test is plotted and an
incorrect test can thus be detected immediately.

The Umbilical is designed to operate in 56 mm boreholes down to
1000 m. Greater diameter holes can be measured by changing the
packers. Maximum borehole diameter tested so far is 165 mm.

The lower measurement limit for "production" measurements is set to
a hydraulic conductivity of 1×10^{-11} m/s in a 20 m packed-off
section.

The great advantages of the system are easy handling and automatic
control of the measurement sequence.

The Umbilical is normally operated by two persons.

5.3 Hydraulic head monitoring system

The hydraulic head monitoring system PIEZOMAC II, see figure 9, is
designed to measure the variation of hydraulic head in five different
sections in a 56 mm borehole. An optional design of the downhole units
enables up to 9 sections to be measured in boreholes with a diameter
of 76 mm. Additional packers may be used to arrange blind intervals
between the monitored sections. Interference from vertical water flow
in the borehole is thereby avoided and measurements in sections of
special interest can be focused. The PIEZOMAC II consists of the
following units:

- Multipressure probe
- Analogic unit

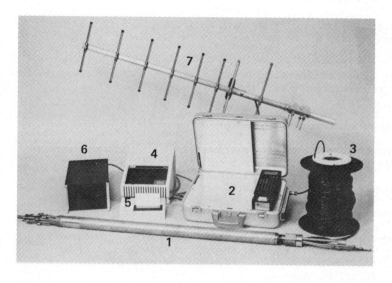

Figure 9. The components of the Piezomac II
system. 1. Multipressure probe, 2. Data collection and control unit, 3. Signal cable, 4. Data
recorder, 5. Matrix printer, 6. Radio, 7. Antenna.

- Data collection and control unit
- Data transfer unit
- Solar cell
- Hoisting device

The borehole is sectioned off by rubber packers equipped with a
feed-through for pressure tubes. The packers are connected to a pipe
string which is used for spacing the packers and also for lowering the
equipment into the borehole.

Each section is hydraulically connected to a valve in the multipressure probe. The valve opens one section at a time and the
pressure in the sections is recorded sequentially by a pressure
transducer. By using one transducer for all sections, errors due to
individual differences between transducers can be avoided. The
transducer signal is converted and transferred in serial form to the
data collection unit on the surface.

In the standard version, the data collection and control unit is
designed for simultaneous control and monitoring of up to eight
individual multipressure probes. The surface unit, including a
microcomputer, is battery operated and built for all weather conditions. The recording intervals can be programmed from one minute to
17 hours.

Communication via radio and a modem to the telephone network makes
remote operations possible, such as reprogramming of sampling times,
as well as transfer of measured data to a central computer. Printouts
can also be made in the field. The batteries of the surface unit may
be backed-up with solar cells, making longer field operations
possible.

A low weight, hydraulically operated hoisting device is used for lowering and lifting of the downhole units. The device is built on a trailer.

The hydraulic head measurement units can be used in field interference tests between boreholes. In this case up to 7 multipressure probes can be connected to each monitoring unit.

5.4 Hydrogeochemical monitoring system

The acquisition of reliable chemical data is of crucial importance in a site characterization study. Great care has to be taken in the groundwater sampling in order to minimize disturbances from the drilling operations. The traditional method of collecting samples in the field and then sending them to a central laboratory for analyses has the disadvantage that the samples may be contaminated or undergo chemical changes during handling and transport. In order to avoid these disadvantages, a mobile field laboratory has been designed for use in the Swedish site investigation program. The laboratory is combined with downhole equipment for in situ recording of the most sensitive chemical parameters.

The system for groundwater sampling and chemical analysis consists of three components:

- Downhole unit containing packers, pump and chemical probe
- Surface chemical probe
- Laboratory for chemical analysis

A schematic drawing of the field laboratory is shown in figure 7 whereas figure 10 shows an interior view.

The downhole unit can be connected in two different ways to the surface units. In the most advanced version an umbilical hose is used. An alternative is to use a pipe string with hoses and cables tied to the string.

Distinct water-bearing sections - single fractures or fracture zones - are chosen for groundwater sampling. The selected section is sealed off by rubber packers at each end of the downhole unit. Forced by the hydraulically operated piston pump between the packers, water passes the chemical probe, where the in situ pH, Eh and sulphide concentration values are registered and transferred to the surface in digital form. Downhole electrodes are calibrated before lowering the probe into the borehole and after completion of the measurement.

A stainless steel vacuum gas sampler is mounted above the electrodes. When the electrode signals have become stable the gas sampler is moved hydraulically towards a needle, which penetrates a rubber membrane of the sampler. The sampler is filled with water of ambient pressure and sealed again through withdrawal of the needle. Thus the collected groundwater can also be analyzed for the content of different gases.

When the groundwater has passed the downhole unit it is pumped to the chemical probe on the surface. This cell is equipped with the same electrodes as the downhole probe and the same kind of recordings are made as a check. Additional recordings are made by a dissolved oxygen and conductivity sensor.

The surface electrodes can be calibrated at any time. As the stabilization of the electrodes takes 5-7 days, calibration is normally done only one or two times during a test period.

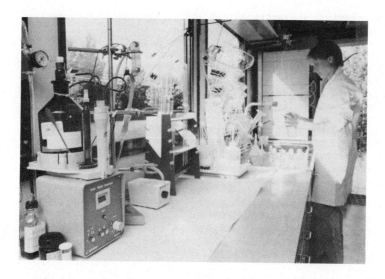

Figure 10. Interior of the field laboratory

The field laboratory for hydrochemical investigations has the following advantages:

- Stable and reliable readings of in situ values of pH and Eh

- Iron (Fe^{+2}/Fe^{+3}) analyses are improved, as oxidation during transport to the laboratory is avoided

- Fast and continuous information is obtained on the chemistry of the sampled groundwater and the tracer content of the drilling fluid Contamination and irregularities are detected at an early stage, and time-consuming and useless sampling of contaminated water can be avoided

- The field laboratory allows for certain chemical experiments in the field, e.g. enrichment of natural organic substances from large volumes of groundwater has been tested

From the extensive field measurements performed within the Swedish site investigation program a comprehensive base of data regarding properties of deep crystalline rock and groundwater has been built up. With the aid of a purpose-designed computerized system for data storage and processing, the researchers of SKB are given easy access to basic information needed in their further work.

REFERENCES

Almén, K-E., O. Andersson, B. Fridh, B-E. Johansson, M. Sehlstedt, E. Gustafsson, K. Hansson, G. Nilsson, K. Axelson & P. Wikberg. Site investigation - Equipment for geological, geophysical, hydrogeologi-

cal and hydrochemical characterization. 1986. Swedish Nuclear Fuel
and Waste Management Co. TR 86-16. Stockholm.

Gustafsson, B., O. Jansson & G. Chevrier. The realization of the
central interim storage facility for spent nuclear fuel, CLAB, in
Sweden. June 1-6, 1986. European Nuclear Conference, ENC'86, Geneva.

Hedman T. & I. Aronsson. The Swedish final repository for reactor
waste (SFR). March 3-7, 1986. International Symposium on the siting,
design and construction of underground repositories for radioactive
wastes. Hannover.

Swedish Nuclear Fuel and Waste Management Co., SKB, 1986.   Handling
and final disposal of nuclear waste - Programme for research
development and other measures. SKB. R&D-Programme 86. Stockholm.

Swedish Nuclear Fuel and Waste Management Co., SKB, 1986. Annual
report 1985. SKB TR 85-20. Stockholm.

2nd International Symposium on Field Measurements in Geomechanics, Sakurai (ed.)
© 1988 Balkema, Rotterdam. ISBN 90 6191 778 6

In situ testing programme related to the mechanical behaviour of clay at depth*

B.Neerdael, D.De Bruyn & M.Voet
Centre d'Etudes de l'Energie Nucléaire (CEN/SCK), Mol, Belgium
R.André-Jehan
Agence Nationale pour la gestion des Déchets RAdioactifs (ANDRA), Paris, France
M.Bouilleau & J.F.Ouvry
Bureau de Recherches Géologiques et Minières (BRGM), Orléans, France
G.Rousset
Laboratoire de Mécanique des Solides, Ecole Polytechnique (LMS), Palaiseau, France

1 ABSTRACT

In the framework of the Hades project (High Activity Disposal Experimental Site) managed by the Belgian Nuclear Research Establishment (CEN/SCK), an in situ research programme in an underground laboratory has been launched in 1980. This facility is built in a deep tertiary clay formation underlying the nuclear site of Mol-Dessel. Various issues related to the characterization of the host rock and to the technological capabilities have been and are being studied.

After the first investigations made in frozen clay during construction works and already reported (Neerdael & al, 1983), efforts were devoted to the behaviour of non frozen clay by digging and lining an experimental shaft and drift, two meters in external diameter, both being instrumented as well as the surrounding clay for long term measurements.

For investigating more accurately the characteristics of "virgin" clay in view of defining tunneling techniques and equipment as well as waste repository concepts, appropriate long term tests at smaller scale are performed around the completed underground facility.

An overview is given of the most relevant field data already obtained allowing comparison and calibration of numerical models.

In the context of the construction of a longer test drift, part of a future pilot facility, a geotechnical investigation programme is also foreseen.

2 INTRODUCTION

The depths generally considered for the construction of disposal facilities in clay and the relatively weak mechanical resistance of these materials even when consolidated, turns the geomechanical aspect into one of the key elements when studying the technical feasibility of such projects.

---

*Research performed in the framework of contracts with the Commission of the European Communities (Brussels), Niras/Ondraf, National Waste Management Authority (Brussels) and Andra (Paris)

This aspect was therefore included from the beginning in the research and development activities on disposal of radioactive waste in argillaceous media set up by the CEN/SCK in the early seventies (De Beer & al, 1977).

The first five-year programme led to the construction in the tertiary Boom clay formation of an underground laboratory to collect data in situ, to confirm laboratory observations and to demonstrate the construction capabilities for underground structures in clay.

The ground freezing technique used to dig the access shaft through water-bearing sands was further applied to build the gallery in clay at 223 m depth by way of horizontal freezing pipes implanted from the bottom of the shaft (Funcken & al, 1983). But the important deformation rates and connected problems raised during digging in the frozen clay body and lining (cast-iron) led rapidly to the conclusion that the study of the behaviour of "virgin" clay at depth was first requested before considering any other ground conditioning technique.

The construction of exploratory excavation works (shaft and gallery, 2 m in diameter, lined with concrete) in non frozen clay was directly started, in parallel with the instrumentation of both clay body and lining (figure 1). One of the most important characteristics of clay in this field being its time dependent behaviour, auscultation was developed in order to allow long term measurements (Manfroy & al, 1984 - De Bruyn & al, 1987).

These digging works provided also for the first time the opportunity for sampling undisturbed clay blocks much more representative than previous core samples for laboratory experiments which are now running in different institutions in Belgium and abroad. The LMS at Palaiseau is concerned for such study on request of the French Agency for Radio-active Waste Management (ANDRA) which contributed also extensively since mid 1983 to the geotechnical experimental in situ programme.

The results of these investigations were very encouraging and, during the last two years, several in situ tests were designed and performed together with ANDRA, LMS and BRGM from the underground excavations in order to calibrate mathematical models describing the rheological behaviour of deep clay (Rousset & al, 1985). One can mention :

- the determination of the in situ deformation moduli of clay (pressure-dilatometer tests),
- the convergence of the walls of unlined boreholes (calipers),
- the pressure build up and stress field in clay, (Gloetzl borehole cell cylinders) ;

as well as more specific test devices and/or testing procedures such as :

- the deformation of instrumented tubes in boreholes simulating at small scale, the lining of galleries, as a result of pressure build up of the clay body,
- long term retracting dilatometer tests as an application of the convergence-confinement theory.

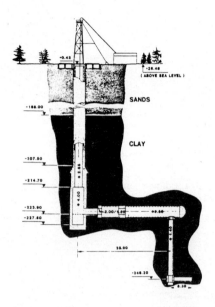

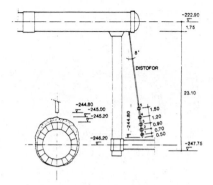

Figure 2 : Location of the
multi-points extensometer
(Distofor)

Figure 1 : Scheme of the as built
underground facility

The decision to extend the underground laboratory to a pilot size
demonstration facility has now been taken. In the first phase to be
started in February 1987, the construction capability in non frozen clay
will be upscaled to a diameter of around 5 m by making a 25 m long test
drift. It will have a 3.5 m useful diameter, be connected to the
presently existing shaft and be lined with 60 cm thick concrete
segments. Auscultation is now developed within the frame of a real
scale mine-by test as a further contribution to geomechanics in deep
clay.

3 SMALL DRIFT INSTRUMENTATION AND INTERPRETATION

The study of the in situ mechanical behaviour of deep clay was first
carried out as mentioned above by reduced scale exploratory works 2 m in
diameter and fully instrumented. In particular, the 8 m long experimental drift was equipped in a central section with convergence measuring
bolts and load cells in the lining as well as with total pressure cells
and a multi-points extensometer in the surrounding clay.

This last measuring system placed in a borehole from the underground
laboratory has allowed the monitoring of the in situ displacements in
clay consecutive to the digging works and for different distances
radially to this profile (see figure 2).

After more than 900 days, the deformation rate and stress evolution
are still observable ; the following interpretations are related to
convergence/confinement aspects including the estimation of the
stiffness of the lining as well as to the determination of the
plastified zone.

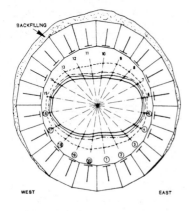

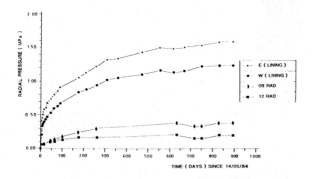

Figure 3 : Convergence diagram of measuring section 17

Figure 4 : Radial pressures around the small drift ; comparative measurements from load and pressure cells

Ten diameters are measured by means of studs in each of the 20 segments constitutive of a ring, using an invar wire system under tension (Telemac). At the end of October 86, the deformations illustrated by figure 3 reached values ranging from 10 mm to 30 mm, depending on the orientation. This ratio of 3 traduces a stress field anisotropy, the main stress being vertical ; however, it must be noted that all diameters are decreasing.

Similar ovalization is observed in the 0.5 m long unlined part of the drift close to the bare clay front except that the order of magnitude of the deformations at this location are eight times higher (80/240 mm) for the same time period.

The values recorded by pressure and load hydraulic cells (Gloetzl) in the same profile are not particularly in good agreement (figure 4). The load or stress measured within the lining has been converted to equivalent radial pressure to be compared on this graph to the pressure

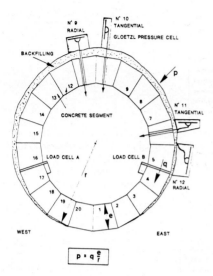

Figure 5 : Hydraulic cells position
for ring 17

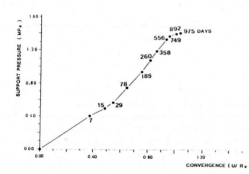

Figure 6 : Convergence/
confinement curve related to
the small drift ; time period
21/05/84 - 14/01/87

cells numbered 9 and 12. The discrepancy can be explained when
considering the emplacement procedures of each sensor (figure 5). The
earth pressure cells are embedded in clay in recesses made at a distance
of 20 or 30 cm from the lining or from the backfill (19 cm sand-based
material at the top) and sealed with remolded clay whereas the load
cells are inserted in the lining itself between truncated segments and
are of course more directly sensitive to stress evolution. For further
interpretation, a radial pressure of 1.5 MPa has been taken into
account.

Considering the following simplifying assumptions :

- hydrostatic stress field in clay,
- plane strain conditions,
- autoclamping type lining.

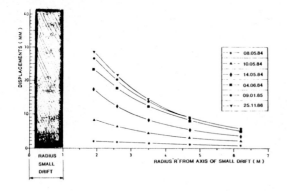

Figure 7 : Extensometric measurements in clay ; displacements versus radius

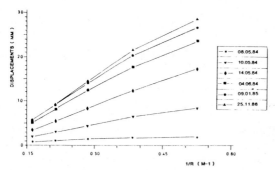

Figure 8 : Extensometric measurements in clay ; displacements versus reciprocal of the radius

it is possible to assume this problem to the convergence of a homogeneous lining under uniform confining pressure. The stiffness of the lining system can then be deduced from the convergence-confinement diagram given in figure 6 showing the ratio of the respective variation of the support pressure to the corresponding convergence. A stiffness of about 200 MPa can be estimated. It also worth to point out that the convergence of the lining is limited to 1,2 %.

The Distofor borehole multi-points extensometer (Telemac) has recorded deformations in clay until a few meters from the excavation wall which already gives information on the disturbed zone caused by digging at such a diameter. After 30 days, cumulative displacements of 23, 17, 12, 7 and 5 mm have been recorded respectively at distances of 0.9, 1.6, 2.5, 3.7 and 5.2 m from the excavated wall ; they have increased slowly with time to reach now, after 975 days, a 20 % higher value expressing the time dependent behaviour of clay. Deformations are now still perceivable at shorter distances (figure 7). A total convergence of 5 % at the excavated wall can be deduced.

By plotting these results in function of the reciprocal of the radius (figure 8), the graph indicates a first estimation of the extension of the plastic zone ; a radius of around 2.6 m (at the maximum) can be deduced. The analytical solution of such a problem using the elastoplastic theory brings similar results for both the deformation distribution and the plastified zone (see point 5).

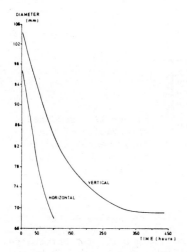

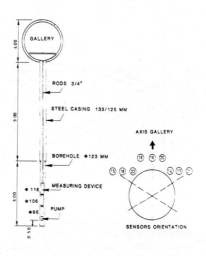

Figure 9 : diametral convergence
of the walls of a borehole

Figure 10 : emplacement of an
instrumented tube in a vertical
borehole

The "Caliper" tests performed from the experimental shaft were made in
one horizontal and one vertical borehole, 104 mm in diameter. Diametral
convergence for the unlined borehole walls of about 30 % within a few
days is shown on figure 9, pointing out once more the influence of the
stress distribution. Similar results were recorded for other tests
performed in 147 mm diameter boreholes.

4 IN SITU EXPERIMENTS IN BOREHOLES.

A measuring system to be used in situ for the simulation at a small
scale of the behaviour of a lining in deep clay has been studied and
designed for ANDRA by BRGM and LMS. It contributes to the experimental
testing of the interaction clay/lining as well as to the development of
the measuring techniques. The principle is based on the measurement of
the deformation of a tube placed in a borehole.

In practice and in a first phase including the emplacement of such a
device in horizontal and vertical directions, each apparatus is composed
of 3 successive steel tubes (E = 220.000 MPa), all of 3 mm thickness,
but with different diameters (96, 106 and 116 mm) allowing the follow up
of 3 confining times. Each 1 m long tube is equipped with 3 inductance
variation probes placed at 120° in a central zone to follow the
diametral displacement. Each system, previously calibrated at the
surface, has been placed between 12 and 15 m far from the gallery lining
at 0.5 m distance of the end of the borehole which is lined along 9
meters as illustrated by figure 10. As far as the vertical orientation
is concerned, previously gained experience (calipers) has shown the
necessity for relatively long term experiments (more than 3 months) to
provide below the device a pumping system to ensure the evacuation of
the pore water which may be accumulated with time.

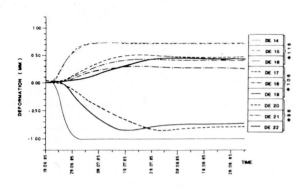

Figure 11 : recorded
deformation of the
instrumented tube in
case of the vertical
configuration

The diametral variations given by the 9 measuring points of the
apparatus placed vertically are given in figure 11 for the first 70
days. The time needed in this configuration (hole ϕ 125 mm) before the
collapsing clay comes into contact with the tube of ϕ 116 mm can be
estimated to 47 hours ; in the horizontal configuration, this delay is
reduced to 6 hours. The excessive stiffness of the tubes didn't allow
the determination of the isotropic component of the stress tensor.
Making several assumptions such as an ellipsoidal-shaped ovalization of
the measuring section with corresponding center points, the deviatoric
stress can be estimated to 0.3 or 0.4 MPa depending on the orientation.
These results are not to be considered as a representative in situ
estimation of the stress field at rest because of the probable
interference caused by the freezing technique previously applied for the
shaft construction.

However, to confirm the validity of the interpretation, it has been
decided to place 2 Gloetzl borehole cell cylinders from the same gallery
section and in the same configuration (distance, direction) to get
information on the stress field in clay at this location using orien-
tated hydraulic pressure cells in a borehole. The flat oil-filled
cells, assembled in different orientations have been enclosed in the
laboratory in a grout on a vibrating desk, repressurized and calibrated
in a pressure vessel ; the grout is made of cement and 6 to 7 %
bentonite in the form of a rod and is characterized by a deformation
modulus of 235/250 MPa slightly higher than the surrounding clay (hard
inclusion).

One borehole cell cylinder consists of :

  - 3 cells inclined at 0,45 and 90° with respect to the vertical
    axis ;
  - 1 cell for measuring stress in the borehole axis ;
  - 1 pore water pressure cell.

Maximal values of 1.75 MPa have been measured after 8 months. The
stress field indicates a slight anisotropy which corroborates the
interpretation already expressed.

In a second experimental phase, further investigations were developed
according to the principle of the instrumented tube with the purpose of

choosing a material with a modulus closer to the stiffness of the
existing small drift in order to obtain comparison of the mechanical
behaviour of clay at different scales around a lining of similar
characteristics. An extensive survey was made to select an homogeneous
elastic material with a low young's modulus (around 200 MPa),
characterized by a linear elasticity in a strain field of 2 or 3 % with
reversible deformations and a dimensional stability with time under
confinement. Other materials with higher young's modulus had finally to
be investigated. In spite of its great stiffness (74.000 MPa) an
aluminium alloy (AU 4G) was chosen. The convergence of this tube will
be recorded along 12 radii and the deformations at the intrados and
extrados can be monitored using strain gauges.

Other applications of the convergence-confinement theory to be started
early in 1987 will involve the simulation of sliding lining systems by
performing long term retracting dilatometer tests in boreholes before
going to large scale in situ testing.

On the other side, self-boring pressuremeter tests performed at
different distances from the excavation will quantify the stress field
anisotropy (coefficient of earth pressure at rest) and in situ moduli
for different configurations.

5 INTERPRETATION AND MODELLING

The research programme in geomechanics aims at modelling the
mechanical behaviour of deep clay when digging and lining galleries.
Taking into account the characteristics of the Boom clay at this depth
(water content, strength, consolidation, plasticity, ...) two approaches
can be envisaged :

- the first one using an elasto-visco-plastic (EVP) model with strain
  softening, will take place at LMS where the model has been
  developed in the frame of a collaboration agreement with ANDRA ;
- the second one, taking into account the responses to external
  solicitation of the clay skeleton as well as of the pore water will
  use one version of the "Cam Clay" model describing the time-
  dependant behaviour as a consequence of pore-water dissipation.

Some numerical results of the behaviour of the small experimental
drift allow the determination of the model's parameters. The clay
response when digging galleries of different diameters will afterwards
be investigated for both approaches to compare their respective
predictions with observed data.

In order to start from basic references for further comparison with
more sophisticated models, a small computer code, called Eplast, was
developed and is operational. It is based on the convergence-confinement principle theory with a perfect, isotropic elasto-plastic model.
The Eplast programme requires as input data the soil characteristics
(cohesion c, angle of shearing strength ϕ, Young's modulus E_s, poisson's
ratio ν_s, dilatancy α), the lining characteristics (internal R_i and
external R_e radii, Young's modulus E_l, poisson's ratio ν_l, uniform soil
convergence occuring before lining e_v) and the in situ initial stress
p_o. It calculates the equilibrium condition (p, u), the radius of the

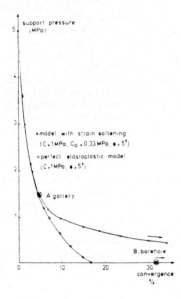

Figure 12 : Long term convergence - confinement curve for the Boom clay

plastic zone if existing and for different distances from the excavation, the corresponding displacements and stresses.

The following applications have been made :

1 Eplast

When applying the geometrical data of the small drift at 246 m with the respective parameters of clay and concrete, the numerical and experimental results are in good agreement for a e_v value of 0.05 m which is acceptable considering the construction procedure and a α value of 0.9985 expressing a contraction of the clay around the cavity probably due to its drying. By the same way, the equilibrium pressure reaches 1.75 MPa and the plastic radius 2.12 m ; both are acceptable values.

Extrapolating to the future test drift construction at 223 m, (R_e = 2.35 m, R_i = 1.75 m) the radius of the plastic zone can be predicted to 3.3 or 3.9 m, depending on the value of e_y (5 or 7 cm) ; the corresponding equilibrium pressures are respectively of 2.45 and 2.05 MPa.

2 EVP model

Applying this model to describe the behaviour of clay with the usual simplifying assumptions (e.g. cylindrical symmetry), computations at LMS have led to the long term convergence-confinement diagram of figure 12, together with the curve resulting from modelling. The figure shows the two points accounting for the long term behaviour of the gallery (A : p = 1 MPa, u = 5 %) and of the unlined borehole - caliper test - (B : p = 0, u = 30 %). For comparison, the convergence-confinement curve corresponding to a perfect elastoplastic behaviour has also been drawn (constant cohesion equal to 1 MPa).

6 CONCLUSION

Access to real in situ conditions by means of an underground laboratory is essential for the geomechanical studies in clay at depth. A good cooperation is further required between the user and the manufacturer of a measuring device to answer to the specific requirements.

As shown by the first computations and in spite of the simplifying assumptions taken into account, the situation observed experimentally can be fairly well reproduced. Comparison of stress and strain evolution using different measuring equipments and if possible at different scales provides a wealth of recorded data and evolues readily into a quite comprehensive in situ programme. This allows, together with clay characterization studies, the definition of tunnelling techniques and equipment as well as the design of waste repository concepts.

In this field, the auscultation planned in the clay and on the lining for the test drift has been made according to the experience gained with the small drift and refined to improve the interpretation by extensive cross-checking.

This mine-by test will mainly consider the follow up of :

- the stress field and pore water pressure evolution during and after construction (total pressure cells and piezometers of different types) ;
- the rate of loading (Gloetzl load cells) and the subsequent convergence of the lining (invar wire under tension) ;
- the deformations in clay at different distances of the excavation (single and multi-points extensometers, clinometer, deflectometer).

7 REFERENCES

De Beer, E. and al. 1977, Preliminary studies of an underground facility for nuclear waste burial in a tertiary clay formation. Rockstore 77, Volume 3 : 771-781.

De Bruyn, D. and al. 1987, Time-dependent behaviour of the Boom clay at great depth- An application to the construction of a waste disposal facility. Computers and Geotechnics, Special issue "Radioactive Waste Management", March 1987.

Funcken, R. and al. 1983, Construction of an underground laboratory in a deep clay formation. Eurotunnel Conference, 22/24 June 83, 79-86.

Manfroy, P. and al. 1984, Experience acquise à l'occasion de la réalisation d'une campagne géotechnique dans une argile profonde. CEC/NEA Workshop on the design and instrumentation of in situ experiments in underground laboratories for radioactive waste disposal, Brussels, May 1984.

Neerdael, B. and al. 1983, Field measurements during the construction of an underground laboratory in a deep clay formation. Int. Symposium FMGM, September, 5/8 1983, Volume 2 : 1419-1430.

Rousset, G. and al. 1985, Mechanical behaviour of galleries in deep
   clay- Study of an experimental case.   Int. Meeting on HLW disposal,
   Pasco, September 1985

2nd International Symposium on Field Measurements in Geomechanics, Sakurai (ed.)
© 1988 Balkema, Rotterdam. ISBN 90 6191 778 6

Geomechanical instrumentation applications at the Canadian underground research laboratory

P.M.Thompson & P.A.Lang
Atomic Energy of Canada Limited, Whiteshell Nuclear Research Establishment, Pinawa, Manitoba

ABSTRACT

A major geotechnical research facility, the Underground Research
Laboratory (URL), is being developed in southeastern Manitoba, Canada.
The URL provides an environment in which engineers and scientists can
conduct experiments to assess the concept of disposal of nuclear fuel
waste deep in stable plutonic rock. This paper outlines the
geomechanical field measurements that have taken place during the
development of the URL and describes experiences with the various
field instrumentation systems that have been used. The instruments
that have performed the best for measurements in the research program
are summarized.

1 INTRODUCTION

The Underground Research Laboratory (URL) is a research facility being
developed by Atomic Energy of Canada Limited (AECL) in the Lac du
Bonnet granite batholith, about 120 km northeast of Winnipeg, Manitoba
(Simmons, 1986). The facility will be used for in-situ experiments
designed to provide data required to assess the concept of nuclear
fuel waste disposal deep in stable plutonic rock of the Canadian
Shield.

Initial development work at the URL includes a 2.8 m x 4.9 m
rectangular shaft excavated to a depth of 255 m below surface with
shaft stations at the 130-m and 240-m depths. About 350 m of
horizontal access tunnels have been developed on the 240-m level. A
number of experimental activities associated with characterization and
excavation have taken place during the course of this development.
In-situ studies and measurements have involved the fields of geology,
hydrogeology, geochemistry, geophysics and geomechanics. This paper
focuses on geomechanical field work done to date at the URL. The
timetable for past geomechanical experimental activities at the URL
showing the work phase and type of measurements performed are
summarized in Figure 1.

The major phases in the development of the URL to date include:
· shaft sinking to 255 m,
· geotechnical characterization of the shaft and shaft stations,
· development of the 240-m level, and
· characterization of Fracture Zone #2 (a major sub-horizontal fault

about 10 m below the base of the present shaft) and preparations for shaft extension to 465 m.

Each of these phases will be discussed in separate sections of this paper, with emphasis placed on the instrumentation and methods used to measure the following parameters:

- excavation response rock displacements,
- excavation response convergence,
- excavation response stress changes, and
- in-situ stresses.

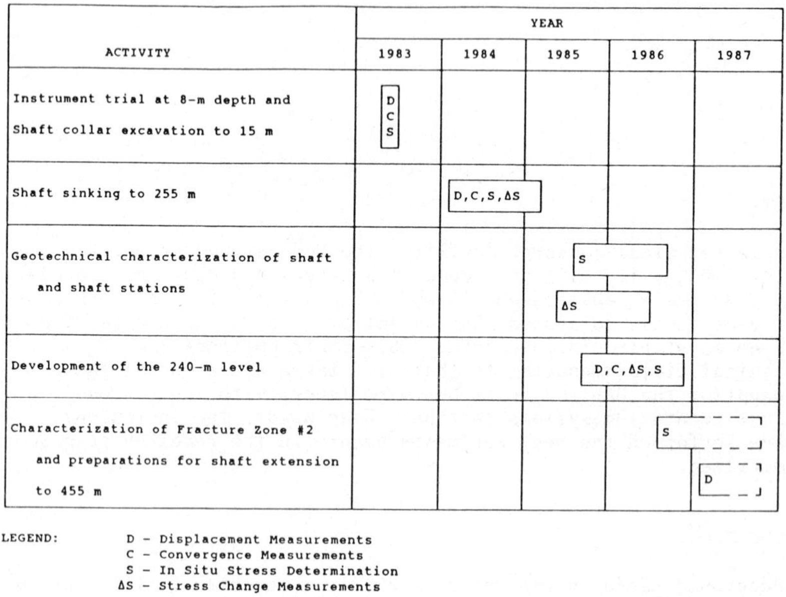

ACTIVITY	YEAR				
	1983	1984	1985	1986	1987
Instrument trial at 8-m depth and Shaft collar excavation to 15 m	D C S				
Shaft sinking to 255 m		D,C,S,ΔS			
Geotechnical characterization of shaft and shaft stations			S ΔS		
Development of the 240-m level				D,C,ΔS,S	
Characterization of Fracture Zone #2 and preparations for shaft extension to 455 m				S	D

LEGEND: D - Displacement Measurements
            C - Convergence Measurements
            S - In Situ Stress Determination
            ΔS - Stress Change Measurements

Figure 1   Geomechanical Measurement Timetable During URL Construction

2 SHAFT SINKING TO 255 m

During sinking of the URL shaft (1984 May to 1985 March), as part of the URL Construction Phase Experimental Program, measurements were made of rock stress changes, rock displacements, and shaft wall convergence in response to removal of shaft bottom support during excavation (Thompson et al. 1984). These measurements were conducted at four elevations during shaft sinking, and convergence alone was measured at eleven additional elevations in the shaft.

Construction was suspended at four separate locations during shaft sinking to drill and instrument arrays of boreholes. These instrumentation arrays were installed at depths of 15 m, 62 m, 185 m and 218 m (275 m, 228 m, 105 m and 72 m Above Mean Sea Level (AMSL), respectively) (Thompson et al. 1984). A vertical section of the URL shaft showing instrument array locations is shown in Figure 2. The instruments were monitored when shaft sinking resumed to measure the excavation-induced changes that occurred. A typical instrumentation array layout is shown in Figure 3. To measure rock displacements at each array four to six IRAD GAGE Sonic Probe multiple-point borehole extensometers (MPBXs) plus six convergence anchors were used. The MPBXs were installed in 15-m-long 60-mm-diameter boreholes. The

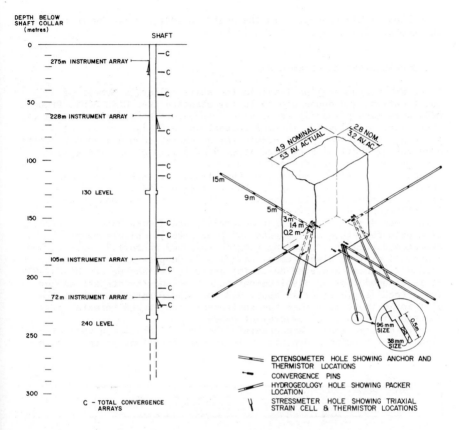

Figure 2   Instumentation Array
Locations

Figure 3   Typical URL Shaft
Instrumentation Array

convergence anchors were grouted into the rock surface at or near the
centre lines of the east and west walls and on either side of the
centre lines of the north and south walls.   Convergence measurements
were made with a Slope Indicator Company (SINCO) Model 51855 tape
extensometer.

Six Commonwealth Scientific and Industrial Research Organization
(CSIRO) hollow inclusion (HI) triaxial strain cells were installed at
each of the instrumentation arrays.   The cells were installed in near-
vertical boreholes 9 m below the shaft bottom and nominally 1.3 m and
2.6 m outside the projected shaft excavation line (see Figure 3).
Each triaxial strain cell measured strains in 9 directions, so enough
information was obtained from each cell to calculate the complete
strain tensor at the cell location.   We assumed an elastic modulus of
50 GPa (based on results from nearby overcore tests) and Poisson's
ratio of 0.20, to convert the strains to stress changes.

All of the instruments and cables were protected from blast damage
by 6-mm-thick steel plates.   To obtain information on the variation of
the rock mass deformation modulus with depth and varying rock
conditions, sets of six tape extensometer convergence pins were

installed at eleven depths in the shaft in addition to the four main
instrumentation arrays as shown in Figure 2.

2.1 Displacement Measurements

After an instrumentation trial in the shaft collar, a number of
modifications and enhancements to the standard IRAD GAGE Sonic Probe
MPBX were made primarily to provide temperature measurements along the
instrument to allow for thermal corrections and also to prevent
slippage of the instrument within the borehole. Nominal anchor depths
used during shaft sinking were 15 m, 9 m, 5 m, 3 m, and 1.4 m, with
the reference collar recessed by 0.2 m.

Quality control procedures and checklists for rock displacement
instrumentation used during URL shaft sinking, (Snider et al. in
prep.), were designed to provide quality control and consistency in
installation and monitoring.

The rock displacement measurements obtained during URL shaft sinking
have been described by Thompson and Lang (1986). Displacements
measured during shaft sinking ranged from 0.09 mm to 1.80 mm. Figure
4 shows a plot of displacement relative to deepest anchor (15 m)
versus distance excavated below the array for extensometer 105-EXT-E1.
It is assumed that no displacement occurs beyond the deepest anchor.
Total rock displacement is about 0.2 mm. The erratic nature of the
plot is an indication that the instrumentation lacked repeatability
and stability in this measurement range.

Where displacements were greater, the lack of precision was not a
significant problem. Overall, with the modifications to the system

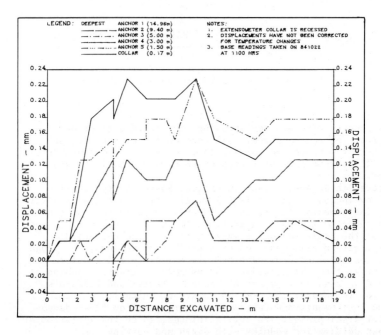

Figure 4   Displacement Versus Depth Excavated for MPBX 105-EXT-E1

made prior to use, the Sonic Probe MPBX system proved to be economical, reliable, and easy to install and monitor. However it did not have the precision needed to validate and calibrate numerical models in stiff granite. In addition the minimum distance between the reference collar and the closest anchor was about 1.2 m due to the length of the staggered magnet rods that are a necessary part of the measurement system. This placed a limit on the displacement information that could be obtained close to the excavation wall, the area of most interest.

From the results of these tests it was decided that, for subsequent displacement measurements at the URL, it would be beneficial to install more measuring points along a borehole to better resolve the rock movements due to joint dilation, particularly in the rock close to the excavation wall. It was also decided that data logging of instrument response would be preferable to manual measurements because it would increase the data quality and reduce the construction delays and associated costs that result from manual readings.

2.2 Convergence Measurements

The SINCO Model 51855 tape extensometer employs a proving ring to accurately set tape tension. The resolution of the instrument is 0.01 mm, and the manufacturer's specified accuracy is 0.08 mm. Removable spherical reference anchors were used during URL shaft sinking.

A number of procedures were introduced to minimize the likelihood of inaccurate measurements:

1. Each reference ball was marked so that it was used in the same place each time.

2. The instrument was allowed to stabilize to ambient air temperature for ten to fifteen minutes prior to taking measurements.

3. Air temperature was measured before and after each set of readings. Thermal corrections were made for each measurement.

4. Double-averaged readings were taken across each pair of convergence pins, with the instrument reversed for each set of readings.

5. Laboratory calibration of the tape extensometer was done before and after each set of readings.

Convergences measured during shaft sinking varied from 0.5 to 2.5 mm (Thompson and Lang 1986). Overall, despite the careful procedures used to make the convergence measurements, the SINCO tape extensometer did not provide high quality data. It was decided to procure more accurate, temperature-insensitive instrumentation for subsequent convergence measurements at the URL. These are described in section 4.2.2.

2.3 Stress Change Measurements

The CSIRO HI cell consists of a molded epoxy cylinder into which 3 strain gauge rosettes are embedded (Worotnicki and Walton, 1976). The rosettes are spaced at 120° around the circumference. Each rosette contains three 120-ohm strain gauges 10 mm long and at 45° to each other relative to the drill hole axis. The HI cell is installed by bonding the cylinder inside an EWG borehole (37.7-mm dia.) with epoxy. The epoxy must cure for 12 to 16 hours before the cell is considered properly installed. The electrical signal cable from the cell is sealed at the cell body and once installed and cured, the device is considered waterproof.

The HI cell can be used to determine in-situ stress changes by monitoring the various strains on the borehole wall measured by the strain gauges. These strains can then be converted to stress changes once the elastic properties of the rock have been determined by laboratory testing. Six independent strain measurements are necessary to provide the complete three dimensional stress tensor. The HI cell provides eight independent values (two are parallel), therefore there is some redundancy. In order to monitor temperature changes which could influence the stress change measurements, each HI cell contained an integral thermistor.

A gas-lift system was developed to dewater the 9-m-deep, near-vertical holes during HI cell installations. A mechanical packer sealed between PVC casing and the borehole wall at the base of the HQ hole. Compressed nitrogen was used to force the borehole water down the casing and up to surface via a centrally located steel water pipe. A check valve at the base of this central pipe prevented water from draining back inside the borehole. Final dewatering was accomplished using lint-free swabs on a cleaning tool.

As with the other instrumentation used during shaft sinking, installation procedures and checklists were closely followed to ensure a high level of quality control and to increase the percentage of successful installations.

To confirm that the HI cells had operated correctly, they were overcored after shaft sinking was complete and all stress changes due to elastic response had occurred. If the sum of the triaxial stresses determined during overcoring plus the triaxial stress changes measured due to shaft sinking equalled the virgin stress conditions as determined from in-situ stress measurements, then the cells were assumed to have operated accurately. Equally important, by overcoring these cells, it allowed the actual Poisson's Ratio and Young's modulus to be determined through biaxial pressure testing. Without the testing only assumed values could be used for these parameters.

An AECL Technical Record summarizing the results of all of the stress change measurements during URL shaft sinking is presently near completion (Ng et al. in prep.). Preliminary results were presented by Lang and Thompson (1985).

The stress changes recorded at one of the strain cells at the 218-m-depth array are shown in Figure 5. The stereo nets give the orientation of the principal directions of stress change, and the plot shows the magnitude of the principal stress changes for each level of the shaft bottom as the shaft was sunk past the strain cell. For example, consider the stress changes measured at the cell when the shaft bottom had reached the level indicated as 18 in the shaft. The stereo net number 18 shows that the maximum principal stress change marked by the circle symbol is horizontal and parallel to the shaft walls, the cross representing the intermediate principal stress is vertical, and the triangle representing the minor principal stress is horizontal and perpendicular to the shaft wall. (The line across the stereo net is the orientation of the shaft wall.) The plot of stress change magnitude versus depth shows that the tangential stress change indicated by the circle has increased by about 6 MPa, the vertical stress change indicated by the cross shows zero change, and the radial stress change indicated by the triangle shows about 13 MPa stress decrease.

In general the HI triaxial strain cells performed well for accurately measuring stress changes in the rock mass. However, to work well, the stress changes should take place over a relatively short time period of about two weeks or less. For longer periods, errors due to glue creep would become significant.

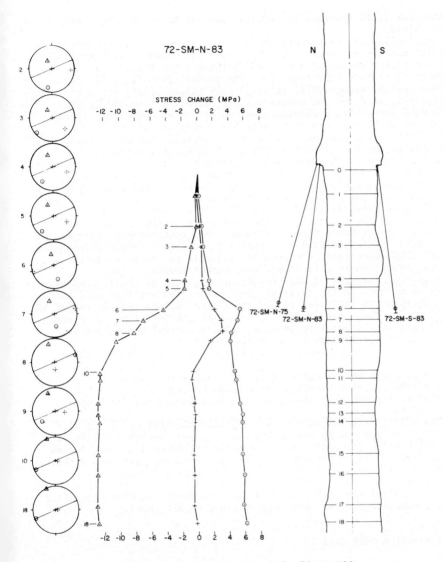

Figure 5   Stress Changes Measured in Borehole 72-SM-N83

3 GEOTECHNICAL CHARACTERIZATION OF THE SHAFT AND SHAFT STATIONS

After construction of the URL shaft and shaft stations, a comprehensive geotechnical characterization program began (Thompson et al. 1984). As part of this program over 300 overcore stress determinations were made at several locations in the 240 Level and 130 Level and at five different elevations in the shaft, including the locations of the four instrumentation arrays used to monitor excavation response. The program ran from 1985 June to 1986 June.

Both near-field (excavation-induced) and far-field stresses were
determined.
Originally when we planned the testing program, we assumed existing
techniques and equipment would be used. The CSIRO HI (triaxial) and
United States Bureau of Mines (USBM) (Hooker and Bickel 1974)
(biaxial) methods were selected. The Council for Scientific and
Industrial Research (CSIR) triaxial cell (Leeman 1968) was not
originally selected for use at the URL primarily because, in its
standard form, strain observations can not be made during drilling.
Of the 28 HI overcore tests attempted early in the geotechnical
characterization program, none were successful. Extensive laboratory
testing of the HI cell and of various epoxies, and extensive
experimentation with installation procedures, led us to the conclusion
that the cell did not work for overcoring stress measurements at
ambient rock temperatures of 8°C with recommended and other available
epoxies. Similar problems with the HI cell in cool near-surface rock
have been previously reported (Garritty et al. 1985).
Because of the difficulties encountered using the HI cell, we
modified the standard CSIR cell to meet our requirements (Thompson et
al. 1986). This modified CSIR cell and the USBM borehole deformation
gauge (BDG) were then both used successfully in the geotechnical
characterization in-situ stress measurement program. Three-
dimensional analysis of USBM results from groups of three holes were
used as a check on the triaxial results from the modified CSIR tests.
Comprehensive installation and testing checklists were developed and
used during the stress measurement program (Snider et al. in prep.).
These provided a suitable level of quality control and helped to
increase the percentage of successful tests.
Other major improvements to standard stress measurement methods
were:
 · a portable data logging system for overcoring tests
 · a temperature-controlled drill water system
 · an environmentally controlled and data-logged biaxial pressure
   chamber.
The effect of residual stress and drill hole size on the in-situ
stresses determined by overcoring was assessed through a series of
tests in which concentric overcore holes of 200-mm diameter, 150-mm
diameter, and 96-mm diameter were drilled consecutively over either a
modified CSIR cell or a USBM BDG. The results of these tests
indicated that the residual stresses at the URL site are only about
1.5 to 2.5% of the applied stress and that 96-mm diameter overcoring
was preferable for the test program (Lang et al. 1986).

3.1 Modified CSIR Cell

The modified CSIR triaxial cell used during the geotechnical
characterization program at the URL differs from the standard cell in
the following ways:
 1. It permits continuous strain observations during overcoring.
 2. It has a thermistor located against the borehole wall to permit
temperature at the cell location to be monitored.
 3. It uses quarter bridge wiring with no temperature compensation.
The modified CSIR cell is installed in a similar way to the standard
CSIR cell but the signal cable remains attached throughout overcoring
(Figure 6).

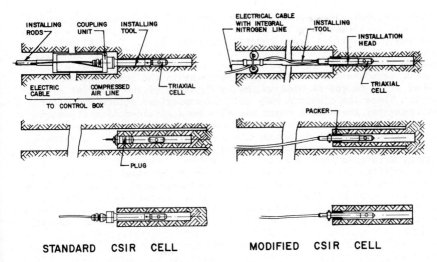

Figure 6 Installation and Overcoring Details for Standard and Modified CSIR Cells

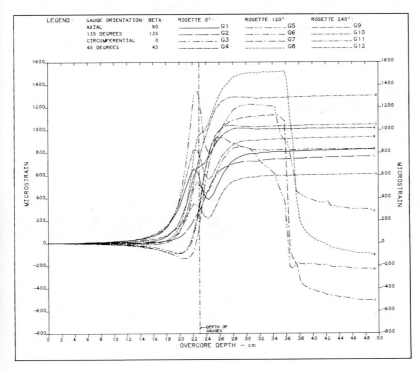

Figure 7 Computer Plot of Modified CSIR Overcore Test Data Showing a Rosette Debonding

The principal advantage of the modified CSIR cell over the standard CSIR cell is illustrated by the overcore test plotted in Figure 7. In this test, one strain gauge rosette (channels 5, 6, 7 and 8) starts to debond at 31 cm into the test, 7 cm past the centre line of the cell. Because the strains are being continually monitored, and the total elastic response has occurred by this point in the test, the strain values at 31 cm can be used in the calculation of the triaxial stress field. Since the standard CSIR cell only provides before and after measurements, these data could not have been obtained if a standard cell had been used. Furthermore a debonded rosette or faulty strain gauge may not even be detected with a standard CSIR cell, in which case erroneous strain values would be used in the stress calculations causing errors in the results.

The modified CSIR cell in conjunction with the other innovations used during the stress measurement program gave good results. Figure 8 shows a stereonet plot of five successive overcore tests conducted in homogeneous, massive grey granite at one location. The consistency in stress magnitudes and orientations is an indication of the quality of data.

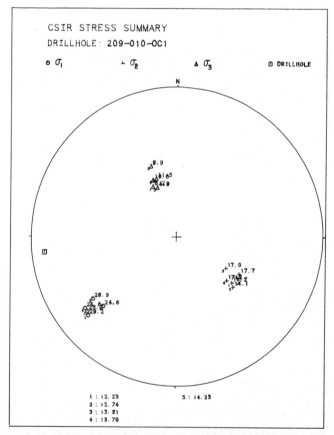

Figure 8   In Situ Stresses Determined Using Modified CSIR Cell in Drillhole: 209-010-0C1

3.2 Data Acquisition

The portable overcoring data logger was developed based on the Fluke 2280 unit. The following parameters were monitored, usually at a 15 second scan rate, during each overcore test:
  · depth of drill penetration (mm),
  · strains on all strain channels (microstrains),
  · temperature of drill supply water (°C),
  · temperature of strain cell (°C),
  · temperature of drill return water (°C),
  · drill water pressure (kPa),
  · drill rotation speed (RPM),
  · drill thrust (kN), and
  · drill torque (N·m).
The data logger was equipped to monitor overcore or biaxial pressure tests on either the USBM BDG, the CSIRO HI cell or on the modified CSIR cell. Compared with the standard manual methods of data measurement and reduction previously used for overcore testing at the URL, the data-logger system:
  · reduced the chance for errors during data transfer,
  · was more accurate,
  · recorded more parameters more frequently,
  · expedited data transfer for plotting,
  · obtained plots faster and of report quality,
  · required less manpower for data entry,
  · provided easier data storage and took up less physical space, and
  · enabled anomalies to be detected more easily because the main parameters that could affect a test were monitored.

3.3 Temperature-Controlled Drill Water

Temperature changes in the rock or strain cell during an overcore test have adverse effects on results, particularly when the stain cell uses quarter bridge wiring (Garritty et al. 1985). The temperature-controlled drill water system used in the stress measurement program reduced temperature changes during the test. Ambient rock temperature at the URL is about 8°C and the drill water supply was maintained at about 1°C below this to compensate for heat from the diamond bit and to maintain temperatures as close to ambient as practical.

3.4 Biaxial Pressure Testing System

The biaxial pressure testing system used during the characterization program consisted of an environmentally controlled cabinet with a Hoek-type biaxial pressure cell inside. Temperature and humidity were maintained at ambient rock conditions of 8°C and 99% respectively. A Fluke 2400 data logger was used to record microstrains, temperatures, and applied biaxial pressure.

4 DEVELOPMENT OF THE 240 LEVEL

The initial development of the 240 Level, when about 100 m of tunnel was excavated, began concurrently with the start of geotechnical characterization. The remainder of the level was excavated after geotechnical characterization of the shaft was completed. This development work provided an opportunity to test different geomechanical instruments to determine their suitability for future field measurements at the URL.

4.1 Traversing Casing Deflectometer

The SINCO Traversing Casing Deflectometer (TCD) was assessed for use in applications where lateral displacements in the horizontal plane were of interest. This instrument is used in standard inclinometer casing that can be oriented in any direction. The device consists of two one-metre-long beams mounted on spring-loaded wheels and hinged at the centre where a strain gauged cantilever is located. By traversing the instrument in one-metre intervals along a horizontal measurement casing, the lateral displacements along the casing can be determined from the strain measurements. A second traverse with the device rotated 90° allows vertical deflections to be measured.

At the URL two installations were made in parallel holes 40 m long and 1.5 m apart. A tunnel was then driven parallel to the two measurement casings, 1.5 m from the closest. Measurements of deflections were made after each excavation blast round.

The instrument was not suited for measuring displacements of less than 1 mm , since repeatability in the horizontal plane is about $\pm$ 0.5 mm. The scatter in the measurements was too great to provide any useful data. Errors were attributed to:

1. Difficulty in locating the instrument in precisely the same location on subsequent traverses.
2. Dirt in the measurement casing grooves.
3. Wear of the plastic measurement casing with time.

In addition the instrument was extremely cumbersome and fragile. It is over 2 m long and must be carefully supported by two technicians when it is installed in the measurement casing to prevent damage to the strain-gauged cantilever. The measurement process is slow and labour intensive and data logging is not possible. Unless the device can be substantially improved, it will not be used for future rock displacement measurements in the URL program.

4.2 Excavation Response Test

An excavation response test was performed in Room 209 of the 240 Level at the URL. One of the main objectives of this experiment was to assess various types of rock displacement instrumentation to help select the most suitable for the shaft extension construction program, which will commence in the summer of 1987. An array of instrumentation was installed as shown in Figure 9. The array consisted of 10 borehole extensometers of four types, a ring of 8 convergence pins, 8 CSIRO HI triaxial strain cells to measure stress changes and hydrogeological instrumentation to monitor hydraulic pressures in a vertical fracture approximately 4 m ahead of the face. The holes were drilled and the instrumentation was installed between 1985 October and 1986 September. A data logger was installed nearby to monitor those instruments that allowed remote monitoring. Excavation commenced on 1986 October 21. Analysis of the measurements is presently underway.

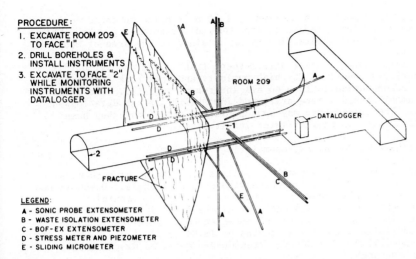

PROCEDURE:
1. EXCAVATE ROOM 209 TO FACE "1"
2. DRILL BOREHOLES & INSTALL INSTRUMENTS
3. EXCAVATE TO FACE "2" WHILE MONITORING INSTRUMENTS WITH DATALOGGER

LEGEND:
A - SONIC PROBE EXTENSOMETER
B - WASTE ISOLATION EXTENSOMETER
C - BOF-EX EXTENSOMETER
D - STRESS METER AND PIEZOMETER
E - SLIDING MICROMETER

Figure 9 Arrangement of Instrument Array Installed in Room 209 of the 240 Level

4.2.1 Displacement Measurements

The four rock displacement measurement systems in the Room 209 instrumentaion array are summarized below:

Irad Gage Sonic Probe Extensometer

The Sonic Probe extensometer measurements confirmed that the instrument lacked the precision for measuring displacements significantly less than 1 mm. The advantages and disadvantages of this instrument are described earlier. It was used in the Room 209 array as a reference for the other instruments; also, some were left over from the shaft program. At the present time the sonic probe transducer cannot be data logged and must be monitored manually.

SINCO Waste Isolation Extensometer

The Waste Isolation Extensometer is an MPBX with 4 borehole anchors. The extensometer rods are tensioned 6-mm-diameter invar steel. The long tensioning springs within the bulky reference head require a 1.5-m-long, 100-mm-diameter borehole collar, with an outer collar about 0.20 m long and 200-mm diameter in which to recess the head. The anchors consist of hydraulically activated, coiled-copper bladders, which are pressurized to 12 MPa against the borehole wall. One of the twelve anchors used in the three waste-isolation extensometers in the array burst before the design pressure was achieved.

Assembly and installation of the waste-isolation extensometer was time consuming and difficult in comparison with the other measurement systems. Several days were needed to assemble and install each unit. In addition, the displacement information provided by the instrument is limited. The collar reference must be recessed about 20 cm for blast protection, therefore the 20 cm of rock beside the excavation wall cannot be monitored. Furthermore, the closest anchor to the reference head must be a minimum of 1.75 m from the borehole collar, because of the long tension springs and collar reference tube. This results in minimal information being obtained in the area of most interest.

The linear potentiometer displacement transducers used with this MPBX performed poorly. The high humidity and cool rock conditions caused condensation to form on all instrument surfaces. The moist environment caused severe drift and instability in the data-logged output. The transducers may have performed better had they been properly sealed to prevent water infiltration.

Roctest Limited Bof-Ex Extensometer

The Roctest Bof-Ex extensometer consists of a series of mechanical screw-type anchors installed line-wise down a borehole. Downhole DC-LVDT linear displacement transducers record the incremental displacements between each anchor. At the trial installation at Room 209, the instrument consisted of seven anchors installed in an NQ borehole (76 mm diameter) at depths of 15.01, 9.01, 5.01, 3.01, 1.41, 0.51, and 0.11 m. Six DC-LVDT transducers measured displacements between the anchors.

A number of modifications to the standard system were incorporated in the Room 209 installation to better assess the potential of this system:

1. Stainless steel construction was used instead of the standard aluminum, to provide greater strength, more corrosion resistance, and reduce the thermal coefficient of expansion.

2. Thermistors were installed at each anchor to monitor down-hole temperatures to allow for temperature compensation.

3. The mechanical anchors incorporated an "active" spring in the three shoes that locked the anchor to the borehole wall to prevent anchor slippage due to blast vibrations.

4. The instrument was data logged.

The instrument performed exceptionally well as can be seen in the plot of results in Figure 10. The total displacement measured during pilot advance was 0.42 mm, and the incremental displacements were monitored with excellent resolution and stability. For example the total displacement measured between 9.01 and 15.01 m in the borehole was 0.021 mm ($\pm$ 0.001 mm stability.)

The advantages of this system for measurements at the URL are:

1. Up to 11 measuring points can be installed in a 76-mm-diameter borehole.

2. Measurement points can be within 300 mm of each other and within 100 mm of the excavation surface, so that more measurements can be made in the zone of rock of most interest, i.e., adjacent to the excavation.

3. The system is easily and quickly installed (about 3 hours).

4. The system can be data logged.

5. The entire instrument is retrievable by unscrewing and removing the mechanical anchors. This has beneficial implications for long-term displacement monitoring since the entire instrument can be removed and recalibrated or replaced within hours. Such a feature will be desirable for instruments required to monitor an actual nuclear waste disposal vault during the approximately 50 years of development and operation before it is closed.

ISETH/Sol Experts Ltd Sliding Micrometer

The sliding micrometer is a probe capable of making highly accurate

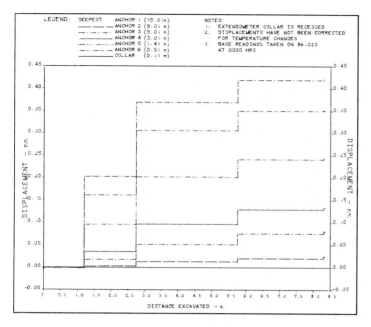

Figure 10   Displacements Measured With Bof-Ex Extensometer in
Borehole209-018-EXT7

axial displacement measurements along an instrumented borehole in one-
metre increments, by means of specially machined measurement marks
installed at one-metre intervals within a PVC casing (Kovari et al.
1979).

In the application at the Room 209 excavation response test, the
sliding micrometer casing was installed from an adjacent excavation to
pass about 4 m ahead of the pre-excavation face of Room 209.
Measurements were taken as the excavation proceeded, and continued
after the instrumented hole was intersected by the excavation.

The instrument performed reasonably well although overall
repeatability was only about ± 0.04 mm. This is considerably less
accurate than the manufacturer's claim of ± 0.003 mm in the field.
The instrument hole was inclined upwards at 1°. Although done with
the intention of keeping the measurement surfaces and the probe dry,
we later learned in discussion with K. Kovari of ISETH that the
following problems can occur in such an installation:

1. Dirt and any other debris within the casing is carried or washed
down onto the seating faces of the measurement heads.

2. Thermal stability within the air-filled casing is much less than
if the casing were filled with water; this can adversely affect
measurement accuracy.

In addition, some grout had penetrated the measurement casing during
grouting and continued to cause problems with a few anchors thoughout
the measurement period despite repeated attempts to clean the casing.

It is expected that in future installations we should be able to
increase reading accuracy by installing the measurement casing in a
downward sloping hole. The instrument appears to be well suited for

measuring small displacements in stiff rock; however, measurements must be taken manually. This can be a problem when making excavation response measurements from inside an excavation heading, since the measurements hinder other activities near the excavation face.

4.2.2 Convergence Measurements

The ISETH/Kern Distometer was selected to monitor convergences. Unlike the tape extensometer used previously at the URL, this device uses a tensioned invar wire to accurately measure convergence. The instrument performed very well as can be seen in Figure 11. Measured convergences were as low as 0.2 mm with relatively little scatter (about ± 0.03 mm). Based on the displacement measurements the convergences measured with this device were as we would expect.

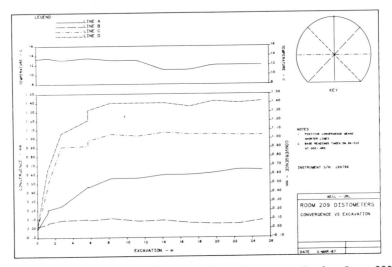

Figure 11   Convergence Measured With Distometer During Room 209 Excavation

4.2.3 Stress Change Measurements

Prior to the Room 209 excavation response test, a comparison test between vibrating wire stressmeters and CSIRO HI triaxial strain cells was conducted. Five of the twelve vibrating wire stressmeters installed failed shortly after adjacent excavation commenced. Furthermore, the ones that survived provided scattered results which did not agree with the stresses measured by the HI cells. Since these instruments only provide stress change data in one direction and did not appear to be as reliable as the HI triaxial strain cells, they are not being included in future instrumentation plans.
The HI triaxial strain cells used in the Room 209 excavation response test were installed in much the same way as during shaft sinking, although only the two cells installed in the floor required the installation boreholes to be dewatered. These cells differ slightly from those used earlier in the program in that they each contain 12 strain gauges, thus providing additional strain

information. The cells have operated well and the measurement results
are presently being assessed.

4.2.4 Data Logging

A Fluke 2400 data logger located adjacent to the Room 209 entrance
(about 10 m from the face) was used to monitor the instrument response
from approximately 200 sensors. These sensors included thermistors,
HI triaxial strain cells, DC-LVDT and linear potentiometer
displacement transducers, and vibrating wire piezometers. During
excavation of Room 209 the normal scan rate between blasts was once
every four hours. Immediately prior to and for at least 10 minutes
following blasts, the scan rate was once every two minutes.
  The data logger performed well, despite its close proximity to the
blasting activities. Superficial damage to the data logger cabinet
was sustained due to concussion, but this had no noticeable effect on
the measurements.
  Data was fed to the host computers on the surface at the URL where
it was stored on hard disk, ready for transfer to the mainframe VAX
computer at nearby Whiteshell Nuclear Research Establishment (WNRE)
for plotting and analysis. Report-quality plots could be produced
quickly at the URL using the link to the WNRE mainframe.

5 SHAFT EXTENSION PREPARATIONS

AECL and the United States Department of Energy (US/DOE) have joined
in a co-operative agreement to extend the URL shaft to 465-m depth.
Plans call for shaft sinking to commence in 1987 July. A major
subhorizontal fault, known as Fracture Zone #2, exists immediately
below the present shaft bottom. As part of the co-operative work
program, this fracture zone is being characterized and grouted, and a
shaft extension probehole is being drilled and tested.
  A total of 23 hydrogeological characterization boreholes have been
drilled to intersect the fracture zone. These holes are instrumented
with multiple zone piezometers. In addition we have drilled and
instrumented nine geomechanical characterization boreholes through the
fracture zone. In-situ stresses were measured by overcoring above and
below the fracture zone in five of these holes. Displacement
instrumentation has been installed in all of these holes.
  A 250 m deep shaft extension probehole has been started. This hole,
which dips at 79°, will pass about 10 m from the shaft extension at
its mid-point and is being used to obtain geological data and for deep
in-situ stress measurements prior to the start of shaft sinking.

5.1 Displacement Measurements During Characterization and
    Stabilization of Fracture Zone #2

Four geomechanical characterization holes located beyond three metres
from the shaft extension have been instrumented with sliding
micrometer casing to measure normal displacements across the fracture
zone related to either:
  1. Water pressure changes due to hydrogeological testing or eventual
drawdown due to leakage into the shaft extension, or
  2. Grouting activities.
The other five boreholes immediately surrounding the shaft extension
have been instrumented with Trivec measurement casing to measure the
lateral displacements due to excavation response as well as the axial

displacements across the fracture zone. Base measurements are currently being taken on these installations.

5.2 In-Situ Stress Measurements

The modified CSIR cell, standard USBM borehole deformation gauge, and the Swedish State Power Board (SSPB) cell (Hiltscher et al. 1979) were used to measure the in-situ triaxial state of stress in five of the characterization boreholes. The CSIR cell is limited to depths of about 15 m and it was necessary to dewater each borehole in order to perform a measurement. The standard USBM BDG was used to 25-m depth. The SSPB cell was used for all other stress measurements in these holes, with measurements as deep as 59 m.

The SSPB cell consists of three strain gauge rosettes spaced at 120° with three 120-ohm strain gauges each. Three strain gauges measure axial strains, three measure circumferential strains and three measure strains at 45° to the core axis. The instrument works in deep waterfilled boreholes (up to 500 m) through the use of an acrylic resin which bonds to wet rock. The instrument is lowered down the HQ (96 mm dia.) borehole into the 0.5-m-long EWG (38 mm dia.) borehole at the base. The strain rosettes are positioned inside a metal glue pot containing the high viscosity resin. A mechanism on the installation tool causes the glue pot to be forced off and the rosettes to be pressed against the EWG borehole wall as the installation tool reaches the bottom of the HQ hole. Strain measurements are taken prior to and after overcoring since the instrument cannot be monitored during drilling. Slight modifications to the cells and installation tool were made to allow the cell to be used in North America where borehole sizes differ slightly from those used in Sweden.

All three stress determination methods worked reasonably well with good agreement between the different instruments used in adjacent tests in the same borehole. Measurements are currently being assessed. An interesting preliminary observation is that normal stresses across the fracture zone are inversely proportional to the permeability, as one might expect.

Overcore stress measurements are planned for approximately ten locations as the hole progresses. At each location a minimum of five tests using each of the following two methods will be performed:
1. SSPB triaxial cell.
2. Deep USBM borehole deformation gauge (developed by AECL).

The SSPB cell has already been described and has operated satisfactorily in the shaft extension probe hole to 86-m depth, the deepest such test performed to date.

The deep USBM borehole deformation gauge (DBDG) is illustrated in Figure 12. This device differs considerably from a standard USBM BDG. The maximum depth of operation of a standard USBM BDG is about 60 m. Beyond this depth, water pressure in the test borehole forces the measurement buttons into the gauge body and prevents strain measurements from being obtained. The DBDG is oil filled and external water pressure is transmitted inside the device by a flexible diaphram thus equalizing internal and external pressures. Special protection is provided for the strain gauges and a pressure bulkhead prevents oil leakage into the back section of the device. The instrument incorporates an RTD temperature sensor and a miniature pressure transducer to record water pressure. An integral signal conditioner multiplexes and transmits signals to surface via a wireline geophysical cable.

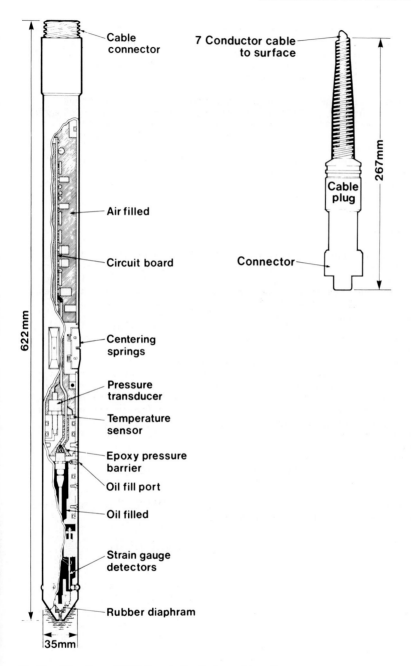

Figure 12   Deep USBM Borehole Deformation Gauge

Installation is performed using a "hammer" to set the instrument at depth. Since the measurements are performed in an inclined hole, orientations are obtained by integrating a heated cylinder filled with oil and containing a ball bearing. The ball remains on the lowest part of the cylinder when the heater is shut off and the oil solidifies.

As expected with such a complex instrument there were numerous problems encountered in initial trials but the device is now debugged and working well. The deepest successful test has been at 120 m.

Overcoring at this depth is being accomplished using wireline techniques pioneered by Longyear Canada Ltd. By using wireline inserts for flattening the HQ hole and drilling the concentric EWG hole through the HQ core barrel the efficiency of overcoring at depth is greatly improved. The drill rods only have to be removed once per test to install the overcoring bit.

6 CONCLUSIONS

The construction phase activities at the URL have provided suitable opportunities to assess the performance of geomechanical instruments to measure displacements, convergences, stress changes, and in-situ stresses in granitic rock. Based on these assessments, we have been able to select the most appropriate instruments and methods for measuring these parameters in future URL experiments.

Of the instruments tested, those most suited to measuring displacements in stiff granite are:
· Bof-Ex extensometer
· Sliding micrometer, (Trivec not yet assessed)
The Bof-Ex extensometer has been selected for the rock mass response instrumentation arrays for the shaft extension. Sliding micrometer/ Trivec casing has been installed across Fracture Zone #2 to measure displacement across the zone. Satisfactory convergence measurements can be obtained by using the Distometer. Triaxial stress change measurements can be made using the CSIRO hollow inclusion triaxial strain cell.

At shallow depths in-situ stresses in cool stiff granite can best be determined using the USBM BDG (biaxial) and modified CSIR cell (triaxial). In deeper boreholes the DBDG (biaxial) and SSPB cell (triaxial) provide good results. Data logging of all pertinent parameters during an overcore test and careful control of temperatures provide a greater level of confidence in results.

REFERENCES

Garritty, P., R.A. Irvin, and I.W. Farmer 1985. Problems associated with near surface in-situ stress measurements by the overcoring method, 16th U.S. symposium on Rock Mechanics, Rapid City.
Hiltscher, R., J. Martna and L. Strindell 1979. The measurement of triaxial rock stresses in deep boreholes and the use of rock stress measurements in the design and construction of rock openings. Proc. 4th International Congress on Rock Mechanics, Vol. 2. pp. 227-334.
Hooker, V.E. and L. Bickel 1974. Overcoring Equipment and Techniques Used in Rock Stress Determination, USBM, IC 8618:32.
Kovari, K., C.H. Anstad, and J. Köppel 1979. New developments in the instrumentation of Underground Openings, Proc. Rapid Excavation and Tunnelling Conference, Atlanta..

Lang, P.A., and P.M. Thompson 1985. Geomechanics experiments during excavation of the URL shaft. 20th Nuclear Fuel Waste Management Information Meeting, Winnipeg, Manitoba. Atomic Energy of Canad Limited Technical Record TR-375.*

Lang, P.A., P.M. Thompson, and L.K.W. Ng 1986. The effect of residual stress and drill hole size on the in-situ stress determined by overcoring. Proc. Int. Symposium on Rock Stress and Rock Stress Measurements, Stockholm, pp. 687-694.

Leeman, E.R. 1968. The determination of the complete state of stress in rock in a single borehole-laboratory and underground measurements, Int. J. Rock Mech. Min. Sci. Vol. 5, pp. 31-56.

Ng, L.K.W., R.A. Everitt, P.A. Lang, and P.M. Thompson. Stress change measurements during URL shaft sinking. Atomic Energy of Canada Limited Technical Record* (in prep).

Simmons, G.R. 1986. "Atomic Energy of Canada Limited's Underground Research Laboratory for Nuclear Waste Management Proceedings of International Symposium on Geothermal Energy Development and Advanced Technology, Tohoku University, Sandai, Japan. Nov 1986.

Snider, G.R., P.A. Lang, and P.M. Thompson. Procedures Used for Installing and Monitoring Geomechanical Instrumentation during Sinking of the URL Shaft, Atomic Energy of Canada Limited Technical Record* (in prep).

Snider, G.R., P.A. Lang, and P.M. Thompson. Procedures used for overcore testing during sinking of the URL shaft. Atomic Energy of Canada Limited Technical Record* (in prep).

Thompson, P.M., P. Baumgartner, and P.A. Lang 1984. Planned construction phase geomechanics experiments at the Underground Research Laboratory. Proc. of OECD/NEA and CEC Workshop on Design and Instrumentation on In Situ Experiments in Underground Labora-tories for Radioactive Waste Disposal; Brussels, Belgium, May 1984.

Thompson, P.M., and P.A. Lang 1986. Rock displacements measured during URL shaft sinking. Proc. 2nd International Conference on Radioactive Waste Management, Winnipeg, Canada, September 1986.

Thompson, P.M., P.A. Lang, and G.R. Snider 1986. Recent Improvements to in-situ stress measurements using the overcoring method. Proc. 39th Canadian Geotechnical Conf., Ottawa, pp. 143-150.

Worotnicki, G. and R.J. Walton 1976. Triaxial "hollow inclusion" gauges for the determination of rock stresses in-situ. Proc. Symp. on Investigation of stress in rock and advances in stress measurement. I.S.R.M., Sydney, supplement, pp. 108.

*   Unrestricted, unpublished report available from SDDO, Atomic Energy of Canada Research Company Limited, Chalk River, Ontario K0J 1J0.

2nd International Symposium on Field Measurements in Geomechanics, Sakurai (ed.)
© 1988 Balkema, Rotterdam. ISBN 90 6191 778 6

Generic technical position on in situ testing during site characterization for high-level nuclear waste repositories

Mysore S.Nataraja
Rock Mechanics Section, US Nuclear Regulatory Commission, Washington, D.C.

1 INTRODUCTION

The United States Nuclear Regulatory Commission (NRC) interacts with the United States Department of Energy (DOE) on a continuous basis to provide timely guidance on the resolution of licensing issues concerning the disposal of high-level nuclear waste in deep geologic repositories. Generic and Site-Specific Technical Positions are prepared by the NRC on a number of topics and provided to the DOE as a part of the early and ongoing prelicensing consultations. This paper summarizes the salient points of the Generic Technical position on In situ Testing during Site Characterization prepared by the Engineering Branch of the Division of Waste Management.

Before submitting a license application, the DOE is required by the Nuclear Waste Policy Act of 1982 and by 10 CFR Part 60 to conduct a program of site characterization. In situ testing is an important element of site characterization. These tests are to be performed from the exploratory shaft(s) and underground openings on surrounding rock and on other materials and components such as the waste package, engineered backfill, linings, and seals. The conditions under which these in situ tests are to be run should represent, as closely as possible, the realistic repository environment (for (example, temperature and stresses). The tests performed under such conditions would provide data to assess the suitability of a particular site and a particular geologic medium to host high-level nuclear waste and realistic input parameters for the design of a geologic repository.

In situ tests can only be conducted for a limited duration compared with the long time span during which the repository must function to isolate the waste. Analytical, experimental, and numerical models must be used to make predictions far into the future; however, models have their own limitations on applicability and are sensitive to the quality of data used as input. Some of the uncertainties in the prediction process can be reduced by conducting appropriate in situ tests on a representative volume of rock and by using appropriate models to account for possible inherent spatial variations of physical, hydraulic, and chemical properties within the rock formation. By comparing in situ test data with modeling results, models can be validated, thereby reducing some uncertainties in the prediction process.

2   PUBLIC LAW AND REGULATORY FRAMEWORK

The regulatory framework governing the in situ testing during site characterization consists of the Nuclear Waste Policy Act of 1982 and the USNRC rule 10 CFR Part 60. The intent of these regulations is that the in situ testing program should (1) obtain data to assess the suitability of a particular site, (2) provide representative parameters for design of a repository, and (3) help resolve before the licensing hearing, important issues that will affect the performance of the repository.

2.1   Nuclear Waste Policy Act of 1982

NWPA, (Public Law 97-425, Section 113(b)1(A)ii), requires DOE to submit a site characterization plan, including a description of onsite testing, which should cover (1) the extent of planned excavations for any onsite testing with radioactive or nonradioactive materials, (2) plans for any investigative activities that may affect the capability of such candidate sites to isolate high-level radioactive waste and spent nuclear fuel, and (3) plans to control any adverse safety-related effects of such site characterization activities. In situ testing is expected to help meet the provisions of the NWPA.

2.2   NRC Rule for Disposal of High-Level Radioactive Waste

The NRC rule 10 CFR Part 60 requires in situ testing as a part of site characterization. Therefore, the construction of exploratory shafts, the excavation of test facilities, and the performance of in situ tests will precede license application by the DOE. Site characterization will take place at a minimum of three sites before DOE submits a license application for one site.

3   TECHNICAL POSITIONS ON IN SITU TESTING

The NRC staff technical positions on in situ testing during site characterization are:
   (A) Before submitting a license application, DOE should perform a necessary and sufficient variety and amount of in situ testing to support, if the facts so warrant, a staff position that the requirements for issuance of a construction authorization (10 CFR Part 60.31) have been met.
   (B) The in situ testing program should be developed with two major objectives: (i) characterization of host rock and in situ measurement of its properties prior to construction and waste emplacement; and (ii) determination of response characteristics of the host rock and engineered components to construction and waste emplacement.
   (C) DOE should present its site specific and design specific in situ test plans in the Site Characterization Plan (SCP).
   (D) Before developing the in situ test plan, DOE should develop a rationale for in situ testing and present this rationale with the test plan in the SCP. The overall goal of the rationale should be to ensure that all important parameters are identified and ranked according to

their relative importance in supporting 10 CFR Part 60 licensing findings.

(E) For successful site characterization, DOE should integrate the data from surface borehole testing and laboratory testing on small-scale samples with the in situ test results.

This technical position is generic and covers in situ testing for all potential repository sites and designs.

4 DISCUSSION

In situ testing is essential to assess the suitability of a geologic repository site for hosting high-level nuclear waste and to provide realistic input parameters for a repository design. In situ testing is expected to significantly reduce uncertainties regarding the ability of the host rock to provide long-term isolation and containment of the high-level nuclear waste. The uncertainty for compliance with the performance requirements of 10 CFR Part 60 is expected to decrease significantly during the site characterization period. However, during the performance confirmation period, DOE efforts to lessen the uncertainties will continue.

As noted earlier, in situ testing is an important and necessary element of site characterization. In spite of being the most effective and direct approach to the characterization of the host rock, any in situ testing program will be, of necessity, limited by practical considerations. For example, the lateral extent of underground excavation for in situ testing purposes will be very limited in comparison to the total volume of rock being characterized. The number and duration of tests will also be limited by practical considerations. However, in the absence of in situ test results, confidence in the predictions based only on borehole and laboratory testing may be limited. It is very important to note that although in situ testing is necessary, it is not sufficient by itself and requires integration with all other testing.

The following unique features make in situ tests an essential element of site characterization and a rational design of the repository.

(a) Scale Effects Can Be Minimized
     Numerous laboratory and in situ tests have shown that many of the measured rock properties (for example, compressive strength and permeability) are influenced by the size of the rock specimen tested. In highly jointed rocks, this dependence on size could be more pronounced. In situ tests that measure crucial design parameters clearly minimize the scale effect as a source of error.

(b) The Rock Mass in Its Natural Condition Can Be Observed and Tested
     The natural conditions of the rock mass cannot be exactly duplicated in the laboratory. Examples are (a) geologic discontinuities such as joints and shear planes, (b) hydrologic conditions such as hydraulic head and pore pressure, (c) loading conditions such as the in situ stress field, and (d) temperature. The in situ tests, by definition, encompass the natural rock conditions; therefore, the rock is tested in its natural state.

(c) Coupled/Interactive Processes Can Be Directly Observed
     Many coupled/interactive processes are likely to occur in the host rock in which the nuclear waste will be disposed. In situ tests

can measure representative properties resulting from coupled/interactive processes, unlike most small-scale laboratory tests. Furthermore, the in situ tests, if conducted properly, can provide for measuring possible nonlinear behavior that is difficult to extrapolate from small-scale laboratory experiments.

(d)   Host Rock Variability Can Be Evaluated
Variability in geology (for example, joint patterns and spacing), hydrology, and geochemistry can only be directly assessed through in situ testing. Estimation of variability and assessment of ability to predict rock behavior in different parts of the repository are necessary for satisfactory design of the repository.

4.1   Rationale for In Situ Testing

The in situ testing program must be site and design specific. However, the following topics should be addressed in the in situ test plan for each site.

(a) All relevant issues requiring resolution by in situ testing and measurements;
(b) the information needs for the license application (identified on the basis of 10 CFR Part 60 requirements and the level of uncertainty in predicting the performance of the repository);
(c) the availability of existing tests to provide all the information needed;
(d) the capabilities and limitations of available tests and measurement methods;
(e) the need, if any, to develop new tests;
(f) the effects of testing on long term repository performance;
(g) the extent of the underground test facility required to assess host rock variability properly and to assess the effect of that variability on design and performance;
(h) the extent of the underground test facility needed in order to minimize or avoid interference among tests;
(i) the sufficiency of subsurface geologic mapping, geophysical testing, and core drilling to assess the characteristics of the host rock and the variability of its properties;
(j) the representativeness of the in situ test location compared to the entire volume of rock, that must be assessed in determining compliance with the U.S. Environmental Protection Agency limits for releases to the accessible environment
(k) the basis for selection of a particular scale and duration of testing;
(l) justification for conducting or not conducting coupled/interactive tests
(m) extent of credit taken for the performance of important components of the barrier system, engineered and natural;
(n) the sufficiency (amount, scale and duration) of geological, hydrological, geomechanical, geochemical, thermal, and coupled/interactive testing, to make findings on compliance with the performance objectives on a scale that is sufficient to realistically

represent the inhomogeneities and discontinuities of the rock being
tested; and
   (o) description of the manner in which data from surface borehole
testing and laboratory testing on small-scale samples will be
integrated with the in situ test results.

4.2 Content of In Situ Test Plan

The in situ test plan should identify all important parameters,
classify them according to their relative importance, and document
their potential variability and its influence on design/performance.
The plan should be flexible to accommodate new issues that might
develop during site characterization.

4.3 Description of In Situ Tests

As a minimum, geological-geophysical, hydrological, geomechanical,
geochemical, thermal, and thermal-hydrological-mechanical-chemical
(coupled/interactive) tests, as well as tests on seals and backfill
materials, should be considered in formulating the in situ test plan.
The extent of each type of testing may vary with the individual site,
the selected repository design, and the site issues that require
resolution by testing. The following is a brief description by
discipline of these in situ tests.

(A) Geological-Geophysical Tests should consist of:

   (a) inspection, examination, and geologic mapping of exposed rock in
exploratory shafts and accessible underground excavations including
characterization of lithology, structural features, and discontinuities
intersecting underground openings;
   (b) drilling of horizontal, vertical, or inclined boreholes from
within the exploratory shaft(s) and underground excavations to obtain
samples for testing;
   (c) lithological logging of cuttings and cores obtained from
boreholes, shafts, and underground excavations to characterize
lithological variations and discontinuities; and
   (d) underground geophysical testing (for example, resistivity and
radar) and borehole geophysical testing and measurement techniques (for
example, sonic, gamma, and electrical logging; strain meters; vertical
seismic profiles; borehole gravity and magnetics; and shear wave
surveys) to assess the properties and distribution of units and their
associated discontinuities.

(B) Hydrological Tests should consist of:

   (a) estimation of the potential for high rates of groundwater influx
into the exploratory shaft and underground excavations, which may be
obtained by (i) means of pilot holes drilled from the exploratory shaft
and underground test facility for detecting the presence of zones of
anomalously high hydraulic conductivity in close proximity to the
underground test facility or the exploratory shaft(s); (ii) correlation
of discontinuities mapped along the drifts with observed zones of

groundwater influx and their corresponding calculated flow rates; and
(iii) a chamber test for determining groundwater influx potential, if
needed;
  (b) monitoring of transient changes in formation pressures in the
volume of rock close to the underground workings and evaluation of
changes induced by in situ hydrological and geomechanical tests,
subsequent construction activities, and surface-based hydrological
testing;
  (c) determination of hydrologic parameters such as directional
hydraulic conductivity, total and effective porosity, specific storage,
pore water pressure, and degree of saturation; measurement of
parameters within the
hydrologically significant units overlying and underlying the
host rock, if appropriate; and
  (d) testing to evaluate the potential significance of fracture
flow and matrix diffusion on radionuclide transport.

(C) Geomechanical Tests should consist of:

  (a) representative volume testing (for example, block tests) to
predict the constitutive behavior (strength and deformability) and
potential failure mechanisms of the host rock;
  (b) where applicable, measurement of the in situ stresses from
underground openings so that they can be compared with the stresses
measured from the surface boreholes and so that the prevailing stress
field around the underground openings can be more accurately estimated;
and
  (c) demonstration of repository construction (for example, mine-by
test), emplacement, and retrievability, and observation of full-scale
response of underground openings and backfill by simulation.

(D) Geochemical Tests* should consists of:

  (a) sampling and in situ testing of retardation and migration
characteristics of host rock to support laboratory studies of all
significant geochemical conditions and phenomena to provide a reliable
data base for modeling studies;
  (b) measurement of physical parameters such as ambient temperature
and pressure;
  (c) chemical analyses of rock cores to determine the pre-waste
emplacement mineralogy and elemental composition of the host rock and
surrounding strata;
  (d) chemical analyses of the groundwater, including analyses of the
compositions of elements and species, pH, redox conditions, gas
components, and organic/colloidal material, to provide an understanding
of possible transport mechanisms;
  (e) mineralogical and petrological analyses to determine mineral
phases, morphologies, distributions, and textures; and
  (f) tracer tests to determine retardation factors and scaling
factors for laboratory tests.

---

*The nature of geochemical tests may require certain samples to be
transported to a laboratory for testing and analysis.

(E) Thermal Loading Tests

Thermal loading tests (for example, small- and large-scale heated block tests) are to support modeling studies of repository scale temperature effects such as thermal expansion and stability effects, and to establish, by simulation, canister scale behavior; special emphasis on the effects of the thermal field to open or close the existing fractures or initiate new fractures.

(F) Tests on Seals and Backfill Materials

Appropriate hydrological, geomechanical, geochemical, thermal, and coupled/interaction testing of seals and backfill materials, which are major components of the repository system, should be conducted to study their
effectiveness, durability, and long-term performance; results of these tests also should be used in performance confirmation.

4.4 Sufficiency of Testing

The design of a geologic repository is governed by the 10 CFR Part 60 requirement of a multiple-barrier approach. The amount of testing required for individual components of the barrier system will depend in part on the amount of credit taken for individual components in meeting the performance requirements. This implies that in developing its plans for testing during the site characterization and engineered component design phases, DOE will identify performance goals for repository system components on a site-specific basis. Such identification of performance goals is a necessary prerequisite to establishing what is a necessary and sufficient level of testing. Thus, tentative goals are needed at the time that the Site Characterization Plans are issued. Because of the large uncertainties that would be associated with DOE's understanding of a candidate site at that time, these tentative performance goals necessarily involve making technical and management decisions and should be conservatively chosen. As site characterization proceeds and a more complete understanding of the site is developed, the initial choice may have to be revised and appropriate changes made to the test plan. Given a particular set of performance goals, little or no testing of a certain component of a barrier may be necessary.

4.5 Planning and Scheduling of Testing

Many phases of in situ testing are required to continue during siting design, construction, and operation stages. Because of the long lead-time of some tests, the schedule of testing will be crucial to the licensing activity. When a construction authorization application is submitted for a particular site, that application must be complete and fully supported by the data and analyses necessary to make a decision on construction authorization findings (10 CFR Part 60.31).
DOE should identify in its test plan which tests will be completed at the time of the construction authorization application, and which tests and long-term monitoring activities will continue after that.

4.5.1 Amount and Variety of Tests

Decisions related to establishing the amount and variety of testing
needed should be made on a site- and design-specific basis. Several
different tests can be used to obtain the same rock parameters. For
example, the plate test, pillar test, and block test can each provide
sufficient information to calculate the material modulus. Each test or
type of test can be repeated a number of times depending on the
required level of confidence. The same test may be repeated at a
number of different locations to assess the inherent variability of the
measured parameter. Also, it may be desirable to conduct tests under a
range of conditions to represent the extremes of the anticipated
environment.

The in situ test plan should include criteria to determine whether
the amount and variety of testing is sufficient. For all tests
important to performance, a general guideline is that testing should
continue or be repeated until the results are technically defensible.

4.5.2 Scale of Tests

Because of the complexities of designing and constructing an
underground repository, testing will have to be performed at different
scales. Laboratory testing on small specimens will provide useful
information for preliminary designs and analyses. However, in many
cases, large-scale testing will be required to yield a realistic and
convincing data base. The need for large-scale testing will have to be
established on a site and design specific basis. The in situ test plan
should discuss the scale of testing and its implications for site
characterization. Moreover, the underground openings should be of
sufficient extent so that the variability in the host rock and adjacent
strata can be properly assessed.

4.5.3 Duration of Tests

Testing of geologic materials for certain design parameters can be of
long duration. On other hand, many design parameters can be obtained
by testing over relatively short periods. When the processes being
observed are slow, complex interactions are involved, or time effects
are important (or predominant), it is extremely important that tests be
of sufficiently long duration so that meaningful and representative
data can be obtained. There is particular uncertainty in the testing
required to analyze coupled/interactive thermal effects of waste
emplacement on the host rock and groundwater. The test plan should
discuss how the data from such long-duration tests (if any) will be
used in the repository design.

4.6 Special Testing

Special tests in this paper refer to the unconventional or non-standard
tests or both, for instance, accelerated tests to simulate long-term
effects in a short-duration test period and tests to assess
interactions among different processes, such as

Thermal/Hydrologic/Mechanical/Chemical (THMC) effects. These types of tests should be conducted after proper determination of their appropriateness, adequacy, and procedures. The test plan should discuss the need for and the rationale behind such complex tests and present details on how the data will be analyzed and used.

Because the coupled/interactive processes will take place following waste emplacement, the test plan should either provide for direct testing of the coupled/interactive behavior or demonstrate that such testing is not needed. The need for coupled/interactive testing should be based on site-specific conditions. If DOE establishes that coupled/interactive testing is not needed, then DOE should provide a technical evaluation identifying the volume of affected rock for which no performance credit is taken.

5.   SUMMARY

The Nuclear Waste Policy Act of 1982 and NRC Rule 10 CFR Part 60 require the U.S. Department of Energy to undertake a program of site characterization for each of the three sites nominated as a potential site for a nuclear waste repository. In situ testing will be an important activity in the site characterization process. In situ testing results will be the basis for (a) evaluating the suitability of each site for a nuclear waste repository, (b) determining representative parameters to be used in the design of the repository, and (c) resolving issues regarding the expected performance of the repository prior to licensing.

Each in situ test plan will be site specific. However, in developing the in situ test plans for the sites being characterized, DOE should follow similar guidelines. First, DOE should plan and perform necessary and sufficient amounts of testing to support the issuance of a construction authorization. Second, the in situ test program should be developed to adequately characterize the host rock prior to construction and waste emplacement as well as determine the post construction and emplacement characteristics of the host rock and engineered barriers. Third, DOE should present the in situ test plans in the site characterization plans. Fourth, a rationale for in situ testing should be developed and presented in the SCP. Finally, a strategy should be developed for the integration of surface test data and in situ test data.

This generic position was developed and presented to the Department of Energy in an effort to provide ongoing pre-licensing guidance. The NRC staff envisions that adequate in situ test plans will be developed by the DOE if the provisions set forth in this GTP are met.

REFERENCES

Code of Federal Regulations, Title 10, Energy (10 CFR Part 60), U.S. Government Printing Office, Washington, D.C., January 1, 1985.
Nuclear Waste Policy Act of 1982, Public Law 97-425, January 7, 1983.
U.S. Nuclear Regulatory Commission, "Disposal of High-Level Radioactive Wastes in Geologic Repositories: Technical Criteria", Supplementary Information, 48 FR 28195, June 21, 1983.

....., NUREG/CR-2983, "Selected Hydrologic and Geochemical Issues in Site Characterization for Nuclear Waste Disposal," Lawrence Berkeley Laboratories, January 1983.

....., NUREG/CR-3065, "In Situ Testing Programs Related to Design and Construction of HLW Deep Geologic Repositories, Task 2," Vols. 1 and 2, Golder Assoc., March 1983.

Observation of the spreading of a fracture during and after hydraulic fracturing tests

W.R.Fischle, G.Kappei, Th.Meyer, M.W.Schmidt, K.Schwieger, E.Taubert &
K.Thielemann
*Gesellschaft für Strahlen- und Umweltforschung mbH München, Institut für Tieflagerung,
Braunschweig, FR Germany*
J.Eisenblätter
Battelle Institut e. V., Frankfurt/M., FR Germany

INTRODUCTION

The geotechnical preliminary exploration of unknown sections of the rock has gained greater significance in the course of the construction of increasingly larger underground workings at greater depths. Besides application of indirect geophysical measurement methods, which, in combination with the geological exploration serve to describe the structure of an area in particular, it is necessary to determine the state of stress according to size and direction, i. e. the stress tensor. Only by these means can shape and layout of caverns, rooms and other cavities be defined, thus enabling a long term prognosis. Critical areas can then be avoided, resp. such constructional methods can be employed that long term stability is guaranteed.

The Gesellschaft für Strahlen- und Umweltforschung mbH has carried out an extensive Research and Development Programme with regard to partial aspects of these general problematics in connection with the disposal of radioactive wastes in salt formations in cooperation with Battelle Frankfurt and with the support of the Bundesministerium für Forschung und Technologie (Federal Ministry for Research and Technology).

GOALS

Research and Development activities on the disposal of radioactive wastes in salt are taking place in the former Asse salt mine.

By using the hydraulic fracturing procedure in combination with acoustic emission measurements in the rock information on the magnitude of main stresses as well as on the position and extension of induced cracks is expected. The produced data are – apart from mining experiences – e. g. the basis for dimensioning of borehole casings as well as for the estimation of mutual stress influences on the supporting elements of mining constructions. Results of model calculations of selected supporting elements can be revised by these measurement data obtained in situ.

MEASUREMENT TECHNIQUE

The known hydraulic fracturing technique to determine stress intensities was tested in the Asse salt mine by Rummel (1979) and Frohn et al. (1982) and was used in connection with the acoustic emission analysis (Erlenkämper, 1982/84). Here reference is made to the corresponding literature while thus dispensing with a detailed description.

The hydraulic fracturing tests were carried out in a vertical borehole with a diameter of 56 mm between 878 m and 924 m depth in the north-eastern field in the Older Halite (see Fig. 1).

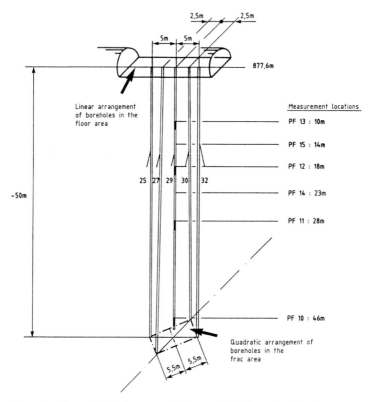

Fig. 1: Test field between 878 m and 924 m level

The area is situated at a large distance from workings in the southern field which terminate at 775 m depth. The packer probe had a length of 2.16 m. Two packers and the injection area took approximately 1/3 of the length. The packers as well as the injection area were pressurized independently with oil (type: Shell Tellus 32). Permeability tests were carried out under stepwise increased injection pressures of 50, 100 and 150 bar to test the leak-tightness of the system. In order to be able to localize acoustic emission (AE) events, the signal sequence was not to be too close. Therefore, special emphasis was laid upon a slow rate of pressure increase of about

2 bar/min contrary to the usual "sudden pressure increase" within a
few seconds. The manual control of the pressure rate permitted an
even increase of pressure only as long as no fracture progress had
set in. At fracture progress the entire pumping capacity had to be
turned on. After hydraulic fracturing haulage was stopped and the
shut-in pressure was measured. Subsequent further load changes served
to determine the refrac-pressure, i. e. reloading until further fracture progress sets in and to control the shut-in pressure.

Four further boreholes of approx. 100 mm ∅ were arranged around
the central borehole for the frac-tests in such a way that a square
of approx. 10 m diagonal length spread out over the centre. According
to the depth of the test acoustic emission transducers, which were
mounted on a support together with a preamplifier, were emplaced in
these boreholes. A pneumatically controlled cylinder pressed the
transducer, an PZT-ceramics element with a resonance frequency of
about 150 kHz, to the borehole wall. The amplified signals were
stored in digital form by a signal analysis system SAS 8, as shown in
Fig. 2.

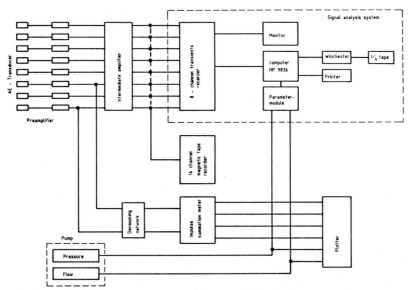

Fig. 2: Measurement arrangement to record acoustic signals during
hydraulic fracturing tests

The system SAS 8 consists of an 8 channel transient recorder KRENZ
TRC 4050, which processes the signals quickly (sample frequency up to
5 MHz, 8 bit resolution, storage capacity per channel 4096 points)
and a desk computer HP 9836. During the test the signals recorded by
the transient recorder were displayed by an oscilloscope and stored
on a hard disk by the computer. The evaluation took place later in
the laboratory. To follow up the acoustic emission activity of six
channels the so-called threshold count, i. e. the number of trans-
gressions of a given threshold was recorded. Moreover, it was possible to plot the parameters packer pressure, injection pressure,
flow rate and amount of flow.

TEST PERFORMANCE

In the first series hydraulic fracturing tests were carried out at
46 m (PF 10), 28 m (PF 11), 18 m PF 12 and 10 m (PF 13) depth. During
the first three tests PF 10, PF 11 and PF 13 three refracturing tests
each, with ensuing back flow measurements, were carried out after the
initial hydro-fracturing. In test PF 12 only one fracture test was
made. While in the case of PF 11 and PF 12 the pressure increase took
place at full pumping capacity (increase rate > 200 MPa/min), fracturing of PF 10 and PF 13 - and in the case of PF 10 also the 1. refracturing - was performed with a slow pressure increase rate of 0.2
MPa/min. In the course of test PF 10, 100 1 of stained $MgCl_2$-satu-
rated brine were injected at a flow rate of 4 l/min, subsequent to
three refracturing tests with back flow measurement. After the first
two hydraulic fracturing tests the positions of the transducers were
adapted to the respective change of depth by means of emplacing them
in a correspondingly higher location. Two transducers got stuck and
could not be recovered. The open boreholes were measured at the end
of the tests, and the coordinates of the transducer locations were
determined.
 For a better consolidation of the hydraulic facturing test data, a
second series of tests took place after 68 days at 23 m and 14 m,
namely frac-tests PF 14 and PF 15.

MEASUREMENT RESULTS

The data stored by signal analysis system SAS8 were processed as follows:
 Plotting of the events, as in Fig. 3; identification of the arrival
times; calculation of source location; recalculation of the arrival
times of the longitudinal (L) and transverse (Tr) waves; assessment
of the quality of the location; storage of the signal data and coordinates with a view to the later projection. Of the approximately
10 000 events which occurred during the test PF 10, 440 were recorded
and 235 of them were located, while most of the events due to obviously external sources were not included in the locating. Frequencies up to approx. 100 kHz have been observed with, however, gradually varying spectra; in general peak frequencies lay between 60 and 70
kHz. Fig. 4 shows a typical spectrum. The digital filtering of this
signal with a bandpass between 0 and 20 kHz shows in Fig 5 that higher amounts of low frequency portions are to be found particularly in
the transversal signal.
 Fig. 6 shows the hydraulic fracturing pressures, shut-in pressures
and refracturing pressures. The number of events do not yet permit
any substantiated statements. Direct differences between slow pressure increase (Pf 10 and PF 13) and fast pressure increase (PF 11, PF
15) as well as with the borehole standing unequipped for a longer period, as in the case of the tests carried out 68 days later (PF 14,
PF 15), are to be clearly recognized as regards the critical fracture
pressures. Upon taking the shut-in pressure to be smallest principal
stress it can be observed that the results from a series, e. g. PF
10, lie closely together. Furthermore, the smallest principal stress
increases with increasing depth. All measurement results are compiled
in Table 1.
 The coloured fracture plane was surveyed approximately 2 months af-

ter the test at mining of the gallery. Five vertical sections are shown in Fig. 9 which were produced in the course of driving of the gallery. The northern boundary is demarcated by the gallery wall. To the south the trace of the fracture is to be seen clearly in the floor of the gallery.

Table 1: Results of fracturing tests at the 878 m level, vertical

Meas-ure-ment loca-tion	Bore-hole depth	Depth	Crit. frac pres-sure	Shut-in pres-sure	1. Re-frac	Shut-in sure	2. Re-frac	Shut-in sure	3. Re-frac	Shut-in press.	Press. incr.
	[m]	[m]	[bar]	[bar]	[bar]	[bar]	[bar]	[bar]	[bar]	[bar]	
PF10	46	924	325	186	295	192	281	189	291	185	2 bar/min.
PF11	28	906	284	168	214	168	214	164	208	162	imm.
PF14*	23	901	320	166	245	168	–	–	–	–	imm.
PF12	18	896	260	177	–	–	–	–	–	–	imm.
PF15*	14	892	318	147	248	152	252	152	–	–	imm.
PF13	10	888	290	170	250	168	243	163	238	160	2 bar/min.

imm. = immediately
*These tests were carried out 68 days subsequent to the others

Surface contour lines have been outlined in Fig. 10, basing on the determined coordinates. It should be observed that the lines refer to 700 m below sea level and that their increasing values are therefore indicative of greater depths. If, starting out from the place of injection as origin of the fracture, the line of dip is illustrated as in vertical section I, then the angle of dipping of the fracture plane can be determined up to a distance of about 2.9 m with 27°, resp. 34°. Casts of the borehole wall, however, displayed traces of fracturing parallel to the borehole axis. In so far this area is not quite specific, and it must be assumed that the fracture has rotated immediately after its development, resp. that it has developed from a number of small cracks. At a greater distance from the place of origin the fracture plane dips at an angle of approx. 10° to 13°, as shown in vertical section II. Both angles and the local arrangements are to be found in the results of the seismic location in the y-z profile, Fig. 7, taking into account the rotated x-axis.

There the calculated coordinates of the selected events from PF 10 were represented graphically in a rectangular system of coordinates. It can be seen in the x-z plane that most events take place above the origin of coordinates as the centre of the injection area. A preferred orientation is clearly recognizable in the y-z plane. Starting out from negative y-values the fracture rises steadily in a northerly direction at an angle of about 33°. At about $y = + 2.5$ m the accumulation of the events is almost parallel to the y-axis. The stagewise

SIGNAL 06 (Diskette 030)   Messung ASSEF (28 Jun 1984)

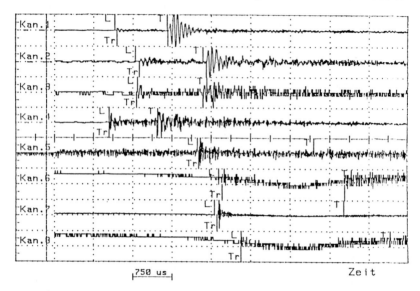

Fig. 3: Typical acoustic emission event during a frac-test,
registered by 8 channels by means of signal analysis system

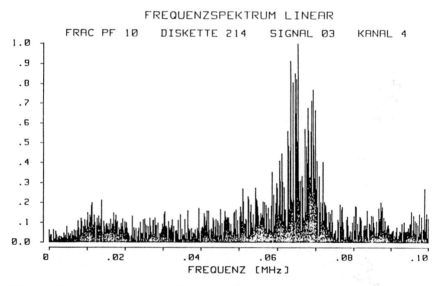

Fig. 4: Frequency spectra of channel 4

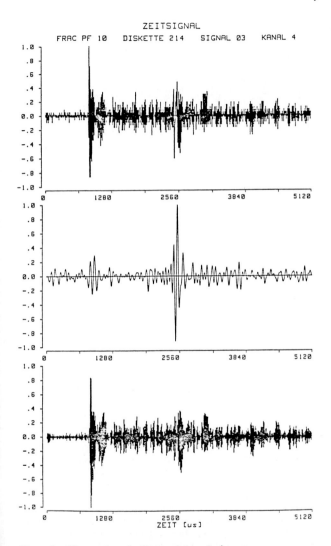

ZEITSIGNAL

FRAC PF 10   DISKETTE 214   SIGNAL 03   KANAL 4

Fig. 5: Time signal from channel 4.
Top: registered signal, Middle: after low frequent filtering
(0 to 20 kHz), bottom: after high frequent filtering (40 to
90 kHz)

progress of the fracture plane is shown in Fig. 8. Subsequent to the
initial hydraulic fracturing the fracture already extends by about
6 m in the N-S-direction and about 7 m in the W-E-direction. After
the injection of 100 l brine a fracture plane of approx. 88 m^2 has
been reached in the x-y-projection. The fracture then extends to approx. 12.5 m in N-S direction and to about 11.9 m in E-W direction.

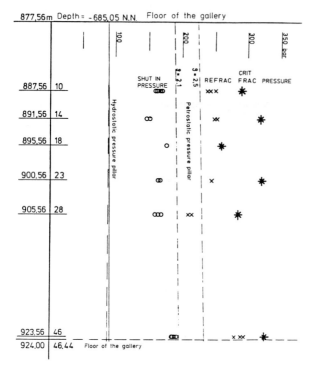

Fig. 6: Results of the hydraulic fracturing tests from the 878 m level

INTERPRETATION

Upon considering the extension of the fracture plane due to the microseismic events as seen in Fig. 8, then the result is not contradictory to the vertical sections of Fig. 9. It needs to be taken into consideration that the final boundaries of the fracture plane were not determined in Fig. 9 and that Fig. 8 does not comprise all AE events. As the AE detection network had been laid out in a square with a diameter of 10 m, the events lying far without this range were not always clearly recognizable and were therefore ignored.

Assuming that the smallest principal stress is normal to the fracture plane, it is noticable that it lies approximately in the direction of the vertical stress.

The reasons for the alterations in the surroundings of the origin of the fracture have not yet been finally explained. The packer cast of the borehole wall renders no information on the spreading of the fracture. Calculations on changes of stress as a result of the hydraulic fracturing process are to explain the possible changes of the direction of the fracture. References to the accuracy of the determined stress directions are given among other things by the overall structure of the salt dome, which extends lengthwise more or less in E—W—direction and has a strong vertical anticlinal fold.

SUMMARY

AE measurements were combined with hydraulic fracturing. The expansion of a purposely produced fracture was recorded and represented by means of the location and evaluation of seismic events. This is an important aid in the interpretation of states of stress in salt. The measurement in boreholes and the registrations in situ permit an investigational radius of action down to approx. 100 m depth. The measurements were confirmed by subsequent registration of the stained fracture with good accuracy. It was not possible to determine the position of the fracture plane by means of packer casts at the injection location. The acoustic emission method also enables the pursual of individual expansion steps and, e. g., the correlation of rates of events with the momentary injection pressure or with the injected amount of liquid. Using this information it is possible to give an account of the processes of origin of fractures.

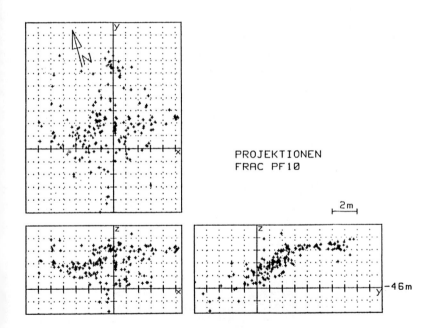

PROJEKTIONEN
FRAC PF10

2 m

−46 m

Fig. 7: Projection of the detected AE sources to the three coordination planes at Fracture PF 10

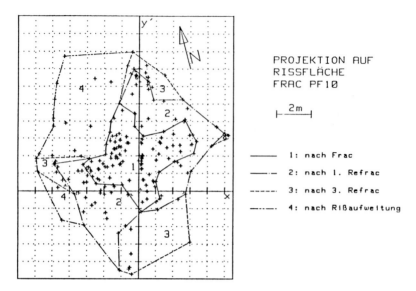

PROJEKTION AUF
RISSFLÄCHE
FRAC PF10

├── 2 m ──┤

——————   1: nach Frac

——·—·—   2: nach 1. Refrac

————————   3: nach 3. Refrac

——··——··—   4: nach Rißaufweitung

Fig. 8: Projection of the locations to the fracture plane showing the extension after several steps at PF 10

Fig. 9: Fracture traced by surveying at mining of the drift

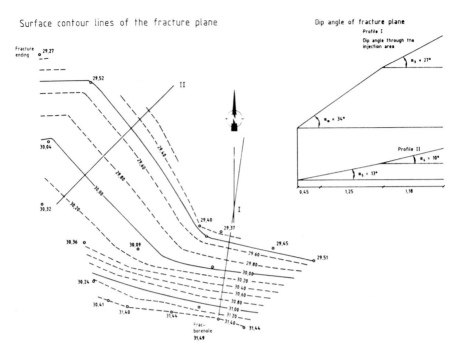

Fig. 10: Surface contour lines and profiles of the fracture plane

REFERENCES

Eisenblätter, J. 1986. Einsatz der Schallemissionsanalyse zur Über-
wachung von Teilbereichen eines Salzstocks während und nach der
Einlagerung radioaktiver Abfälle, Stufe II. Bericht BIeV-R-65.241-2
zum Vorhaben KWA 5209. Frankfurt: Battelle Inst.

Erlenkämper, S. 1982. Einsatz der Schallemissionsanalyse zur Über-
wachung von Salzstöcken während und nach der Einlagerung radioaktiver Abfälle, Stufe I. Bericht BIeV-R-64.288-1 zum Vorhaben KWA
2110. Frankfurt: Battelle Inst.

Erlenkämper, S. 1984. Einsatz der Schallemissionsanalyse zur Über-
wachung von Salzstöcken während und nach der Einlagerung radioaktiver Abfälle, Stufe II. Bericht BIeV-R-65.241-1 zum Vorhaben KWA
5209. Frankfurt: Battelle Inst.

Flach, D., Frohn, C., Hente, B., Schmidt, M. W. and Taubert, E. 1983.
Hydraulic fracturing experiments in rock salt with seismic frac location. Proc. 1. Symposium Field Measurements in Geomechanics, Zürich.

Frohn, C., Grafe, V., Schmidt, M. W. and Urff, A. 1982. Geotechnische
in situ Untersuchungen zum Frac-Verhalten von Salzgebirge. Proc.
ISRM Symposium, Aachen.

Rummel, F. 1981. Hydraulic Fracturing Spannungsmessung in der Schacht-
anlage Asse. Az. 7084809-1. Bochum: Ruhr Universität.

Instrumentation for nuclear waste studies in salt

Frank S.Shuri & Gordon E.Green
Shannon & Wilson, Inc., Seattle, Wash., USA

1 INTRODUCTION

The United States plans to extensively characterize three sites for
possible storage of high-level radioactive waste. At the present
time, one of these potential sites is in bedded salt. An underground
facility consisting of two 800-m-deep shafts and approximately 1,500 m
of horizontal openings will be constructed to evaluate the thermal,
hydrologic, and geomechanical properties of the salt and surrounding
rock types. The shafts and underground openings will be
comprehensively instrumented to provide data for evaluating design
assumptions, monitoring actual performance, and validating
constitutive models of salt behavior. In addition, several major
thermomechanical tests will be performed to obtain basic data on the
performance of various repository components, such as the waste
package and backfilled rooms, as well as additional data for model
validation under elevated temperature conditions. These tests will
also be heavily instrumented. A more detailed description of the
underground testing program in salt is given by Kalia et al. (1986)
and Golder Associates, Inc. (GAI) (1986). The currently planned type,
number, and purpose of the various proposed instruments is presented
in Table 1. Other instruments and equipment associated with
geophysical, geologic, and hydrologic characterization testing are
included in the program but are not discussed here.
    In this context, it is clear that satisfactory instrument
performance is necessary to achieve the objectives of the underground
testing program. However, it is often difficult to obtain accurate,
reliable geotechnical measurements even in conventional mining and
civil engineering applications. Success is difficult to evaluate,
because in many instances, instrument calibration is minimal, and
field performance is measured by the user's confidence in the data,
based on whether the data fit a particular theory or whether the
results seem plausible. Measurement problems are more severe for
nuclear waste repository studies in salt because

1. The accuracy requirements for repository studies are generally
higher.

2. In many tests, measurements must be made at significantly
elevated temperatures, up to 150°C.

3. Corrosion problems in salt are more severe than in most other
geologic materials.

4. The required lifetime for instruments in repository studies is
often substantially longer than in many conventional applications, for
example 3 to 5 years or more versus 6 months to 1 year.

Table 1. Summary of Instrumentation for Underground Testing in Salt

Instrument	Number	Function
Multiple Position Borehole Extensometer	148	Measure rock mass deformation parallel to borehole
Strain Gage	484	Measure strain in concrete and steel of shaft liners
Convergence Equipment	239 pairs of studs	Measure closure of underground openings
Jointmeter	40	Measure displacement between preliminary and final shaft liners
Borehole Closure and Shear Gages	44	Measure radial borehole closure and offset along bedding planes, respectively
Inclinometer	900 m of casing	Measure borehole position and rock mass deformation, transverse to borehole.
Deflectometer	150 m of casing	Measure borehole position and rock mass deformation, transverse to borehole, in horizontal plane
Slot Cutting Tool	200 tests	Measure absolute magnitude of stress in rock mass
Biaxial Borehole Stressmeter/ Hydraulic Pressure Cell	217	Measure stress change in rock mass
Embedment Stress Cell	326	Measure stress change in concrete and backfill
Piezometer	266	Measure pore fluid pressure in geologic materials
Temperature Sensor	2895	Measure temperature in rock mass, concrete, other instruments, and hydrologic test fluids

5. Quality Assurance and licensing processes require that instrument performance be clearly demonstrated in a formal, defensible way. All data may be subject to intense public scrutiny, debate, and legal action.

For these reasons, few geotechnical parameters can be measured in the field with a high degree of confidence unless the most appropriate instruments and techniques, tailored to the specific rock types and site conditions, are employed. This paper presents the results to date of an ongoing study to identify the best instruments. The information presented here is based on (1) an extensive literature review; (2) visits to test facilities and discussions with researchers, manufacturers, and consultants; and (3) the authors' own experiences. It is recognized that a study of this type can never be complete, and that much useful information has not yet been identified or incorporated. Nonetheless, the material in this paper is presented (1) to familiarize the geotechnical community with several innovative approaches to in situ measurements, and (2) to elicit comments on the advantages and pitfalls of the proposed techniques.

2 PERFORMANCE REQUIREMENTS

Performance is a general term encompassing several aspects of an instrument's ability to obtain the desired measurement. For this study, the most important criteria are accuracy, temperature range, longevity, and capability for automation.

Accuracy is defined as the difference between the measured value and the true value of a parameter at the measurement location. Accuracy requirements for the various parameters pertinent to the underground testing program are summarized in Table 2; ranges of accuracy have been included where different applications have different requirements. Appropriately, no accuracy requirements have been formally specified for U.S. repository studies, because overall repository performance depends on a complex interaction of many factors which in turn are functions of many parameters; the overall accuracy of performance predictions depends on the accuracy of parameter measurements in such a way that no individual measurement must be made to an absolute accuracy. In this context, the accuracy requirements presented in Table 2 were determined on the basis of (1) accuracy requirements of other nuclear waste studies in both salt and hard rock, (2) the perceived overall level of accuracy required to meet the needs of the program, and (3) the known capabilities of existing geotechnical instrumentation. It is recognized that these accuracy figures are preliminary and may be revised as performance assessment and repository design requirements become more quantitatively defined.

Table 2. Estimated Measurement Requirements in Salt

Parameter	Application	Accuracy, + or -	Max. Temp. Limit, °C
Deformation	Rock mass, parallel to borehole	0.03 to 0.1 mm	60 to 150
	Rock mass, transverse to borehole	0.08 to 0.4 mm per meter of borehole	60 to 80
	Convergence of of underground openings	0.1 mm	60
	Borehole closure and shear offset	0.03 mm	60 to 150
	Strain in concrete and steel components	5 microstrain	60 to 250
	Relative displacement of shaft liners	0.03 mm	60
Stress	Absolute	10%	60
	Change in rock mass, concrete, or backfill	10%	60 to 200
Pressure	Pore water pressure	1 kPa	60
Temperature	Instrument corrections	1°C	150
	Fluid property corrections for hydrologic tests	1°C	60 to 200
	Rock mass	0.5 to 1°C	60 to 200
	Heater surface	1°C	250

Temperature range requirements are summarized in Table 2 as the maximum operating temperature of the instrument. Function at elevated temperature includes not only physical survival, but also the ability to accurately characterize thermally-induced changes in instrument output that are not related to changes in the parameter being measured.

Longevity refers to the ability of the instrument to function and maintain accuracy for a specified time. The term is almost synonymous with reliability, but emphasizes the time aspect more strongly. Longevity is achieved primarily through design and materials selection, once the instrument has satisfied the accuracy and temperature requirements. The desired lifetime of most instruments for the underground testing program in salt is 3 years, based largely on the duration of the major thermomechanical tests. It is recognized that some instruments which are inaccessible for repair or replacement, such as those in the concrete/steel shaft liners, will need to function longer than 3 years. These will be engineered for the maximum possible lifetime subject to practical considerations, and redundant instruments or supplemental measuring techniques have been incorporated in the monitoring program to compensate for potential limits of longevity.

The capability to collect and store instrument readings automatically is required because of the large number of instruments. A computer-based automatic data acquisition system is being designed with a capacity of 10,000 channels. For purely logistical reasons, it is not feasible to manually collect and analyze this amount of data. More importantly, in some large thermomechanical tests, each of which has several hundred instruments, it is highly desirable that parameters be measured under similar conditions (i.e., as simultaneously as possible). Instruments that can be automated have been selected once the accuracy, temperature range, and longevity criteria have been satisfied.

3 DEFORMATION MEASUREMENTS

Multiple Position Borehole Extensometer (MPBX)

Deformation of the rock mass will be measured primarily with MPBXs. The U.S. Department of Energy, Salt Repository Project Office (DOE/SRPO), recently funded development of an MPBX designed specifically for use in salt. A prototype instrument, see Figure 1, has been designed, fabricated, and tested to a limited extent by Geokon, Inc., and Soil and Rock Instrumentation (SRI). At the present time, this extensometer is the leading candidate for use at the salt site, although additional characterization work needs to be done, and the final design may contain significant modifications. The innovative features of this MPBX include

1. Lengthening the body of the C-ring anchor to improve stability, and fabricating it from compressed graphite for reduced weight (easier installation) and increased corrosion resistance.

2. Identifying a rod material (carbon fiber reinforced vinyl ester resin composite) with an extremely low coefficient of thermal expansion. Initial test results indicated a value of 0.04 ppm/$^{\circ}$C. If this material proves to be suitable, the low thermal expansion would eliminate the need for temperature measurements along the rods to provide corrections for rod expansion. Absence of creep and warping, as well as evaluating the variability of the coefficient of thermal expansion between rods, still need to be investigated.

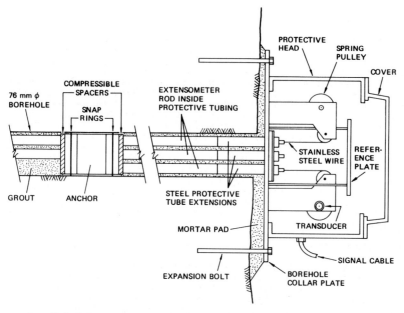

Figure 1.   Multiple Position Borehole Extensometer

3. Using a rotary resolver as the displacement transducer. This
sensor is a rotary transformer whose voltage output changes as the
coils are displaced relative to each other. The output is converted
to a digital form by signal conditioning electronics. The primary
advantage of this transducer is stability; calibration tests at
temperatures up to 120°C showed no measurable zero offset or change in
calibration factor. Stability tests at 90°C for several weeks also
showed no measurable change in output. The resolver will divide a
circle into 1,000 parts, so that the resolution and range of the
extensometer depends on the mechanical linkage between the measuring
rods and the resolver. At present, this is accomplished by connecting
the end of the rod to a thin stainless steel aircraft cable, which is
then wrapped around a pulley and tensioned with a constant force
spring.

Apart from the considerations noted above, the major uncertainties
relate to head design and use of grout in the borehole. The MPBX head
is a relatively complex assembly of pulleys and other mechanical
components that may be susceptible to friction, particularly if it
cannot be perfectly sealed. With respect to grouting, it seems
desirable to fill the borehole with a strong grout, as shown on Figure
1, to resist radial borehole closure that could eventually deflect the
rods and introduce significant uncertainty in the measurement.
However, this approach also increases the vulnerability of the
extensometer to shear offsets along clay seams in the bedded salt that
intersect the borehole, as observed at the Waste Isolation Pilot Plant
(WIPP) in New Mexico (Bechtel, 1985). A better understanding of this
problem is needed before the installation technique is finalized.

Strain Gages

Strain gages will be installed on steel liner plate, in concrete, and on/in reinforcing bar in the shaft liners. Vibrating wire instruments will be used because (1) they exhibit good long-term stability; (2) they are rugged and have a relatively long lifetime, which is desirable since they cannot be replaced; and (3) the output signal has low susceptibility to electrical noise, which will allow wires to be run several hundred meters to a location suitable for automatic data acquisition equipment (Bordes and Debreuille, 1985; DiBiagio, 1986). All strain gages will be of the "pluck and read" type to avoid signal conditioning electronics in nonaccessible locations. On the liner plate, strain gages will be spot welded in place and protected with layers of rubber or epoxy compounds and metal foil. Strain gages embedded in concrete will have supports and anchors designed in such a way that intimate contact will be maintained with the concrete as it shrinks during curing, for example by using convex surfaces and multiple holes in the anchor, removable support wires, etc. For reinforcing bar, strain gages will probably be installed in a small axial hole in a section of bar.

Convergence Equipment

A conventional tape extensometer will be used to make convergence measurements. Hook-and-eye attachment points such as those manufactured by the Slope Indicator Co. will be used; these points are manufactured from stainless steel bar on a numerically-controlled machine, so that the dimensions are very precise and the repeatability of tape attachment is high. These points will be anchored in the rock or concrete shaft liner by grouted bar or expansion shell anchors.

Manual convergence measurements in the shafts will have a significant impact on facility construction and operation. To avoid this problem, an automatic non-contacting monitoring system is desirable. However, no system has yet been identified with the required accuracy (0.1 mm) across the shaft diameter (4 to 6.5 m). Potential approaches include optical and acoustic systems, but considerable evaluation work remains to be done.

Jointmeters

Jointmeters will be installed in the 15-cm-wide gap between the preliminary and final liners in the upper 300-m section of the shaft, which passes through several major aquifers. This gap will eventually be filled with bitumen to provide a water barrier. Jointmeters will also be used to monitor chemical seals installed in the lower section of the shaft to prevent water movement along the liner/rock interface. In both applications, these instruments will utilize conventional field-grade linear displacement transducers packaged in a bellows or O-ring sealed tube for protection. The primary concerns relate to survivability, as the bitumen is poured hot at 150°C, and the chemical seal is a potentially corrosive, high-pressure environment.

Borehole Closure and Shear Gages

Borehole closure gages will consist of 3 or more linear displacement transducers, such as linear variable differential transformers (LVDTs), mounted across several borehole diameters. A packer or mechanical anchor system will support the transducers. The components of this technology are used in other geotechnical instruments and can probably be assembled to form a closure gage without extensive modification.

Borehole shear gages will also use packers or mechanical anchors,
one on each side of the actively-shearing feature. The linkage
between the two anchor points can be accomplished in several ways, for
example by direct connection using one or more linear displacement
transducers, or by lateral movement of a reference surface inclined
with respect to the shearing feature (Sun et al., 1985). As with the
closure gage, the major task is to adequately package existing
components.

Both gages will be as retrievable as possible, to allow the
instruments to be removed for calibration. It will also facilitate
moving the gages to new locations in the boreholes where the most
significant deformations or offsets are occurring, which may not be
apparent at the beginning of the monitoring program. The gages will
also be designed to allow multiple installations in a single borehole,
if possible.

Inclinometer and Deflectometer

Borehole inclinometers are routinely used in geotechnical
applications. Both vertical and horizontal inclinometers have been
successfully used at WIPP. The maximum operating temperature,
however, is lower than the expected rock mass temperature of some
thermomechanical tests where it is desirable to use the inclinometer.
Consequently, modifications to the instrument or operating technique
are required for these applications. At the present time, circulating
cool air around the probe while readings are taken is the preferred
alternative.

Deflectometers have not been widely used in the geotechnical
industry, and the accuracy of data from existing instruments has been
relatively low. Unlike inclinometers, where each measurement is
independently referenced to vertical, each deflectometer measurement
is referenced to the preceding measurement. Consequently, errors
accumulate rapidly. The accuracy of existing deflectometers is
limited by the quality of the mechanical linkage between the two
halves of the instrument, sensor design and performance, and geometry
of the wheel assembly. To obtain a deflectometer that can satisfy the
requirements of this project, these factors must be improved. To
further increase accuracy, casings will be installed with access at
both ends, so that traverses can be run from both ends, and the
movement of each end can be independently measured by optical
surveying techniques.

4 STRESS

Absolute Stress Measurement

None of the presently available techniques for measuring absolute
in situ stress is adequate for use in salt. Techniques relying on
analytical methods, such as stress relief and borehole dilatometers,
have low accuracy in salt because the deformational behavior of this
material is complex and only partially understood. Geophysical
techniques are far too inexact and developmental to be useful.
Hydrofracturing may provide supplemental data, but recent laboratory
studies attempting to quantify the relationship between fracturing
pressure and in situ stress in salt have not been encouraging (Boyce
et al., 1984).

Passive borehole stress gages will theoretically measure stress as
the salt creeps on to them, but the time required to achieve
equilibrium may be unacceptably long. Techniques involving several
hydraulic cells with different initial setting pressures show some
promise (Lu, 1981), but further evaluation of this concept is
required.

Cancellation techniques are particularly attractive for use in salt, because they require minimal knowledge of constitutive behavior and material properties. To avoid the depth limitations of conventional flatjack equipment, a borehole cancellation technique using high pressure abrasive fluidjets to cut a slot has been developed to a conceptual level by GAI. As presently conceived, the tool will operate in a 50-mm-diameter borehole. At the test location, straddle packers will stabilize the tool in the borehole and isolate the test interval. High pressure fluid jets containing abrasive garnet powder will cut two planar slots along the borehole axis, each about 20 cm deep and 30 cm long, into opposing walls of the borehole. The resulting borehole closure perpendicular to the slot will be measured with displacement transducers mounted on the tool. The slot will then be pressurized to cancel the deformation. If necessary, it may be possible to control the pressure in the test interval during slot cutting so that no deformation whatsoever occurs. The fluid jet cutting principle has been successfully demonstrated in salt blocks by Flow Industries, Inc. The next development activity involves packaging the fluid jet nozzles and displacement sensors in a compact configuration that will fit in the borehole. The prototype tool will then be tested in concrete and salt blocks in the laboratory.

Stress Change

Measuring stress change in salt has generally been unsuccessful because of instrument characterization problems. Recognizing this, DOE/SRPO recently funded development of a stress change instrument specifically for use in salt. A prototype instrument (see Figure 2) consisting of a thick-walled stainless steel tube with vibrating wire deformation sensors mounted in duplicate across 3 diameters has been developed by Geokon, Inc. and SRI. This instrument also contains longitudinal vibrating wires to provide data to correct for the effects of longitudinal extension, and temperature sensors so that temperature effects can be compensated for. Work to date has consisted mainly of characterizing the instrument itself and verifying the data reduction algorithm in short-term tests in granite and salt blocks where elastic behavior was predominant. Results from this phase of the testing indicated accuracies on the order of $\pm$ 10% to $\pm$ 15%. Preliminary testing of a stressmeter grouted into salt blocks was also done to determine the time required for the stressmeter to register the total applied stress following a temperature change in the system. In this test, the contact between the stressmeter and the borehole wall is temporarily lost because (1) salt has a higher coefficient of thermal expansion than the stressmeter body, and (2) the boundaries of the test block were stress-controlled. Results from this phase of the testing were ambiguous, with the time required for the stressmeter to re-equilibrate with the surrounding stress field ranging from a few days to several weeks. Further work is needed to characterize the behavior of this instrument under simulated field conditions. However, the biaxial stressmeter has performed better than other instruments under laboratory conditions and may potentially satisfy the requirements of the underground testing program.

Because the biaxial stressmeter is still unproven, other concepts for stress change measurement in salt are also being evaluated. Hydraulic pressure cells have been used in salt, with the results ranging from poor to ambiguous. This is probably related to inadequate cell stiffness or overly complex installations which cannot be adequately characterized. Nevertheless, the concept of the hydraulic pressure cell is attractive, and experience to date provides useful guidance for future efforts. Designing a stiffer cell and improving installation techniques, possibly by cutting a close-fitting slot in the end of the borehole with a downhole chainsaw, will be investigated. Another type of flat borehole pressure cell that

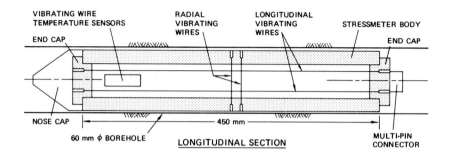

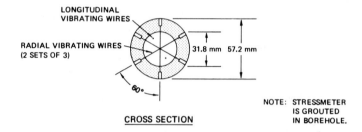

Figure 2. Biaxial Stressmeter

will be evaluated is the AWID cell developed in Germany for use at the Asse test facility in salt; a description of this instrument and some initial results are given by Kessels (1986).

Stress change in concrete and crushed salt backfill will be measured with embedment stress cells. Long-term reliability and, in concrete, high cell stiffness are major design requirements for these applications. Existing concrete stress cells and earth pressure cells all have some limitations, and it is believed that a more satisfactory cell (see Figure 3) can be developed by combining the best features of these instruments. Like the Carlson concrete stressmeter, this cell uses a thin film of mercury with a thin diaphragm on one side and a thick (rigid) diaphragm on the other. The new stress cell has two separate active faces to provide two stress measurements, which may improve confidence of data interpretation and provide redundancy. The vibrating wire pressure transducers are set off to one side, and repressurizing tubes, as used in the Gloetzl cell, may be provided. However, it is presently unclear whether a cell with a double active face can be built that is sufficiently stiff. Additionally, although diaphragm-type earth pressure cells with two active faces have demonstrated the concept, the usefulness of this approach for liquid-filled cells can be questioned; it may be argued that each face must experience the same pressure if the free-body equilibrium of the cell is maintained, and that this pressure may not reflect actual conditions on either side. If the double-face cell concept proves to be unworkable, a cell with a single active face and a vibrating wire pressure transducer mounted on the back plate, similar to the Carlson concrete stressmeter, can be used.

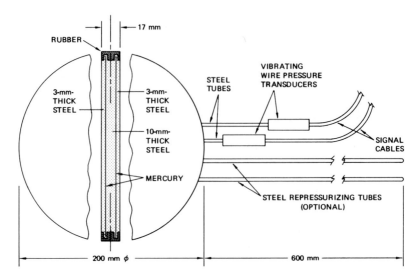

Figure 3.   Embedment Stress Cell

5 PORE PRESSURE

Piezometers in several different configurations will be used to
measure pore pressure. Where possible, such as installations in
relatively permeable strata below the test facility, the pressure
transducer will be mounted at the borehole collar, connected by a
small stiff metal tube to the completion zone. With this approach,
the transducer can be removed and recalibrated when necessary; for
this reason, the operating principle of the transducer is of secondary
importance, so long as it has the necessary accuracy and can be
automated. A variation of this approach, which may be required in
formations which are less permeable and require a minimum-volume
completion zone, involves access to the completion zone through a
plastic pipe and installing the transducer in the zone with a packer.
This approach also allows the transducer to be periodically removed
and recalibrated. A third approach will be used in situations where
the piezometer is inaccessible, for example, behind the shaft liners.
In this case, a vibrating wire transducer will be used because of its
ruggedness and good long-term stability. A manually-read pneumatic
transducer will also be installed in the same package to provide (1)
the capability for periodic verification of the vibrating wire
measurement, and (2) a backup should the other transducer fail.

6 TEMPERATURE

Temperature measurement technology in harsh environments is well
established from industrial applications, and has been successfully
demonstrated in earlier nuclear waste studies (Rogue and Binnall,
1983). A variety of potentially suitable transducers are available.
In the shaft liners, vibrating wire temperature sensors will be used
where very long lead lengths are required. For the major thermomechanical tests, thin-film platinum resistance temperature detectors
(RTDs) in Inconel 600 sheaths will be used to measure rock mass
temperature. Although thermocouples have generally been employed for
this application, it is difficult to obtain instrument accuracies
better than about $\pm$ 2°C unless laborious and time-consuming system

calibrations are performed. RTDs, on the other hand, are
significantly more accurate and are expected to eliminate the need for
system calibrations. These transducers also have excellent long-term
stability. As shown on Figure 4, RTD arrays will be installed in the
rock mass by grouting, with individual sensor tips positively
positioned against the borehole wall with mechanical spring clips.
The assembly will include an open central tube containing a removable
thermal baffle for access with a manual probe to periodically verify
measured temperatures.

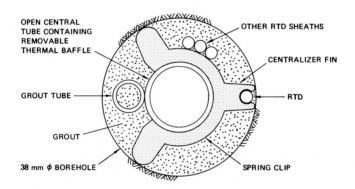

OPEN CENTRAL TUBE CONTAINING REMOVABLE THERMAL BAFFLE

OTHER RTD SHEATHS

CENTRALIZER FIN

GROUT TUBE

RTD

GROUT

38 mm ∅ BOREHOLE

SPRING CLIP

Figure 4. Resistance Temperature Detector Array Installation

7 SUMMARY

Various instruments have been evaluated for use in nuclear waste
repository studies in salt. With respect to satisfying the
requirements of this program, these instruments can be classified in
three groups:

1. Some instruments are clearly suitable with little or no
modification, and only a minor amount of characterization.
Temperature sensors, strain gages, manual convergence equipment, and
inclinometers fall into this category.

2. Some instruments have been identified that are expected, with a
high degree of confidence, to satisfy the desired performance
requirements. However, basic mechanical configurations must be
developed or substantial modifications must be made to the existing
design. Once the instrument is in prototype form, a moderate to large
effort will be required to characterize its behavior, and further
modifications may be required. Instruments in this category include
the MPBX, jointmeter, borehole closure and shear gages, embedment
stress cell, and piezometer.

3. The third category includes instruments that are not presently
capable of meeting the desired performance criteria. Although in all
cases promising and innovative approaches have been identified, some
are only at a conceptual stage, and all will require substantial
development effort. At the present time, it is not possible to
predict with a high degree of confidence that these approaches will be
successful. Instruments in this category include automated
convergence monitoring equipment, biaxial stressmeter, hydraulic
pressure cell, borehole slot cutting tool, and deflectometer.

ACKNOWLEDGEMENTS

This review of instrumentation capabilities and assessment of
development needs was performed by Shannon & Wilson, Inc., a
subcontractor to GAI, under Contract E512-09900 with Battelle Memorial
Institute, Project Management Division, Office of Nuclear Waste
Isolation (ONWI), under Contract DE-AC0283CH10140 with the U.S. DOE.
The interpretations and conclusions presented in this paper are solely
those of the authors and do not necessarily reflect those of ONWI or
DOE. The support of these organizations, and in particular the
assistance and encouragement of Dr. H.N. Kalia of ONWI and Dr. R.J.
Byrne of GAI is gratefully acknowledged. The authors also wish to
thank the many individuals in research, manufacturing, and consulting
organizations who generously contributed their time and expertise to
this study. A special note of appreciation is due Mr. J. Dunnicliff
for his comprehensive review of this material and many useful
suggestions.

REFERENCES

Bechtel, Inc. 1985. Quarterly Geotechnical Field Data Report, December
  1985. WIPP-DOE-221. San Francisco: Bechtel National, Inc.
Bordes, J.L., and Debreuille, P.J. 1985. Some Facts About Long-term
  Reliability of Vibrating Wire Instruments. Transportation Research
  Record 1004. Washington D.C.: National Research Council.
Boyce, G.M., Doe, T.W., and Majer, E. 1984. Laboratory Hydraulic
  Fracturing Stress Measurements in Salt. Proceedings of the 25th U.S.
  Symposium on Rock Mechanics. Baltimore: Port City Press.
DiBiagio, E. 1986. Comments on the Reliability and Performance of
  Vibrating-Wire Type Instruments. NGI Internal Report 55100-5.
  Oslo: Norwegian Geotechnical Institute.
Golder Associates, Inc. 1986. Draft Underground Test Plan for Site
  Characterization and Testing in an Exploratory Shaft Facility in
  Salt. Redmond: Golder Associates, Inc.
Kalia, H.N., J.F. Kircher, and T.I. McSweeney 1986. Underground
  Testing for Characterization of a Salt Site for the Repository. R.G.
  Post (ed.), Waste Management 86, Vol. 2, p. 193-200. Tucson:
  University of Arizona.
Kessels, W. 1986. Operational Principle, Testing, and Applications of
  the AWID-Flat Jack for Absolute Stress Determinations Using Voltage
  Measurements. Rock Mechanics and Rock Engineering, Vol. 19, p.165-
  183.
Lu, P.H. 1981. Determination of Ground Pressure Existing in a
  Viscoelastic Rock Mass by Use of Hydraulic Cells. Proceedings of the
  International Symposium on Weak Rock, Tokyo, Japan. Rotterdam:
  A.A. Balkema.
Rogue, F., and Binnall, E.P. 1983. Reliability of Geotechnical,
  Environmental, and Radiological Instrumentation in Nuclear Waste
  Repository Studies. NUREG/CR 3494. Livermore, CA: Lawrence
  Livermore National Laboratory.
Sun, Z., Gerrard, C., and Stephansson, O. 1985. Rock Joint Compliance
  Tests for Compression and Shear Load. Int. J. Rock Mech. Min. Sci. &
  Geomech. Abstr., Vol. 22, No. 4.

FE coupled process analysis of buffer mass test in Stripa project

Yuzo Ohnishi
Kyoto University, Japan
Akira Kobayashi
Hazama-gumi, Aoyama, Tokyo, Japan

1 INTRODUCTION

The general objective of the Buffer Mass Test (BMT) conducted for the Stripa Project, was to check the functions of highly compacted Na-bentonite as a canister overpack and of sand/bentonite mixtures as a tunnel backfill. The test arrangement consisted of six large boreholes for the setting of electrical heaters, surrounded by tightly fitting blocks of highly compacted sodium bentonite. The power of the heaters simulated the heat production of waste canisters. The holes were covered by the sand/bentonite backfill that was compacted on site in the tunnel which was located at a depth of 340 m in the Stripa mine. The temperatures, swelling pressures and water contents of the buffer materials were measured. The BMT involved the prediction of temperature distributions, moisture distributions and swelling pressure developments in the heater holes. The predictions were performed by Push,et al.(1985) and Knutsson(1983). Their analyses, however, were too simple to examine the influence of the change in water content and temperature on heat transfer and seepage in the buffer materials. Moreover, because of the difficulty in measuring the coupled effects, the complicated phenomena did not seem to be sufficiently grasped.

In this paper, by a newly developed FE code for the analysis of thermo-hydro-mechanical behavior, we tried to understand such complicated coupled phenomena, specifically the behavior of the highly compacted bentonite. The fundamental functions of this code have been verified in comparison with analytical solutions and experimental results.(Ohnishi, et al.(1985))

2 NUMERICAL METHOD

The governing equations used in the FE code were derived with the fully coupled thermal-hydraulic-mechanical relationships as shown in Fig.1. In the formulations, Fourier's law was employed for constitutive relationships for the energy flux, and the effect of temperature change on stress-strain relationships was taken into account by the extension of Biot's classical consolidation equations with Duhamel-Neuman form of Hooke's law.

The Galerkin method was used to formulate the finite element discretization. Linear quadrilateral elements for temperature and water pressure, and quadratic quadrilateral elements for displacement were

employed in the code.
The detailed derivation of these field equations and the general set up of this initial boundary value problem, along with a numerical solution approach is given in Ohnishi, et al.(1985).

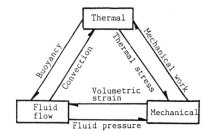

Figure 1. Interaction mechanisms of the fully coupled thermo-hydro-mechanical analysis.

3 MATERIAL NONLINEARITIES

Highly compacted clay (HCC) was used as a buffer material in the heater holes at the BMT. The dependency of hydraulic conductivity in the HCC on void ratio change, suction and degree of saturation, and that of kinematic vicosity of water on temperature were considered in the code.

3.1 Hydraulic conductivity of HCC

To investigate analytically the function of the HCC as an artificial barrier, it is necessary to know the characteristics of the unsaturated permeability change in the HCC during the process of swelling. The unsaturated properties in the HCC, however, were not investigated intensively in the Stripa project program. So, we assumed the following water retention curve dependent on the void ratio change on the basis of the report by Push,et al.(1985),

$$k=10^{-13}\exp(\frac{e-0.26}{1.09/2.30})\times S_r^{14.27} \quad (cm/s) \tag{1}$$

$$\psi=\exp(-1.76e+7.0)\times S_r^{-2.7} \tag{2}$$

where Sr designates a degree of saturation,and ψ denotes suction.
Eq.(1) is composed of the equation for relative permeability proposed by Irmay (1954), of which coefficients are deduced from the equation proposed by Nishigaki(1985) and the relationships between permeability and void ratio shown by Lambe and Whitman (1969). Eq.(2) is composed of the relationships between saturation and suction proposed by Brooks and Corey (1966), of which a coefficient is deduced from Mualem's work(1978).
It should be noted that these equations do not take the evaporation effect into consideration.

3.2 Kinematic viscosity

Kinematic viscosity is obtained from the interpolation in Table 1, in which it is shown as a function of temperature.

4 SWELLING PRESSURE

The HCC was made up of highly compacted Na-bentonite which had a high swelling capacity. The HCC absorbs water from the surrounding rock. The HCC swells with the water and has a tight contact with the confining rock and embedded waste canister.

In this paper, for the finite element computation, swelling pressure Ps is assumed as a prescribed force. While the degree of saturation Sr is less than 100%, Ps=ψ ,and while Sr equals 100%, Ps is obtained by interpolating the values shown in Table 2 as a function of temperature and density.

5 ANALYSES OF BMT

Numerical analyses have been performed with the axisymmetric finite element mesh as shown in Fig.2. The model contains the slot that separates the HCC from rocks. Case 1 is for a bentonite powder placed in the slot, Case 2 is for water placed in the slot, and Case 3 is same as Case 1 except for no thermal expansion. Material properties of the HCC are given in Table 3. The power of the electric heater is assumed to be 600 W.

We examined the following items:
1)the change of heat conductivity in the HCC with seepage and swelling,
2)the change of permeability in the HCC with seepage and swelling,
3)the effects of thermal expansion on heat conductivity and swelling capacity in the HCC,
4)the effects of the materials placed in the slot on heat transfer, seepage, and swelling in the HCC.

5.1 The change of heat conductivity in the HCC during seepage and swelling

Fig.3 shows the distribution of porosity,n,degree of saturation,Sr, heat conductivity, K_t, and permeability,k, with time. Fig.4 shows the comparison of the numerical results from Cases 1 and 2 with the experimental results of the BMT. The heat conductivity of the HCC is obtained from the equation $(1-n)K_{ts}+nS_rK_{tf}$, where K_{ts} is the heat conductivity of the clay particle and K_{tf} is that of water.

As shown in Fig.3, swelling causes an increase of porosity in the HCC. Sr goes up to 100% once after reduction of its value, depending on the state of the porosity and the permeability. As a result, the heat conductivity in the HCC gradually decreases with the increase of porosity.

The temperature distribution obtained in the field measurement is well predicted by the computation as shown in Fig.4.

Fig.5 shows the change of calculated temperature distributions with time. Though Fig.3c) shows the difference between the heat conductivity distribution in Case 1 after 3 weeks and after 4 months, Fig.5a) shows that the temperature distribution in Case 1 after 3 weeks is same as that after 4 months.

Moreover, Fig.5a) shows that the temperature increases in the beginning, but then and settles down to steady state after 7 weeks. This tendency is the same as the real phenomena shown in Table 4, which shows the comparison of the calculated temperature change with the measured one. This is probably because the heat conductivity in the HCC decreased temporarily at the 7th week, since the HCC was not saturated enough as shown in Fig.7a). This low water content is caused by an abrupt increase in porosity in the HCC as shown in Fig.6a). Fig.6 shows the change of calculated porosity distributions with time and Fig.7 shows the change of calculated water content distributions with time.

It is concluded that the heat conductivity distribution is dependent on the porosity change more effectively than on the water contents change, and that the temperature distribution is not influenced much by the change in heat conductivity distribution.

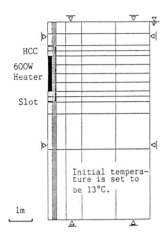

Figure 2. Finite element model of BMT. Right,top and bottom boundaries are set to be fixed temperature condition.

Table 1. Kinematic viscosity μ versus temperature.

°C	μ (m²s⁻¹)	°C	μ (m²s⁻¹)
	×10⁻⁶		×10⁻⁶
0	1.792	60	0.475
10	1.307	70	0.413
20	1.004	80	0.365
30	0.801	90	0.326
40	0.658	100	0.296
50	0.554		

Table 2 . Swelling pressure P_s versus bulk density ρ of HCC at different temperatures.

ρ(tf/m³)	P_s (tf/m²)		
	20°C	70°C	90°C
2.15	4592.	4082.	3571.
2.10	3061.	2041.	1735.
2.05	1531.	1020.	816.
2.00	714.	510.	408.
1.95	459.	306.	255.

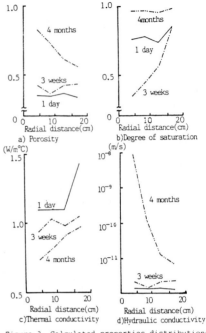

Figure 3. Calculated properties distributions at mid-height heater in hole.

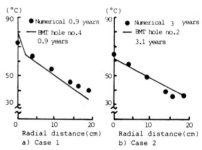

Figure 4. Comparison of recorded temperatures at mid-height heater in holes of BMT and calculated ones by numerical analyses.

5.2 The change of permeability in the HCC during seepage and swelling.

The permeability in the HCC is calculated by Eqs.(1) and (2). As shown in Fig.3, the permeability distribution is influenced more by the porosity than by the degree of saturation, and the permeability increases with the porosity.

Fig.8 shows the comparison of the calculated water content distribution with the measured one. However, it indicates that the calculated distribution of water content is higher than the measured one. This is possibly because the evaporation of water is neglected in

Table 3. Data of HCC used for analysis.

Property	Value
mass density	2.15 tf/m^2
porosity	0.42
Young's modulus	1.0×10^6 tf/m^2
Poisson's ratio	0.3
thermal expansion coef.	6.0×10^{-6} °C^{-1}
specific heat	1220. J/kg°C
thermal conductivity	1.46×10^{-3} kJ/ms°C
permeability	1.0×10^{-19} m^2

Table 4. Comparison of the temperatures at mid-height heater in hole no.1 recorded at BMT and calculated by numerical analysis.

Time	Heater surface		Rock/HCC interface	
	BMT	Numerical	BMT	Numerical
1 weeks	66	64.8	32	31.2
10weeks	70	78.6	33	38.8
1 year	65	73	34	39.9
2.4years	65	73.9	35	39.9 (°C)

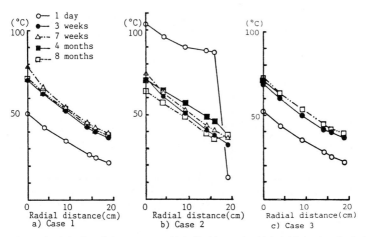

Figure 5. Calculated temperature distributions at mid-height heater in hole.

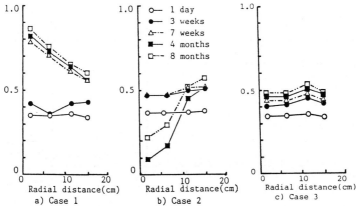

Figure 6. Calculated prosity distributions at mid-height heater in hole

our analyses. It is concluded that the water content may be strongly influenced by the water movement due to evaporation.

Figs.6a) and 7a) show a complicated unsaturated seepage behavior such that the saturation decreases temporarily at the 7th week because of the abrupt increase in porosity by swelling, and then the HCC is saturated gradually with the increase of permeability.

Fig.3d) shows that the permeability distribution in the HCC is not uniform. This is because the porosity distribution is not uniform by the

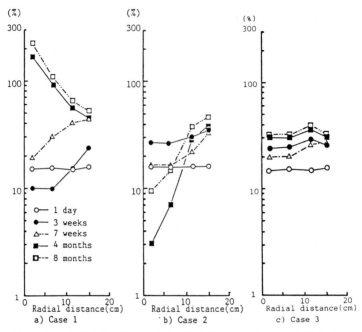

Figure 7. Calculated water contents distributions at mid-height heater in hole.

partial swelling.

As mentioned above, because of the strong dependency of the permeability on the porosity change, the seepage behavior is very complicated, depending on the condition of swelling. Moreover, the real behavior is expected to be more complex with the additional effects of evaporation.

5.3 The effects of thermal expansion on heat conductivity and swelling capacity.

Figs.6a) and 6c) indicate that in Case 1, the HCC swells near the heater more than near the rock. In Case 3, on the other hand, the HCC swells uniformly everywhere. Fig.9 shows the swelling pressure developments at the middle-height of the heater hole wall, and indicates that the value of the swelling pressure with thermal expansion is about half that

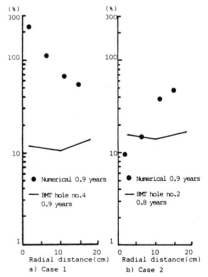

Figure 8. Comparison of water contents distributions at mid-height heater in hole and calculated ones by numerical analyses.

without thermal expansion.

We may overestimate the thermal expansion because of the assumption of linear relationships between thermal stress and temperature increase. But it is expected that thermal expansion has too much effect on the swelling behavior to be neglected.

Fig.9 shows that the calculated initial swelling pressure is larger than the measured one. This may be caused by the overestimation of suction at the initial state which is calculated by Eq.(2).

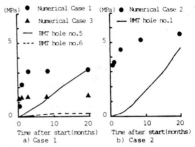

Figure 9. Comparison of recorded swelling pressure developments at mid-high part of hole and calculated ones by numerical analysis.

Fig.5 indicates no difference between temperature distributions at the steady state in Cases 1 and 3, and no temporary increase in the temperature at the 7th week in Case 3. This is probably because the temporary decrease in the water content at that time is too small for the heat conductivity to decrease much. It is concluded that the effects of thermal expansion on swelling are neglectable and that nonuniformity of the porosity distribution in the HCC is caused by thermal expansion when swelling by seepage is uniform like in Case 3.

Though the swelling in Case 3 is caused only by seepage, a temporary decrease in saturation after 7 weeks is shown in Fig.7c). This is possibly because the degree of saturation calculated by Eq.(2) is a little low.

5.4 The effect of filling material in the slot on heat conductivity, permeability and swelling capacity.

Fig.5 indicates that the temperature in Case 1 rises gradually, and the temperature in Case 2 settles down to steady state after an increase at the initial stage. This is probably because in Case 2, the heat is confined in the HCC at the initial stage, since it is not transferred by the high heat capacity of water in the slot. After a while, the temperature in the HCC begins to decrease.

Fig.6b) shows that porosity near the heater is smaller than that near the slot after 4 months. This is because the part in the HCC near the heater is compressed by swelling near the slot. In other words, swelling by seepage from the slot is larger than the thermal expansion by a heat increase of the heater. This is different from the phenomena in Case 1. Thus, it is expected that the swelling phenomena in the HCC is dependent on the material in the slot.

Fig.7b) shows that the water content decreases temporarily after 7 weeks. This is probably because water does not seep into the HCC enough to fill the void which has been expanded by swelling.

6 CONCLUSIONS

This paper describes a coupled thermo-hydro-mechanical analytical method and its application to the BMT. The conclusions are as follows:
1)The distribution of heat conductivity and permeability is more affected by the porosity change than by the water content change. And the distributions in the HCC become nonuniform with swelling.

2)The temperature distribution is not so dependent on the complicated change of heat conductivity distribution in the HCC.
3)The water retention curve Eq.(2), which takes the void ratio change into account, gives a slightly low degree of saturation. The suction at the initial saturated state is estimated to be higher, and the swelling at the initial stage is given to be larger than the real value.
4)The evaporation has much effect on the water content distribution in the HCC.
5)When water is placed in the slot that is the space between the HCC and rock, the temperature rises during the initial stage after the canister has been set up.
6)The swelling process is influenced by the filling material in the slot.
  It is expected that the real seepage process becomes more complicated in consideration of the additional evaporation effects.
  Although these analyses are based on many assumptions, they are helpful in understanding the phenomena which may not be observed in experiments.   The phenomena will be understood better by the defined experiments based on the results of these analyses.
  Further research is required on the unsaturated properties dependent on the void ratio change like Eqs.(1) and (2), and the constitutive law which includes swelling with seepage .

REFERENCES

Brooks, R.H. and Corey, A.T. 1966. Properties of porous media affecting fluid flow, ASCE, IR(92), pp.61–88.
Irmay, S. 1954. On the hydraulic conductivity of unsaturated soil. Trans. Amer. Geophys. Union, Vol.35(3), pp.463–467.
Knutsson, S. 1983. Buffer mass test-thermal calculation for the high temperature test, SKBF/KBS Internal Report 83–03.
Lambe,T.W. and Whitman,R.V. 1969. Soil Mechanics, John Wiley & Sons,Inc.
Mualem, Y. 1978. Hydraulic conductivity of unsaturated porous media generalized macroscopic approach, Water Resources Research, Vol.14, No.2, pp.325–334.
Nishigaki, M. 1983. Some aspects on hydraulic parameters of saturated-unsaturated regional ground water flow. Journal of the Japanese Society of Soil Mechanics and Foundation Engineering. Vol.23, No.3, pp.165–177.
Ohnishi,Y.,Shibata,H. and Kobayashi,A. 1985. Development of finite element code for the analysis of coupled thermo-hydro-mechanical behaviors of saturated-unsaturated medium, Int. Symp. on Coupled Process Affecting the Performance of a Nuclear Waste Repository.
Push,R. and Borgesson,L. 1985. Final report of the buffer mass test – Volume 11; test results, SKB Technical Report 85–12.

2nd International Symposium on Field Measurements in Geomechanics, Sakurai (ed.)   1027
© 1988 Balkema, Rotterdam. ISBN 90 6191 778 6

A new technique for measuring evaporation from the ground surface

Kunio Watanabe
Faculty of Engineering, Saitama University, Urawa, Japan
Hidehiro Tamaki
Sumitomo Cement Co. Ltd, Kanda-Mitoyocho, Tokyo, Japan
Shoichi Matsushima
Tokyo Keisoku Co. Ltd, Tachikawa, Japan

1 INTRODUCTION

It has been planned to dispose a huge amount of low radiation waste in many underground pits which are constructed in the shallow part of the ground. A suitability analysis based on the detailed investigation of the groundwater flow around these pits is indispensable before any disposal operation is begun. Groundwater flow under both saturated and unsaturated conditions can be fairly well analyzed by means of some calculation techniques such as the finite element method and the finite difference method, if the boundary conditions of flow and some hydrogeologic parameters of soil are properly given. However, many unsolved problems in the evaluation of the boundary conditions still remain. One of the largest problems is the estimation of the influence of evaporation on the groundwater flow passing near the ground surface. An exact measurement of the evaporation rate is a basic need for the estimation.

Evaporation from the ground surface has been a subject of great concern in some academic fields, e.g., agriculture, hydrology and meteorology. Many efforts have been made to create a practical technique which would enable on exact evaluation of the evaporation rate. Some techniques for this evaluation have already been proposed by previous authors (Penman (1948), Prevol et al.(1984)). Moreover, the sensitivity of the formula proposed by Penman has been studied by some authors (Saxton (1975), Cavillo & Gurney (1984)). These techniques are mainly concerned with the estimation of the average evaporation rate in a wide rural area. The evaporation rate is strongly affected by many factors, e.g., small scale geomorphology, soil structure, wind velocity near the surface, temperature, sunlight, land use and moisture content of soil. Consequently, the rate varies from place to place, and moreover, it considerably changes with time, even within a day. Recently, some studies have been done to clarify the influence of soil structure or sunlight on the evaporation rate (Hartman & De-Boodt (1984), Nakano & Cho (1983)). Generally speaking, however, it is very difficult to properly estimate the rate from a narrow and/or artificially changed area, such as the vicinity of the disposal pits, by some formulas. A practical technique to directly measure the rate is surely needed. For this reason, the authors of thia paper present a fundamental study on a practical technique, which enables precise evaluation of the rate in the field.

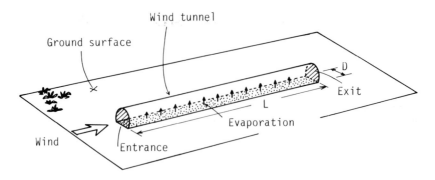

Figure 1. Schematic view of the measuring system

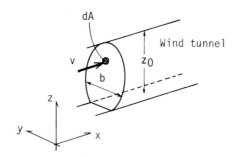

Figure 2. Wind passing through a cross-section of the tunnel

2 MEASURING TECHNIQUE

Vapor coming from the ground surface due to evaporation may be upwardly trnsported in air by wind motion and/or molecular diffusion.   If the air is standing and the movement of the vapor is in a steady state, vapor is transported one-dimensionally by molecular diffusion and the evaporation rate is essentially identical to the transportation rate in air.   There-fore, the evaporation rate from a horizontal ground surface can be eval-uated from the humidity gradient over the surface.   A measuring apparatus based on this idea has already made (Tokyo Keisoku Co. Ltd.: ETH 2001). However, there is seldom such an ideal condition in natural circumstances. Wind is more or less blowing under natural circumustances, so that the vapor may be mainly transported by the wind motion.   Also, the direction and velocity of wind frequently change near the ground surface.   Moreover, the flow condition of wind may often becomes turbulent, except for the thin layer formed immediately above the surface.   The mechanism of vapor transportation under an intense wind is quite different from that under a no-wind condition.   Furthermore, it must be noted that the wind increases the evaporation rate, because the humidity immediately above the surface is decreased by its motion.   Another technique which makes it possible to measure under natural conditions must be developed.
    The measuring apparatus proposed here is schematically shown by Figure 1. A small wind tunnel made of a transparent pipe is put on the ground

surface. The lower part of the pipe is cut off so that vapor can freely enter into the tunnel from its base. The length of the tunnel and the width of its base are refered to as L and D, respectively. Now, let's assume that the wind is blowing from the left hand side (entrance) to the right hand side (exit) of the tunnel. As vapor is continuously supplied from the tunnel base, the total flux of vapor passing through the cross-section of the exit (Fv_{out}) must be larger than that through the entrance section (Fv_{in}). The evaporation rate from the unit area of the tunnel base (Ev) can be calculated from these Fv_{out} and Fv_{in} values with following equation:

$$Ev = (Fv_{out} - Fv_{in})/L/D \qquad (1)$$

The total flux passing an arbitrary section perpendicular to the tunnel axis (Fv) can be presented with notations in Figure 2 as:

$$Fv = \int_A v \cdot \theta \cdot S_r(t)\ dA \qquad (2)$$

where, A is the cross-sectional area, $v(y,z)$ is the wind velocity passing through a small area of dA, $\theta(y,z)$ is the relative humidity measured at this area and $S_r(t)$ is the absolute humidity at a temperature of t°C. When the wind velocity, relative humidity and temperature depend only on the height z from the ground surface, equation (2) can be rewritten as:

$$Fv = \int_0^{z_0} v \cdot \theta \cdot S_r(t) \cdot b\ dz \qquad (3)$$

where, z_0 is the tunnel height and $b(z)$ is the tunnel width at the height of z (see Figure 2). By combining equation (1) with either equation (2) or (3), the evaporation rate Ev can be calculated by putting distributions of relative humidity, wind velocity and temperature into these equations.

It should be noted that the wind tunnel intercepts the velocity component perpendicular to the tunnel axis. Therefore, the wind velocity measured in this tunnel is usually less than the net velocity around the tunnel. Consequently, the tunnel axis must be set in a direction identical to the flow direction of wind.

The largest problem to be solved is the proving of the accuracy and sensitiveness of the device for measuring the relative humidity. Some laboratory and field tests were conducted with the aim of fundamentally proving this point.

3 LABORATORY TESTS

Four cases of laboratory tests were firstly carried out to prove the applicability of this technique. The measuring system used is schematically shown by Figure 3(a). A wind tunnel 4m long and 17.4cm high was prepared. The diameter of the tunnel was 20.4cm and the width of the base 15cm. Wind was supplied by use of an electric fan. Three wire nettings were set in front of the entrance to the tunnel to make the wind uniform. This tunnel was put on wooden boards, except for the 60cm long test section, which was placed near the mid part of the tunnel. A toploader was set under the test section and two soil boxes of 30cm long, 15cm wide and 4cm high were put on the toploader as shown by Figure 3(b). These two soil boxes were filled with a mixture of fine sand (Toyoura standard sand) and water. The average diameter of the sand grain was about 0.18mm. The hydraulic conductivity and the porosity of the sand layer were about 0.006cm/s and 0.41, respectively. The surface of the sand layer in these boxes was set of the same level as the surface of the wooden boards. Evaporation occurs from the surface of the sand layer. The evaporation rate can be directly measured by the toploader. Distri-

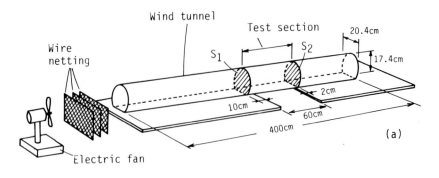

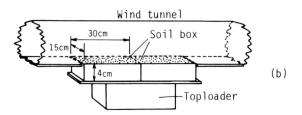

Figure 3. Measuring system used in laboratory tests

butions of relative humidity, wind velocity and temperature were measured
in two cross-sections of S_1 and S_2, which were located 10cm ahead and 2cm
behind the test section, respectively.   Relative humidity and temperature
were measured by two high accurate sensors utilizing a semiconductor.
The sensor was about 1.5cm long, 1.0cm wide and 4mm high.   The response
time of the sensor was about 3 minutes long.   As the response time is not
very short, every test had to be done under a constant temperature due to
the fact that the relative humidity was greatly affected by temperature.
Wind velocity was measured by use of a hot wire anemometer.   Conditions
given for each run are summarized in Table 1.
    The wind velocity distributions in the two sections, namely, S_1 and S_2
were measured prior to every run.   Examples of the velocity distributions
in these sections are shown by Figure 4.   It can be seen that the low
velocity zone is commonly formed at the left hand corner of the lower part
of these sections.   Figure 5 shows relative humidity distributions in S_1
(θ_1) and S_2 (θ_2) sections, and the difference (dθ) between θ_1 and θ_2 measured in Exp-1.   Figure 6 shows results observed in Exp-4.   Figure 5(c)
and 6(c) clearly show that the humidity difference is extremely large at
the left hand corner of the lower part of these sections.   The humidity
distribution is closely related to the velocity distribution.   It was
found that the humidity difference becomes large when the wind velocity
decreases.   Figure 7(a) and 7(b) show temperature differences between
these two sections observed in Exp-1 and Exp-4, respectively.   Evaporation needs latent heat, so that the temperature in S_2 section is lower
than that in S_1.   Figure 8 shows humidity differences observed in Exp-2
and Exp-3.

Table 1. Conditions given in every test

Test case	Average temperature $\bar{t}$	$S_r(\bar{t})$	Water in soil boxes	Wind velocity*
Exp-1	18.8 °C	16.1 gf/m^3	720.6 gf	0.7 m/s
Exp-2	19.1 ″	16.4 ″	355.8 ″	0.6 ″
Exp-3	17.5 ″	14.9 ″	240.3 ″	0.7 ″
Exp-4	17.1 ″	14.6 ″	123.6 ″	0.7 ″

*Velocity at the center of S2 section

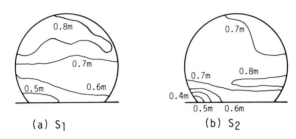

(a) S_1 (b) S_2

Figure 4. Wind velocity distributions in sections S_1 and S_2 (dimensions are in m/s)

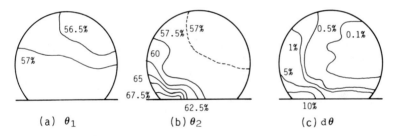

(a) θ_1 (b) θ_2 (c) dθ

Figure 5. Relative humidity distributions measured in Exp-1. (a) and (b) are distributions in sections S_1 and S_2, respectively. (c) is the humidity difference between these sections. dθ means $\theta_2 - \theta_1$.

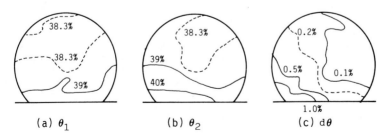

(a) θ_1 (b) θ_2 (c) dθ

Figure 6. Relative humidity distributions measured in Exp-4. Every figure has the same meaning as those in Figure 5.

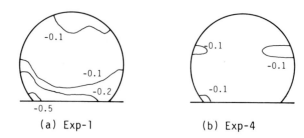

Figure 7. Temperature difference between sections S_1 and S_2 (dimensions in °C). Values mean $Ts_2 - Ts_1$, where Ts_1 and Ts_2 are temperatures in S_1 and S_2, respectively.

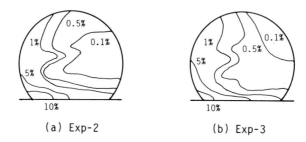

Figure 8. Humidity differences (dθ) observed in Exp-2 and Exp-3.

Table 2. Comparison between calculated evaporation rates (Ev_{cal} and Ev'_{cal}) and measured rates (Ev_{obs}).

Test case	$Ev_{obs}* (10^{-3})$	$Ev_{cal}* (10^{-3})$	$Ev'_{cal}* (10^{-3})$
Exp-1	15.7	14.8	7.5
Exp-2	10.0	10.7	9.9
Exp-3	11.8	13.7	7.5
Exp-4	2.3	2.2	1.4

* Dimension : $gf/cm^2/hour$

Evaporation rates in every test are tested through calculation by putting the measured distributions of relative humidity, wind velocity and absolute humidity into equations (1) and (2). Hereinafter, the rate calculated by these equations and that directly measured by the toploader are refered to as Ev_{cal} and Ev_{obs}, respectively. Equation (2) must be used to estimate the evaporation rate because relative humidity and wind velocity considerably vary along any horizontal line drawn in these sections. Equation (3) can not be used. However, as the response time of the humidity sensor is not short, much time is required to measure the humidity distribution in each section. Under conditions where the wind velocity changes frequently, a precise measurement of the distribution is very difficult to obtain. On the contrary, it is easier and often said more practical to measure only the distribution along the vertical line passing

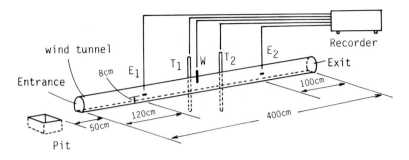

Figure 9. Measuring system used in the field test. E_1, E_2 : humidity and temperature sensors, T_1, T_2 : tensiometers, W : anemometer

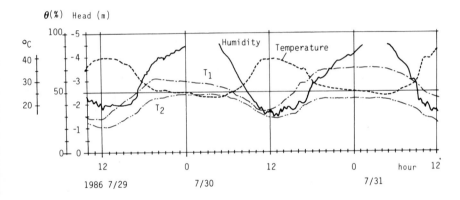

Figure 10. Transient changes of average temperature, average humidity and suction heads observed in the field test.

through the tunnel center.   Taking this point into consideration, the evaporation rate Ev'_{cal} can also be calculated by equations (1) and (3). Ev_{cal}, Ev'_{cal} and Ev_{obs} obtained in every run are summarized in Table 2. It can be seen that Ev_{cal} values are in good agreement with Ev_{obs}.   However, as the wind velocity and relative humidity distributions greatly incline toward the lower left corner of the tunnel, Ev'_{cal} values are lower than Ev_{obs}.   Although a few problems remain to be overcome, it can be concluded that this technique, all in all, has a high availability for use in the evaluation of the evaporation rate in the field.

4 FIELD TEST

A field test was carried out in the yard of Saitama University's campus during the period of 7/28 - 7/31, 1986.   Main attention was centered on the examination of the sensivity of the humidity sensor.   There had been no rain during one full week preceeding the above mentioned period. Therefore, the moisture content near the ground surface was considerably

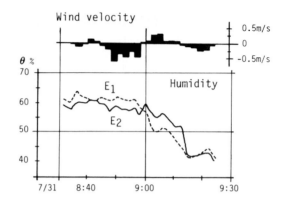

Figure 11. Relation between wind velocity and humidities measured

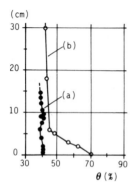

Figure 12. Vertical profiles
of humidity distribution.
(a) bared area, (b) grass
field

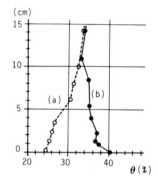

Figure 13. Vertical profiles
of humidity distribution at (a)
the entrance and (b) the exit
sections.

low during the field test.   The measuring system is schematically shown
by Figure 9.   A wind tunnel which was the same as that used for the laboratory test was installed on the ground surface.   Grass growing at the
tunnel base was completely pulled out before the start of the test.   Two
humidity sensors (E_1 and E_2) were set at a height of 8cm from the ground
surface in the tunnel.   The distance between these two sensors was 1.8m.
Temperature at these points was also measured by these sensors.   A small
pit was made 50cm in front of the tunnel entrance for the installation of
an electric fan.   However, the fan was not actually used during the test
at all.   A hot wire anemometer was fixed at the middle point between the
two humidity sensors.   The suction head of soil near the ground surface
was measured by the use of two tensiometers (T_1 and T_2). T_1 and T_2 measured suction heads at a depth of 7cm and 11cm, respectively.   All data
of humidities, temperatures, wind velocity and suction heads were automatically recorded.
   Transient changes of mean relative humidity and temperature between the

two sensors and suction heads measured in the period of 7/28 to 7/31 are
shown in Figure 10.   Temperature rises in the morning and drops in the
evening.   On the contrary, relative humidity decreases in the daytime and
increases at night.   As the humidity near the surface reached 100% at
midnight, dew was created on both tunnel base and wall at this time.   Unfortunately, humidity over about 85% can not be measured by these sensors.
Because of this, the data taken at midnight is insufficient.   The evaporation rate becomes extremely high in the morning because dew is rapidly
evaporated.   The change pattern of suction heads shows a similar tendency
as that for temperature.   However, this data for the suction heads may
lack reliability because the tensiometer is strongly influenced by temperature.

Figure 11 shows the relation between the wind velocity and humidities
observed by both sensors on the morning of July 29.   Mean velocity, averaged in 2 minutes, is presented in this Figure.   The positive value of
the velocity means that wind is passing from the tunnel entrance to the
exit.   Wind direction frequently changed in the morning.   As the evaporation rate in the morning is high as mentioned above, the influence of
wind direction on the humidity differences between these two points is
clearly seen in this Figure.   When wind velocity is possitive, relative
humidity measured by the sensor E_2 becomes larger than that measured by
E_1.   On the other hand, when the velocity is negative, the humidity at
the location of E_1 is larger than that at E_2.

The vertical distribution of relative humidity near the ground surface
measured on the afternoon of July 30 is presented in Figure 12.   Two profiles of the distribution measured on (a) bared area and (b) grass field
are presented.   The average height of the grass was about 6cm.   The
average wind velocity and temperature were about 0.5 m/s and 42 ℃, respectively.   Humidity becomes high near both of these ground surfaces.   The
humidity over the grass field was considerably higher than that over the
bared area.   Figure 13 shows two vertical profiles of relative humidities
measured along vertical lines passing through the centers of the tunnel
entrance and exit.   The average wind velocity and temperature at this
time were 0.5 m/s and 41°C, respectively.   It can be seen from the profile in the entrance section that the humidity becomes low near the ground
surface.   On the other hand, the humidity at the lower level in the exit
section is relatively high.   The humidity profile in the entrance section
seems to be influenced greatly by the complicated wind motion formed by
the pit located in front of the entrance.   At any rate, the difference
between the two humidity distributions can be clearly seen.   From this
fact, although the field tests are not totally sufficient, it can be said
that this technique is applicable for practical purposes.

5 CONCLUSIONS

A practical technique for measuring the evaporation rate from the ground
surface is proposed in this paper.   The technique is essentially based
on the fact that evaporation must increase the humidity in the air above
the surface.   A wind tunnel is used to measure the rate under conditions
when the wind is blowing.   The evaporation rate can be evaluated by calculating the vapor fluxes passing through two different cross-sections of
the tunnel.   The applicability of this technique has been successfully
proved by some laboratory and field tests.

ACKNOWLEDGEMENTS

The authors are grateful to Dr. T. Asaeda of the University of Tokyo, Mr. K. Masumoto of Power Reactor & Nuclear Fuel Development Corporation, and Mr. H. Nomoto of Saitama University for many helpful suggestions during the course of this work.

REFERENCES

Cavillo, P. J. & R. J. Gurney 1984. Asensitivity analysis of a numerical model for estimating evapotranspiration. Water Resour. Res. 20: 105–112.
Hartman, R. & M. De-Boodt 1984. Infiltration and subsequent evaporation from surface aggregated layered soil profiles under simulated laboratory conditions. Soil Sci. 137: 135–140.
Nakano, Y. & T. Cho 1983. Effect of transient, partial-area shading on evaporation from a bare soil. Soil Sci. 135: 282–295.
Penman, H. L. 1948. Natural evaporation from open water, bare soil and grass. Proc. R. Soc. London A193: 120–145.
Prevol, L., R. Bernard, O. Taconet & D. Vidal-Madjar 1984. Evaporation from a bare soil evaluated using a soil water transfer model and remote sensed surface soil moisture data. Water Resour. Res. 20: 311–316.
Saxton, K. E. 1975. Sensitivity analysis of the combination evapotranspiration equation. Agric. Meteorol. 15: 343–353.

2nd International Symposium on Field Measurements in Geomechanics, Sakurai (ed.)
© 1988 Balkema, Rotterdam. ISBN 90 6191 778 6

Field measurement techniques applied to slurried fine grained coal mine tailings

David J.Williams & Peter H.Morris
Department of Civil Engineering, University of Queensland, Brisbane, Australia

1 INTRODUCTION

The expanded production of coal in Australia over the last decade has led to an escalation in the volume of fine grained coal mine waste produced, which has created disposal problems. The fine grained coal mine waste is generally deposited as a slurry in tailings dams, formed in disused open pits, in ponds, or in disused underground workings. Existing disposal areas are becoming depleted, and there is a need to study the performance of tailings deposited by conventional means in order to make more efficient use of remaining and future disposal areas.

An ongoing study of tailings dams at New Hope, Aberdare and Rhondda Collieries, located in the Bundamba District of the West Moreton Coalfields in South East Queensland, Australia, has involved the application of a variety of field measurement techniques. These have included survey levelling to monitor surface settlement, crusting and beaching, vane shear strength testing, dynamic cone penetrometer testing and the use of a nuclear probe for determining moisture content and density profiles with depth. The field measurements have been accompanied by laboratory classification testing including moisture content determination, Atterberg limit testing and particle size distribution analysis. During the course of the work it became apparent that some estimation of the accuracy and reliability of the data collected was essential. A relationship appeared to exist between the peak shear strength obtained using the vane shear apparatus and the corresponding moisture content of the tailings determined in the laboratory. Even plotted to logarithmic scales, the data cloud displayed significant scatter and it was necessary to account for the randomness of individual data points in determining the most appropriate form of regression analysis to be applied.

Calibration of the nuclear probe became a major obstacle. Initial attempts to calibrate the probe for moisture content determination, by obtaining corresponding moisture contents in the laboratory, proved to be too unreliable. It was also found that the decay of the sources was sufficiently rapid that this factor could not be ignored. Attempts were then made to calibrate the probe in the laboratory under carefully controlled conditions. The calibration of the nuclear probe was found to depend, among other things, on the material and dimensions of the access tube, on the chemical composition of the tailings under investigation and on the interdependence of moisture content and density determination. In the field, it is also dependent on the

method of installation of the access tube. This paper highlights the
importance of determining the accuracy and reliability of measured
data, with particular reference to the shear strength versus moisture
content relationship suggested for slurried fine grained coal mine
tailings and the calibration of a nuclear probe for use in coal tailings.

2 SHEAR STRENGTH VERSUS MOISTURE CONTENT RELATIONSHIP

2.1 Base data

In order to examine the relationship between the shear strength and
moisture content of fine grained coal mine tailings initially deposited
as a slurry, a number of disused tailings dams located in the Bundamba
District of the West Moreton Coalfields in South East Queensland,
Australia, was investigated. The investigation involved advancing
boreholes by hand augering from the crusted surface of the tailings.
Each 0.2 m length of a borehole was bag sampled after obtaining an
estimate of the peak undrained shear strength using a 110 mm long by 55 mm
diameter vane shear device operated by hand using a torque wrench.
Only one moisture content determination was made for each bag sample
and only one vane shear test was conducted at each depth interval. The
rate of vane shear testing was such that the peak shear strength was
obtained after about 30s of rotation at an angular rotation of the
order of 45°. Since only a single value of moisture content and shear
strength was obtained at each depth interval, no information describing
the randomness of the distributions of either variable was obtained.
However, it was discovered that the relationship between shear strength
and moisture content varied with the mean particle size of the tailings.
As the mean particle size increased, the shear strength at a particular
value of moisture content decreased. The data presented in this paper
is restricted to 64 data points from New Hope Colliery for which the
mean particle size of the tailings was reasonably constant at about
0.01 mm.
   The shear strength and moisture content data from New Hope Colliery
are plotted to logarithmic scales in figure 1. The data cloud shows
up to a 6-fold spread in peak shear strength at a given value of
moisture content and up to a 1.7-fold spread in moisture content at a
given value of peak shear strength.

2.2 Linear regression analyses

The most common application of linear regression to a set of data
involves the assumption that one variable is non-random and that all
randomness or error is in the other variable. The regression line
is then obtained by minimising the sum of the squares of the differences
between the random variable and its corresponding value on the regression
line. In some instances, where one variable can be precisely
determined, this approach is perfectly valid. However, in most instances
both variables are randomly distributed and account must be taken of
this in the regression analysis. A general solution for the linear
regression equation (1) is available, a detailed derivation for which
is given in York (1966).

(1) $y = a + bx$

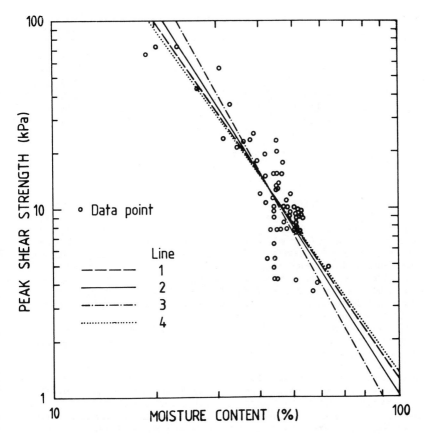

Figure 1. Shear strength versus moisture content relationship for New Hope Colliery tailings

where x and y are 2 dependent variables related linearly, and a and b are constants defining the dependence between x and y. In the general solution, the randomness of one variable is assumed to be independent of the randomness of the other.

In order to be able to apply the general solution, the variances of each variable over their full range of values must be determined. The reciprocals of the variances, which may or may not differ over the range of measured values, give the weights to be applied in the regression analysis. The base data obtained from New Hope Colliery provided no information about the variability of individual data values and it was necessary to carry out further work. It was decided to carry out this work under controlled laboratory conditions, so that the full range of values of both variables could be readily obtained. A bulk tailings sample was obtained from New Hope Colliery and to assess the randomness of moisture content determinations, 16 samples were made up by combining known weights of oven dried tailings and water to cover the range of moisture content from 3 to 70%. After 24 hours curing, 14 to 16 determinations of moisture content were made from each sample. The

Table 1. Summary of results of regression analyses

Line	a	b	CV^*_τ (%)	CV_w (%)	CV_τ/CV_w
1	12.74	-2.719	15.1	1.8	8.4
2	13.75	-2.989	15.1	4.0	3.8
3	15.47	-3.447	40	22	1.8
4	12.36	-2.619	all error in τ		∞

* CV is the coefficient of variation

results gave a coefficient of variation (standard deviation expressed as a % of the mean) for moisture content of 1.8% and an approximately constant variance (mean value of 3.58×10^{-4}) over the full range.

   To assess the randomness of shear strength determinations, 12 samples covering the strength range 1 to 102 kPa were tested using a laboratory vane with 13 to 18 determinations per sample. The results gave a coefficient of variation for shear strength of 15.1% and an approximately constant variance (mean value of 0.0267) over the full range. The weighted regression line corresponding to the 2 laboratory determined variances is given as line 1 on figure 1.

   Limited further work in the field suggests that the coefficient of variation for moisture content determination of field samples is about 4.0% and the coefficient of variation for field determined shear strength is similar to that obtained in the laboratory. A possible explanation for this latter suggestion is that the small volume of soil effected by a small laboratory vane (measuring 12.6 mm long by 10 mm in diameter roughly compensates for the larger natural field variability which is measured by the use of a larger vane. The weighted regression line corresponding to these values for the coefficients of variation is given as line 2 on figure 1.

   Harr (1977) suggested that for natural clays coefficients of variation of 22% and 40% could be expected for moisture content and shear strength, respectively. These values give line 3 on figure 1. Taking the extreme case that all randomness or error is in the shear strength gives line 4 on figure 1. The equation to the regression lines is given by:

(2) $Ln (\tau) = a + b\,Ln (w)$

where τ is the peak shear strength, and w is the moisture content. The results of the regression analyses are summarised in Table 1. A sample correlation coefficient of -0.833 applies to the analyses.

   The ratio of the observed maximum spreads in peak shear strength and moisture content values, observed on figure 1, is 3.5. This prompts the choice of line 2 as the best representation of the functional relationship between shear strength and moisture content for New Hope Colliery tailings based on the work to date.

3 CALIBRATION OF THE NUCLEAR PROBE

3.1 Description of apparatus

The nuclear probe used was a Campbell Pacific Nuclear Corporation 501 Depthprobe Moisture/Density Gauge, which employs backscatter techniques. The probe has 2 radiation sources and detectors. In the Americium 241 - Beryllium neutron source, the fast neutrons emitted by the source are moderated by collisions with surrounding nuclei, until they slow to thermal level. Collisions with hydrogen nuclei (protons) are most efficient at slowing fast neutrons and as water is the most common source of hydrogen in soils, the neutron source may be used to measure mositure content. The boron tri-flouride detector is sensitive to neutrons slowed to the thermal level. Unfortunately, the reading is also sensitive to the total chemistry of the soil (especially the presence of boron and chlorine), soil density, the material and dimensions of the access tube and the decay of the source. All of these factors must be accounted for in the calibration.
     In the Caesium 137 gamma source, the gamma rays are scattered by interaction with the electrons of the surrounding atoms. At the energy levels associated with the Caesium source, Compton scattering and the photoelectric effect are dominant. The scattered photons are countered by Geiger Muller tubes. As with the neutron source, a number of factors comes into play and all must be accounted for in the calibration.
     Previous attempts by others to use backscatter gauges for research into natural soils have failed because of the large variability of the soils and particularly their variable boron and iron contents. However, since coal tailings are relatively uniform greater success with the gauge could be expected in this material. In view of the interdependence of moisture content and density readings an iterative procedure is required to arrive at calibration relationships.

3.2 Compensation for source decay and random effects

The half lives of the neutron and gamma sources are 458 years and 30 years, respectively. The source decay follows an exponential relationship with time and the probe reading is proportional to the activity of the source at the time of the reading. It is necessary to determine the standard count and its variation with time. To obtain a standard count the source is placed in its shield and the shield placed on the fibreglass carrying case in an open area. A minimum of 30 readings is required for a standard count and 60 readings was adopted. The mean of these 60 readings is then the standard count. For periods of observation very much less than the half-life, the readings are randomly distributed about the mean according to the Poisson distribution. The logarithm of the standard counts plotted against time should be linear.
     To date 22 standard density counts have been taken over 134 days and 21 standard moisture content counts have been taken over 133 days. The standard counts are plotted on figure 2. Weighted regression lines are shown for each set of data, which indicate cummulative drops of 4.05% in the standard density count and 2.05% in the standard moisture content count. However, based on the known half-lives of the sources, only 0.84% and 0.06% drop for density and moisture content, respectively, can be accounted for by source decay over the monitoring period.

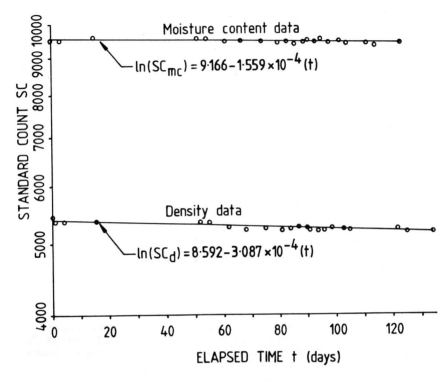

Figure 2.   Standard counts versus elapsed time

Monitoring over an extended period is expected to shed more light on
this aspect.
   The drop in the standard count is used to correct readings taken at
a given time and standard counts will be taken during the entire period
over which the probe is used.

3.3   Preliminary calibration for moisture content

A field calibration of the moisture gauge was attempted using 3 PVC
access tubes and 1 aluminium access tube. Two of the PVC tubes and
the aluminium tube were installed in deposited tailings at New Hope
Colliery and a single PVC tube was installed in finer grained deposited
tailings at Aberdare Colliery. The PVC tubing had an outside diameter
of 48.3 mm and an inside diameter of 40.5 mm. The corresponding
dimensions of the aluminium tubing were 48.4 mm and 39.0 mm, respectively
The tubes were installed closed-ended by hand augering using a 45 mm
diameter auger, which provided a close fitting hole for the tubes and
also provided samples for moisture content determination (mean value
of several determinations per sample). The moisture content ratios
(= moisture gauge reading at time t/standard moisture content count
at time t) obtained for PVC tube no. 1 in New Hope tailings are compared
with the measured moisture contents on figure 3. Even discarding the

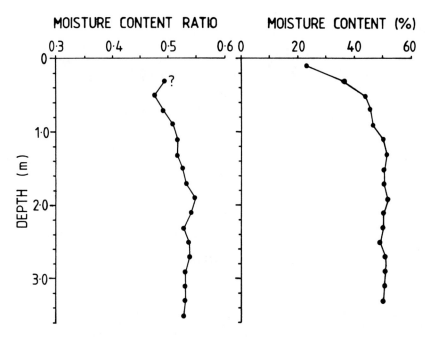

Figure 3. Moisture content ratio and moisture content profiles with depth for PVC tube No. 1 in New Hope tailings

uppermost moisture content ratio, which may be affected by closeness to the surface of the tailings and possible hole oversize, the shapes of the 2 curves show only poor agreement.

Plots of moisture content versus moisture content ratio for each of the access tubes are presented on figure 4. This shows very much higher moisture content ratios for the aluminium tube compared with those for the PVC tubes due to the high absorption of neutrons by the high levels of chlorine in the PVC. The data for the aluminium tube also show a lower sensitivity to moisture content. The New Hope data for PVC tubes show a higher trend line than the Aberdare data, and for PVC tube No. 2 in New Hope tailings there is a large amount of scatter and no trend line. This may be due to the installation hole for PVC tube No. 2 being oversize, but may also reflect greater variability in the chemical composition and particle size distribution of the tailings at that location.

The New Hope tailings were deposited progressively, while the Aberdare tailings were pumped from one location in the main tailings dam at Aberdare Colliery and deposited immediately within bunds. Greater material variability would therefore be expected in the tailings investigated at New Hope Colliery. The difference in the trend lines for New Hope and Aberdare tailings is due to the Aberdare tailings being finer grained, and hence having more mineral matter and less coal, and a different density. Clearly, separate calibrations are required for different access tubes, and account must be taken of the grain size and chemical composition of the tailings and of the interdependence of moisture content and density.

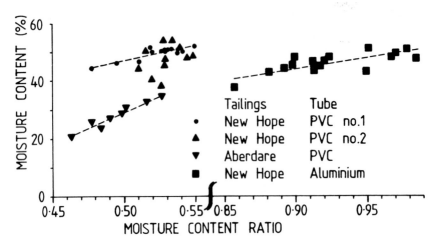

Figure 4.   Preliminary plots of moisture content versus moisture content ratio.

Table 2.   Calculated radii of the sphere of importance

Material	RI_{99} (mm)	RI_{96} (mm)
vacuum	∞	∞
water	167	113
water + salt (NaCl)	<167	<113
dry sand (SiO$_2$)	1648	1136
saturated sand	284	197
Aberdare tailings (40% ash)		
- dry	461	317
- saturated	219	119

3.4   Laboratory calibration for moisture content

Before attempting a laboratory calibration, it is necessary to define the sphere of importance of the probe in order to decide the sample size required. The sphere of importance was defined by Zuber and Cameron (1966) as the sphere around the source yielding 99% of the count rate obtained in an infinite medium. The calculation of the radius of the sphere of importance RI_{99} is based on three group diffusion theory with the neutrons split according to their energies, into 2 fast groups and 1 thermal group. The values of RI_{99} calculated for a range of materials are given in table 2.

A 44 gallon drum (285 mm radius) is therefore adequate for calibration in water, water with dissolved salt, saturated sand and saturated Aberdare Colliery tailings. Calibration in air should be carried out in a large open space with the source located inside a length of the appropriate tube. Semmler (1963) derived equation (3) for the moisture content count for a point source in a spherical cavity radius R_0:

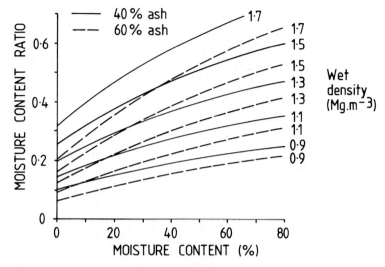

Figure 5. Calculated calibration curves for Aberdare Colliery tailings with 40% and 60% ash for the CPN 501 Depthprobe in a PVC tube

(3) moisture content count = $\dfrac{K_2}{4\pi \sum_a (L_1 + R_0)(L_2 + R_0)(L_1 + L_2)}$

where $\sum_a$, L_1 and L_2 are functions of the material density and chemistry and R_0 and K_2 are constants to be determined experimentally for a given probe and tube. Results from a minimum of 2 materials is required to estimate R_0 and K_2, although better estimates are obtained using more than 2 materials. Using distilled water and distilled water with 6.6% salt by weight gave values of 0.712 cm and 37.15 cm^2 for R_0 and K_2, respectively, for the CPN 501 Depthprobe in a PVC tube. Using these estimates of R_0 and K_2, the moisture content count for this probe/tube combination may be calculated for material of any density and chemistry. The calculated calibration curves for Aberdare Colliery tailings with 40% and 60% ash, which covers the range expected in the field, are given on figure 5. A measure of the wet density of the material is provided by the density count.

3.5 Calibration for wet density

Figure 5 cannot be used without a corresponding calibration for wet density. The 2 calibrations would then need to be used iteratively in order to converge on the best estimate of both moisture content and density for a given material. Christensen (1972 and 1974) describes a mathematical model for the calibration of a gamma probe, which is based on the Monte Carlo method. However, the necessary approximations make the model too inaccurate for the purposes of calibration and fitting to experimental results is required. A simpler phenomenological model described by Ballard and Gardner (1965) accounts for Compton scattering and the photoelectric effect which are the dominant features.

This model requires results from a minimum of 4 materials of known
density and chemistry to define the 4 constants involved. In addition,
the model assumes an infinite medium and its application to a finite
laboratory sample may require some modification. The determination of
calibration curves for wet density for coal mine tailings has yet to
be achieved and is the subject of ongoing work.

4   CONCLUSIONS

The application of field measurement techniques to slurried fine grained
coal mine tailings has highlighted the importance of determining the
accuracy and reliability of the measured data. Reference is made to
the shear strength versus moisture content relationship suggested for
coal tailings and the effect of adopting different values for the
coefficients of variation of the 2 parameters. The calibration of a
nuclear moisture/density probe for use in coal tailings is also
discussed. It has proved extremely difficult to obtain a reliable
calibration. However, considerable progress has been made, the
problem areas and required procedure have been defined and further
work should see a reliable calibration emerge.

5   ACKNOWLEDGEMENTS

The work presented in this paper has been made possible through the
co-operation of the managements of New Hope and Aberdare Collieries
in South East Queensland, Australia, which is gratefully acknowledged.

REFERENCES AND BIBLIOGRAPHY

Ballard, L.F. and Gardner, R.P. 1965. Density and moisture content
  measurements by nuclear methods. NCHRP Rep. 14, Highway Res. Board.
Christensen, E.R. 1972. Monte Carlo calculations for the subsurface
  gamma density gauge. Nucl. Eng. Des. 24:431 - 439.
Christensen, E.R. 1974. Use of the gamma density gauge in combination
  with the neutron moisture gauge. Proc. Symp. on Isotopes and
  Radiation Techniques in Soil Physics. IAEA, Vienna: 27-41.
Gardner, R.P. and Ely, R.L. 1967. Radioisotope measurement applications
  in engineering. Reinhold, New York.
Haahr, V. and Olgaard, P.L. 1965. Comparative experimental and
  theoretical investigations of the neutronic method for measuring the
  water content in soil. Proc. Symp. on Isotopes and Radiation in
  Soil - Plant Nutrition Studies, Ankara. IAEA, Vienna: 129-146.
Harr, M.E. 1977. Mechanics of particulate media. McGraw-Hill, New York
Neutron moisture gauges. 1970. Cameron, J.F., Ed. IAEA Tech. Rep.
  No. 112, Vienna.
Operator's manual: 501 depthprobe moisture/density gauge. 1981.CPN
  Corp, Pacheco, California.
Semmler, R.A. 1963. Neutron - moderation moisture meters: analysis of
  application to coal and soil. USAEC Rep. COO-712-73, Univ. of
  Chicago.
York, D. 1966. Least - squares fitting of a straight line. Can. Jnl.
  of Physics, 44: 1079-1086.
Zuber, A. and Cameron, J.F. 1966. Neutron moisture gauges. Atomic
  Energy Rev. 4 4 : 143-167.

6 Interpretation of field measurement results

Interpretation of plate-loading tests in tunnels

T.Rotonda
Department of Structural and Geotechnical Engineering, Rome, Italy

1 INTRODUCTION

One of the methods most widely used in determining the deformability of
rock masses is the plate-loading test. A uniform pressure is applied on
the rock surface in a circular area; the displacement can be measured on
the surface (above or around the loaded area) and/or at various depths
along the loading axis.

Often the test is carried out in exploratory drifts; the scheme which
is generally used in Italy is shown in Figure 1. Two plate-loading tests
are symmetrically carried out at two diametrically opposed areas of a
tunnel. The diameter of the plate, a, is usually 0.5 m and sometimes
1.0 m; the diameter of the tunnel, R, varies between 1.8 m and 3.0 m.
The displacements are measured near the boundary of the plate (at a
distance about 0.2a) and below the loading plate at various depths up
to 3.0 m.

The interpretation of the test is usually based on the theoretical
relationships which are valid for a uniform circular load on an elastic
homogeneous half-space (Figure 2). The displacements at the surface are
related to the applied load, q, and to the elastic parameters of the
rock E and ν by means of the relation

$$(1) \qquad u = k_x \left(\frac{x}{a}\right) \frac{qa}{E} (1 - \nu^2)$$

Likewise, the displacements at depth along the loading axis are given by

$$(2) \qquad u = K_h \left(\frac{h}{a}, \nu\right) \frac{qa}{E}$$

When measuring the surface displacements an "average" elastic modulus
can be determined from eq. (1). With in depth measurements the local
moduli between two reference points at depths h_A and h_B are obtained
for eq. (2)

$$(3) \qquad E = \frac{qa}{u_A - u_B} \left[K_h \left(\frac{h_A}{a}\right) - K_h \left(\frac{h_B}{a}\right) \right]$$

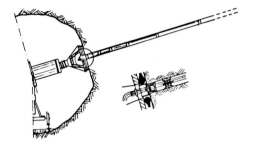

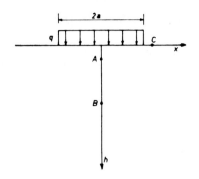

Figure 1.   Scheme of plate-
loading tests in an explora
tory tunnel.

Figure 2.   Plate-loading test
on the surface of a seminfinite
medium.

The values of the coefficients K_x and K_h are given for instance   by
Giroud (1972).

   As is discussed in another paper at this conference (Ribacchi, 1987),
the real situation can be somewhat different from the idealized model
and therefore various errors can be introduced. They may   derive for
instance from the anisotropy of the rock, the presence of a superficial
layer of loosened rock below the loading plate, the influence of   the
finite radius of the test tunnel. This paper is devoted to a detailed
analysis of this latter factor.

   Some preliminary results can be derived from a paper by Booker   and
Carter (1984); they can be utilized however only for the central part
of the loaded area and for a loading configuration which does not fully
correspond to that which is really adopted in practice.

2 METHOD OF ANALYSIS

The problem to be solved is that of a load applied to a portion of the
internal surface of a hollow infinite cylinder of radius R in a homogeneous elastic medium. Cylindrical coordinates r, θ and z (Figure 3)
are utilized. The displacements along these axes are indicated by u, v,
and w. Normal and shear components of the applied load at the tunnel
wall are indicated by τ_{rr}, $\tau_{r\theta}$, τ_{rz}.

   The solution to the elastic problem is found by applying a Fourier
transform with respect to the z axis and a Fourier series in the angle θ.
The elasticity equations are then reduced to a set of ordinary differential equations in the variable r. By solving these equations the transforms of the displacements (U, V and W) can be obtained in a Fourier
series as

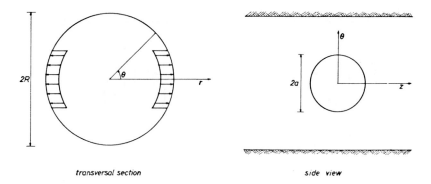

transversal section   *side view*

Figure 3.   Plate-loading test in tunnel.

(4)
$$\left[U^{(n)}, V^{(n)}, W^{(n)}\right]^T = \left[\Phi^{(n)}\right]\left[T_{rr}^{(n)}, T_{r\theta}^{(n)}, T_{rz}^{(n)}\right]^T$$

n indicates the general term of the Fourier series, $T_{rr}^{(n)}$, $T_{r\theta}^{(n)}$, $T_{rz}^{(n)}$
indicate the transforms of loads and the terms of the matrix $\Phi^{(n)}$
contain Bessel functions of order n. The displacement field is subsequently obtained by inverting the Fourier transforms. In a similar way
the strains can also be obtained.

The number of n terms in the Fourier series required to obtain a good
approximation depends on the load configuration and on the position of
the points in which the displacements are required; a number of terms
varying from 50 to 120 was utilized in this study. The inversion of the
Fourier transforms was carried out by a modified Clenshaw-Curtis procedure for the integration of oscillating integrals (Piessens and Branders,
1975).

Three load configurations were investigated. The first correspond to
the case discussed by Booker and Carter, that is a "square" loading area
having a width 2b along the z axis and a curved width $2b = 2\theta R$ in a
circumferential direction. The second case correspond to a radial loading on a circular surface. The solution was obtained by a numerical
integration: the circular area was subdivided into 36 rectangular strips
at a distance b_i from the loading axis, each having a width Δb_i and a
length $2\theta_i R$.

Finally, a third analysis was carried out for a situation similar to the
preceding one, but with a load acting along the axis of the circular
plate (uniaxial loading) (Figure 4).
In this case

(5)
$$\begin{aligned}
\tau_{rr} &= q \cos \theta \\
\tau_{r\theta} &= q \sin \theta \\
\tau_{rz} &= 0
\end{aligned}$$

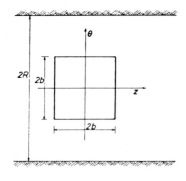 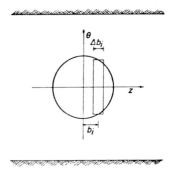

Figure 4. Loading test on square plate (left) and on circular plate (right).

with $|\theta| < \theta_i$ and $|\theta-\pi| < \theta_i$. The third situation more closely corre-
sponds to the configuration adopted in the in situ tests (although the
load is really applied on a flat surface) and therefore only these results will be presented here. It is to be noted however, that the solutions for the case of radial loading and those for uniaxial loading
are only slightly different, expecially for low values of the a/R ratio.

3 ANALYSIS OF THE RESULTS

The results have been obtained for values of plate radius/tunnel radius
ratios equal to 0.0, 0.1, 0.2 and 0.3. The former case corresponds to
the half-space condition; in situ test conditions will almost always
fall within the investigated range of a/R values. The analysis was
carried out for a Poisson's coefficient equal to 0.25.
    Figure 5 shows the surface displacements of the tunnel wall along a
circumferential direction. The scaled surface displacements obviously
decrease as the ratio a/R increases; for a ratio a/R = 0.2 (a typical
value for in situ tests) the displacement at the centre is about 0.89
with respect to the half-space condition. The decrease of the displacements, outside the loaded area, is sharper than in the half-space case.
These results can be utilized when surface measurements are carried out
in the in situ plate-loading test.
    Figure 6 shows the displacements of the points along the loading axis.
The depth h is obviously equal to r-R. The relative displacements of the
various points with respect to that at the surface are shown in Figure 7:
this graph is more convenient for practical utilization.
    The figures show that a conventional half-space analysis of relative
displacements below the plate would give overestimated values of the
true moduli. The error is slight when the displacements are measured
near the tunnel wall, and it increases with depth.
    For an evaluation of the errors in the estimation of the modulus be-
tween two measurement points at a short distance from each other in

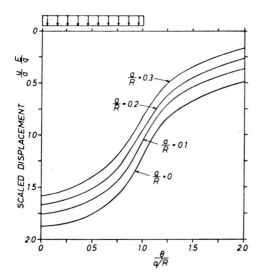

Figure 5. Scaled displacements at the surface of the tunnel in a circumferential direction.

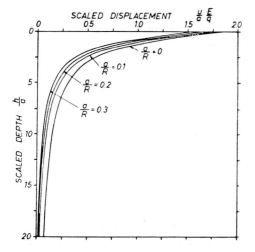

Figure 6. Scaled displacements below the center of the plate.

depth Figure 8 can be helpful. The radial strain at depth is lower than in the half-space condition, except in a small zone near the loading surface. Near the tunnel wall the differences are small, but they become increasingly important with depth and a conventional analysis would give unreliable results.

For a quick interpretation of the in situ tests with the configuration commonly utilized in Italy (Figure 1), it was deemed useful to prepare the graphs of Figures 9, 10, 11, each of which refers to a different value of the a/R ratio. Each solid line curve indicates the difference in displacement between the points at the beginning of the curve and any other point at greater depth; this displacement is scaled down with

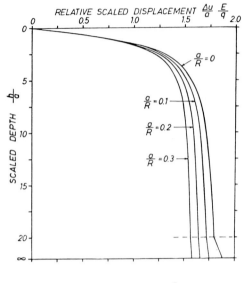

Figure 7.   Relative scaled displacements between a point below the center of the plate and the point on the surface.

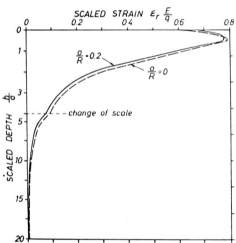

Figure 8.   Scaled strains below the center of the plate.

respect to the corresponding value given by the half-space theory. This ratio obviously represents the correction coefficient η by which the moduli $\underset{\sim}{E}$ obtained from a conventional half-space interpretation are to be multiplied to obtain the true moduli of the rock

(6) $E = \eta \underset{\sim}{E}$

The line passing through the initial points of the various lines obviously gives the ratio between the true strains (given in Figure 8) and those corresponding to the half-space conditions.

    An example will show the application of these graphs. Let us suppose

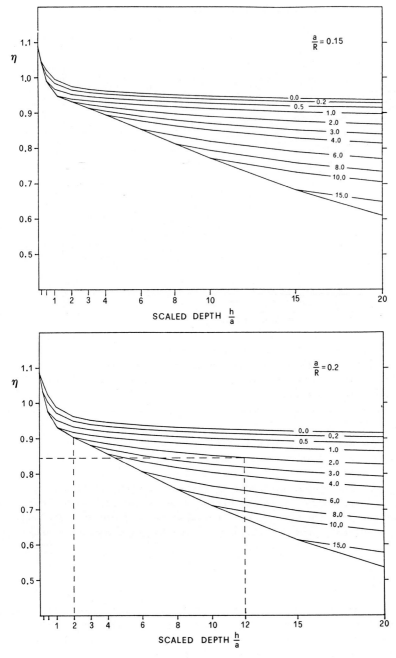

Figures 9 and 10.   Ratio between the true rock modulus and that obtained
with a conventional half-space analysis, on the basis of the relative
displacements of two points below the plate on the loading axis. Plate
radius/tunnel radius = 0.15 and 0.2

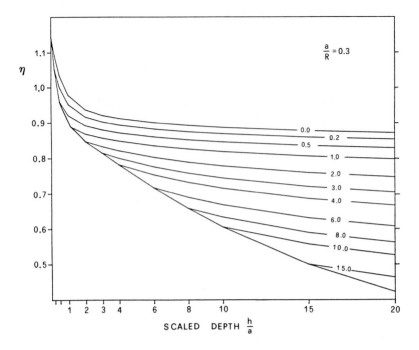

Figure 11.   Ratio between the true rock modulus and that obtained with
a conventional half-space analysis, on the basis of the relative   displacements of two points below the plate on the loading axis. Plate
radius/tunnel radius = 0.3

that a plate-loading test with a plate diameter of 0.5 m is carried out
in a tunnel with a diameter 2.5 m and that a modulus of 10 GPa has been
determined on the basis of the relative displacements of points 0.5 m
and 3.0 m below the plate. In Figure 10 we trace a vertical line from
abscissa 2.0 (0.5/0.25) up to the dashed line and we follow the solid
line starting from this point up to abscissa 12.0 (3.0/0.25). The corresponding ordinate, equal to 0.84 will give the true modulus of the
rock, that is 8.4 GPa.

4 CONCLUSIONS

On the basis of the Fourier transform method the displacements deriving
from the application of loads on the walls of circular tunnels can be
determined. This allows a more accurate interpretation of plate-loading
tests which are carried out in exploratory drifts.
    The influence of the curvature of the tunnel walls depends on the
ratio between the plate radius and the tunnel radius, which typically
falls in the range 0.15-0.25. In these conditions the interpretation of
the test by the conventional relationships which are valid   for   half-

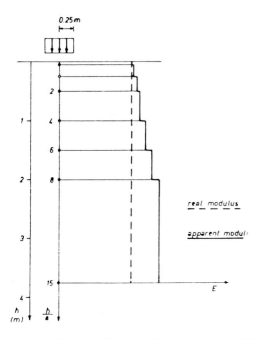

Figure 12. Profile of the apparent moduli (solid line) as determined with a conventional analysis in a homogeneous rock (modulus indicated by the dashed line). The block date indicate the position where displacements are measured. Plate radius/tunnel radius = 0.2

space conditions could introduce some errors in the evaluation of the rock modulus.

When the displacements are measured on the tunnel wall at the centre of the plate the error is small, that is of the order of 11%, but it increases up to 22% when the displacements are measured near the border of the plate, as commonly occurs in practice.

The displacements are often measured at some depth below the loading plate in order to avoid the influence of a loosened layer of rock near the tunnel wall. In this case the rock moduli could be sometimes grossly overestimated by the conventional analysis, but such a drawback could easily be avoided by applying the correction factors given in this paper.

Finally, it must be pointed out that when the displacements are measured in various points below the plate, the conventional analysis would indicate a gradual increase in the rock quality with distance from the surface, even when the moduli are really uniform. This effect is clearly shown in Figure 12 with reference to a typical scheme of measurements, which is normally applied in various engineering sites in Italy.

REFERENCES

Abramowitz, M. Handbook of mathematical functions with formulas, graphs
   and mathematical tables.
Booker, J.R. & Carter, J.P. 1984. The analysis of deformations caused
   by loading applied to the walls of a circular tunnel. Int. J. Numer.
   Anal. Methods Geomech. 8(5): 445-455.
Giroud, J.P. 1972. Table pour le calcul des fondations. Paris: Dunod
Piessens, R. & Branders, M. 1975. Computation of oscillating integrals.
   J. comp. Appl. Math. 1: 153-164.
Ribacchi, R. 1987. Rock mass deformability: in situ tests, their inter-
   pretation and typical results in Italy. 2nd Int. Symp. on "Field
   measurements in Geomechanics".

ACKNOWLEDGEMENT

This research was supported by a M.P.I. contribution.

The back-analysis method from displacements for viscoelastic rock mass

Wang Sijing & Yang Zhifa
Institute of Geology, Academia Sinica, Beijing, People's Republic of China
Xue Ling
Qingdao Institute of Architectural Engineering, People's Republic of China

1 INTRODUCTION

When engineers make design for an underground opening within
a weak rock mass, they should determine some important parameters, such as modulus of elasticity, initial stresses of
the rock mass and coefficient of viscosity. Certainly, there
are some traditional methods that can be used in a field.
But those tests are costly, and may take too much time and
labor. Sometimes, for economic reason, engineers have to
abandon some tests and make the design by means of only their
experience. Obviously, it is necessary to develop a new efficient, simple and inexpensive method. Probably, the method
to back-analyse various important parameters of a weak rock
mass from displacements which are caused by excavation is
promising.
    In this paper we will discuss a back-analysis method for
a rheological rock mass on the basis of the theory of rheological body. In concrete terms, three rheological bodies, namely, Maxwell, Kelvin and Poyting-Thomson body are dealt
with. For convenience, we also make program XMKT. In order
to put the back-analysis method into practice, the authors
advocate that one should reduce the numbers of parameters to
be analysed as far as possible, and also suggest to combine
the rheological back-analysis method in plane strain problem
with the back-analysis method in three dimentional problem
of elasticity.
    Theoretically, since the displacements caused by excavation
are related to various parameters, it is possible to back-
analyse them from those displacements. But experience from
practice has shown that it is unrealistic and not necessary
to try to determine all of parameters by means of back-analysis. A better way, perhaps, is to find and back-analyse
only main parameters which may have a great influence upon
stability of the underground opening and determined by the
other method difficultly. For instance, we could consider
only the horizontal component, perpendicular to the axis of
a tunnel. Obviously, the way will contribute not only to reduce amount of work but also to increase reliability and
accuracy of back-analysis. So we prefer to back-analyse less
parameters even though we are able to analyse more theoreti-

cally. In practice, putting emphasis on the principal problems and neglecting secondary problem, as mentioned above,
we could reduce the numbers of parameters to be back-analysed. Besides, we should also consider to use more reliable
results from the other tests which were excuted before and
determine some parameters according to the reliable laws.
For terrain with a more or less horizontal surface it is
reasonable to assume that the vertical normal stress in the
initial state is approximately equal to the overburden stress
of the overlying rock mass; In addition, some further knowledge is important. For instance, if the elevation of the
bottom of a tunnel is much higher than that of two deep gul-
lies which are near the engineering work, we could not ima-
gine that the horizontal initial stress is very high. Finally, some examples will be given to explain the idea further
and check the program.

2 A COMBINATION OF TWO METHODS OF BACK-ANALYSIS

Considering the degree of difficulty of the rheological back-
analysis in the three dimentions, we choose the back-analysis
in the plane strain problem, here. But it will bring about
two problems:(1) Generally, instruments are used on the face
of a tunnel to measure the displacements, so we will miss a
part of displacements for the plane strain problem; (2) Because the excavation needs time, it is difficult to determine the beginning of the creep deformation.
  In view of those facts, the following proposals are to be
made: As shown in Fig. 1, the rheological back-analysis in
the plane strain problem is combined with the back-analysis
in three dimensional problems of elasticity. The latter is
special for determination of the horizontal initial stress
and modulus of elasticity from the instantaneous elastic
displacement; The former — for determination of rheological
parameters from the creep displacements. Here, it seems to

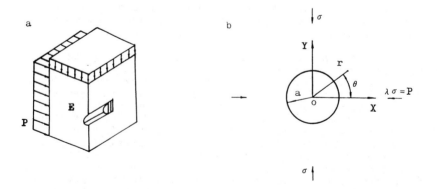

Fig. 1. a. For elastic back-analysis in three dimensions;
b. For rheological back-analysis in plane strain problem.

be necessary to mention the latter very simply. Because the calculation in three dimensions is very costly and may take too much time and labor, the standard pattern method based on the finite element method (Yang Zhifa, Liu Zhuhua, 1982) is used for analysis of the displacements. The analysis can be made by microcomputer or by hand which is very cheap, simple and convenient to the engineers in field. Provided the displacements calculated by the standard pattern method and displacements measured in a field are put into a microcomputer, the results can be got by program YPE: elastic modulus E and horizontal initial stress P, perpendicular to the axis of a tunnel. If necessary, a special diagrammatic back-analysis can also be used without computer. Obviously, when the rheological back-analysis is excuted further, P and E will be treated as the known values. Thus, it will save us much time, labour and computation expenses.

Thus, there are two ways for rheological back-analysis: One is only rheological back-analysis in the plane strain problem from the creep displacements; The other is a combination of the rheological back-analysis in the plane strain problem from the creep displacemenrs with the three-dimensional back-analysis from the instantaneous elastic displacements.

3 VISCOELASTIC SOLUTIONS FOR THREE RHEOLOGICAL MODELS

The idea to get the viscoelastic solution is that the viscoelastic could be got form the elastic solution by means of Laplace's transformations. It is justifiable to assume that the initial stress is as homogeneous provided the overburden is thick enough. The horizontal initial stress is determined by the product of the vertical initial stress σ and the coefficient of lateral pressure λ . Here, we also assume the rock mass to be elastic, homogeneous and isotropic.

1.1 The elastic solution for the radial displacements of a round opening

Because the radial displacements are caused by excavation, they should be equal to the difference between values of displacement after excavation and initial displacements. The expression of the radial displacements u_r is as follows (Fig. 1b):

(1) $$u_r = \frac{1+\mu}{2E} \sigma \frac{a^2}{r}((\lambda+1)+(\lambda-1)(4(1-\mu)- \frac{a^2}{r^2})\cos 2\theta)$$

where E is modulus of elasticty of the rock mass; μ is Poisson's ratio; a — radius of a round opening; r and θ — polar coordinates. Considering that extensometer and convergence indicator are usually used to measure the displacements in the field, the elastic solution should be changed from formula (1) into expressions (2) and (3): For the convergence indicator (Fig. 2a), have

(2) $$u_{co}= 2u_{r=a}$$

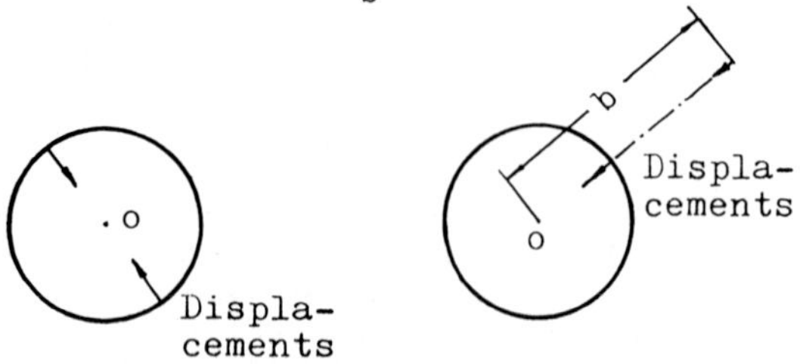

a   b

Fig. 2 a. The method using a convergence indicator; b. The
method using an extensometer.

$$= \frac{1+\mu}{E}\,\sigma a((\lambda+1)+(\lambda-1)(3-4\mu)\cos2\theta)$$

For the extensometer (Fig. 2b), its formula is as follows:

(3) $u_{ex}= u_{r=a} - u_{r=b}$

$$= \frac{1+\mu}{2E}\,\sigma a((\lambda+1)(1-\alpha)+(\lambda-1)(4(1-\mu)(1-\alpha)$$

$$-(1-\alpha^3))\cos2\theta)$$

where b is the distance between the centre of a circle and
the point fixed at the bottom of the bole hole; And $\alpha=a/b$.

1.2 Viscoelastic solution

On the basis of the elastic solution above, the viscoelastic
solution can be determined by meas of Laplace's transformations.
    For Maxwell body as shown in Fig. 3a, we have

(4) $u_{co}= \sigma a((\lambda+1)(t_M+t)\frac{1}{\eta} +(\lambda-1)((t_M+t)\frac{1}{\eta}$

$$+ \frac{1}{K}(1- \frac{\eta}{6Kt_M+\eta}\,\exp(- \frac{6Kt}{6Kt_M+\eta})))\cos2\theta)$$

(5)   and $u_{ex}= \sigma a((\lambda+1)(1-\alpha)(t_M+t)\frac{1}{2\eta} +(\lambda-1)((\alpha^3-2\alpha+1)(t_M$

$$+t)\frac{1}{2\eta} +(1-\alpha)(1- \frac{\eta}{6Kt_M+\eta}\,\exp(- \frac{6Kt}{6Kt_M+\eta}))$$

$$\frac{1}{2K})\cos2\theta)$$

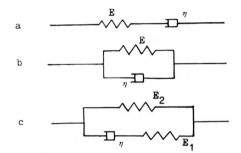

Fig. 3 Three rheologi-
cal bodies. a. Maxwell
body; b. Kelvin body;
c. Poyting-Thomson body.

where, η is viscousity coefficient; $t_M = \eta/E$ is relaxation
time; And t is time.

For Kelvin body given in Fig. 3b, we can obtain

(6)
$$u_{co} = \sigma a((\lambda+1)(1 - \exp(-\frac{E}{\eta}t))\frac{1}{E} + (\lambda-1)((1$$

$$-\exp(-\frac{E}{\eta}t))\frac{1}{E} + (1-\exp(-\frac{E+6K}{\eta}t))\frac{6}{E+6K})\cos2\theta)$$

(7)   and $$u_{ex} = \sigma a((\lambda+1)(1-\alpha)(1 - \exp(-\frac{E}{\eta}t))\frac{1}{2E} + (\lambda-1)$$

$$((\alpha^3-2\alpha+1)(1-\exp(-\frac{E}{\eta}t))\frac{1}{2E} + (1-\alpha)(1$$

$$-\exp(-\frac{E+6K}{\eta}t))\frac{3}{E+6K})\cos2\theta)$$

For Poyting-Thomson body shown in Fig. 3c, we suppose $t_P = \eta/E_1$ and $E = E_1+E_2$, then

(8)
$$u_{co} = \sigma a((\lambda+1)(1+\frac{E_2-E}{E}\exp(-\frac{E_2 t}{Et_P}))\frac{1}{E_2} + (\lambda-1)$$

$$(\frac{1}{E_2} + \frac{6}{6K+E_2} + \frac{E_2-E}{EE_2}\exp(-\frac{E_2 t}{Et_P}) + \frac{6(E_2-E)}{(6K+E)(6K+E_2)}$$

$$\exp(-\frac{6K+E_2}{t_P(6K+E)}t))\cos2\theta)$$

(9)   and $$u_{ex} = \sigma a((\lambda+1)(1-\alpha)(1+\frac{E_2-E}{E}\exp(-\frac{E_2 t}{Et_P}))\frac{1}{2E_2} + (\lambda-1)$$

$$((1-\alpha)(1+\frac{E_2-E}{6K+E}\exp(-\frac{6K+E_2}{t_P(6K+E)}t))\frac{3}{6K+E_2} + (\alpha^3-2\alpha$$

$$+1)(1+\frac{E_2-E}{E}\exp(-\frac{E_2}{Et_P}t))\frac{1}{2E_2})\cos2\theta)$$

where, E is the instantaneous elastic modulus; E_2 is the long-term elastic modulus.

4 ETHOD TO BACK-ANALYSE PARAMETERS OF THE RHEOLOGICAL ROCK MASS AND ITS PROGRAMME

For Maxwell, Kelvin and Poyting-Thomson body it is possible to back-analyse their parameters according to the formulas from (4) to (9). Here, except Kelvin body the creep displacements u^c could be got from difference between the total displacements u and the instantaneous elastic displacements u^e is considered as follows:

For Maxwell body,

(10)　　　$u^c_{co}=\sigma a((\lambda+1)\frac{t}{\eta}+(\lambda-1)(\frac{t}{\eta}+\frac{\eta}{K(6Kt_M+\eta)}(1-$

$$\exp(\frac{-6Kt}{6Kt_M+\eta})))\cos2\theta)$$

(11)　and　　$u^c_{ex}=\sigma a((\lambda+1)(1-\alpha)\frac{t}{2\eta}+(\lambda-1)((\alpha^3-2\alpha+1)\frac{t}{2\eta}+(1$

$$-\alpha)\frac{\eta}{2K(6Kt_M+\eta)}(1-(\exp\frac{-6Kt}{6Kt_M+\eta})))\cos2\theta)$$

For Poyting-Thomson body,

(12)　　　$u^c_{co}=\sigma a\frac{E-E_2}{EE_2}((\lambda+1)(1-\exp(-\frac{E_2t}{Et_P}))+(\lambda-1)((1-$

$$\exp(-\frac{E_2t}{Et_P}))+\frac{6EE_2}{(6K+E)(6K+E_2)}(1-$$

$$\exp(-\frac{6K+E_2}{t_P(6K+E)}t)))\cos2\theta)$$

(13)　and　$u^c_{ex}=\sigma a\frac{E-E_2}{2EE_2}((\lambda+1)(1-\alpha)(1-\exp(-\frac{E_2t}{Et_P}))+(\lambda-1)$

$$((\alpha^3-2\alpha+1)(1-\exp(-\frac{E_2t}{Et_P}))+(1-\alpha)\frac{6EE_2}{(6K+E)(6K+E_2)}$$

$$(1-\exp(-\frac{6K+E_2}{t_P(6K+E)}t)))\cos2\theta)$$

But for Kelvin body, there are no instantaneous elastic displacements, i. e. $u^c_{co} = u_{co}$ and $u^c_{ex} = u_{ex}$.

The authors had made program XMKT according to those formulas to back- analyse Maxwell, Kelvin and Poyting-Thomson body. In the program, the least squares procedure, the golden section method and the simplex method are used. The application of programme XMKT is very convenient. One can choose any one from three bodies arbitrarily.

5 SOME EXAMPLES

It is necessary to give some examples to explain how to use programme XMKT to back-analyse the parameters of a rheological rock mass. Because our purpose is only to discuss its principle and concrete method, we might as well assume some examples and their 'real values' of the parameters to be back-analysed.

Example 1: A 250m deep round tunnel is excavated within a homogeneous rock mass. Its radius is 1.8m. Here, we suppose that Maxwell body can represent mechanical behaviour of the rock mass and $\eta = 10^{17}$ P. The unit weight $\gamma = 2.5$ T/m^3. The problem is how to back-analyse elastic modulus E and coefficient of lateral pressure λ from measured displacements. The positions of measuring points of six extensometers are shown in Fig. 4. In order to examine programme XMKT we might as well assume in advance 'real values' of the parameters to be E = 90000T/m^2 and $\lambda = 0.7$. Thus, according to the expression (11) the relative displacements in various measuring points at various time are calculated by a computer, which are listed in Table 1. Here, we regard those value as 'the displacements measured' and back-analyse E and λ from them with program XMKT. Certainly, there are no measuring error. Generally, the optimum seeking ranges can be determined by our experience to some extent. For example, the range of E is taken from 40000 to 120000T/m^2; As to λ its

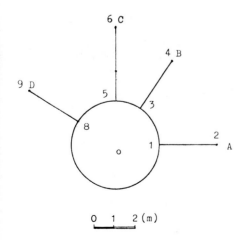

0 1 2 (m)

Fig. 4 A round opening with four extensometers.

Table 1. `Displacements measured´(mm).

Exten-	Hours						
someter	6	12	24	48	72	120	168
A 1-2	4.912	4.924	4.946	4.992	5.037	5.128	5.219
B 3-4	7.524	7.537	7.565	7.619	7.643	7.781	7.889
C 5-6	8.394	8.409	8.437	8.494	8.551	8.666	8.780
C 5-7	4.432	4.440	4.445	4.484	4.541	4.574	4.633
D 8-9	5.783	5.795	5.819	5.867	5.916	6.012	6.109

Table 2. Results to back-analyse E and λ.

Parame- ters	A 1-2	B 3-4	C 5-6	C 5-7	D 8-9	Ave- rage
$E(T/m^2)$	90050	89500	90920	89860	90140	90090
λ	0.703	0.694	0.690	0.698	0.707	0.698

range is supposed from $\lambda_a=0.43$, which is determined by the theory of elasticity, to 1.0 tentatively. Putting those `measured displacements´ and the other known values into a computer, one will obtain results and show them in Table 2. the relative errors of E and λ are only 0.1% and 0.3% as compared with the valure assumed at the beginning.

Example 2: There is a round opening within a homogeneous rock mass. The thinckness of the overlying rock mass is about 210 m. The example has the same positions of four extensometers as Example 1 in Fig. 4. Supposing that a law of the rock mass accords with Poyting-Thomson body and $\gamma = 2.65$ T/m^3, we try to back-analyse λ, η, E_1 and E_2 from displacements by means of the combination of both methods above-mentioned. Here, the `real values´ $\lambda = 0.6$, $\eta = 10^{18}$ P, $E_1=120000$ T/m$^2$ and $E_2=200000$ T/m$^2$ are also given in advance. At the first step, using the elastic back-analysis in three dimensions we can obtain $\lambda$ and E from the instantaneous `displacements measured´ which are calculated according to `real values´ mentioned above. The `displacements measured´ mean those `measured´ within a short time after a blasting. Because this method has already been introduced in detail in another paper (Yang Zhifa et al., 1983), we intend to make no mention of it. At the second step, considering $E=E_1+E_2$, we will analyse only η and E_2 from the creep displacements by means of the rheological back-analysis in the plane strain problem because E and λ have already been got in the first step. Therefore, according to expression (13) `the creep displacements measured´ are also

Table 3. `The displacements measured´(mm).

Exten-someter		Days				
		1	5	9	13	15
A	1-2	0.807	0.809	0.811	0.813	0.814
B	3-4	1.945	1.948	1.950	1.953	1.954
C	5-6	2.325	2.327	2.330	2.333	2.335
C	5-7	1.207	1.208	1.210	1.211	1.212
D	8-9	1.186	1.189	1.191	1.193	1.194

Table 4. The results of η and E_2.

Exten-someter	A 1-2	B 3-4	C 5-6	C 5-7	D 8-9	Ave-rage
$\eta(\times 10^{18} P)$	1.02	0.96	1.04	0.98	1.00	1.00
$E_2(T/m^2)$	200100	199500	198900	198900	200300	199500

calculated from the `real values´ by a micro-computation, which are shown in Table 3. Their optimum seeking ranges are given in accordance with our experience: The range of E_2 is from 100000 T/m^2 to 150000 T/m^2; For η it is from 2×10^{17} P to 2×10^{18} P.

As a results, η and E_2 can be calculated in the light of programme XMKT (Table 4). Their relative error are less than 0.2%.

6 CONCLUSION

Through the examination of two examples, it is clear that the methods to back-analyse various parameters of rock mass, including the rheological parameters, and some components of initial stress are more useful and cheap for engineers. In terms of those methods they are able to extract more information about a rock mass from the displacements measured in field for design and construction of an underground opening.

REFERENCES

Yang Zhifa, Liu Zhuhua, Wang Sijing 1983. A practical back-analysis method from displacements to estimate some parameters of a rock mass for design of underground openings. Proceedings of International Symposium on Field Measurements in Geomechanics, vol. 2, p. 1267-1276. Zürich.

Yang Zhifa, Liu Zhuhua 1982. The application of the standard
    pattern method in the underground construction, Underground
    Construction, No. 11 (In Chinese).
Xue Lin, Qie Yuting, Yang Zhifa 1986. Back-analysis method
    from displacements for determining parameters of rheologi-
    cal rock mass and initial stress, Scientia Geological Si-
    nica, No. 4 (In Chinese).

2nd International Symposium on Field Measurements in Geomechanics, Sakurai (ed.)
© 1988 Balkema, Rotterdam. ISBN 90 6191 778 6

Determination of rock mass strength through convergence measurements in tunnelling

Chikaosa Tanimoto
Department of Civil Engineering, Kyoto University, Japan

Toshio Fujiwara, Hisaya Yoshioka & Koji Hata
Research Institute, Ohbayashi Corporation, Tokyo, Japan

Kazutoshi Michihiro
Department of Civil Engineering, Setsunan University, Osaka, Japan

1 INTRODUCTION

The most popular and practical measurement for ground deformation caused by tunnelling is the convergence measurement. It is a simple measurement which can be easily carried out during construction, and provides serial information on the displacement of the tunnel wall and change in the radial stress corresponding to the displacements. When the ground behaves as an elastic body with no time dependency, the curve of convergence reaches a constant level within the range of a double distance of the tunnel diameter in the driving direction.

On the other hand, when the ground is subject to the non-elastic behaviour such as the strain-softening and/or plastic flow, deformation does not reach a constant level within a double length of the tunnel diameter, and converges in the distance of 3 to 10 times longer than the tunnel diameter. As the ground arch formation based on the stress redistribution is mobilized along the elastic/nonelastic boundary around the tunnel opening, it has been verified through several case histories that the convergence curve suggests the development of the nonelastic zone.

Once a field measurement provides us with some pieces of information concerning the magnitude of tunnel wall displacement and the location of the elastic/nonelastic boundary in the surrounding rock, we can analyze the relationship between the competence factor and the available support pressure. Besides, geologic information such as rock type, frequency and orientation of joints/faults, and quantity of inflow which are obtained from a contractor's daily report on the state of driving face gives us Barton's Q-value comparatively easily. Tanimoto & Goodman (1984) found that there exists some empirical rule between Hoek & Brown 's estimation of rock mass strength and Barton's Q. Through a parametric study on the initial stress field, rock mass strength, and support pressure, we can estimate the strength of rock mass with reasonable precision.

2 OBSERVED DEFORMATIONS IN TUNNELLING

In most of cases the results of convergence measurements are plotted on a deformation-time graph, which gives the apparent deformation rate. However, because of the dependence of deformation on the position of the mining face, it is more informative to plot the results of the convergence survey relative to the distance to the face and tunnel

walls.   There are many uncertainties involved in the interpretation of
geological conditions   for   tunnelling   from   exploratory   borings.
Terzaghi(1946)   and   Peck(1969)   proposed the use of   an   "observational
construction method"   for tunnelling in difficult   ground.   Under   such
conditions,   a   convergence   survey may serve the purpose of   providing
information   on the deformation rate at the early stages of   excavation,
from which   it is possible to estimate the approximate support load   and
timing for the installation of the support elements.

The optimum goal in tunnelling is to advance the face in such a manner
as   to maintain the original stress state of the ground.   However,   the
stress   distribution is altered by the removal of the rock from   within
the opening, and the displacement of the surrounding rock with   the loss
of the inner confining pressure.It has been found that by allowing   some
displacement of the rock around the opening, the shear strength   of   the
rock is mobilized so that it is capable of supporting some   of   its   own
load by the ground arch effect.

The   largest deformation occurs in the vicinity of the advancing face
and is subject to the advance rate and mechanical behaviour of the rock.
As   the face advances,   artificial supports take the place of the   half-
dome action of the face.   To fully mobilize the bearing capacity of the
rock   mass,   the   supports   must be installed within a distance   of   one
tunnel diameter from the face(Terzaghi,1946;   Mueller,1977;   AFTES,1978;
Tanimoto,1980;   Tanimoto et al.,1981a,b).   Experience and   observations
from   convergence   surveys have shown that the deformation rate is   most
influenced   by the position of the face relative to the installation of
the support elements and that time-dependent behaviour of the rock is of
little consequence(Tanimoto et al.,1980).   In the case of some   tunnels
through tertiary mudstone which exhibit strong time-dependent behaviour,
the deformation rate decreases to almost zero with a halt in the advance
over an extended holiday.

Tanimoto et al.(1983;1986)   summarized the experience and   data   from
tunnel   constructions which had been executed in Japan   and   the   United
States from 1975 to 1985,   and proposed a method to find the approximate
range of the maximum allowable convergence   of rock   based   on   tunnel
convergence measurements.   On the other hand, using data from a number
of case histories, Pacher(1972) defined the rock classification for tunnelling based on the support loads associated with the deformation   rate
obtained   from convergence measurements.   The classification   includes:
(I) low rock pressure,   (II) moderate rock pressure,   (III) strong   rock
pressure and (IV) very strong pressure.   The severest case (IV) of very
strong rock pressure   in   Pacher's   classification   corresponds to   a
deformation rate of 3 mm per day.

In Japan, large deformations and expansive behaviour of rock have been
observed   in tunnels   driven through rocks of the Tertiary Green   Tuff
Region.   The deformation of   these rocks often exceeds   10 % of   the
original   tunnel diameter and rocks deform at rates exceeding   3 mm/day,
which   corresponds to the severest case in Pacher's classification.

As stated previously,the early and quick interpretation of convergence
measurements is   essential   in tunnelling,   and consequently the   authors
propose the rock load classification for tunnelling based on convergence
measurements   as shown   in Table 1.(Tanimoto et al.,1983;1986)   The
deformation rate which is defined by such as the deformation per day
(e.g. mm/day) is not appropriate, and it is better to define it as the
ratio of deformation (in mm) to face advance (in m) from the results of
convergence measurements relative to the distance to the face. When   the
measurement section is very close to the mining face (the initial
relative distance should be less than one meter in the driving

Table 1. Classification of support load based on convergence (D=10 m)

Class	Support Load	Initial Deformation Rate (mm/m)	Observed Deformation ΔD/D (%)	Estimated Support Pressure Pi(MPa)	cf Terzaghi's Rock Class & Load Y H(MPa)
I	Slight	less than 0.3	less than 0.05	less than 0.1	1-3 ; 0-0.1
II	Medium	0.3 - 2	0.05 - 0.3	0.1 - 0.3	4-5 ; 0.1-0.3
III	Heavy	2 - 7	0.3 - 1	0.3 - 0.6	6-7 ; 0.3-0.7
IV	Very Heavy	7 - 15	1 - 2	0.6 - 1	8-9 ; 0.7-1.0
V	Extremely Heavy	over 15	over 2	over 1	10 ; 1.0-2.2

direction),the initial deformation rate can be defined by   deformation versus   face advance during early excavation such as 3 to 5 m.   Its unit is mm per meter. For example, if the initial deformation rate is observed to be between 7 and 15 mm/m, the final deformation described by the ratio of the deformation to the original diameter of a tunnel (denoted as $\Delta D/D$) and the magnitude of final support pressure are estimated to be 1 to 2 % and 0.6 to 1 MPa, respectively.

3 MEANING OF DEVELOPING NON-ELASTIC ZONE

It seems that the relationship between the formation of a rock arch and the development of a non-elastic zone around a tunnel has not been clar- ified sufficiently in past literature. Here, let us consider the mean- ing of developing a non-elastic zone such as strain-softening and plastic flow.

Based on many observations, it has been suggested that actual deforma- tion caused by tunnelling does not converge within a double length of the tunnel diameter (which is denoted as 2^*D where D is the tunnel dia- meter) and does in the distance of 3 to 10^*D. The execution of multiple headings is one of the reasons for it. Even if this influence is taken into account for further analysis, it should be concluded that the ground shows rather remarkable non-elastic behaviour in many cases and convergence is strongly influenced by the development of a non-elastic zone around the tunnel opening.

Fig. 1 shows the change of the non-elastic zone in association with the decrease of inner pressure given by a support system for the case in which the ground has a competence factor of 0.5 and is in the state of being overstressed.(Tanimoto et al.,1983;1987)   It can be seen that a small change in the supporting effect (as an inner pressure) such as 0.1 MPa has a large influence on the extent of the non-elastic zone. And, the state of stress on the elastic/non-elastic boundary does not vary; only the thickness of the non-elastic zone around an opening changes in correspondence with the magnitude of inner pressure. This means that the ground around a tunnel trends to equilibrate itself by developing a non- elastic zone which is capable of providing a constant inner pressure toward the outer ring in the state of elastic.   This can be obtained by allowing a certain deformation, but on the other hand a magnitude of allowable deformation is limited in practice.

Although many discussions on the minimum support load have appeared, there are few papers in which the relationship between the minimum sup- port load and the corresponding magnitude of displacement has been clear- ly defined. Using experience and data from recent tunnel construction in Japan and the United States,   Tanimoto et al.(1983) proposed a method to find the appropriate range of maximum allowable convergence of rock

based on tunnel convergence measurements. and also Tanimoto et al.(1986)
showed that, when a change (ΔD) of the original tunnel diameter or
height (D) is over 1.0 % (as ΔD/D), a very heavy support pressure such
as 0.6 - 1.0 MPa has been observed. Then, to reconsider the tendency of
the convergence in several cases where the effect of the support system
was clearly confirmed, it has been clarified that the outer diameter of
the non-elastic zone (hereafter, it is denoted as D') corresponds with
the tendency of convergence. Namely, the distance of 2*D' agrees with
the point which reaches a constant displacement. It behaves as if a
tunnel with a diameter od 2*D' has been driven instead of a tunnel with
a diameter of 2*D. It was clearly observed in the construction of the
Orizume Highway Tunnel, in which a deformation reached constant within 4
- 5*D (D = 10 m) and the observed non-elastic zone was 5 - 6 m wide
around an opening. 2*D' was equal to 40 -45 m as shown in Fig. 2.
    In order to verify this matter from a theoretical point of view, a
numerical analysis was conducted. out. Fig. 3 shows some convergence
curves obtained from the analysis which assumed three different
competence factors (denoted as C_f), namely 1.5, 1.0 and 0.5,
corresponding with circled ① , ② and ③ , respectively. As stated
previously, the magnitude of convergence is strongly subject to the
support pressure. In the figure, the shaded areas correspond with the
range of support pressure between zero and 0.15 MPa, which is classified
into the medium support load, Class II in Table 1. Inversely the behaviour of convergence suggests the development of a non-elastic zone in
terms of the distance of D', as well as the deformation of tunnel wall.
a tunnel depends on the timely installation of support elements and the
skillfulness of mobilizing effective ground arch around a tunnel.

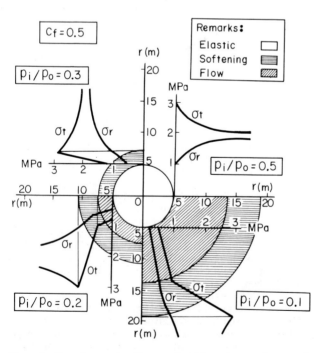

Fig. 1　Development of non-elastic ring and ground arch

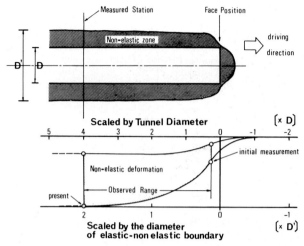

Fig. 2   Convergence in association with non-elastic deformation

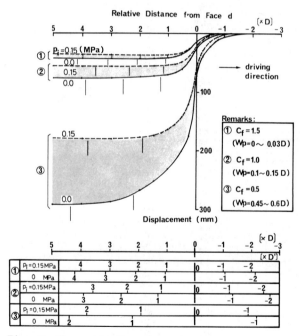

Fig. 3   Theoretical convergence curve for several competence factors

Needless to say, it is very remarkable for cases with a small competence factor. Unless a client or consultant is capable of evaluating this matter, "field measurement" or "monitoring" does not make practical sense for construction management. Then, the estimation of support pressure must be discussed next.

4 ESTIMATED SUPPORT PRESSURE

Regardless of the type of support/lining elements to be employed, it is possible to estimate an equivalent support pressure, the change in the half-dome action and timing for the installation of the support elements from the strain-softening analysis (Tanimoto et al.,1983). The analysis procedure for fully bonded rockbolts is outlined by Tanimoto et al.(1981a,b).

The basic concept for the support pressure has been described in the publications, e.g. Hoek et al.(1980), Tanimoto et al.(1982a,b;1983) etc. The problem on the estimation of support pressure comes from the difference between the theoretical values and observed ones. An assumed cross-section in design is apt to vary greatly from a realistic one, which is quite irregularly shaped at excavation in most cases, and also it is very difficult to evaluate the realistic stiffness of a support system in-situ after installation. The authors consider that variation in installation timing and skillfulness in construction result in the wide range of support/lining stiffness, with a rather different maguni-tude of 1/2 to 1/20 less than the theoretical value. For instance, a thin shotcrete layer such as 5 cm thick, which is applied onto the ideally circular opening, shows very high stiffness and bearing capacity, but many measurements aimed at monitoring the magnitude of stresses in a shotcrete lining by means of pressure cells indicate quite low values compared with the theoretical ones. The same tendency can be seen on a rockbolt system too.

Since it does not make any sense to compare much data obtained under different conditions at the same level, the authors have tried to esti-mate the approximate values for various support systems by employing the data which has been obtained directly by the authors and is considered

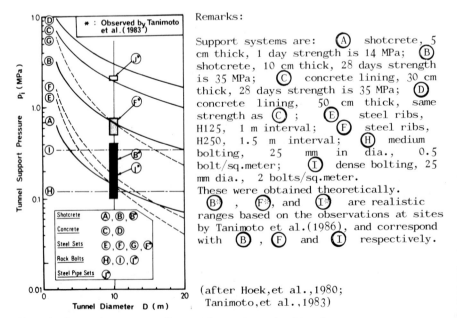

Remarks:

Support systems are: Ⓐ shotcrete, 5 cm thick, 1 day strength is 14 MPa; Ⓑ shotcrete, 10 cm thick, 28 days strength is 35 MPa; Ⓒ concrete lining, 30 cm thick, 28 days strength is 35 MPa; Ⓓ concrete lining, 50 cm thick, same strength as Ⓒ; Ⓔ steel ribs, H125, 1 m interval; Ⓕ steel ribs, H250, 1.5 m interval; Ⓗ medium bolting, 25 mm in dia., 0.5 bolt/sq.meter; Ⓘ dense bolting, 25 mm dia., 2 bolts/sq.meter. These were obtained theoretically. Ⓑ′, Ⓕ′, and Ⓘ′ are realistic ranges based on the observations at sites by Tanimoto et al.(1986), and correspond with Ⓑ, Ⓕ and Ⓘ respectively.

(after Hoek,et al.,1980; Tanimoto,et al.,1983)

Fig. 4 Support pressure for various support elements

reliable with the proper measurement.   They are plotted onto the Hoek's
diagram as shown in Fig. 4.   Circled A to H with thin curves are given
by Hoek et al.(1980), and circled alphabets with asterisks show   the
observed ranges for the case of D = 10 m by Tanimoto et al.(1983;1986).

5 ESTIMATION ROCK MASS STRENGTH BASED ON CONVERGENCE

As the deformation behaviour of ground associated with tunnelling is
highly influenced by the strength of the surrounding rock (q_u), primary
stress field (p_o) and reaction of support system (p_i), the introduction
of the concept of "support intensity" is useful.   It is defined by the
ratio of support pressure (p_i) versus major primary stress (p_o), namely
p_i/p_o (Tanimoto et al.,1986).[1] Generally speaking, the value of support
intensity is considered to be 0.02 to 0.2 in past cases of rock tunnels
constructed by conventional methods, and nearly equal to 1.0 for earth
tunnels with a thin cover.   There are few cases corresponding with the
range of 0.3 to 0.8 for support intensity.   This is because most projects
have been executed in rock with a competence factor higher than 1.5.
However, recent social demands force us to drive tunnels under severe
conditions with a low competence factor such as below 1.0.   Tunnel projects in Japan which have been constructed since 1970, especially in the
Tertiary Green Tuff Region, are good examples.   Because of the difficulty
 of installing a support system with a high support intensity, too large
deformation has been accepted, and sometimes it has yielded the remarkable expansive behaviour of the surrounding rock.   As mentioned previously, the magnitude of available support pressure is in the range of 0.1
to 0.4 MPa for many cases.
   For tunnelling through rock with a low competence factor (it may be
called weak rock in general) discussion on an appropriate support intensity must be taken into account for better tunnel design based on the
allowable limit of deformation.   Fig. 5 shows one example of the relationship between support intensity and competence factor.   In the figure,
w/E corresponds to the negative gradient for the state of strain-softening in the stress-strain relation, and a high value of w/E indicates a
rapid drop of peak strength.   The line marked by the circled p
expresses the boundary of elastic and non-elastic bahaviours.

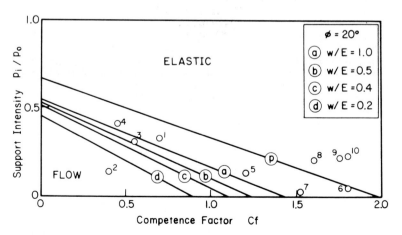

Fig. 5   Support intensity and competence factor

Table 2. Case study for estimating rock mass strength

Site No.	Geologic Condition	Intact Rock σ(MPa)	In-situ Stress Po(MPa)	Support Intensity Pi / Po	Observed Deformation ΔD/D(%)	Non-elastic Zone Wp(m)	Rock Mass σ'(MPa)	Qindex	Rock Mass σ'ₐ(MPa)	Remarks
①	Lapilli-tuff	0.34	0.50	0.33	2.35	6.6	0.07 ~ 0.15	0.03	0.20	SR,Mu-29 SC,20cm RB,21bolts(l=6m)
②	Tuff	0.56	1.55	0.14	4.30	unknown	——	0.04	0.28	Fracturing in the roof and sidewalls (Mu-29, 20cm, 50bolts)
③	Tuff-breccia	0.74	1.36	0.31	1.69	4.0	0.41 ~ 0.61	0.09	0.55	SR,H-200 SC,20cm RB,65bolts(l=3~9m)
④	Schalstein	0.71	1.55	0.41	3.41	6.6	0.06 ~ 0.31	0.03	0.75	"
⑤	Limestone	58.60	2.28	0.14	0.72	5.9	0.23 ~ 0.57	0.3	1.57	SR,H-150 SC,15cm RB,19bolts(l=4m)
⑥	Slate/Chert alt. Layer	46.45	2.94	0.047	0.27	2.1	2.00 ~ 2.35	4.2	3.15	SR, 125 SC,10cm RB, 15bolts(l=3m)
⑦	"	46.45	3.41	0.029	0.24	0.85	4.10 ~ 4.40	4.2	2.90	SC,10cm RB,15bolts(l=3m)
⑧	Mudstone	1.86	1.15	0.21	0.22	0.0	1.46 ~ 1.58	1.3	0.69	SR,H-150 SC,15cm RB,18bolts(l=4m)
⑨	"	1.86	1.06	0.22	0.14	0.0	1.33 ~ 1.43	1.0	0.61	"
⑩	"	1.86	1.01	0.23	0.35	0.0	1.23 ~ 1.32	0.7	0.59	"
⑪	Slate	59.29	0.83	0.28	0.07	0.0	0.75 ~ 0.87	2.9	3.20	SR,H-150 SC,t=15cm RB,18bolts(l=4m)
⑫	"	59.29	0.98	0.15	0.09	0.0	1.39 ~ 1.47	2.5	2.50	SR,H-125 SC,t=10cm RB,15bolts(l=3m)
⑬	Sandstone	115.84	1.18	0.13	0.08	0.0	1.76 ~ 1.84	2.9	4.90	"
⑭	Slate	59.29	1.40	0.10	0.02	0.0	2.28 ~ 2.35	5.9	3.60	"
⑮	Sandstone	115.84	1.52	0.09	0.02	0.0	2.51 ~ 2.58	7.5	7.75	"
⑯	"	115.84	0.86	0.17	0.05	0.0	1.15 ~ 1.23	7.5	7.80	"

The farther left a plot approaches, the easier plastic flow occurs.   The area between the line of the circled p and that of a circled alphabet is subject to the state of post-peak behaviour, which does not reach a state of complete plastic flow.   The plotted point with a number corresponds to Site Number in Table 2.

Based on the above mentioned concept, let us try to analyze the approximate strength of rock mass in terms of a convergence survey.   In order to make it simpler now, it is assumed that a circular tunnel with a radius, a. is driven through the ground whose stress-strain relation is subject to elasto-plastic (elastic-perfectly-plastic) behaviour in the hydrostatic primary stress field (p_0).   When Coulomb's yield criterion is employed, the width of the plastic zone Wp can be expressed by the following equation. (Tanimoto,et al.,1980)

$$\overline{W}_p = a\left[\frac{2\{P_0(\varsigma-1) + q_u\}}{(1+\varsigma)\{(\varsigma-1)P_i + q_u\}}\right]^{\frac{1}{\varsigma-1}} - a \qquad \text{Eq.(1)}$$

$$\text{where} \quad \varsigma = \frac{1 + \sin\phi}{1 - \sin\phi}, \quad \mu = \frac{E}{2(1+\nu)}$$

and q_u : unconfined compressive strength of rock mass, ϕ : internal friction angle of rock mass, and p_i: support pressure.

Next, as the competence factor C_f is defined by q_u/p_o, it can be expressed in the following form by deforming Eq.1:

$$C_f = \frac{(\varsigma^2 - 1) A \cdot (P_i/P_o) - 2(\varsigma - 1)}{2 - (1 + \varsigma) A} \qquad \text{Eq.(2)}$$

$$\text{where} \quad A = \left(\frac{\overline{W_p} + a}{a}\right)^{\varsigma - 1}$$

Fig. 6 shows the convergence curves corresponding with Site No.3 in Table 2. The upper and lower half sections were separately excavated and heading for the lower half was done 25 m behind the upper half. The plots by open circles and triangles corresponding with the headings for the upper and lower halves respectively. It can be seen that the deformation of the tunnel wall converged at the relative distance of 3.5*D from the mining face. As the diameter of the excavated cross sectional area (D) is 12 m, the approximate thickness of the annular ring formed by the non-elastic zone is considered to be 4.0 m (W_p = 4.0 m).

Assuming ϕ equals to 10° to 15° for Site No.3 in Table 2, it is obtained that q_u/p_o is 0.3 to 0.45. Therefore, the strength of rock mass (q_u : denoted as σ' in Table 2) at Site No.3 can be estimated to be 0.41 to 0.61 MPa.

In the same manner we can obtain the approximate value for the rock mass strength as shown in Table 2.

Hoek & Brown(1980) showed the approximate estimation of rock mass strength by the following equation:

$$\sigma_1' = \sigma_3' + \sqrt{m \cdot \sigma_c \cdot \sigma_3' + s \cdot \sigma_c^2} \qquad \text{Eq.(3)}$$

where σ_1' and σ_3' are major and minor principal stresses at failure; σ_c is the unconfined compressive strength of intact rock; and, m and s are the empirical constants for the rock mass. When a tunnel is in a critical state, σ_1' and σ_3' correspond with q_u and p_i respectively in the previous discussion. Also, if Barton's Q is known, constants m and s can be obtained from Q-value. Due to the limited pages, only the figures for

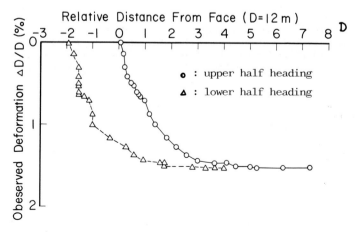

Fig. 6 Convergence curve obtained at Site No.3

the rock mass strengths obtained from Barton's Q-values and Eq.3 are listed as σ_q' in Table 2.

There are some disagreements between the authors' estimations and those based on Q-values for the cases with no non-elastic zone or slight support load corresponding with Class I in Table 1.

It must be emphasized that, so long as the interaction between surrounding rock and support elements remains unknown, it is hard to estimate rock mass strength. Available support pressure and stiffness of a support system at installation show quite a wide range of scattering in reality. This subject is still in consideration, and further comments will appear in the near future.

REFERENCES:

AFTES(1978):General Report,Conference on Convergence Confinement Method, Paris,Oct.26;J.of Tunnels et Ouvrages Souterrains,No.32(1979) (in French); Underground Space,4-4,pp.225-232(1980)(in English).

Barton,N,V.Chouby(1977): The Shear Strength of Rock Joint in Theory and Practice,Rock Mech.,Springer,No.1/2,pp.1-54.Also,NGI-Publ.119,1978.

Hoek,E.,E.T.Brown(1980): Underground Excavation in Rock,The Institute of Mining and Metallurgy,London,Ch.8.

Mueller,L.(1977): The Use of Deformation Measurements in Dimensioning the Lining of Subway Tunnels, Proc.Int.Sympo. on Field Measurements in Rock Mechanics: edited by Kovari,Balkema(1979),pp.451-472.

Pacher,F.(1972): Erfahrungen mit Gebirgsdruckmessungen bei osterreichi-schen Verkehrstunnelbauten,Int.Sympo.f.Untertagebau,Luzern,S.381-391.

Peck,R.B.(1969): Deep Excavation and Tunnelling in Soft Ground,State-of-the-Art Report,7th ICSMFE,Mexico,State of the Art Volume,pp.225-290.

Tanimoto,C.,S.Hata(1980): Fundamental Concept of Designing Tunnel Supports in Consideration of Elasto-plastic and Strain Softening Behaviour of Rock, Memoir of the Faculty of Engineering, Kyoto University,Vol.42,pp.349-376.

Tanimoto,C.,et al.(1981a): Interaction between Fully Bonded Bolts and Strain Softening Rock in Tunnelling, Proc.22nd U.S.Sympo. on Rock Mech.,pp.347-352.

Tanimoto,C.,et al.(1981b): Interaction between Rockbolts and Weak Rock in Tunnelling, Proc.Int.Sympo. on Weak Rock,Tokyo,Sept.,Vol.3,pp.157-162.

Tanimoto,C.,et al.(1982a): Interpretation of Characteristic Line for Tunnel Stability, 14th Sympo.on Rock Mech.,JSCE,pp.86-90(in Japanese).

Tanimoto,C.,S.Hata(1982b): Fundamental Concept of Tunnel Supports to be Placed in the Vicinity of Mining Face, Proc.of JSCE, No.325,pp.93-106 (in Japanese).

Tanimoto,C.,et al.(1983): Allowable Limit of Convergence in Tunnelling, 24th U.S.Sympo.on Rock Mech.,pp.251-263.

Tanimoto,C.,R.E.Goodman(1984): Technical Discussion on the application of Steel Ribs in the VAT Tunnel Project at the request of Bureau of Reclamation,U.S.A.

Tanimoto,C.,et al.(1986): Allowable Limit of Deformation Rates in Tunnelling, 18th Sympo.on Rock Mech.,JSCE,pp.431-435(in Japanese).

Tanimoto,C.,et al.(1987): Relationship Between Deformation and Support Pressure in Tunnelling through Overstressed Rocks, 6th ISRM Congress, Montreal,1987.

Terzaghi,K.v.(1946): Rock Defects and Loads on Tunnel Supports, Publ. in "Rock Tunnels with Steel Supports" by Proctor and White, Commercial Shearing,Inc.

2nd International Symposium on Field Measurements in Geomechanics, Sakurai (ed.)
© 1988 Balkema, Rotterdam. ISBN 90 6191 778 6

Displacements and strains around a non-flat-end borehole

S.Kobayashi, N.Nishimura & K.Matsumoto
Kyoto University, Japan

1 INTRODUCTION

Knowledge of in-situ stress state is of basic importance for designing rock structures. Several techniques have been developed for measuring in-situ rock stresses. Among them, the stress-relief technique using over-coring is one of the most efficient methods. For the measurement of relieved stresses, strain-gauge rossettes are usually placed on the flat-end of a bore hole. However, the method is not accurate enough to evaluate the stress along the axial direction of the borehole.

Recently, an efficient improvement for the method has been proposed (Sugawara and Obara 1986) by introducing a hemispherical-ended borehole instead of the flat-end. This technique is very promising for a more accurate measurement of in-situ rock stresses. Motivated by the improvement, we hereby propose another version of the technique. The idea is the same, but we have simply replaced the hemispherical-end with a conical-end.

In the present paper, we show some fundamental data useful for implementing the technique.

2 OBSERVATION EQUATIONS

In-situ stresses can be expressed by a matrix

$$\{\sigma\}^T = \{\sigma_{11}, \ \sigma_{22}, \ \sigma_{33}, \ \sigma_{12}, \ \sigma_{23}, \ \sigma_{31}\}$$

$$= \{\sigma_1, \ \sigma_2, \ \sigma_3, \ \sigma_4, \ \sigma_5, \ \sigma_6\} \tag{1}$$

referring to the rectilinear cartesian coordinate system $(O:x_1,x_2,x_3)$, where $\{\sigma\}^T$ means the transpose of $\{\sigma\}$. The induced displacement components and strain components on the wall of the borehole by over-coring are, respectively,

$$\{u\}^T = \{u_{11}, \ u_{12}, \ u_{13}, \ \ldots\ldots\ldots, \ u_{M1}, \ u_{M2}, \ u_{M3}\} \tag{2}$$

$$\{\varepsilon\}^{T} = \{\varepsilon_{11}, \ \varepsilon_{12}, \ \cdots\cdots\cdots, \ \varepsilon_{N1}, \ \varepsilon_{N2}\} \tag{3}$$

where the first subscripts M and N stand for the position (node) numbers, at which displacements and strains are considered, and the second subscripts imply their components. In Eq.(3) subscripts 1 and 2 stand for components along generators and those orthogonal to them, respectively.

For a linear isotropic elastic body, for example, the induced strains can be expressed by

$$\{\varepsilon\} = \frac{1}{E}[A]\{\sigma\} \tag{4}$$

where E is Young's modulus, and element A_{ij} implies strain ε_{ni} (n:$1 \le n \le N$, i=1,2), which is induced by the stress $\sigma_{j}=1$ provided that E=1. It is noted that A_{ij} is dependent on Poisson's ratio ν.

For displacements, we have a similar relation

$$\{u\} = \frac{1}{E}[B]\{\sigma\} \tag{5}$$

In this paper, we list [A] and [B] for some typical cases, which can be accurately evaluated by the use of the boundary element method. Using some selected numbers ($6 \le 3M$, 2N) of [A] or [B], we can easily obtain the observation equation for determining the in-situ stress state.

For example, using the least-squares technique, we have for the measured strain $\{\varepsilon\}$ (more than 6 measurements)

$$[A]^{T}[A]\{\sigma\} = E[A]^{T}\{\varepsilon\} \tag{6}$$

and thus stress is determined by

$$\{\sigma\} = E[C]\{\overline{\varepsilon}\} = E[D]\{\varepsilon\} \tag{7}$$

where [C] is the inverse of $[A]^{T}[A]$, $\{\overline{\varepsilon}\} = [A]^{T}\{\varepsilon\}$, and $[D] = [C][A]^{T}$.

3 MATRICES [A] AND [B]

An integral equation for the linear isotropic elastic body is expressed by using the indicial notation with the summation convention rule, as follows;

$$C_{ki}(\mathbf{x})u_{i}(\mathbf{x}) = \int_{\partial D} U_{ki}(\mathbf{x},\mathbf{y})t_{i}(\mathbf{y})dS_{y}$$
$$-\oint_{\partial D} T_{ki}(\mathbf{x},\mathbf{y})u_{i}(\mathbf{y})dS_{y} \tag{8}$$

where u_i and t_i stand for displacement and traction on the boundary ∂D, respectively, $\mathbf{x}$ and $\mathbf{y}$ are position vectors, C_{ki} the free term ($\delta_{ki}/2$ for a smooth surface, δ_{ki} : the Kronecker's delta), and U_{ki} and T_{ki} are fundamental solutions which are given by

$$U_{ki}(\mathbf{x},\mathbf{y}) = \frac{(1+\nu)}{8\pi E(1-\nu)r}\{(3-4\nu)\delta_{ki} + \frac{r_k r_i}{r^2}\} \qquad (9)$$

and

$$T_{ki}(\mathbf{x},\mathbf{y}) = \frac{1}{8\pi(1-\nu)r^2}\{(1-2\nu)[n_k(\mathbf{y})\frac{r_i}{r} - n_i(\mathbf{y})\frac{r_k}{r}]$$

$$+ [(1-2\nu)\delta_{ki} + 3\frac{r_k r_i}{r^2}]n_s(\mathbf{y})\frac{r_s}{r}\} \qquad (10)$$

where n_i stands for the outward unit normal vector to the boundary surface, $r_i = x_i - y_i$, $r = |r_i r_i|$, and $\oint_{\partial D} (\quad) dS_y$ denotes the Cauchy's principal value integral.

The integral equation is solved numerically by descretizing the boundary and approximating boundary displacements and tractions by use of the isoparametric elements as used in the finite element method.

In our numerical treatment, we used quadratic isoparametric elements. Displacements are obtained first with such boundary conditions as homogeneous stress state $\sigma_j = 1$ at infinity and free of traction on the wall of the borehole.

Elements of [A] and [B] are easily calculated from these displacements with Young's modulus $E = 1$. They are listed in Tables 1 and 2, in which the label of displacement and strain components should be referred to in Fig. 1.

4 APPLICATIONS

When the induced strains or displacements on the wall of the borehole are measured, we can easily estimate the in-situ stress state by Eq.(7), for instance, provided that Young's modulus and Poisson's ratio are estimated in advance, which may be obtained from the bored core.

Using strain components

$$\{\varepsilon\}^T = \{\varepsilon_{11}, \; \varepsilon_{12}, \; \varepsilon_{31}, \; \varepsilon_{32}, \; \varepsilon_{51}, \; \varepsilon_{52}, \; \varepsilon_{71}, \; \varepsilon_{72},$$

$$\varepsilon_{91}, \; \varepsilon_{92}, \; \varepsilon_{11,1}, \; \varepsilon_{11,2}\} \qquad (11)$$

Table 1. Strain matrix [A].

label	ν=1/3						ν=1/6					
11	-0.814	2.684	-0.402	-0.000	-0.000	-0.123	-0.877	2.760	-0.235	-0.000	-0.000	-0.105
12	0.211	-0.478	0.662	0.000	0.001	-1.634	0.326	-0.294	0.662	0.000	0.001	-1.520
21	0.061	1.811	-0.403	-3.029	-0.061	-0.107	0.033	1.851	-0.235	-3.148	-0.052	-0.090
22	0.036	-0.305	0.661	0.597	-0.816	-1.410	0.168	-0.138	0.660	0.537	-0.760	-1.312
31	1.810	0.060	-0.401	-3.029	-0.107	-0.061	1.850	0.032	-0.234	-3.148	-0.090	-0.052
32	-0.305	0.037	0.661	0.597	-1.409	-0.817	-0.138	0.169	0.660	0.537	-1.312	-0.761
41	2.685	-0.814	-0.402	-0.000	-0.123	0.000	2.760	-0.877	-0.235	0.000	-0.105	0.000
42	-0.478	0.211	0.662	-0.000	-1.634	-0.001	-0.294	0.326	0.662	-0.000	-1.520	-0.001
51	1.810	0.060	-0.401	3.029	-0.107	0.061	1.851	0.033	-0.234	3.148	-0.090	0.052
52	-0.305	0.036	0.661	-0.597	-1.410	0.816	-0.138	0.169	0.660	-0.537	-1.312	0.760
61	0.060	1.811	-0.402	3.029	-0.061	0.107	0.033	1.851	-0.234	3.148	-0.052	0.090
62	0.037	-0.305	0.661	-0.597	-0.817	1.409	0.169	-0.138	0.660	-0.537	-0.761	1.312
71	-0.814	2.684	-0.402	-0.000	0.000	0.123	-0.877	2.760	-0.235	-0.000	0.000	0.105
72	0.211	-0.478	0.662	0.000	-0.001	1.634	0.326	-0.294	0.662	0.000	-0.001	1.520
81	0.061	1.811	-0.403	-3.029	0.061	0.107	0.033	1.851	-0.235	-3.148	0.052	0.090
82	0.036	-0.305	0.661	0.597	0.816	1.410	0.169	-0.138	0.660	0.537	0.760	1.312
91	1.810	0.060	-0.401	-3.030	0.107	0.061	1.851	0.033	-0.234	-3.148	0.090	0.052
92	-0.305	0.037	0.661	0.597	1.409	0.817	-0.138	0.169	0.660	0.537	1.312	0.761
10,1	2.685	-0.814	-0.402	0.000	0.123	-0.000	2.760	-0.877	-0.235	0.000	0.105	-0.000
10,2	-0.478	0.211	0.662	-0.000	1.634	0.001	-0.294	0.326	0.662	-0.000	1.520	0.001
11,1	1.810	0.060	-0.401	3.030	0.107	-0.061	1.850	0.033	-0.234	3.148	0.090	-0.052
11,2	-0.305	0.036	0.661	-0.597	1.410	-0.816	-0.138	0.169	0.660	-0.537	1.312	-0.760
12,1	0.060	1.810	-0.402	3.029	0.061	-0.107	0.033	1.851	-0.234	3.148	0.052	-0.090
12,2	0.037	-0.305	0.661	-0.597	0.817	-1.409	0.169	-0.138	0.660	-0.537	0.761	-1.312
13,1	-0.833	2.687	-0.347	0.000	0.000	0.004	-0.892	2.764	-0.190	0.000	0.000	0.004
13,2	-0.854	-0.644	1.051	-0.000	0.000	-0.049	-0.624	-0.314	1.038	-0.000	0.000	-0.043
14,1	0.046	1.807	-0.348	-3.049	0.002	0.003	0.022	1.850	-0.191	-3.166	0.002	0.004
14,2	-0.801	-0.696	1.051	-0.182	-0.024	-0.042	-0.547	-0.392	1.038	-0.268	-0.022	-0.037
15,1	1.807	0.046	-0.347	-3.049	0.003	0.001	1.850	0.022	-0.191	-3.166	0.003	0.001
15,2	-0.696	-0.802	1.051	-0.182	-0.042	-0.024	-0.392	-0.547	1.038	-0.268	-0.037	-0.022
16,1	2.687	-0.833	-0.347	0.000	0.004	0.000	2.764	-0.892	-0.190	0.000	0.004	0.000
16,2	-0.644	-0.859	1.051	0.000	-0.049	-0.000	-0.314	-0.624	1.038	0.000	-0.043	-0.000
17,1	1.807	0.046	-0.348	3.049	0.003	-0.002	1.850	0.022	-0.191	3.166	0.004	-0.002
17,2	-0.696	-0.801	1.051	0.182	-0.042	0.024	-0.392	-0.547	1.038	0.268	-0.037	0.022
18,1	0.046	1.807	-0.347	3.049	0.001	-0.003	0.022	1.850	-0.191	3.166	0.001	-0.003
18,2	-0.802	-0.696	1.051	0.182	-0.024	0.042	-0.547	-0.392	1.038	0.268	-0.022	0.037
19,1	-0.833	2.687	-0.347	0.000	0.000	-0.004	-0.892	2.763	-0.190	-0.000	0.000	-0.004
19,2	-0.854	-0.644	1.051	-0.000	-0.000	0.049	-0.624	-0.314	1.038	-0.000	-0.000	0.043
20,1	0.047	1.807	-0.348	-3.049	-0.002	-0.003	0.022	1.850	-0.191	-3.166	-0.002	-0.003
20,2	-0.801	-0.696	1.051	-0.182	0.024	0.042	-0.547	-0.392	1.038	-0.268	0.022	0.037
21,1	1.807	0.046	-0.347	-3.049	-0.003	-0.001	1.850	0.022	-0.191	-3.166	-0.003	-0.001
21,2	-0.696	-0.801	1.051	-0.182	0.042	0.024	-0.392	-0.547	1.038	-0.268	0.037	0.022
22,1	2.687	-0.833	-0.347	0.000	-0.004	0.000	2.764	-0.892	-0.190	0.000	-0.004	0.000
22,2	-0.644	-0.854	1.051	0.000	0.049	0.000	-0.314	-0.624	1.038	0.000	0.043	0.000
23,1	1.807	0.046	-0.348	3.049	-0.003	0.002	1.850	0.022	-0.191	3.166	-0.003	0.002
23,2	-0.696	-0.801	1.051	0.182	0.042	-0.024	-0.392	-0.547	1.038	0.268	0.037	-0.022
24,1	0.046	1.807	-0.347	3.049	-0.001	0.003	0.022	1.850	-0.191	3.166	-0.001	0.003
24,2	-0.801	-0.696	1.051	0.182	0.024	-0.042	-0.547	-0.392	1.038	0.268	0.022	-0.037

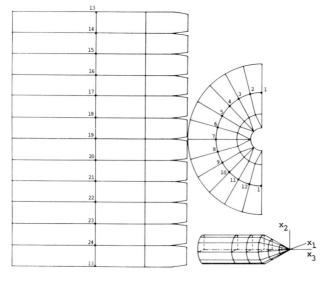

Fig. 1. Surface elements of the conical-end borehole.

Table 2. Displacement matrix [B].

label	ν=1/3						ν=1/6					
13,1	2.637	-0.655	-0.372	0.000	0.000	-1.897	2.717	-0.715	-0.204	0.000	0.000	-1.657
13,2	0.000	0.000	0.000	3.295	-1.901	0.000	0.000	0.000	0.000	3.416	-1.662	0.000
13,3	0.508	0.588	-1.584	0.000	0.000	0.177	0.282	0.327	-1.567	0.000	0.000	0.135
14,1	2.285	-0.568	-0.322	1.645	0.002	-1.898	2.350	-0.616	-0.177	1.721	0.002	-1.658
14,2	-0.329	1.320	-0.186	2.852	-1.900	0.002	-0.352	1.353	-0.102	2.966	-1.661	0.002
14,3	0.528	0.568	-1.584	-0.069	0.088	0.153	0.293	0.316	-1.567	-0.039	0.068	0.117
15,1	1.320	-0.329	-0.186	2.852	0.002	-1.900	1.352	-0.352	-0.102	2.966	0.002	-1.661
15,2	-0.568	2.285	-0.322	1.645	-1.898	0.002	-0.616	2.350	-0.177	1.721	-1.658	0.002
15,3	0.568	0.528	-1.584	-0.069	0.153	0.088	0.316	0.293	-1.567	-0.039	0.117	0.068
16,1	0.000	0.000	0.000	3.295	0.000	-1.901	0.000	0.000	0.000	3.416	0.000	-1.662
16,2	-0.655	2.637	-0.372	0.000	-1.897	0.000	-0.715	2.717	-0.204	0.000	-1.657	0.000
16,3	0.588	0.508	-1.584	0.000	0.177	0.000	0.327	0.282	-1.567	0.000	0.135	0.000
17,1	-1.320	0.329	0.186	2.852	-0.002	-1.900	-1.352	0.352	0.102	2.966	-0.002	-1.661
17,2	-0.568	2.285	-0.322	-1.645	-1.898	-0.002	-0.616	2.350	-0.177	-1.721	-1.658	-0.002
17,3	0.568	0.528	-1.584	0.069	0.153	-0.088	0.316	0.293	-1.567	0.039	0.117	-0.068
18,1	-2.285	0.568	0.322	1.645	-0.002	-1.898	-2.350	0.616	0.177	1.721	-0.002	-1.658
18,2	-0.329	1.320	-0.186	-2.852	-1.900	-0.002	-0.352	1.353	-0.102	-2.966	-1.661	-0.002
18,3	0.528	0.568	-1.584	0.069	0.088	-0.153	0.293	0.316	-1.567	0.039	0.068	-0.117
19,1	-2.637	0.655	0.372	0.000	0.000	-1.897	-2.717	0.715	0.204	0.000	0.000	-1.657
19,2	0.000	0.000	0.000	-3.295	-1.901	0.000	0.000	0.000	0.000	-3.416	-1.662	0.000
19,3	0.508	0.588	-1.584	0.000	0.000	-0.177	0.282	0.327	-1.567	0.000	0.000	-0.135
20,1	-2.285	0.568	0.322	-1.645	0.002	-1.898	-2.350	0.616	0.177	-1.721	0.002	-1.658
20,2	0.329	-1.320	0.186	-2.852	-1.900	0.002	0.352	-1.352	0.102	-2.966	-1.661	0.002
20,3	0.528	0.568	-1.584	-0.069	-0.088	-0.153	0.293	0.316	-1.567	-0.039	-0.068	-0.117
21,1	-1.320	0.329	0.186	-2.852	0.002	-1.900	-1.352	0.352	0.102	-2.966	0.002	-1.661
21,2	0.568	-2.285	0.322	-1.645	-1.898	0.002	0.616	-2.350	0.177	-1.721	-1.658	0.002
21,3	0.568	0.528	-1.584	-0.069	-0.153	-0.088	0.316	0.293	-1.567	-0.039	-0.117	-0.068
22,1	0.000	0.000	0.000	-3.295	0.000	-1.901	0.000	0.000	0.000	-3.416	0.000	-1.662
22,2	0.644	-2.637	0.372	0.000	-1.897	0.000	0.715	-2.717	0.204	0.000	-1.657	0.000
22,3	0.588	0.508	-1.584	0.000	-0.177	0.000	0.327	0.282	-1.567	0.000	-0.135	0.000
23,1	1.320	-0.329	-0.186	-2.852	-0.002	-1.900	1.352	-0.352	-0.102	-2.966	-0.002	-1.661
23,2	0.568	-2.285	0.322	1.645	-1.898	-0.002	0.616	-2.350	0.177	1.721	-1.658	-0.002
23,3	0.568	0.528	-1.584	0.069	-0.153	0.088	0.316	0.293	-1.567	0.039	-0.117	0.068
24,1	2.285	-0.568	-0.322	-1.645	-0.002	-1.898	2.350	-0.616	-0.177	-1.721	-0.002	-1.658
24,2	0.329	-1.320	0.186	2.852	-1.900	-0.002	0.352	-1.352	0.102	2.966	-1.661	-0.002
24,3	0.528	0.568	-1.584	0.069	-0.088	0.153	0.293	0.316	-1.567	0.039	-0.068	0.117

Origin of the coordinates is taken at the center of the circular right-section.

for example, we have [D] as listed in Table 3. In general, more numbers of measured strains are taken into account, the more accurate estimation of the in-situ stress state is achieved.

It should be noted that over-coring with the same diameter as that of the borehole may be possible when we apply the technique proposed in this paper.

5 CONCLUDING REMARKS

In the present paper, we have showed fundamental data for estimating the in-situ stress state with the aid of a conical-end borehole and an over-coring stress-relief technique. The proposed method is advantageous in two points;

1) we can estimate the in-situ stress state using one borehole, and

2) the over-coring size can be the same as that of the measuring borehole.

Table 3. Stress matrix [D].

label	11	12	31	32	51	52	71	72	91	92	11,1	11,2
ν=1/3	0.006	0.077	0.143	0.050	0.143	0.050	0.006	0.077	0.143	0.050	0.143	0.050
	0.189	0.041	0.052	0.068	0.052	0.068	0.189	0.041	0.052	0.068	0.052	0.068
	0.040	0.276	0.039	0.276	0.039	0.276	0.040	0.276	0.039	0.276	0.039	0.276
	-0.000	0.000	-0.079	0.016	0.079	-0.016	-0.000	0.000	-0.079	0.016	0.079	-0.016
	-0.000	0.000	-0.013	-0.176	-0.013	-0.176	0.000	-0.000	0.013	0.176	0.013	0.176
	-0.015	-0.203	-0.008	-0.101	0.008	0.101	0.015	0.203	0.008	0.101	-0.008	-0.101
ν=1/6	-0.001	0.046	0.132	0.024	0.133	0.024	-0.001	0.046	0.133	0.024	0.132	0.024
	0.177	0.016	0.043	0.039	0.043	0.039	0.177	0.016	0.043	0.039	0.043	0.039
	-0.004	0.251	-0.004	0.251	-0.004	0.251	-0.004	0.251	-0.004	0.251	-0.004	0.251
	-0.000	0.000	-0.077	0.013	0.077	-0.013	-0.000	0.000	-0.077	0.013	0.077	-0.013
	-0.000	0.000	-0.013	-0.190	-0.013	-0.190	0.000	-0.000	0.013	0.190	0.013	0.190
	-0.015	-0.218	-0.007	-0.109	0.007	0.109	0.015	0.218	0.007	0.109	-0.007	-0.109

REFERENCES

Sugawara, K. & Y. Obara 1986. Measurement of in-situ rock stress by hemispherical-ended borehole technique. Mining Science and Technology. 3:287-300.
Watson, J. O. 1979. Advanced implementation of the boundary element method for two- and three-dimensional elastostatics, Chapt. 3 in P. K. Banerjee & R. Butterfield (eds.), Developments in boundary element methods - 1, p.31-63. Applied Science Publishers, London.

2nd International Symposium on Field Measurements in Geomechanics, Sakurai (ed.)
© 1988 Balkema, Rotterdam. ISBN 90 6191 778 6

A preliminary assessment of correct reduction of field measurement data: Scalars, vectors and tensors

C.G.Dyke, A.J.Hyett & J.A.Hudson
Department of Mineral Resources Engineering, Imperial College of Science and Technology, South Kensington, London, UK

1. INTRODUCTION

It is well known that the factors of accuracy, precision and resolution are crucial to the success of a field measurement programme in geomechanics. It is, of course, also important that the measured data are correctly reduced and interpreted. One major aspect of such a programme is that the average values and scatter of the data are correctly obtained and presented. The conventional presentation of the mean of a set of measurements, together with the standard deviation, is a straightforward operation for any scalar quantity (i.e. a quantity with magnitude only), such as the density of rock. However, it is not readily apparent how to obtain, and present unambiguously, measures of the average and scatter of vector and tensor quantities, such as ground movement components and in situ stress, respectively. How, for example, does one present the mean and standard deviation of ten measured stress tensor?

We develop here, preliminary ideas on the determination and presentation of the average and scatter of vector and tensor data; this is fundamental to any field measurement programme in geomechanics. Because, in three dimensions a vector has three independent components and a tensor has six, the means and standard deviations must be considered very carefully. In fact, special vector and tensor statistical methods, based on multivariate analysis, must be utilized. These will be explained with reference to the parameters commonly measured in geomechanics, such as displacements, permeability and in situ stress.

Considering this latter example, the stress tensor, the search for improved statistical analysis is illustrated by the work of Chambon and Revalor (1986). The necessity for improved methods is highlighted by the presentation in the literature of non-perpendicular sets of mean principal stresses, which can be caused by the application of least squares regression analysis to the calculation of principal directions.
Futhermore, correct analysis of data enables a better understanding of the scatter of field measurements. In the case of in situ stress there is an increased awareness that the scatter and variability of results is partially a property of the rock (Brady, Lemos and Cundall, 1986; Fairhurst, 1986). By analysing the variance of a tensor in a scalar way, one is either assuming that a measure of the scatter is scalar

and hence isotropic, or forfeiting a wealth of information about the response of the rock. Similarly univariant variance analysis of the stress tensor is not particularly clear, let alone rigorous. An illustrative example of magnitudes and trend/plunge data presentation is;

$$\sigma_1 \quad 10 \pm 4 \text{ MPa} \qquad 010/00 \pm 5^{\circ}$$
$$\sigma_2 \quad 4 \pm 2 \text{ MPa} \qquad 090/90 \pm 5^{\circ}$$
$$\sigma_3 \quad 2 \pm 1 \text{ MPa} \qquad 100/00 \pm 5^{\circ}$$

Within the magnitude scatter, σ_1, σ_2 and σ_3 can be interchanged as major, intermediate and minor principal stresses, therefore changing the direction of eg the major principal stress.

2. TERMINOLOGY

There is some ambiguity over the terms relating to field measurement data. The following definitions are included as a preface to the subsequent discussion on the means and variances of vectors and tensors.

2.1 Type of Measurement: Scalar, Vector , Tensor

Fundamental to the correct analysis of data is an understanding of the directional nature of the property. The basic difference between a scalar, vector and tensor, is that a scalar has magnitude only (e.g. mass), a vector has magnitude, and direction (e.g. displacement), and a tensor has magnitude direction and the plane which is being considered (e.g. stress). However, to manipulate the mathematics for the statistical considerations in this paper, it is necessary to consider a more rigorous definition of a tensor: " A tensor is an invariant multilinear function of direction, i.e., it is a function of a number of directions which takes the form:

$$S_{\alpha\beta\gamma} \quad \cdots \qquad l_{\alpha} \; m_{\beta} \; n_{\gamma} \ldots \ldots$$

in any particular basis, (l_{α}), (m_{α}), (n_{α}) .... being the direction cosines of the directions in the basis and which have the same numerical value in every basis. The number of directions involved is the rank of the tensor" (Temple, 1960). A scalar can be regarded as a tensor of rank zero, and vectors are tensors of rank one. What is commonly called a tensor in geotechnics is specifically a second rank tensor. These will be discussed below. It is not just the quantities themselves that conform to this description, but also their corresponding means and variances. A list of frequently used geotechnical properties and their corresponding multilinear rank or order is presented in Table 1.

2.2 Sampling parameters

We can only utilize a value if we can assess the variability of the measured data. The single value itself is not sufficient, because we do not know its accuracy, precision or measurement resolution (these are often assumed). The scatter within a sample is modified from that of the population scatter by the following instrument measuring characteristics (Figure 1)

Table 1. Examples to illustrate the classification of some Rock Mechanics Properties.

Scalar	Vector	Tensor (2nd order)
Temperature	Displacement	Stress
Time	Force	Strain
Mass	Discontinuity frequency	Permeability
Volume		Moment of Inertia

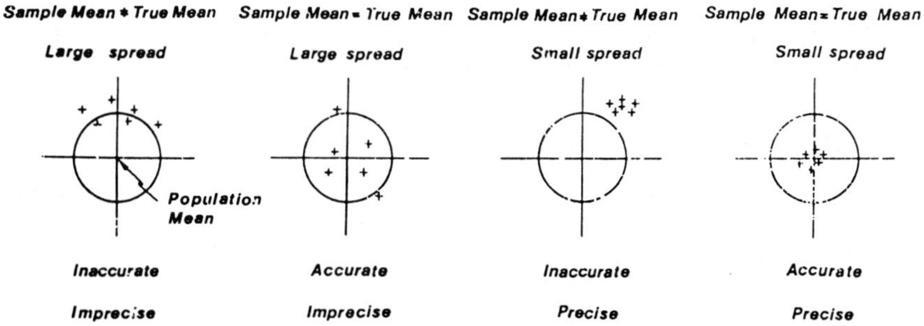

Sample Mean ≠ True Mean   Sample Mean = True Mean   Sample Mean ≠ True Mean   Sample Mean = True Mean

Large spread   Large spread   Small spread   Small spread

Population Mean

Inaccurate   Accurate   Inaccurate   Accurate

Imprecise   Imprecise   Precise   Precise

Figure 1. A graphical representation of accuracy and precision.

Accuracy

"If the expected value of an estimator does not equal the true value being estimated, the difference between the expected value and the true value is known as the bias of the estimator. If an estimator has zero bias it is said to be accurate; if an estimator has a large bias it is said to be inaccurate." (Ostle, 1963)

Precision

"Precision refers to the spread of measurements about the estimated mean. Thus, the precision of an estimator is a measure of the re-peatability of the estimator. Therefore, precision may be expressed in terms of the variance of an estimator, with a large variance repre-senting low precision, and a small variance signifying high precision. " (Ostle, 1963)

Resolution

The resolution is the smallest measurable difference between any two measurements. If the measuring instrument can differentiate between

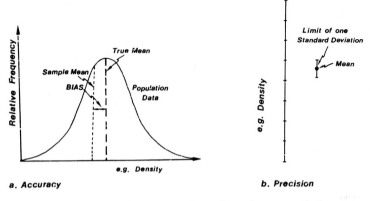

a. Accuracy   b. Precision

Figure 2.   Accuracy and precision of scalar quantities.

two closely spaced points then it is said to possess high resolution;
if only between widely spaced points, then it has low resolution.

3. REDUCTION AND PRESENTATION OF SCALAR VALUES

No directions are associated with each data member, and likewise
for the mean and variance.   The analysis is straight forward and is
included for completeness.   The scalar mean and variances are summarised in Figure 2 and defined below;

Mean, $\bar{X} = 1/N \sum_{i}^{N} X_i$

Variance $S^2 = 1/(N-1) \sum_{i}^{N} (X_i - \bar{X})^2$

where N is the sample size and X_i is the ith observation value.

4. REDUCTION AND PRESENTATION OF VECTOR QUANTITIES.

The analysis of vectorial data is a subset of trivariate data analysis, with the characteristic that the sum of the squares of the components along each axis equals the square of the magnitude.   The
scatter cloud of vectorial data can be represented as in Figure 3a.
Relative orientation of this ellipsoid enclosing the scatter depends
on the level of dependency between the three variables.   For instance,
in 2D if X and Y are independent then the ellipse axes will be orien-
tated parallel to the X and Y directions.   If they are interdependent,
then the ellipse axes are inclined at an angle corresponding to the
degree of dependency (Figure 4).   Which of the above two occurs for a
particular set of data is, of course, primarily controlled by the
initial choice of co-ordinate system.

The mean and measures of accuracy, precision and resolution of vectors
are themselves all directional parameters.   The mean is a vector and
the variance is a second order tensor.   However, during preliminary
data analysis, it is convenient at times to calculate scalar summaries
of these quantities.

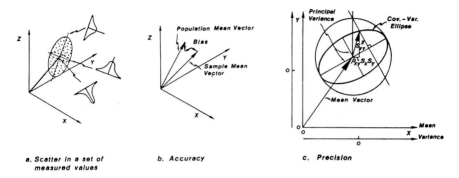

a. Scatter in a set of measured values

b. Accuracy

c. Precision

Figure 3. Accuracy and precision of vector quantities.

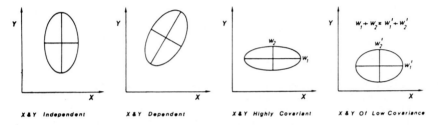

X & Y Independent

X & Y Dependent

X & Y Highly Covariant

X & Y Of Low Covariance

Figure 4. The effect of dependency and covariance on the relative geometry of the variance ellipse.

4.1 The Mean

For a series of n vectors: $v = a_i \bar{X} + b_i \bar{Y} + c_i \bar{Z}$,

the mean equals: $\bar{V} = \left[\sum_1^n a_i \ / \ n \right] \bar{X} + \left[\sum_1^n b_i \ / n \right] \bar{Y} + \left[\sum_1^n c_i \ / n \right] \bar{Z}$

of magnitude R $= (\ a^2 + b^2 + c^2)^{\frac{1}{2}}$

Around this vector the sum of the squared deviations is minimised. The above degenerates for unit vectors into the standard technique used, for instance, in the geometrical analysis of discontinuity planes. The orintation of the normal is represented by a unit vector, the components of each unit vector are summed and subsequently divided by the sample size (Priest,1985).

4.2 Accuracy

Accuracy, the difference between the population and sample means, can be represented by the line separating the two vector means in Figure 3(b). Hence, the bias vector is of the form:

$$B = b_1 \bar{i} + b_2 \bar{j} + b_3 \bar{k}.$$

Total bias, a scalar, equals the square root of the sum of the squares of the bias components.

4.3 Precision

To compare means of multivariate observations an analogue to the variance of univariate observations is required. This analogue is the variance–covariance matrix.

One approach to generate such information for the analysis of unit vectors is through the use of Bingham's distribution (Bingham, 1974). The three parameters k_1, k_2, k_3, which define the degree of clustering of the data in orthogonal directions, can be obtained; in essence these are similar to the constant k associated with the Fisher distribution. The mean is always parallel to the direction of k_3.

However, the approach cannot be used for the analysis of data of varying magnitudes such as ground displacements, even though the variance is a three parameter property. Because L, the column vector of the direction cosines, has unit length and U_1, U_2, U_3, the vectors associated with k_1, k_2, k_3 form an orthogonal set, it follows that:

$$(L' U_1)^2 + (L' U_2)^2 + (L' U_3)^2 = L'(U_1 U_2 U_3)(U_1 U_2 U_3)' L = 1$$

where L' is the transpose of L.

As developed by Mardia (1972, p235) and discussed by Kelker and Langenberg (1976), if this relation is used to replace $(L' U_3)^2$ in the probabilty density function, it is seen that the distribution is a function of $k_1 - k_3$ and $k_2 - k_3$ only. Thus only two of the parameters are independent. This can be envisaged by considering the spread of the unit vectors on a sphere. As the angular deviations away from the mean are increased, the component parallel to the mean decreases. It is this variation that the third dependent variable measures.

When the magnitude of the vector varies as well, there are four variables, three of which are independent. A suitable approach is multivariate analysis; the analogy of these two analyses was noted by Mardia (1972, p225).
If X, Y, Z is a trivariate observation, each variable has a sample variance S_i^2 and each pair of variables has a sample covariance defined as $\rho_{ij} S_i S_j$ where ρ_{ij} is the sample correlation coefficient of the paired variable. The variance–covariance matrix is the symmetric matrix

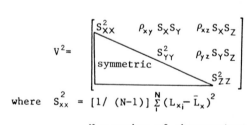

$$V^2 = \begin{bmatrix} S_{XX}^2 & \rho_{xy} S_X S_Y & \rho_{xz} S_X S_Z \\ & S_{YY}^2 & \rho_{yz} S_Y S_Z \\ \text{symmetric} & & S_{ZZ}^2 \end{bmatrix}$$

where $S_{xx}^2 = [1/ (N-1)] \sum_{i}^{N} (L_{x_i} - \bar{L}_x)^2$

N = number of observations.
L = X component of the observation.
$\bar{L}$ = Mean of the X component.

This can be regarded as a tensor as it transforms accordingly. To decompose the six independent components of covariance and variance into the three orthogonal components representing the variance of the independent parameters, principal component analysis is utilised. Davis (1973, p 478-482) discusses this variance transformation and a graphical representation identical to the stress ellipse manipulation is given in figure 3c. The ellipse is centred about the mean with semi-axes equal to the principal variance components, the direction of which are not necessarily parallel and perpendicular to the mean vector. Likewise, confidence limit ellipsoids can also be drawn about the mean; however, in calculating these limits one must take into account variate dependency by using Hotelling's T^2 test.

The total variance, a scalar measure of precision, is defined as the sum of the individual variances (the first invariant).

5 REDUCTION AND PRESENTATION OF TENSOR QUANTITIES

Just as vectors were treated by methods of trivarate data analysis, the statistics of a second order tensor can be calculated by hexavariate data analysis methods. Results lead to a mean tensor, and measures of accuracy and precision. In turn the problem is simplified for visualization in three dimensional space, by using principal component analysis to eliminate the non leading diagonal terms e.g. the analogue of the shear stress components. For example, individual stress observations cannot be plotted on a scatter diagram, as was the case with vectors, but must be represented as ellipsoids with axes of length corresponding to the principal stresses and oriented in the directions through which they act. Scatter can be represented by the superposition of a set of ellipsoids centred about the origin (Figure 6a).

5.1 The Mean

Tensors cannot be averaged regardless of base. In Cartesian co-ordinates, the method outlined in Figure 5 is used, which always yields a set of three orthogonal principal stresses. The mean is a second order tensor.

5.2 The Variance

Likewise, variance calculations must be performed with regard to the correct base (figure 5), producing a nine by nine fourth order variance-covariance tensor. The independent variances can be calculated and presented as below;

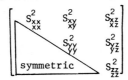

5.3 Accuracy

The bias or inaccuracy of a tensor measurement is the six dimensional vector:

$$\text{Bias} = a\vec{i} + b\vec{j} + c\vec{k} + 2d\vec{l} + 2e\vec{m} + 2f\vec{n}$$

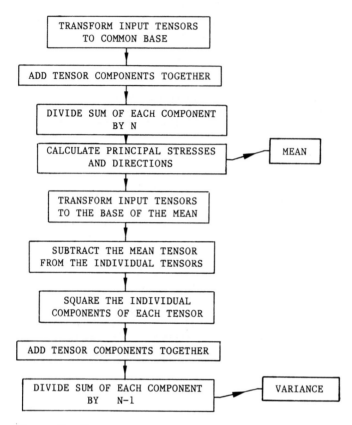

Figure 5. Flow chart of tensor mean and variance calculation.

where each is the difference between the relevant term of the sample and population means, in the base of the population mean. Total bias, a scalar summary, equals the square root of the sum of the squares of the components:

$$\text{Total Bias} = (a^2 + b^2 + c^2 + 2d^2 + 2e^2 + 2f^2)^{\frac{1}{2}}$$

5.4 Precision

Precision is estimated by the total variance, a summation of nine terms measuring the precision within each tensor component:

$$\sum_{1}^{9} S_{ij}$$

This yields an overall measure of the scatter within a data set, which depends on rock properties, instrument error and operator error.

For analysis of the overall consequences of variance within a data set, all nine terms need to be analysed. A graphical representation

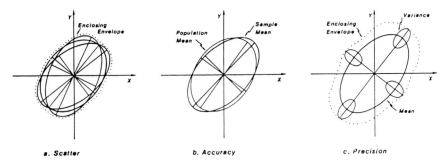

Figure 6. Illustration of accuracy and precision of 2nd order tensor quantities.

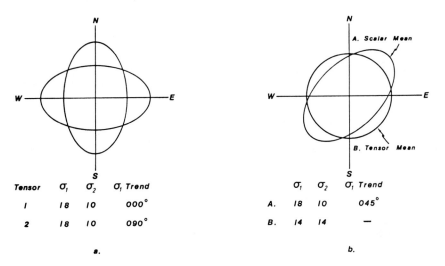

Tensor	σ_1	σ_2	σ_1 Trend
1	18	10	000°
2	18	10	090°

	σ_1	σ_2	σ_1 Trend
A.	18	10	045°
B.	14	14	—

a.   b.

Figure 7. Comparison of scalar and tensor means.

(Figure 6c) plots the variance ellipsoid of each principal stress around the principal mean stress vector. Whereas the representation surface of a second order tensor is an ellipsoid, that of the mean stress tensor including one standard deviation of variability, has the geometry of the enclosing envelope of Figures 6a and 6c. Similar surfaces have been noted for the fourth order stiffness tensor of zinc crystals by Goens (1933), (Nye, 1957).

6. ILLUSTRATIVE EXAMPLES

Two examples of averaging second order tensors by a conventional scalar least squares method, and the method outlined above, are given. In each case, the quantity represented is stress.
Firstly, a two dimensional problem for visual conceptualization (Figure 7a) is analysed. A least squares analysis treating each

Table 2. Comparison of scalar and tensorial averaging techniques.

| | | Magnitude (MPa) | | | Trend / | Plunge (Degs) | |
		σ_1	σ_2	σ_3	σ_1	σ_2	σ_3
Scalar Analysis	Mean	39.9	21.1	13.6	045/18	153/41	298/39
	Stand. Dev.	5.9	4.3	2.6	14	16	16
Tensor Analysis	Mean	39.0	21.0	14.5	046/19	154/42	298/42
	Stand. Dev.	7.5	5.1	3.9	σ_{12} 2.5	σ_{13} 1.6	σ_{23} 1.0

parameter as a scalar results in solution A (figure 7b). However it is intuitive, even with only a basic understanding of tensors, that the correct solution is B - hydrostatic stress.

Secondly, a series of twelve typical in situ stress determinations is analysed (Table 2). Similarly to Figure 6c, the variance can be used to construct a three dimensional representation of mean and associated scatter.

Of particular interest in each case are the differences in the computed deviatoric stress components.

7. CONCLUSIONS

As rock mechanics and rock engineering advance both in theory and in practice, there is an increasing requirement for rigorous analysis of field data. The parameters associated with rock properties and behaviour are not all scalars. Therefore, the data reduction methods must be matched to the type of quantity in question, i.e. scalar, vector or tensor quantities. It has been shown in this preliminary study that vector and tensor quantities can not be reduced in exactly the same way as scalar quantities,ie by using the statistical techniques that are most familiar. These methods have to be extended for the special analysis of vectors and tensors, otherwise the data reduction methods will be incorrect, as has been demonstrated for the tensor quantity, stress.

The discussion in this paper emphasises the complexities of vector and tensor data reduction, and presents possible methods of advance. These methods are not merely esoteric statistical tools, purely of academic interest; they are essential to the success of all field measurement programmes in geomechanics.

8. ACKNOWLEDGEMENTS

The work presented in this paper is part of an in situ stress research programme funded by B.P International and the U.K. Science and Engineering Research Council. The authors are extremely grateful to Dr M.V. Barr, and Mr S.W. Morrison of B.P International for providing them with the opportunity to conduct the research.

BIBLIOGRAPHY

Bingham, C. 1974. An antipodally symmetric distribution on the sphere. Annals Stat.,2: 1201 - 1225

Brady, B.H.G., J.V. Lemos & P.A. Cundall. 1976. Stress Measurement schemes for jointed and fractured rock. Proc Int. Symp. Rock Stress and Rock Stress Measurement. Stockholm. 167-176.

Chambon, C. & R. Revalor. 1986. Statistic analysis applied to rock stress measurements. Proc. Int. Symp. Rock Stress and Rock Stress Measurements Stockholm. 397-410.

Davis, J.C. 1973. Statistics and Data Analysis in Geology, New York: John Wiley.

Fairhurst, C. 1986. In situ Stress Determination - an appraisal of its significance in rock mechanics. Proc. Int. Symp. Rock Stress Measurements. Stockholm. 3-17

Goens, E. 1933. Uber eine verbesserte Apparatur zur Statischen Bestimmung des Drillingmodulus von Kristallstaben und ihre Anwendung auf zink-einkristalle.Ann. d. phys. 16: 793-809.

Kelker, D. & C.W. Langenberg. 1976. A mathematical model for orientation data from macroscopic cylindrical folds. Mathematical Geology. 8: 549-559

Mardia, K.V. 1972. Statistics of directional data. London. Academic Press.

Nye, J.F. 1957. Physical Properties of Crystals. Oxford. Clarendon Press.

Ostle, B. 1963. Statistics in Research. 2nd Ed. Iowa State University Press.

Priest, S.D. 1985. Hemispherical Projection Methods in Rock Mechanics. London. George Allen & Unwin.

Temple G. 1960. Cartesian Tensors. London. Methuen.

2nd International Symposium on Field Measurements in Geomechanics, Sakurai (ed.)
© 1988 Balkema, Rotterdam. ISBN 90 6191 778 6

Determination of elasto-plastic parameters by an inverse analysis method as a boundary control problem

Y.Ichikawa
Department of Civil Engineering, Nagoya University, Japan
T.Kyoya & T.Kawamoto
Department of Geotechnical Engineering, Nagoya University, Japan

1 INTRODUCTION

Since geomaterials show nonlinear characteristics in response to applied loads, engineers are required to determine an appropriate stress-strain model when designing geotechnical structures. It is considerably difficult, however, since the behaviour of geomaterials is affected by a lot of factors. Furthermore, only a small amount of data is measured in the whole structure. Thus, it is suitable that an inverse analysis method be used in order to determine a material model from the measured data.

Inverse analysis using a linear elastic model has been used for the observational construction procedure of geological structures (Sakurai and Takeuchi 1983, Sakurai,Shimizu and Matsumuro 1985), while an iden-tification method for hypoelastic material behaviour is proposed by Idings, Pister and Taylor (1974). The idea of Idings et al.'s method is to apply finite element approximations for both the space of the equation of equilibrium and for the constitutive equation (two stage finite element scheme). They get excellent results for rubber.

We hereby present a new procedure of the inverse analysis method, and determine elasto-plastic parameters from the results of block shear tests. The elaso-plastic model employed is based on the flow theory with a Drucker-Prager type yield function including dilatancy characteristics. The two stage finite element scheme is also applied.

Notations and terminologies used henceforth are listed here. g : stress tensor, $\bar{g} = \text{tr}(g)\underline{I}$: mean stress tensor $(\text{tr}(\underline{A}) = A^i{}_i$; trace of tensor $\underline{A}$), $\bar{\sigma} = \text{tr}(g)/\sqrt{3}$: mean stress, $\underline{s} = g - \bar{g}$: deviatoric stress tensor, $s = \|\underline{s}\| = (\underline{s} \cdot \underline{s})^{1/2}$: deviatoric stress (norm), ε : strain tensor, $\bar{\varepsilon} = \text{tr}(\varepsilon)\underline{I}$: mean strain tensor, $\bar{\varepsilon} = \text{tr}(\varepsilon)/\sqrt{3}$: mean strain, $\underline{e} = \varepsilon - \bar{\varepsilon}$: deviatoric strain tensor, $e = \|\underline{e}\| = (\underline{e} \cdot \underline{e})^{1/2}$: deviatoric strain (norm). Note that the total strain is the direct sum of the elastic and plastic strains: $\varepsilon = \varepsilon^e + \varepsilon^p$, where the superscript e denotes the elastic component and p the plastic one. Here we treat only the hydro-statically symmetric materials: variables of the response function of these materials are only mean and deviatoric components of stress and strain (see details Ichikawa, Kyoya and Kawamoto 1985).

2 INVERSE ANALYSIS AS A BOUNDARY CONTROL PROBLEM

The inverse analysis method is classified into the inverse formulation and the direct formulation (Gioda 1985). Here we present a direct formulation method as a boundary control problem.

Let Ω be a domain with a boundary of $\partial\Omega$. The direct problem in incremental form is formulated as follows:
(equation of equilibrium)

$$\underset{\sim}{\nabla}\cdot d\underset{\sim}{\sigma} = \underset{\sim}{0} \qquad \text{in} \quad \Omega \tag{1}$$

(constitutive law)

$$d\underset{\sim}{\sigma} = \underset{\sim}{\psi}\, d\underset{\sim}{\varepsilon} \tag{2}$$

("direct" boundary conditions)

$$d\underset{\sim}{u} = \hat{\underset{\sim}{g}} \qquad \text{on} \quad \partial\Omega_t \tag{3}$$

$$d\underset{\sim}{t} = d\underset{\sim}{\sigma}\,\underset{\sim}{n} = \hat{\underset{\sim}{f}} \qquad \text{on} \quad \partial\Omega_u \tag{4}$$

Note $\partial\Omega = \partial\Omega_u + \partial\Omega_t$, and $\underset{\sim}{n}$ is an outward unit normal on $\partial\Omega$.

As a boundary control problem, we are observing dual data to the direct boundary conditions (3) and (4), and are adjusting parameters involved in the constitutive law (2). That is, on the displacement boundary $\partial\Omega_u$, we are going to "control" the reaction force $d\underset{\sim}{t}=\bar{\underset{\sim}{f}}$, while on the traction boundary $\partial\Omega_t$, the displacement $d\underset{\sim}{u}=\bar{\underset{\sim}{g}}$ is kept under control:
("control" boundary conditions)

$$d\underset{\sim}{t} = \bar{\underset{\sim}{f}} \qquad \text{on} \quad \partial\Omega_u \tag{5}$$

$$d\underset{\sim}{u} = \bar{\underset{\sim}{g}} \qquad \text{on} \quad \partial\Omega_t \tag{6}$$

In a weak formulation, direct boundary conditions (3) and (4) are fixed as penalty terms, thus we get the following boundary control equation:

$$\int_\Omega d\underset{\sim}{\sigma}\cdot d\underset{\sim}{\varepsilon}'\, dV \;-\; \int_{\partial\Omega_u}\bar{\underset{\sim}{f}}\cdot d\underset{\sim}{u}'\, dS \;-\; \int_{\partial\Omega_t} d\underset{\sim}{t}'\cdot\bar{\underset{\sim}{g}}\, dS$$

$$-\; \frac{1}{\varepsilon_1}\int_{\partial\Omega_u}(d\underset{\sim}{u}-\hat{\underset{\sim}{g}})\cdot d\underset{\sim}{u}'\, dS \;-\; \frac{1}{\varepsilon_2}\int_{\partial\Omega_t}(d\underset{\sim}{t}-\hat{\underset{\sim}{f}})\cdot d\underset{\sim}{t}'\, dS \;=\; 0 \tag{7}$$

where $0<\varepsilon_1, \varepsilon_2 <<1$ and $d\underset{\sim}{\varepsilon}' = [\underset{\sim}{\nabla}(d\underset{\sim}{u}') + \underset{\sim}{\nabla}(d\underset{\sim}{u}')]/2$. Furthermore, it is noted that on the traction boundary $\partial\Omega_t$, we have

$$d\underset{\sim}{t} = d\underset{\sim}{\sigma}\,\underset{\sim}{n} = (\underset{\sim}{\psi}\, d\underset{\sim}{\varepsilon})\underset{\sim}{n} = \underset{\sim}{\psi}(\,\underset{\sim}{\nabla}(d\underset{\sim}{u}) + \underset{\sim}{\nabla}(d\underset{\sim}{u})^T\,)\underset{\sim}{n}/2$$

$$d\underset{\sim}{t}'= d\underset{\sim}{\sigma}'\underset{\sim}{n} = (\underset{\sim}{\psi}\, d\underset{\sim}{\varepsilon}')\underset{\sim}{n} = \underset{\sim}{\psi}(\,\underset{\sim}{\nabla}(d\underset{\sim}{u}') + \nabla(d\underset{\sim}{u}')^T\,)\underset{\sim}{n}/2 \tag{8}$$

By introducing FE approximations for $d\underset{\sim}{u}$ and $d\underset{\sim}{u}'$ in Ω

$$d\underset{\sim}{u} \approx \underset{\sim}{N}\, d\underset{\sim}{U}\;, \qquad d\underset{\sim}{u}'\approx \underset{\sim}{N}\, d\underset{\sim}{U}' \tag{9}$$

where $\underset{\sim}{N}$ is the shape function, and $d\underset{\sim}{U}$ and $d\underset{\sim}{U}'$ the incremental node displacement vectors. Strains are then discretized as

$$d\underset{\sim}{\varepsilon} \approx \underset{\sim}{B}\, d\underset{\sim}{U}\;, \qquad d\underset{\sim}{\varepsilon}'\approx \underset{\sim}{B}\, d\underset{\sim}{U}'$$

$$\underset{\sim}{B}:\text{ displacement-strain matrix.}$$

Boundary displacements are written as

$$d\underset{\sim}{u} \approx \hat{\underset{\sim}{N}}\, d\underset{\sim}{U}\;, \qquad d\underset{\sim}{u}'\approx \hat{\underset{\sim}{N}}\, d\underset{\sim}{U}' \qquad \text{on} \quad \partial\Omega_u \tag{10}$$

where $\hat{\underset{\sim}{N}}=\underset{\sim}{N}|_{\partial\Omega_u}$ is the restriction of $\underset{\sim}{N}$ on $\partial\Omega_u$. Since tractions $d\underset{\sim}{t}$ and $d\underset{\sim}{t}'$ on $\partial\Omega_t$ are related with $d\underset{\sim}{u}$ and $d\underset{\sim}{u}'$ by Eqn(8), we discretize them as

$$d\underset{\sim}{t} \approx \underset{\sim}{\psi}[(\underset{\sim}{\nabla}\underset{\sim}{N})d\underset{\sim}{U} + (d\underset{\sim}{U})^T(\underset{\sim}{\nabla}\underset{\sim}{N})^T]\underset{\sim}{n}/2 \xrightarrow{\text{rewrite}} \underset{\sim}{\psi}\,\underset{\sim}{B}\, d\underset{\sim}{U}$$

$$d\underset{\sim}{t}'\approx \underset{\sim}{\psi}[(\underset{\sim}{\nabla}\underset{\sim}{N})d\underset{\sim}{U}'+ (d\underset{\sim}{U}')^T(\underset{\sim}{\nabla}\underset{\sim}{N})^T]\underset{\sim}{n}/2 \xrightarrow{\text{rewrite}} \underset{\sim}{\psi}\,\underset{\sim}{B}\, d\underset{\sim}{U}' \tag{11}$$

Substituting these into the weak formulation (8), the discretized control equation

$$\underset{\sim}{K} \, d\underset{\sim}{U} - \overline{\underset{\sim}{F}} - \overline{\underset{\sim}{G}} = \underset{\sim}{\Phi} \tag{12}$$

is obtained where

$$\underset{\sim}{K} = \int_{\Omega} \underset{\sim}{B}^{T} \psi \, \underset{\sim}{B} \, dV - \frac{1}{\varepsilon_{1}} \int_{\partial \Omega_{u}} \hat{\underset{\sim}{N}}^{T}\hat{\underset{\sim}{N}} \, dS - \frac{1}{\varepsilon_{2}} \int_{\partial \Omega_{t}} \hat{\underset{\sim}{B}}^{T} \psi^{T} \hat{\underset{\sim}{B}} \, dS$$

$$\overline{\underset{\sim}{F}} = \int_{\partial \Omega_{u}} \hat{\underset{\sim}{N}}^{T}\hat{\underset{\sim}{f}} \, dS$$

$$\overline{\underset{\sim}{G}} = \int_{\partial \Omega_{t}} \hat{\underset{\sim}{B}}^{T} \psi \, \overline{\underset{\sim}{g}} \, dS$$

$$\underset{\sim}{\Phi} = - \frac{1}{\varepsilon_{1}} \int_{\partial \Omega_{u}} \hat{\underset{\sim}{N}} \, \hat{\underset{\sim}{g}} \, dS - \frac{1}{\varepsilon_{2}} \int_{\partial \Omega_{t}} \hat{\underset{\sim}{B}}^{T} \psi^{T} \hat{\underset{\sim}{f}} \, dS$$

It should be noted that $d\underset{\sim}{U}$ is the (incremental) displacement vector defined in the interior region Ω , while $\overline{\underset{\sim}{F}}$ and $\overline{\underset{\sim}{G}}$ are given on the boundaries $\partial \Omega_{u}$ and $\partial \Omega_{t}$, respectively.

Next, let a parameter $\underset{\sim}{P}$ in the constitutive equation (2) be changed:

$$\underset{\sim}{\Psi} = \underset{\sim}{\Psi}(\underset{\sim}{P}) \tag{13}$$

Then, the displacement in Ω and the values on the control boundaries (5) and (6) are changed at the same time. This implies that $\underset{\sim}{P}$ involved in $\underset{\sim}{K}$, $\overline{\underset{\sim}{F}}$ and $\overline{\underset{\sim}{G}}$ are forced to vary in Eqn(12). The nonlinear equation (12) is iteratively solved, so that it is expanded by Talor series. Its first term is written as

$$\underset{\sim}{K}(\underset{\sim}{P})\delta(d\underset{\sim}{U}) + (\frac{\partial \underset{\sim}{K}}{\partial \underset{\sim}{P}} d\underset{\sim}{U})\delta\underset{\sim}{P} - \delta\overline{\underset{\sim}{F}} - \delta\overline{\underset{\sim}{G}} = \underset{\sim}{\Phi} - \underset{\sim}{K}(\underset{\sim}{P})d\underset{\sim}{U} + \overline{\underset{\sim}{F}} + \overline{\underset{\sim}{G}} \tag{14}$$

or a matrix form at the i-th iteration stage is

$$\underset{\sim}{G} \, \underset{\sim}{X} = \underset{\sim}{R} \tag{15}$$

where $\underset{\sim}{G} = [\underset{\sim}{K}(\underset{\sim}{P}^{i}) \quad \underset{\sim}{I} \quad \underset{\sim}{I} \quad \underset{\sim}{A}^{i}]$, $\underset{\sim}{A} = \frac{\partial \underset{\sim}{K}}{\partial \underset{\sim}{P}} d\underset{\sim}{U}$

$$\underset{\sim}{X} = \begin{bmatrix} d\underset{\sim}{U}^{i+1} - d\underset{\sim}{U}^{i} \\ \overline{\underset{\sim}{F}}^{i+1} - \overline{\underset{\sim}{F}}^{i} \\ \overline{\underset{\sim}{G}}^{i+1} - \overline{\underset{\sim}{G}}^{i} \\ \underset{\sim}{P}^{i+1} - \underset{\sim}{P}^{i} \end{bmatrix}$$

$$\underset{\sim}{R} = [\underset{\sim}{\Phi} - \underset{\sim}{K}(\underset{\sim}{P}^{i})d\underset{\sim}{U}^{i} + \overline{\underset{\sim}{F}}^{i} + \overline{\underset{\sim}{G}}^{i}]$$

Since Eqn(14) is an over-determined system, we apply a least square method to Eqn(15), and get the normal equation

$$\underset{\sim}{G}^{T}\underset{\sim}{G} \, \underset{\sim}{X} = \underset{\sim}{G}^{T}\underset{\sim}{R} \tag{16}$$

This equation is iteratively solved until the error norm becomes sufficiently small:

$$\underset{\sim}{X}^{T}\underset{\sim}{X} < \varepsilon^{2} \, , \qquad 0 < \varepsilon \ll 1 \tag{17}$$

3 NONASSOCITED FLOW THEORY USING DILATANCY FACTOR AND INVERSE ANALYSIS

We present a nonassociated flow theory of plasticity which can estimate the dilatacy phenomenon of rock-like materials correctly. Details are shown in Ichikawa (1985).

The material treated here is assumed to be hydrostatically symmetric, thus the Drucker-Prager yield function

$$f(\underset{\sim}{\sigma}, \underset{\sim}{\varepsilon}^{P}) = \alpha\overline{\sigma} + s - K(e^{P}, \overline{\varepsilon}^{P}) \tag{18}$$

is used where α is a material constant and $K(e^{P}, \overline{\varepsilon}^{P})$ the hardening function (that is, an isotropic hardening model with strain hardening parameter). For dilatant materials, the hardening function $K(e^P, \bar{\varepsilon}^P)$ is schematically drawn in Fig.1 where the softening phenomenon is disregarded.

For determining the subsequent yield surface, Prager's consistency condition

$$df = \frac{\partial f}{\partial \underset{\sim}{\sigma}} \cdot d\underset{\sim}{\sigma} + \frac{\partial f}{\partial K} \frac{\partial K}{\partial \underset{\sim}{\varepsilon}^P} \cdot d\underset{\sim}{\varepsilon}^P = 0 \quad (19)$$

is required.

Assuming a plastic potential function g, an increment of the plastic strain is given as

$$d\underset{\sim}{\varepsilon}^P = \lambda \frac{\partial g}{\partial \underset{\sim}{\sigma}} \quad (\text{ flow law })$$

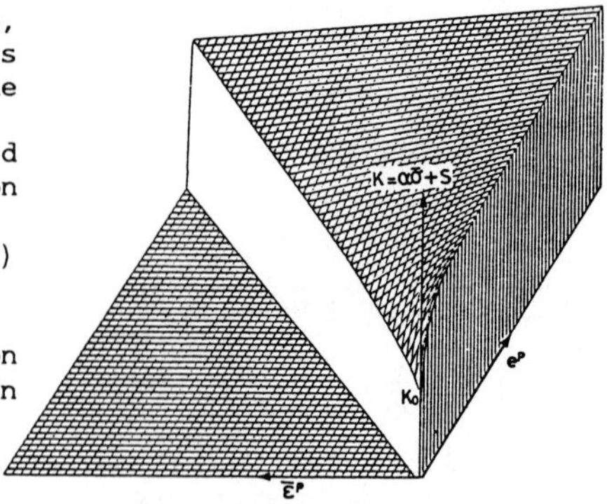

If the plastic strain increment $d\underset{\sim}{\varepsilon}^P$ is coaxial with the stress $\underset{\sim}{\sigma}$, we have

$$d\underset{\sim}{e}^P = \lambda \frac{\partial g}{\partial s} \underset{\sim}{m} \qquad (\underset{\sim}{m} = \underset{\sim}{s}/s)$$

$$d\bar{\varepsilon}^P = \lambda \frac{\partial g}{\partial \bar{\sigma}} \underset{\sim}{n} \qquad (\underset{\sim}{n} = \underset{\sim}{\sigma}/\sigma)$$

Fig.1   Schematic diagram for hardening function K.

Then, using the dilatancy factor defined by

$$\beta = \frac{d\bar{\varepsilon}^P}{de^P} = \frac{\partial g/\partial \bar{\sigma}}{\partial g/\partial s} \tag{20}$$

the incremental constitutive equation is obtained as

$$d\underset{\sim}{\sigma} = \underset{\sim}{D}^{ep} d\underset{\sim}{\varepsilon} \tag{21}$$

$$\underset{\sim}{D}^{ep} = \underset{\sim}{D}^e - \frac{\underset{\sim}{D}^e(\underset{\sim}{m}+\beta\underset{\sim}{n}) \otimes \underset{\sim}{D}^e(\underset{\sim}{m}+\alpha\underset{\sim}{n})}{h' + (\underset{\sim}{m}+\alpha\underset{\sim}{n}) \cdot \underset{\sim}{D}^e(\underset{\sim}{m}+\beta\underset{\sim}{n})}$$

$$h' = \frac{\partial K}{\partial e^P} + \beta \frac{\partial K}{\partial \bar{\varepsilon}^P}$$

Note that the dilatancy factor is represented as a function of the mean stress $\bar{\sigma}$.

Next, we introduce FE discretizations for the functions $K(e^P, \bar{\varepsilon}^P)$ and $\beta(\bar{\sigma})$:

$$K(e^P, \bar{\varepsilon}^P) \simeq \sum_i K^i \phi^i(e^P, \bar{\varepsilon}^P)$$

$$\beta(\bar{\sigma}) \simeq \sum_i \beta^i \psi^i(\bar{\sigma}) \tag{22}$$

where $\phi^i(e^P, \bar{\varepsilon}^P)$ and $\psi^i(\bar{\sigma})$ are interpolation functions in the space $(e^P, \bar{\varepsilon}^P)$ and $(\bar{\sigma})$, respectively (see Fig.2 and Fig.3). We call the finite elements introduced by Eqn (22) as constitutive elements, and the finite elements used in approximating the equation of equilibrium as structural elements. Nodal values K^i and β^i in Eqn (22) are the parameters $\underset{\sim}{P}$ which must be identified by Eqn (16). Then, substituting Eqn (22) into the elasto-plastic constitutive equation (21), we have

$$\underset{\sim}{D}^{ep} = \underset{\sim}{D}^e - \frac{\underset{\sim}{D}^e(\underset{\sim}{m}+\sum\beta^i\psi^i\underset{\sim}{n}) \otimes \underset{\sim}{D}^e(\underset{\sim}{m}+\alpha\underset{\sim}{n})}{h' + (\underset{\sim}{m}+\alpha\underset{\sim}{n}) \cdot \underset{\sim}{D}^e(\underset{\sim}{m}+\sum\beta^j\psi^j\underset{\sim}{n})}$$

$$h' = \sum K^i \left(\frac{\partial \phi^i}{\partial e^P} + \sum \beta^j \psi^j \frac{\partial \phi^i}{\partial \bar{\varepsilon}^P} \right)$$

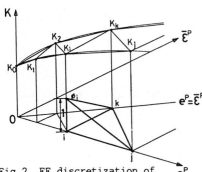

Fig.2 FE discretization of hardening function $K(e^P, \bar{\varepsilon}^P)$.

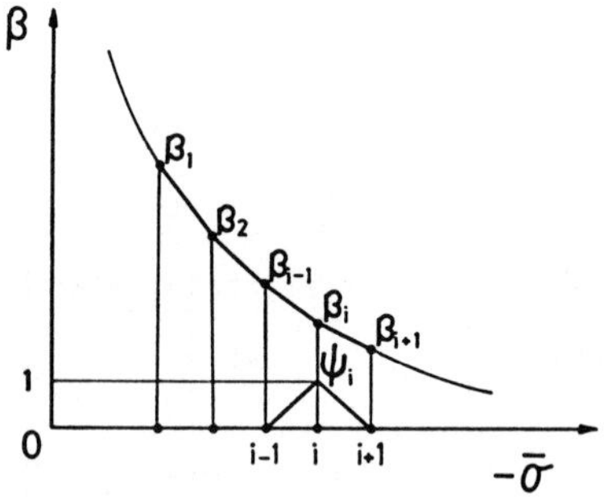

Fig.3 FE discretization of dilatancy factor $\beta(\bar{\sigma})$.

Comprehensive procedures identifying the elasto-plastic parameters for conventional triaxial tests (σ_3=constant) and for block shear tests are as follows:
1) Identify elastic constants by an inverse formulation method (Gioda 1985).
2) Constants for initial yielding α and $K_0 = K(e^P = 0, \bar{\varepsilon}^P = 0)$ are determined. For this purpose, we find stress distributions at the stage of initial yielding under several confining pressures for triaxial tests or under several normal pressures for rock shear tests. We plot these stress states in ($\bar{\sigma}$, s)-space, and get an envelop line which defines the initial yield surface (Fig.4).

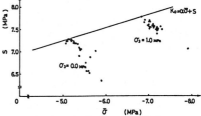

Fig.4 Determining initial yield surface.

3) Plasticity parameters K^j of Eqn(22)$_1$ are identified for the load increase ΔF_i (Fig.5) by using iteration schemes (16) and (17). If CST are used for constitutive elements as shown in Fig.6, two nodal values are identified under one load step.

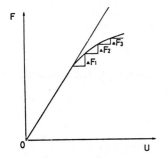

Fig.5 Load-displacement curve and load steps.

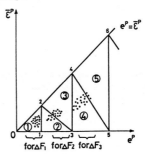

Fig.6 Distribution of plastic strains (dots are values in structural elements).

4) Similarly, the dilatancy parameter β^i of Eqn(22)$_2$ is identified for this load step ΔF_i.
5) Repeat the procedures 3) and 4).
   However, these procedures involve an intrincic difficulty due to the flow theory. Let us recall the stage to identify $\underline{P} = [K^l, K^{l+1}, \ldots, K^m]$ for the load increment ΔF_j. The second term in Eqn(14) is calculated as

$$\frac{\partial \underline{K}}{\partial \underline{P}} \, d\underline{U} = \underline{C} \, \frac{\partial h'}{\partial \underline{P}} \tag{23}$$

for every element Ω_e that has been yielded. Here

$$\underline{C} = (\int_{\Omega_e} \underline{B}^T \frac{\partial \underline{D}^{ep}}{\partial h'} \underline{B} \, dV \,) \, d\underline{U} \tag{24}$$

is constant for any yielded element. Then, terms in $\underline{G}^T\underline{G}$ of Eqn(16) are

$$\underline{A}^T\underline{K} = \frac{\partial h'}{\partial \underline{P}} \, \underline{E} \qquad \text{where} \quad \underline{E} = \underline{C} \, \underline{K}$$

$$\underline{A}^T\underline{A} = \frac{\partial h'}{\partial \underline{P}} \, \underline{F} \qquad \text{where} \quad \underline{F} = \underline{C}^T\underline{C} \, \frac{\partial h'}{\partial \underline{P}} \tag{25}$$

If we set the n-th column or n-th row of $\underline{G}^T\underline{G}$ as

$$\underline{G}^T\underline{G} = \begin{bmatrix} \underline{K}^T\underline{K} & \underline{K}^T & \underline{K}^T & \underline{K}^T\underline{A} \\ & \underline{I} & \underline{I} & \underline{A} \\ & & \underline{I} & \underline{A} \\ \text{Symmetric} & & & \underline{A}^T\underline{A} \end{bmatrix} \begin{array}{l} \\ \\ \text{n-column} \end{array} \tag{26}$$

$$\text{n-row}$$

Eqn(24) implies that every (n+i)-th term (i=2,3,...) is constant times to (n+1)-th column. Thus, the rank of $\underline{G}^T\underline{G}$ is no more than n+1. This concludes that as long as the flow theory is used, that is, the hardening is described by a scalar function h', e^p is not independent from $\bar{\varepsilon}^p$, so that the function cannot be identified in the two dimensional space ($e^p, \bar{\varepsilon}^p$). Preceding schemes 3) and 4) are modified as follows:

3)' Give the dilatancy factor as a constant (β =constant) for a given confining or normal pressure.

4)' The constitutive elements for $K(e^p, \bar{\varepsilon}^p)$ are set in the direction β =constant (one dim. element; see Fig.7), then for given $\Delta \underline{F}_i$ a node value K^i is identified.

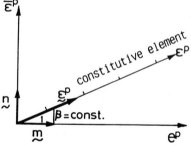

Fig.7 Direction of plastic strain for β =const.

4 APPLICATIONS FOR TRIAXIAL TEST AND ROCK SHEAR TEST

We here present two examples: simulations of uniaxial and triaxial tests, and of rock shear tests.

We have presented several experimental data for uniaxial and triaxial tests of rock materials in which lateral displacements are measured at three points using O-ring type gauges (Ichikawa, Kyoya, Nitta and Kawamoto 1985, Mitsuo and Ichikawa 1986). In this sense, we give all the measured displacements as control boundary data. In the inverse analysis of a uniaxial test, stresses at the lateral surface are set to be zero by the penalty term. In addition, stresses at the top surface are confined, for which results of a direct analysis are employed. The direct analysis is also used in order to calibrate results of the proposed numerical procedures. Material constants of Oya tuff which are used for the direct analysis are

Young's modulus $E=2,200Mpa$, Poisson's ratio $\nu =0.12$
initial yield surface parameters $\alpha =0.29$, $K_0 =5.8Mpa$
dilatancy factor $\beta =0.4$ for $\sigma_3 =0.0$

hardening function

$$K = K_0 + \sum_{i}^{2} a_i \{1 - \exp(-e^P / \tau_i)\} + \sum_{i}^{2} b_i \{1 - \exp(-\bar{\varepsilon}^P / \omega_i)\}$$
$$+ \sum c_i [\{1 - \exp(-e^P / \tau_i)\} \{1 - \exp(-\bar{\varepsilon}^P / \omega_i)\}]$$

$\tau_1 = 5.576 \times 10^{-4}$, $\omega_1 = 0.331 \times 10^{-4}$, $\tau_2 = 1.30110^{-3}$, $\omega_2 = 1.987 \times 10^{-4}$
$a_1 = 8.198$ MPa, $a_2 = -3.551$ Mpa, $b_1 = 1.442$ MPa, $b_2 = -6.690$ MPa
$c_1 = -1.541$ MPa, $c_2 = 7.437$ MPa

Structural and constitutive elements are shown in Figs.8 and 9, respectively. The problem is of course axisymmetric. Results computed by the proposed inverse analysis are plotted in Fig.10, compared with the direct analysis.

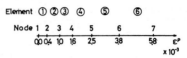

Fig.9 Constitutive FE mesh.

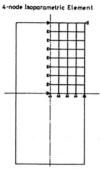

Fig.8 Structural FE mesh for uni/triaxial tests (axisym.).

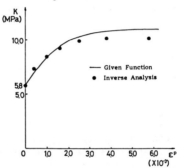

Fig.10 Hardening function computed by inverse analysis and originally given.

The same procedure is applied to a rock shear test. The problem is under plane strain condition. The rock mass is also assumed to be a tuff. Constants α and K_0 for initial yielding are determined by changing the normal stress. However, the stress distributions are different between the uni/triaxial test and the rock shear test. That is, the stresses in a specimen of a uni/triaxial test are mostly uniform, while in the rock mass of a rock shear test the stresses are strongly distributed. This makes the numerical procedure of the rock shear test fairly sensitive or unstable. The structural FE mesh is shown in Fig.11, and the hardening function calculated is in Fig.12. The proposed method gives fair results, however as mentioned, the computation is so unstable that we get only three steps of results.

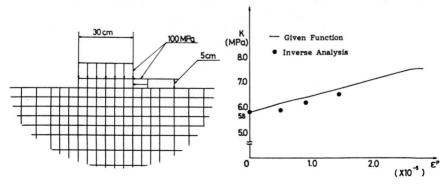

Fig.11 Structural FE mesh for rock shear test (plane strain).

Fig.12 Hardening function computed and original.

5 CONCLUSIONS

We have presented an inverse analysis method to determine elasto-plastic parameters. The problem is formulated as a boundary control equation in which under certain given boundary data (displacement or force; we call these as direct boundaries), the dual data, that is, force for the direct displacement boundary and displacement for the direct force boundary, are observed. The constitutive equation is iteratively determined by controlling parameters in the constitutive equation and the observed boundary data. Finite element approximations are used for both the equation of equilibrium and for the constitutive equation. Node values for the constitutive equation are just the controlling parameters mentioned above.

The procedure is applied for the conventional uniaxial and triaxial tests of a rock specimen, and for the rock shear test. By using this method we can determine the nonlinear constitutive model equivalent to the rock mass. Now, engineers must choose or construct an appropriate material model under each engineering circumstance, and for this purpose, they require an even deeper knowledge of mechanics.

ACKNOWLEDGEMENT

The authors would like to thank Messrs. K.Yoshikawa, T.Ito, Y.Nakamura and H.Ida for their help with experiments and computations.

REFERENCES

Gioda, G. 1985. Some remarks on back analysis and characterization problems. Proc. 5th Int. Conf. Numer. Meth. Geomech., Nagoya, Vol.1, pp.47-61.

Ichikawa, Y. 1985. Viscoelasticity and elastoplasticity problems and finite element programming, Text of 3rd Seminor of Numerical Method in Geomechanics. Union of Japanese Scientists and Engineers, pp.23-96 (in Japanese).

Ichikawa, Y., T. Kyoya and T. Kawamoto 1985. Incremental theory of plasticity for rock. Proc. 5th Int. Conf. Numer. Meth. Geomech., Nagoya, Vol.1, pp.451-462.

Ichikawa, Y., T. Kyoya, H. Nitta and T. Kawamoto 1985. Microcomputer measurement determining plasticity spectrum in incremental theory. Proc. Int. Conf. Education, Practice and Promotion of Comp. Meth. in Eng. Using Small Computers, Macau, Vol.3B, pp.329-341.

Idings, R.H., K.S. Pister and R.L. Taylor 1974. Identification of nonlinear elastic solids by a finite element method. Comp. Meth. Appl. Mech. Eng., Vol.4, pp.121-141.

Mitsuo, J., and Y. Ichikawa 1986. Incremental theory of elasticity and plasticity for rock materials under cyclic loading. Proc. 1st Japan Symp. Numer. Meth. Geotech. Eng., pp.59-64 (in Japanese).

Sakurai, S. and K. Takeuchi 1983. Back analysis of measured displacements of tunnels, Rock Mechanics, Vol.16, pp.173-180.

Sakurai, S., N. Shimizu and K. Matsumuro 1985. Evaluation of plastic zone around underground openings by means of displacement measurements. Proc. 5th Int. Conf. Numer. Meth. Geomech., Nagoya, Vol.1, pp.111-118.

2nd International Symposium on Field Measurements in Geomechanics, Sakurai (ed.)
© 1988 Balkema, Rotterdam. ISBN 90 6191 778 6

Determination of stresses and strain modulus in a rock mass by an experimental-analytical method

N.P.Vlokh, A.V.Zoubkov & Yu.P.Shoupletsov
Institute of Mining, Sverdlovsk, USSR

1 INTRODUCTION

The choice of parameters for design elements of mining systems when mining mineral deposits is often determined by the considerations of rock mechanics. The major mechanical characteristics of a rock mass include the values of original stresses acting beyond the zone of influence of an excavation ($\sigma_1^o, \sigma_2^o, \sigma_3^o$), strain modulus E_M of a rock mass and compressive strength $[\sigma]_c$ of large portions of a rock mass. It is practically impossible to estimate loads on the elements of underground structures without knowing original stresses and strain modulus.

Application of techniques of "forced" rock mass deformation such as stamp method, method of thrust adits and others require use of cumbersome installations, large time and labour in order to determine only strain modulus. Furthermore it is not always possible to obtain the characteristics of a rock mass over large areas. Also for calculations of the pressure on the ground support in an opening one must know the strain modulus of a rock mass over an area many times greater than the diameter of that opening, and determination of the load acting on the backfill in the mined out area requires knowledge of strain modulus for an area measuring tens and hundreds meters. The forced deformation of such rock volumes is naturally impossible.

Some information on mechanical characteristics of a rock mass without any aid from a researcher except general management can be obtained near mined out areas for example in the process of mining or during driving an opening.

2 DESCRIPTION OF THE EXPERIMENTAL-ANALYTICAL METHOD FOR STUDYING THE STRESS-STRAIN STATE (SSS) OF A ROCK MASS

In the process of change in dimensions and configuration of any opening a change in SSS of a rock mass also take place. These changes may be registered by suitable devices: for example, stress change can be measured using photoelastic sensors, and displacements in a rock mass can be

measured by observing movements of datum marks installed
the walls of an opening and in boreholes. The first step
of the method is the in-situ measurements of stress increments and displacements.

The second step is an analytical solution of the problem
of SSS change in a rock mass surrounding the opening being
considered and this problem is solved separately for each
original load and for their values being equal to one.
Strain modulus also may be assumed equal to 1.

From analytical solutions for each original stress i in
the point being considered the stress increments near excavation in the j-direction are:

$$\Delta \sigma_{j(i)p} = \Delta K_j(i) \cdot \sigma_i^o ,\qquad (1)$$

where σ_i^o - original stresses in the i-direction,
$\Delta K_j(i)$ - stress concentration factor in the j-direction
for $\sigma_i^o = 1$ when the values of other components
are equal to 0.

According to calculations the combined values of stress
increments from all components in the point involved in
the j-direction will be

$$\Delta \sigma_{jp} = \sum_1^3 \Delta K_j(i) \cdot \sigma_i^o \qquad (2)$$

Equalizing the observed in-situ stress increments $\Delta \sigma_{jH}$
and the estimated ones we obtain an equation of the following type:

$$\Delta \sigma_{jH} \cong \sum_1^3 \Delta K_{j(i)} \cdot \sigma_i^o \qquad (3)$$

Since calculated model of a medium practically always
does not correspond to a real rock mass the equation (3)
may be strictly true only occasionally. Differences between
the estimated and actual stresses are inevitable due to
the blocky structure of a rock mass and due to other factors of inhomogeneity. It is necessary to develop a redetermined set of equations and to solve it using least
squares method.

When determining original stresses one can make the following generally accepted assumptions:
- original vertical stresses are determined by overlying
  rock weight that is justified when considering large
  mined out areas;
- original horizontal stresses act according to the strike
  of geological structures, mountain ranges and valleys.

These assumptions allow to reduce the number of unknowns
in the equation (3) to two unknowns and to determine the
directions of action for original stresses at the deposit
being considered.

Similar equation may be obtained when considering displacements. In this case the estimated displacement in the
j-direction due to original stress σ_i^o is:

$$\Delta U_{j(i)p} = \frac{1}{E_M} \Delta A_{j(i)} \cdot \sigma_i^o ,\qquad (4)$$

where $\Delta A_{j(i)}$ corresponds to estimated displacement increments in a rock mass when the values of strain modulus and of original stresses are $E_M = 1$ and $\sigma_i^o = 1$, respectively.

If we proceed as when deriving equations (2) and (3) then we obtain the following equations:

$$E_M \cdot \Delta U_{jH} \cong \sum_1^3 \Delta A_{j(i)} \cdot \sigma_i^o \qquad (5)$$

If all parameters E_M, σ_i^o included into (5) are unknown, then the set of equations becomes uniform i.e. the set which have infinite number of solutions. The above mentioned assumption in relation to the equality of vertical original stress components to the overlying rock pressure in the case of large areas of a rock mass makes it possibble to obtain the free term in (5) and therefore to provide unambiguous solution for the compiled set of equations which also has to be statically representative in order to be solved using least squares method.

When we consider a worked out area with a sufficient height then for the solution at average horizontal cross section we obtain a plane problem and the set of equations (5) becomes uniform.

In this case it is necessary to compile a set of equations (3) and then we find E_M at known σ_i^o .Here for performing in-situ observations we require installation of stations of the "displacement measuring-stress increment measuring" type by means of which the strain modulus of a rock mass can be evaluated without analytical estimations provided that stress increments are measured in the range corresponding to the range of displacement measurements.

Thus by compiling redetermined sets of equations (3)or (5) one can find unknown values for the components of original horizontal stresses and for the strain modulus of the rock mass.

3 EXAMPLE OF PRACTICAL APPLICATION

Specific procedure for practical application of this method will be discussed here by way of an example of mining high chamber at Sokolovsky underground mine, North Kazahstan, USSR (Vlokh et al., 1980). This deposit of iron ore lies in the zone of powerful submeridianal tectonic break. Stability of the rock mass in the mine field is variable particularly across the strike of ore deposits (or normally to the break). In the considered case observations have been carried in the most stable deposit under the highwall of the whole property at the depth of 200 m.

Six photoelastic sensors and two datum-mark lines have been installed (Fig.1). Displacements were measured using geodesic tape-measure between datum marks grouted in boreholes. Errors in results of measuring of length parameters does not exceed 0.5 m.

Measurements have been performed three times in the course of mining ore resources in this chamber. During the period of observations some sensors have become failured.

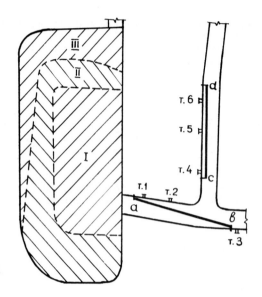

Figure 1. Layout of photoelastic sensors (𝝅) and datum-
mark lines (ab, cd) near the chamber. I,II,III –
positions of the chamber contour at the moment
of observations.

Table 1. Equations of the type (3) for determination of
original horizontal stresses

Position of chamber contour	Sensor number	Stress increments, MPa	
		Estimated	Measured
I	1	$-0.94\,\sigma_y^o - 0.02\,\sigma_x^o$	$= 4.4$
	2	$-0.82\,\sigma_y^o - 0.05\,\sigma_x^o$	$= 1.6$
	3	$-0.55\,\sigma_y^o - 0.08\,\sigma_x^o$	$= 2.5$
	4	$-0.04\,\sigma_y^o - 0.18\,\sigma_x^o$	$= -1.7$
II	2	$-0.9\,\sigma_y^o - 0.04\,\sigma_x^o$	$= 1.8$
	3	$-0.65\,\sigma_y^o - 0.07\,\sigma_x^o$	$= 2.7$
	4	$-0.11\,\sigma_y^o + 0.24\,\sigma_x^o$	$= -2.2$
	5	$-0.09\,\sigma_y^o + 0.24\,\sigma_x^o$	$= -0.7$
III	4	$-0.19\,\sigma_y^o + 0.29\,\sigma_x^o$	$= -1.7$
	5	$-0.18\,\sigma_y^o + 0.29\,\sigma_x^o$	$= -2.7$

Table 2. Equations for determination of strain modulus of
a rock mass

Position of chamber contour	Datum-mark line	Displacements ($m \cdot E_M$)	
		estimated	measured
I		$1.456\,\sigma_x^o + 0.653\sigma_y^o =$	$-0.0043 E_M$
		$-1.658\,\sigma_x^o - 10.84\,\sigma_y^o =$	$0.0065 E_M$
II		$1.862\,\sigma_x^o + 1.570\sigma_y^o =$	$-0.0036 E_M$
		$-1.818\,\sigma_x^o - 9.429\sigma_y^o =$	$0.0061 E_M$
III		$2.618\,\sigma_x^o + 1.954\sigma_y^o =$	$-0.0051 E_M$

Estimated and actual stress increments for each position
of the chamber contour are given in Table 1. The solution
has been performed for a two-dimensional problem from the
theory of elasticity (horizontal section).

When solving the derived set of equations using least
square method the values of $\sigma_x^o = -8.6$ MPa and $\sigma_y^o = -2.8$ MPa
have been obtained. The correlation coefficient between
estimated and actual stress increments has been equal to
0.96. The straight line representing regression of measured stress increments is expressed by the equation $\Delta\sigma_H =$
$= 1.05\,\Delta\sigma_P + 0.64$ (MPa) which is close to the ideal theoretical equation $\Delta\sigma_H = \Delta\sigma_P$ for elastic medium. And in other
cases of determining the original stresses using the method
considered the correlation coefficient has not been less
than 0.9.

For determination of strain modulus E_M in the considered
example a set of equations of the type (5) has been composed at Poisson's ratio $\nu = 0.3$.

The negative sign (-) in the values of stresses and displacements represent compression.

For stresses which are found from Table 2 the average
value of the strain modulus $E_M = 5.5 \cdot 10^3$ MPa at root-mean-
square deviation of 1.45 MPa, i.e. about 25%. Calculations
of E_M for other values of Poisson's ratio indicated that
the effect of Poisson's ratio is insignificant: for $\nu = 0.2$
the value of E_M was defined as $5.2 \cdot 10^3$ and for $\nu = 0.4$
its value was $5.7 \cdot 10^3$, i.e. for practical calculations the
value of Poisson's ratio may be approximately assumed.

4 DEPENDANCE OF STRAIN MODULUS OF A ROCK MASS ON ITS CATEGORY OF STABILITY

The study of the rock mass SSS at Sokolovsky underground
mine has proved that the value of the rock mass strain modulus depend on the category of stability of the considered
rock mass. Similar studies have been also performed at two

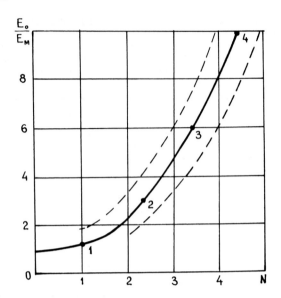

Figure 2. **Relationship between the ratio of strain modulus of a specimen (E_O) to the strain modulus of a rock mass (E_M) and the category of rock mass stability (N according to Table 3):** point 1 – Severopeschanskaya and Yestuninskaya mines (the Urals); points 2,3,4 – Sokolovsky underground mine (Kazakhstan); = = = – the range of possible errors in estimating N.

other mines in the Urals. As a result we have found certain relationship for rock masses as it is given in Figure 2.

The categories of stability are given in Table 3. They provide qualitative descriptive characteristic of a rock mass but in the range of one category the stability of a rock mass can be always identified by those data on fracturing of the rock mass, that are given in the table. These data have been obtained in the course of visual observations in rock masses where in-situ experiments have been performed.

The dotted line in Fig.2 shows the range of possible errors in determining the category of stability for rock masses.

5 SUMMARY AND CONCLUSIONS

i) The method of experimental-analytical study of the stress-strain state (SSS) of a rock mass presented here combines two approaches – a method of in-situ measurements and an analytical method – thereby increasing the effectiveness of both methods: firstly, the confidence level of the results of analytical estimating

Table 3. Categories of stability of rock masses

Category of stability	Description of cracks	Linear fracturing factor (cracks per meter)	Protodyakonov hardness
1. High stable	Cracks are closed or tough due to the presence of tough infill	up to 2	> 10
2. Stable	Almost all cracks are closed without presence of friction clay infill and slip plane	from 2 to 4	> 8
3. Medium stable	Cracks generally do not have friction clay infill and are partly opened without presence of tectonic cracks and slip planes	from 4 to 10	> 6
4. Low stable	Cracks are filled with mylonite and friction clay infill. Tectonic cracks and slip planes are present	-	-
5. Unstable	Crushed zones, zones of mylonitization, powder ores, large tectonic joints	-	-

of the SSS for a rock mass around mined out areas is improved and, secondly, the results of in-situ measurements become more general in its nature that was unachievable earlier and their interpretation is improved.
In view of relative simplicity, low labour consumption in performing in-situ experiments and due to usage of present analytical methods of rock mechanics and computers for the solution of three-dimensional problems such experimantal-analytical methods may be considered as the most perspective ones for studying characteristics and state of rock masses.
ii) Investigations have been performed by authors at some iron ore mines of the Urals and Kazakhstan. Although in all cases the most simple model of the medium (elastic, isotropic, homogeneous) has been assumed the correlation coefficient between measured and estimated stresses has not been less than 0.9 thereby indicating the validity of the assumed model for hard rocks. The

depth at which the measurements have been made range
from 200 to 600 meters.
The average error of the described method of determining original stresses which has been found by a method
of evaluating errors of the functions (2) for equations
(3) do not exceed 30%.
iii) The relationship between the ratio of rock specimen
strain modulus to rock mass strain modulus and the category of stability of a rock mass that is presented
in Fig.2 for 10-20m sections of a rock mass is logical-
ly justifiable: in less stable rock masses, i.e. more
fractured with weak infill, this ratio is greater than
in more stable rock masses.

REFERENCES

Vlokh N.P., Oushkov S.M., Shoupletsov Y.P. et al. 1980.
   Determination of elastic properties and original stress
   components of a rock mass. Mechanics of rock failure.
   Frunze:Ilim - p.325-334.
Tourchaninov I.A., Medvedev R.V., Panin V.I. 1967. Modern
   methods of complex determination of physical properties
   of rocks. Leningrad:Nedra - 198 p.

A method of measuring and computing the deformation of a rock block with large displacement and strain

Guo Jing Zhao
Beijing Graduate School of China, Institute of Mining Technique

1 INTRODUCTION

Many problems such as the underground excavation of mines, crustal motion in tectonics and other topics in geomechanics require analysis of the deformation and stress of rock masses. The small displacement and strain theory of solid mechanics, as everyone knows, has been applied well to analyze and interpret some of the above problems. However, for rock masses which undergo large strain accompanied with large rigid rotation and translation (for instance, a tension of 70 - 100% and a large rotation in foldings of rocks is ordinary) the above theory is no longer applicable. Large strain and rotation occur frequently in the stratas overlying the excavated mining work areas.

A deformable rock mass of a finite size will be regarded as an object of research. It can be named a deformable rock block or block for short. The kinematics of a block with large strain and accompanied by large rigid rotation and translation is to discover the relations between the displacement (relative to an outer reference) and the strain of the rock. It is important to separate the strain and rigid motion for the kinematics of the block. The polar decomposition theory of rational continuum mechanics (Truesdell 1977) provides a principle for carrying out the separation. Since the interpretation of a block's motion and its strain from a record of displacements at several points of the block can be simplified only in the situation of homogeneous deformation, we will discuss emphatically the homogeneous deformation of a block like one with an infinitesimal volume at a point in continuum. In the field or in the laboratory, if some monitoring system has been set up to measure the data of displacements at several points of a block, a method based on the polar decomposition theorem to analyze and compute the strain and rigid motion from the measured data or the inverse is provided in the following. Under this circumstance, the peculiarity of analysis is that all of the mechanics formulas are restricted to kinematics only. If the material properties and constitutive relation of a block are specified, we can also compute the stresses and forces applied to the block. We will give a brief formulation and simple examples to illustrate the function and significance of the method put forward in this paper.

2   BASIC FORMULATION USED IN A FINITE DISPLACEMENT FIELD

In attempting to make a simple representation of the finite displacement field based on the polar decomposition, we have to use the tensor
calculus. The matrix calculus should be involved for convenience in
the next paragraphs.

2.1   The reference of frame

Introduce the following two referenced frames to represent the motion
of a block.
    The outer fixed frame: A rectangular Cartesian coordinates system
is often chosen as the fixed frame. The base vectors $\underline{e}_k$ are unit
vectors.
    The convected frame: It may be an arbitrarily curvilinear frame;
the covariant base vectors $\underline{g}_i$ deformed with the block are, in general,
not unit vectors. At the initial stage of motion, the convected base
vectors are denoted by $\overset{\circ}{\underline{g}}_i$. Note that affine frames $\underline{g}_i$ and $\overset{\circ}{\underline{g}}_i$
are definited at every point of the block.

2.2   Motion transformation

Consider a moving block in which a point P moves to a new position P',
and u is the displacement vector. If $\overset{\circ}{\underline{r}}$ and $\underline{r}$ are the position vectors
of P and P' referred to in the origin of the fixed frame, then it is
apparent that,

$$\underline{r} = \overset{\circ}{\underline{r}} + \underline{u} \qquad\qquad d\underline{r} = d\overset{\circ}{\underline{r}} + d\underline{u} \qquad\qquad (2.2.1)$$

where $d\overset{\circ}{\underline{r}}$ is an infinitesimal line element at P, and it is transformed to
$d\underline{r}$ at P'. $d\underline{u}$ is an increment of $\underline{u}$, $d\underline{r}$, $d\overset{\circ}{\underline{r}}$, $d\underline{u}$ in different frames
which can be notated at

$$d\underline{r} = dX^k \underline{e}_k = d\bar{x}^i \underline{g}_i = dx^i \underline{g}_i$$

$$d\overset{\circ}{\underline{r}} = dx^k \underline{e}_k = dx^i \overset{\circ}{\underline{g}}_i \qquad\qquad (2.2.2)$$

$$d\underline{u} = u^i|_j \, dx^j \overset{\circ}{\underline{g}}_i$$

where $u^i|_j \overset{\circ}{\underline{g}}_i$ is the covariant derivative of $\underline{u}$ referred to frame $\overset{\circ}{\underline{g}}_i$.
Upon inserting (2) into relation (1b), we obtain the relations of
transformation as

$$d\bar{x} = (\ \delta^i_j + u^i\big|_j\ )\ dx^j , \ \underline{g}_i = (\ \delta^i_j + u^i\big|_j\ )\ \overset{\circ}{\underline{g}}_j \qquad (2.2.3)$$

where δ^i_j is Kronecker delta.
    When the frame $\underline{g}_i$ and $\underline{e}_k$ are homeomorphic, then $x^k = X^k$, $x^i = X^i$
and (3) becomes

$$dX^k = F^k_i \, dx^i , \quad \underline{g}_i = F^k_i \, \underline{e}_k , \quad F^k_i = X^k,_i = \partial X^k / \partial x^i \qquad (2.2.4)$$

As $\overset{\circ}{\underline{g}}_i$ is arbitrarily a curvilinear frame in general cases, we introduce

$$\overset{\circ}{\underline{g}}_j = \overset{\circ}{F}^k_j \underline{e}_k \qquad\qquad (2.2.5)$$

and by (5) and (3) we may write the transformation matrix F_i^k in (4) as

$$F_i^k = (\delta_i^j + u^i|_j) \overset{o}{F}{}_j^k \qquad (2.2.6)$$

Obviously, when $\underline{g}_i$ and $\underline{e}_k$ are homeomorphic, $\overset{o}{F}{}_i^k = \delta_i^k$, (6) can be reduced to

$$F_i^k = \delta_i^k + u^k_{,i} \qquad (2.2.7)$$

2.3 Polar decomposition of matrix F_i^k

Consider the computation of metric tensor first. On substituting (2.2.4b) and (2.2.5) into the metric tensor $g_{ij} = \underline{g}_i \cdot \underline{g}_j$ and $\overset{o}{g}_{ij} = \underline{\overset{o}{g}}_i \cdot \underline{\overset{o}{g}}_j$, we obtain

$$g_{ij} = \bar{g}_{kl} F_i^k F_j^l \qquad \overset{o}{g}_{ij} = \bar{g}_{kl} \overset{o}{F}{}_i^k \overset{o}{F}{}_k^l \qquad (2.3.1)$$

where $\bar{g}_{kl} = \underline{e}_k \cdot \underline{e}_l = \delta_{kl}$

From the polar decomposition theorem, any invertible matrix F_i^k can uniquely be factorized as a matrix or tensor product of the orthogonal tensor R_m^k with a positive symmetric tensor U_i^m:

$$F_i^k = R_m^k U_i^m \qquad (2.3.2)$$

where R_m^k and U_i^m are called the rotation tensor and the right stretch tensor, respectively. Letting $F_i^k = R_i^k$ be a rotation tensor, the metric tensor must be an invariable: $g_{ij} = \overset{o}{g}_{ij}$. Therefore, the orthogonality of R_i^k can be written as

$$\overset{o}{g}_{ij} = \bar{g}_{kl} R_i^k R_j^l \qquad \text{or} \qquad \bar{g}^{kl} = \overset{o}{g}{}^{ij} R_i^k R_j^l \qquad (2.3.3)$$

Upon inserting (2) into (1) and paying attention to the above orthogonality, we have

$$g_{ij} = \overset{o}{g}_{kl} U_i^k U_j^l \qquad (2.3.4)$$

It is a valuable relation. On the basis of it, we will suggest a method for computing the stretch tensor U_j^i in the next paragraph. By using the stretch tensor U_j^i, we define the strain at a certain point as

$$E_j^i = U_j^i - \delta_j^i \qquad (2.3.5)$$

The strain tensor E_j^i characterizes such a deformation state of the small neighboring domain at a point, that if a rigid rotation arising from the rotation tensor R_m^k is followed, we will obtain the final deformation state. If unit vectors $\underline{g}_i$ are the proper vectors of tensor U_j^i, they represent one orthogonal triad of principal axes which is called the principal axes of strain. The stretch tensor U^i_j can be reduced to a unique diagonal form: $\text{diag}(U^i_j) = \text{diag}(\lambda_i)$ referred to in the frame formed from the principal axes of strain. Here the λ_i are proper numbers of U^i_j and are the stretch of a unit length in the direction of $\underline{e}_i$ before deformation. In the final deformation state, the principal axes of strain will be rotated about a certain axis with certain angle θ .

3   COMPUTATION OF TENSOR U^m_i AND R^k_m

A method for computation of stretch tensor U^m_i and rotation tensor R^k_m is discussed in this paragraph. For convenient calculation, tensors of rank two are regarded as a 3 x 3 matrix, for example $g_{ij} = [g]$, etc.

3.1   Computation of tensor U^i_j

According to formula (2.3.4), if one uses some orthogonalized procedure on the matrix $[g]$ and $[\underline{g}]$ so as to reduce them to a diagonal from $\text{diag}(g_{ii})$ and $\text{diag}(\underline{g}_{ii})$, then the matrix $[U]$ must be diagonalized at the same time. Thus, we have

$$\text{diag}(U_{ii}) = \text{diag}(\lambda_i) = \text{diag}(g_{ii} / \underline{g}_{ii})^{1/2} \qquad (3.1.1)$$

In order to determine the components of $[U]$, in practice, we need to solve the following eigenproblem.

$$[g]\,\underline{e}_i = \lambda^2\,[\underline{g}]\,\underline{e}_i \qquad (3.1.2)$$

Letting $\underline{e}_i$, corresponding to the proper numbers λ^2, be the unit proper vectors solved from (2), it is easy to verify that the transformation matrix $[P]$ $= [\ \underline{e}_1\ \underline{e}_2\ \underline{e}_3\ ]$ is orthonormal, and by which the matrix $[U]$ can be diagonalized to the diagonal form:

$$[P]^T[U][P] = \text{diag}(\lambda_i) \qquad [U] = [P]\,\text{diag}(\lambda_i)\,[P]^T \qquad (3.1.3)$$

It is apparent that

$$[U]^{-1} = [P]\,\text{diag}(1/\lambda_i)\,[P]^T \qquad (3.1.4)$$

Note that the matrix $[P]$ gives the orientation cosine matrix of the principal axes of strain (before the rigid rotation arising from $[R]$) referred to the frame $\underline{e}_k$, namely,

$$\underline{e}_i = \cos\alpha_i\,\underline{e}_1 = \cos\beta_i\underline{e}_2 + \cos\gamma_i\underline{e}_3 \qquad (3.1.5)$$

3.2 Computation of rotation tensor R^k_m

Formula (2.3.2) can be written as the matrix form

$$[F] = [R] [U] \qquad\qquad [R] = [F] [U]^{-1} \qquad (3.2.1)$$

A substituting of (1.4) into (1b) yields the formula for the computation of [R]

$$[R] = [F] [P] \, diag(1/\lambda_i) \, [P]^T \qquad (3.2.2)$$

It is worth noting that the matrix [R] is, in general, not orthogonormal. We are interested in finding out what the orientation of the axis about which the principal axes of strain rotate rigidly is, and what the angle of this rigid rotation arising from the rotation matrix [R] is. Consider an orthonormalized matrix [R], and let $[\tilde{R}]=([R]-[R]^T)/$ be the antisymmetric part of the matrix [R], and according to the Gibbs theorem (Gibbs 1901), it can be written as

$$[\tilde{R}] = sin\theta \begin{bmatrix} 0 & -n_3 & n_2 \\ n_3 & 0 & -n_1 \\ -n_2 & n_1 & 0 \end{bmatrix} \qquad (3.2.3)$$

Thus, the principal axes of strain rotate about the axis so that their orientation can be determined by the unit vector

$$\underline{n} = [n_1 \; n_2 \; n_3]^T \qquad (3.2.4)$$

and with an angle of the rigid rotation:

$$\theta = sin^{-1} \pm (\Sigma n_i^2)^{1/2} \qquad (3.2.5)$$

And the rotated principal axes are $\underline{\tilde{e}}_i = [R] \, \underline{8}_j$.

4 DISCUSSION ON THE HOMOGENEOUS DEFORMATION OF A BLOCK

As of the present time, the statement of the finite displacement field is a general theory based on the polar decomposition theorem. It is a geometric theory of motion transformation applicable for every point of a considered domain in continuum mechanics. As mentioned above, a finite size block undergoing the homogeneous deformation can be interpreted as the same state occurring in the infinitesimal domain at a point in continuum. The size of the domain considered makes no difference, since the strain and rigid rotation at every point are the same for homogeneous deformation.

4.1 The motion of a block undergoing homogeneous deformation

Assume that the components of the displacements at four points (or three for the plane problem) in a block are measured, and frame is referred to as

$$\underline{u}_s = [\ u_s^1 \ u_s^2 \ u_s^3 \ ]^T \qquad \underline{d} = [\ \underline{u}_1^T \ \underline{u}_2^T \ \underline{u}_3^T \ \underline{u}_4^T \ ]^T \qquad\qquad (4.1.1)$$

where $\underline{d}$ is the total displacement vector of the block. The displacement vector of any point in the block may be represented as

$$\underline{u} = [u^1 \ u^2 \ u^3 \ ]^T \quad = \ [N] \ \underline{d} \qquad\qquad (4.1.2)$$

$$[N] = [\ IN_1 \ IN_2 \ IN_3 \ IN_4 \ ] \quad N_s = (a_s + b_s x^1 + c_s x^2 + d_s x^3)/6v \qquad (4.1.3)$$

where I is a 3 x 3 identity matrix, and a_s, b_s, c_s, d_s and v are the
constants to be dependent only on the coordinates at points P_s (see
Appendix). The transformation (2) will give a unique solution provided
that the four points are not in the same plane. Since the displacement
field represented by (2) is a linear function of convected coordinates
x^i, that is to say (2) determines a state of homogeneous deformation
of the block. In this case, any affine frame $\overset{\circ}{g}_i$ chosen for the total
block must remain its affine property during deformation.
   Consider a transformation matrix $[T_3]$ (see Appendix) and from $\underline{d}$
calculate a new vector

$$\underline{a} = [T_3] \ \underline{d} \qquad\qquad (4.1.4)$$

and by using the new vector

$$\underline{a} = [\ a_0^1 \ a_1^1 \ a_2^1 \ a_3^1 \ a_0^2 \dots \ a_3^3 \ ]^T \qquad\qquad (4.1.5)$$

the displacement field function can be simplified as

$$u^i = a_0^i + a_1^i x^1 + a_2^i x^2 + a_3^i x^3 \qquad\qquad (4.1.6)$$

On inserting $u^k{}_{,i} = a_i^k$ into (2.2.7) for the case $\overset{\circ}{g}_{ij} = \delta_{ij}$, note
the block size is finite we obtain

$$F_i^k = \delta_i^k + a_i^k \qquad x^k = F_i^k \ x^i \qquad\qquad (4.1.7)$$

Thus, the coefficients a_0^i in (6) representing the components of the
displacement at the origin of frame $\overset{\circ}{g}_i$, are not concerned with the
strain and rigid rotation of the block.
   If the data of the total displacement vector $\underline{d}$ of a block is measured, we may calculate the strain and rigid rotation of the block by
means of the formulas above based on the polar decomposition theorem.
Two simple examples will be illustrated in the following. On the other
hand, the inverse problem of determining the motion of a block undergoing homogeneous deformation is an interesting one in geomechanics.
Note that it is possible to compute the inverse problem provided that
all of the data of the strain and rigid rotation are measured.

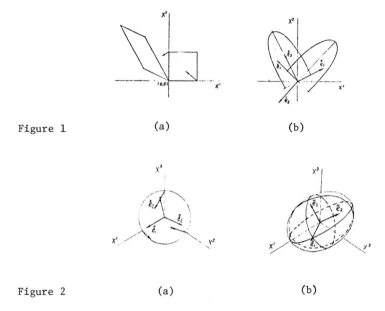

Figure 1   (a)   (b)

Figure 2   (a)   (b)

4.2 Simple examples

1   Consider a plane problem. The large displacements at three points of a square and its rigid rotation and strain in the final state are illustrated in figure 1.

2   The sphere of unit radius undergoing homogeneous deformation becomes an ellipsoid. The displacements at four points of the sphere and its final state are illustrated in figure 2.

CLOSING REMARKS

The main objective of this paper was to draw attention to an analysis method for the finite displacement field which is quite common phenomena met in geomechanics and engineering. There is a significant difference between the theory of finite and small displacement fields. Although the theory is fairly complex, the method put forward in this paper makes the analysis and computation possible to perform. This conclusion has been shown by two simple examples. The specific application of it will be discussed in other articles.

REFERENCES

Truesdell, C. A first course in rational continuum mechanics, Vol. 1, New York, San Franscisco, London, 1977.
Gibbs, J.W. Vector analysis, Scribner, New York, 1901, Chap. IV.

APPENDIX

$$a_1 = \begin{vmatrix} x_2^1 & x_2^2 & x_2^3 \\ x_3^1 & x_3^2 & x_3^3 \\ x_4^1 & x_4^2 & x_4^3 \end{vmatrix} \qquad b_1 = \begin{vmatrix} 1 & x_2^2 & x_2^3 \\ 1 & x_3^2 & x_3^3 \\ 1 & x_4^2 & x_4^3 \end{vmatrix} \qquad c_1 = \begin{vmatrix} 1 & x_2^1 & x_2^3 \\ 1 & x_3^1 & x_3^3 \\ 1 & x_4^1 & x_4^3 \end{vmatrix}$$

$$v = \begin{vmatrix} 1 & x_1^1 & x_1^2 & x_1^3 \\ 1 & x_2^1 & x_2^2 & x_2^3 \\ 1 & x_3^1 & x_3^2 & x_3^3 \\ 1 & x_4^1 & x_4^2 & x_4^3 \end{vmatrix} \qquad d_1 = \begin{vmatrix} 1 & x_2^1 & x_2^2 \\ 1 & x_3^1 & x_3^2 \\ 1 & x_4^1 & x_4^2 \end{vmatrix}$$

The other constants a_s, b_s, c_s, d_s (s=2,3,4) may be computed by alternation of the indices, thus, 1-2-3-4-1.

$$[T_3] = \begin{bmatrix} a_1 & 0 & 0 & a_2 & 0 & 0 & a_3 & 0 & 0 & a_4 & 0 & 0 \\ b_1 & 0 & 0 & b_2 & 0 & 0 & b_3 & 0 & 0 & b_4 & 0 & 0 \\ c_1 & 0 & 0 & c_2 & 0 & 0 & c_3 & 0 & 0 & c_4 & 0 & 0 \\ d_1 & 0 & 0 & d_2 & 0 & 0 & d_3 & 0 & 0 & d_4 & 0 & 0 \\ 0 & a_1 & 0 & 0 & a_2 & 0 & 0 & a_3 & 0 & 0 & a_4 & 0 \\ 0 & b_1 & 0 & 0 & b_2 & 0 & 0 & b_3 & 0 & 0 & b_4 & 0 \\ 0 & c_1 & 0 & 0 & c_2 & 0 & 0 & c_3 & 0 & 0 & c_4 & 0 \\ 0 & d_1 & 0 & 0 & d_2 & 0 & 0 & d_3 & 0 & 0 & d_4 & 0 \\ 0 & 0 & a_1 & 0 & 0 & a_2 & 0 & 0 & a_3 & 0 & 0 & a_4 \\ 0 & 0 & b_1 & 0 & 0 & b_2 & 0 & 0 & b_3 & 0 & 0 & b_4 \\ 0 & 0 & c_1 & 0 & 0 & c_2 & 0 & 0 & c_3 & 0 & 0 & c_4 \\ 0 & 0 & d_1 & 0 & 0 & d_2 & 0 & 0 & d_3 & 0 & 0 & d_4 \end{bmatrix}$$

2nd International Symposium on Field Measurements in Geomechanics, Sakurai (ed.)
© 1988 Balkema, Rotterdam. ISBN 90 6191 778 6

Prediction of dam leakage with Kalman filtering and a safety assessment for dams

Akira Murakami & Takashi Hasegawa
Kyoto University, Japan

1. INTRODUCTION

Dam leakage is one of the most significant indices in a safety assessment for dam performances after construction. It requires the following three factors using such an index to assess the stability of dams: 1)satisfactory prediction of dam leakage; 2)in-situ analysis; 3)evaluation of the behavior of dam leakage.

Most of the attempts to predict dam leakage focus on the use of the finite element method (FEM). It is a great advantage that such a numerical procedure has enormous flexibility in the approximation of the complex domain. In recent decades, several authors have developed finite element solution procedures for predicting leakage in dams by the application of external action, i.e., changing water level and rainfall.

However, for practical cases as in the current problem, conventional finite element predictions have some limitations in light of a large amount of computational effort, and often these predictions are not necessarily in good agreement with the results obtained from observation. Several factors can be used to explain the poor numerical performances. These are: (1) FEM requires an understanding of overall boundary conditions, initial conditions and in-place material properties. However uncertainties in the knowledge of these numerical implementations may exist, (2) FEM only allows deterministic predictions and cannot consider observed data in the analysis.

The study shown herein examines the Kalman filtering prediction as an alternative approach to overcome the above mentioned difficulties. Firstly, we employ a simple predictive model where the water level is taken into account instead of a spatially discretized governing differential equation. Then we identify unknown parameters in the model by Kalman filtering. Once parameters in the model are determined by Kalman filtering using successively observed data, the model readily provides one time-step ahead prediction.

The other aspect to be noted is a methodology of a safety assessment for dams using the above prediction, because there are few previous works evaluating the stability of dam leakage behavior at present.

We show here a new method of safety assessment for dams based upon two judgemental criteria which consider the above predictions and the trend of dam leakage behavior as a time series. The numerical application of these criteria reveal the feasible region of dam leakage one time-step ahead.

We report some experiences of numerical performance using site-specific data.   In the second section we state the preliminary of the computational procedure.   The third section contains the problem formulation to identify the parameters in the prediction model.   The prediction results using observed data at certain actual dams are described in the fourth section.   Finally, a method for safety assessment considering both observed and predicted leakage is discussed.

2. Preliminary

Use of Kalman filtering (Kalman and Bucy 1961) is well-established in the field of hydrology (Chiu 1978).   In recent years, considerable progress in the formulations has also been made in other branches of civil engineering (Hoshiya and Saitoh 1984, Bittanti et al. 1983), although few applications to geotechnical work (Hoshiya and Saitoh 1986, Murakami and Hasegawa 1985) appear to have been achieved.

Before the current problem formulation, we will briefly review the algorithms of Kalman filtering in order to formulate the problem to follow.   The details of these algorithms are given in references: Arimoto 1977, Katayama 1983, Thornton and Bierman 1980.

2.1 Kalman filtering

Kalman filtering has an on-line data acquisition algorithm which offers successively an optimal presumption to system variables using such information as follows:
(1) dynamic characteristics of the system which generates the signals;
(2) statistical characteristics of noises;
(3) prior information on initial values of system variables;
(4) observation data.

Two equations constitute the above filter: One is the linear system equation which denotes the aiming field; The other is the algebraic Riccati equation with which the estimate error covariance matrix is satisfied. The filter gives filter gain (Kalman gain) which is determined uniquely as a solution of the Riccati equation.

Three different problems can be solved by Kalman filtering, i.e., identification, estimation, and the hybrid identification-estimation problem (Table 1).   In this paper, the first problem is dealt with.

Table 1 Three different applications of Kalman filtering

Problem	Observation eq.	State eq.
Identification	Physical model	Stationary condition for parameters
Estimation	Observed condition	Physical model
Identification-estimation	Observed condition	Physical model and Stationary condition for parameters

Now, assume that the following linear system equations denote a field in geomechanics.

$$x_{t+1} = F_t x_t + G_t w_t \tag{1}$$

$$y_t = H_t x_t + v_t \tag{2}$$

where, x_t is a state vector (n×1), y_t is an observation vector (p×1), w_t is a system noise (m×1), v_t is an observation noise (p×1), F_t is a state transition matrix (n×n), G_t is a driving matrix (n×m), H_t is an observation matrix (p×n), and both noises are Gaussian and White.

Next, the following stochastic characteristics on the above noises are assumed:

$$E(w_t) = E(v_t) = 0 \tag{3}$$

$$E\left\{ \begin{pmatrix} w_t \\ v_t \end{pmatrix} (w_s^T \ v_s^T) \right\} = \begin{pmatrix} Q_t & 0 \\ 0 & R_t \end{pmatrix} \delta_{ts} \quad , \ R_t > 0 \tag{4}$$

$$E\{w_t x_s^T\} = E\{v_t x_s^T\} = 0 \quad \text{for } t \geq s \tag{5}$$

where, $E\{\bullet\}$ is an average operator, and δ_{ts} is the Kronecker delta.

When $\hat{x}_{t/t}$ is an estimate for this state vector, the average of error vector, $e_t = \hat{x}_{t/t} - x_t$, should be zero and the quadratic form of $a_t^T[E(e_t e_t^T)]a_t$ should be minimized where a_t is an arbitrary vector (n×1). In this way, the algorithms of Kalman filtering for eqs.(1)-(5) can be formulated as follows.

Filter equation:

$$\hat{x}_{t+1/t} = F_t \hat{x}_{t/t} \tag{6}$$

$$\hat{x}_{t/t} = \hat{x}_{t/t-1} + K_t[y_t - H_t \hat{x}_{t/t-1}] \tag{7}$$

Kalman gain:

$$K_t = P_{t/t-1} H_t^T [H_t P_{t/t-1} H_t^T + R_t]^{-1} \tag{8}$$

Estimate error covariance matrix:

$$P_{t+1/t} = F_t P_{t/t} F_t^T + G_t Q_t G_t^T \tag{9}$$

$$P_{t/t} = P_{t/t-1} - K_t H_t P_{t/t-1} \tag{10}$$

Initial condition:

$$\hat{x}_{0/-1} = \bar{x}_0 \ , \ P_{0/-1} = \Sigma_0 \tag{11}$$

If we know the discretized form of F_t, H_t and select the state variables according to our objective, we can find a new estimate using the above algorithms each time we make an observation.

3. Problem Formulation and Numerical Procedure

It is necessary to specify a predictive model when applying the algorithm appearing in Chapter 2 to the prediction of dam leakage. Assumptions used herein for the development of such a model are listed below:

(1) All information is available at equally-spaced time intervals;
(2) Water level is the most important factor affecting the leakage behavior. The other factors can be ignored in the modelling;
(3) A linear relationship exists between the derivation of dam leakage from its moving average value and the derivation of water level from its moving average value.

These assumptions prescribe the model structure as follows:

1124 *Interpretation*

$$p(k+1) - \bar{p} = \alpha(p(k) - \bar{p}) + a(u(k) - \bar{u}) \quad (12)$$

where, α and a are unknown parameters to be identified from observation, $p(k+1)$ is dam leakage at time $t=(k+1)\Delta t$, $\bar{p}$ is the moving average of dam leakage up to time $t=k\Delta t$, $u(k)$ is water level at time $t=k\Delta t$, $\bar{u}$ is the moving average of water level up to time $t=k\Delta t$. Eq.(12) seems to be rather heuristic without considering the mechanism of flow through the dam. It is analogous to the discretized form of the governing equation for seepage problems.

Identification of unknown model parameters in real-time yet remains to be done in order to get one time-step ahead prediction. Such requirement introduces the use of Kalman filtering, since it is the most suitable recursive estimator which gives the unbiased optimal parameters based on available data.

For such a purpose, eq.(13) in matrix form, to which eq.(12) is rewritten, corresponds to the observation equation (2):

$$p(k+1) - \bar{p} = [p(k)-\bar{p},\ u(k)-\bar{u}]\left\{\begin{matrix}\alpha\\a\end{matrix}\right\}_k + v \quad (13)$$

where, added term v is the observation noise with zero mean and variance R.

On the other hand, the following stationary condition for model parameters is contributed to the state equation (1):

$$\left\{\begin{matrix}\alpha\\a\end{matrix}\right\}_{k+1} = \begin{bmatrix}1 & 0\\0 & 1\end{bmatrix}\left\{\begin{matrix}\alpha\\a\end{matrix}\right\}_k + \left\{\begin{matrix}w_1\\w_2\end{matrix}\right\}_k \quad (14)$$

where, w_i is the system noise with zero mean and variance Q_i. This equation is also called the "random walk" model. State transition matrix in this formulation is a unit matrix.

The above set of notations (eqs.(13)-(14)) constitutes the basic equation for Kalman filtering, which is equivalent to the model representation. The predictive model (eq.(12)) becomes deterministic, when the identified parameters have good convergency in the numerical iteration.

Kalman filtering algorithm, including U-D observation update theorem, offers an optimal estimate for parameters successively from input-output data. The necessary numerical implementation in this case is: 1)initial variances R and Q_i; 2)initial values of the parameters;

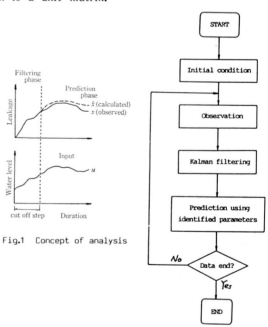

Fig.1 Concept of analysis

Fig.2 Flowchart of computation

Table 2   Duration of the observation

Dam	Duration	Example No.
A	Dec. 26, 1980 - Nov. 10, 1981	1
A	Mar. 27, 1982 - July 28, 1982	2
B	Oct. 2, 1972 - Dec. 24, 1972	3

3)elements in the observation matrix, i.e., leakage and water level at time $t=k\Delta t$ from which those moving average values are subtracted. Fig.1 schematically shows the concept of analysis, and the flowchart of computation is drawn in Fig.2.

4. Prediction Results

Identification of the model structure is performed using a continuous leakage and water level record measured at two actual dam sites; One dam is a filltype dam (called A-dam in this paper), the other dam is a combined dam (called B-dam). Data are sampled with daily intervals Δt.

Table 2 lists the observational duration at each dam and the example number. A-dam has two kinds of measurements over different durations. B-dam has also two kinds of measurements at different zones; the concrete gravity part and the fill part.

To verify that the proposed procedure is valid, its performances by use of the above observation are demonstrated here.

Example 1: Fig.3 shows the observed leakage and water level at A-dam. Each measurement over the whole duration up to the current time is used for the filtering, and the parameter identification is carried out as the performance of the procedure described in the preceding chapter. Fig.4 illustrates that the results of one step ahead prediction using identified parameters in each step agree well with the observation. It can be seen from Fig.5 that the step-by-step improved parameters contribute to high accuracy of the prediction.

Example 2: Fig.6 shows the observed leakage and water level at A-dam which are measured over different durations from Example 1. Fig.7 compares the prediction with the observation in the same manner as in Example 1. Fig.8 illustrates the behavior of the parameter identification process. Also in this example, the well-worked parameter identification operation has good convergency. Thus, comparison between the prediction results and the observation shows a reasonably close agreement.

Example 3: B-dam has two zones where each leakage is observed; Fig.9 contains the water level record and leakage monitored at the concrete gravity part and the fill part. It is seen from Fig.10 and Fig.11 that the prediction in the fill part didn't provide better performance, whereas the reasonable prediction is obtained in the concrete gravity part. The reason for the result in the fill part is that a sudden increase and decrease of observed leakage was found.

5. Safety Assessment and Discussion

Increasing attention is given to evaluating the dam behavior after construction and to developing a method of safety assessment for dams.

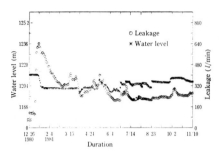

Fig.3 Water level and total leakage
measured at a fill type dam (A-dam)

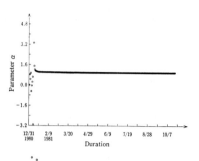

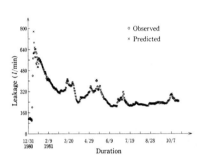

Fig.4 Comparison of the observed
leakage with the predicted one

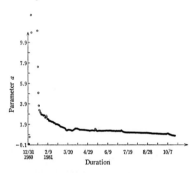

Fig.5 Parameter estimation for the
observed data

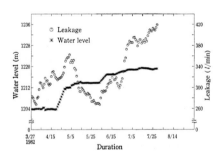

Fig.6 Water level and total leakage
measured at a fill type dam (A-dam)

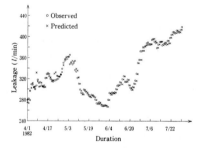

Fig.7 Comparison of the observed
leakage with the predicted one

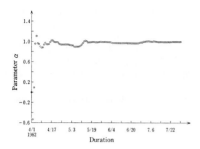 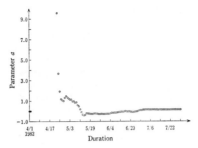

Fig.8 Parameter estimation for the observed data

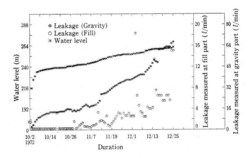

Fig.9 Water level and leakage measured at a combined dam (B-dam)

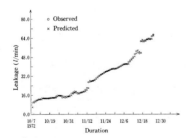

Fig.10 Comparison of the observed leakage with the predicted one in the gravity part of B-dam

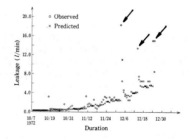

Fig.11 Comparison of the observed leakage with the predicted one in the fill part of B-dam

In this chapter, we suppose two steps for the safety assessment which consider the prediction results under the situation that there are few related studies (Hasegawa and Murakami 1986) on such subject.　For use in engineering practice, the following observational procedure is proposed.

Special emphasis is placed on the definition of the following two judgemental criteria as the first step of dam safety assessment after construction:

(a) the current observation x_k is within a certain confidence band written in the next form

$$\left| x_k - \bar{x} \right| < \lambda \cdot s \tag{15}$$

where, $\bar{x}$ is the average leakage up to the current time, $t = k\,\Delta t$, s is the standard deviation up to that time;

(b) the current observation x_k is in a certain error tolerance from the predicted value

$$\left| \frac{x_k - \hat{x}_{k/k-1}}{x_k} \right| < \varepsilon \tag{16}$$

where, λ and ε are significant. Those values should be determined by experienced engineers.

If both criteria are not satisfied at each step of observation, warning is given concerning the leakage behavior in view of a short term as shown in Fig.12.

Further examination of the leakage behavior should be carried out by comparison with past situations as the second step of assessment.　When it is necessary to assess the leakage behavior over a longer term, another judgemental view (Iida and Yonemichi 1971) gives us useful knowledge.　If necessary, engineers should consider a counter-plan, for instance, a drawdown of the water level, on the basis of their engineering experiences.

Step-by-step judgement based on the above criteria in the trial case of $\lambda = 3.0$ and $\varepsilon = 0.10$ is applied to the whole observed and predicted leakage which appeared in Chapter 4. From this data, warning is given to data with an arrow in Fig.11 at the first step of assessment.　However, in this case, the leakage is thought to be nothing worthy of special attention at the second step of assessment, because the value of the referred data is in rather small in range.

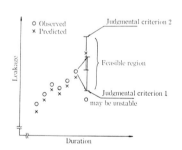

Fig.12　Judgement of the stability of the leakage behavior

A sudden increase in leakage may be attributed to piping or expansion of the cracks in the dam material, and a sudden reduction in leakage may be attributed to the closing or plugging of the cracks, which causes excess water pressure. Thus, synthetic examination considering various circumstances is required when a warning appears.

6. Conclusions

This paper pursues a simple alternative approach to one step ahead prediction for dam leakage. A new method of safety assessment utilizing such a prediction is proposed, and the trial applications of this method to two dam projects are also discussed. Main conclusions throughout the present study can be summarized as follows:

(1) Kalman filtering is used to successively identify a priori unknown parameters of the assumed predictive model. Successfully identified parameters will make the model deterministic.

(2) Applications to site-specific data confirm to us that the simple model in conjunction with Kalman filtering gives high numerical accuracy for dam leakage forecasting.

(3) Considering the prediction, the proposed judgemental criteria allows us to discriminate between significant or non-significant deviations of current observation from the behavior thought to be normal.

(4) The whole process of the computation can be readily completed on many microcomputers while allowing the accumulation of observations on their floppy disks or hard disks.

(5) The fact that the data used herein is limited to a normal range implies that a wider range of data is required to confirm the applicability of the proposed programme. This is left to future investigation.

Acknowledgements

The authors wish to acknowledge Messrs. Hitoshi Inada and Yasushi Sasaki who are former students of Kyoto University, for their assistance in the computation involved in this investigation. Part of the work presented here was supported by the Japanese Ministry of Education, Science and Culture under Grant-in-Aid for General Research (A) No.59420042.

REFERENCES

Arimoto, S. 1977. Kalman Filter. Tokyo: Sangyo-Tosho(in Japanese).
Bittanti, S., G.Maier & A.Nappi 1983. Inverse Problems in Structural Elastoplasticity: A Kalman Filter Approach, Proc. of Int. Symp. on current trends in Plasticity, held at the International Centre for Mechanical Sciences, Udine, Italy:311-329.
C.Chiu(ed.) 1978. Applications of Kalman Filter to Hydrology, Hydraulics, and Water Resources, Proc. of AGU Chapman Conf. held at Univ. of Pittsburgh.
Hasegawa, T. & A.Murakami 1986. Application of the Kalman Filtering to the Prediction of Dam Leakage - Safety Assessment for Dams -, Trans. of Japanese Society of Irrigation, Drainage and Reclamation Eng., 126:1-8(in Japanese).
Hoshiya, M. & E.Saitoh 1984. Structural Identification by Extended Kalman Filter, Jour. of Eng. Mech., ASCE, Vol.110, No.12, Dec.: 1757-1770.
Hoshiya, M. & E.Saitoh 1986. Linearized Liquefaction Process by Kalman Filter, Jour. of Geotech. Eng., ASCE, Vol.112, No.2, Feb.:155-169.
Iida, R. & H.Yonemichi 1971. Manual of Safety Assessment for dams, Revised Edition, Technical Memorandum No.692, Public Works Research Institute(in Japanese).

Kalman, R.E. & R.S.Bucy 1961. New Results in Linear Filtering and Prediction Theory. Trans. of ASME. Jour. of Basic Engg. Vol.83D, No.1:95-108.

Katayama, T. 1983. Advanced Kalman Filter. Tokyo: Asakura-Shoten (in Japanese).

Murakami, A. & T.Hasegawa 1985. Observational Prediction of Settlement Using Kalman Filter Theory, Proc. 5th Int. Conf. on Num. Meth. in Geomechanics, Nagoya, 3:1637-1643.

Thornton, C.L. & G.J.Bierman 1980. UDU^T Covariance Factorization for Kalman Filtering, Control and Dynamic Systems, 16:177-248.

A comparative evaluation of some back analysis algorithms and their application to in situ load tests

Giancarlo Gioda
Department of Theoretical and Applied Mechanics, University of Udine, and Department of Structural Engineering, Politecnico di Milano, Italy

Anna Pandolfi & Annamaria Cividini
Department of Structural Engineering, Politecnico di Milano, Italy

1 INTRODUCTION

Back analysis, or calibration, procedures are frequently adopted in rock engineering, see e.g. [Sakurai and Takeuchi, 1983; Sakurai et al., 1985], for determining the mechanical parameters of rock masses on the basis of the field measurements performed during in situ tests, or during excavation or construction works.

In general terms a back analysis requires the choice of an objective function representing the discrepancy between the quantities measured in the field and the corresponding data obtained by the stress analysis of the in situ test. This function is minimized with respect to the unknown parameters that govern the results of the stress analysis, thus leading to the "optimal" values of the mechanical properties for the rock mass. Note that, depending on the situation to be examined, calibration problems can be solved through a large variety of different methods. In fact, the stress analysis procedure could be based on simple closed form solutions or on sophisticated finite element calculations, and the discrepancy minimization could range from traditional trial-and-error procedures up to sophisticated mathematical programming algorithms.

Confining our attention to complex calibration problems (concerning e.g. tunnels or large underground openings), it can be observed that the stress analysis procedures suitable for their solution are in practice limited to either the finite element or the boundary integral equation methods. On the contrary, the choice of the optimization procedure is not straightforward. In fact, many factors influence the performance of these algorithms (like e.g. the mechanical nature of the problem, the number of unknown parameters, the size of the stress analyses, etc.) and, consequently, it is difficult to define a priori the most convenient among them.

This last aspect of the numerical approaches for back analysis is considered in the present work. An attempt is made, in particular, to compare some minimization algorithms applicable to the calibration of elasticity constants of rock masses and to assess quantitatively their relative performance. Some comments will be also presented concerning the use of these techniques for interpreting the results of plate load tests on layered rock masses. For sake of simplicity, among the mentioned stress analysis tools applicable to calibration problems, only the finite element method will be considered.

2 BACK ANALYSIS BY STANDARD MINIMIZATION ALGORITHMS

Consider an in situ load test during which a set of displacements of points within the rock mass, grouped in vector $\underline{u}^*$, is measured. Denote with $\underline{u}(\underline{p})$ the corresponding displacements evaluated by means of a finite element analysis based on a certain set of material constants $\underline{p}$.

The optimal values of the material parameters for the rock mass can be obtained by minimizing with respect to vector $\underline{p}$ the following objective function F, representing a possible definition of the discrepancy between measured and calculated quantities.

$$\min_{\underline{p}} \left(F = \{ \Sigma_i \; [u_i^* - u_i(\underline{p})]^2 \}^{1/2} \right) \qquad (1)$$

Note that the $F(\underline{p})$ is a nonlinear function even when linearly elastic material behaviour is assumed.

Among the various mathematical programming methods applicable to the above minimization problem, two are briefly described here and will be subsequently applied to the solution of some benchmark examples.

2.1 Rosenbrock's method

The so called direct search techniques for non linear function minimization reach the minimum only by a sequence of evaluations of the objective function $F(\underline{x})$, without requiring the determination of its gradient. The simplest direct search method consists in performing a one dimensional minimization with respect to each variable in turn, keeping the remaining variables fixed. Unfortunately this relaxation procedure turns out to be efficient only when there is a long "valley" almost parallel to one axis. In order to overcome this drawback Rosenbrock's method [Rosenbrock, 1960] performs relaxation moves according to changing orthogonal axes, trying to find at each step the direction of the "valley" and to place the first axis for the next step along this direction.

At the beginning of calculations it is necessary to choose an initial trial vector $\underline{x}_o$ of the free variables. The current vector $\underline{x}_1$ is set equal to $\underline{x}_o$ and the corresponding value of the objective function $F(\underline{x}_1)$ is determined. Then, the minimization process is initiated with a first "exploratory" stage.

Vector $\underline{x}_1$ is perturbed by adding to each one of its components in turn a suitably chosen "perturbation length" δ_1. Each perturbation leads to a new trial vector of free variables that can be expressed as $\underline{x}_1 + \delta_1 a^j \underline{v}_1^j$, where $\underline{v}_1^j$ and a^j are, respectively, the unit vector along the j-th direction and a scalar parameter initially set equal to 1.

The value of the objective function corresponding to the perturbed vectors is evaluated and compared with $F(\underline{x}_1)$. A "success" is obtained in the j-th direction if the following condition is fulfilled

$$F(\underline{x}_1 + \delta_1 a^j \underline{v}_1^j) \leq F(\underline{x}_1) \qquad (2)$$

In this case, vector $\underline{x}_1$ is replaced by vector $\underline{x}_1 + \delta_1 a^j \underline{v}_1^j$, and a^j is set equal to a positive number (e.g. 3). A "failure" is obtained if eq.(2)

is not fulfilled. Then vector $\underline{x}_1$ remains unchanged and a^j is set equal to a negative number (e.g. -0.5).

The exploratory stage ends when all the components of the free variable vector have been perturbed. Further exploratory stages are performed until a "success" is followed by a "failure" in all directions. When this circumstance occurs, the first step of the minimization process terminates.

It is now possible to modify the directions of the axes along which the linear searches are performed, i.e. to define new unit vectors for the second minimization step. The first axis is assumed to be that joining the points that correspond to the initial and final free variable vectors of the first step. The remaining directions are uniquely defined by generating a set of mutually orthogonal unit vectors.

The second and subsequent steps of the minimization process are carried out according to the same rules seen for the first one. At the beginning of the i-th step the current free variable vector $\underline{x}_i$ is set equal to the one obtained at the end of the preceding step $\underline{x}_{i-1}$. The perturbation length δ_i can be equal either to δ_{i-1} or to a chosen percentage of the distance between $\underline{x}_{i-2}$ and $\underline{x}_{i-1}$, that represents the "length" of the preceding (i-1) step. The minimization process ends when the lengths of several successive steps are smaller than a specified lower limit.

2.2 Conjugate gradient method

Among the minimization techniques requiring the determination of the function gradient, the conjugate direction algorithms can be viewed as intermediate between the steepest descent and Newton's methods. In fact, from the one hand they are characterized by a convergence rate faster than that of the steepest descent procedures but, on the other hand, they avoid the computational burden of evaluating and inverting the Hessian matrix, which is required by the more sophisticated Newton's methods. The so called conjugate gradient method [Fletcher and Reeves, 1964], successfully applied to back analysis problems by [Arai et al. 1983], is perhaps the most popular of these procedures.

At the beginning of the i-th iteration the gradient vector $\underline{g}_i$ of the objective function is evaluated at point $\underline{x}_i$ representing the approximation of the minimum obtained at the end of the preceding iteration.

$$\underline{g}_i = \{\partial F(\underline{x})/\partial \underline{x}\}_{\underline{x}=\underline{x}_i} \qquad (3)$$

When dealing with elastic material behaviour it is possible to work out the analytical expression of the gradient of function F defined by eq. (1). However, it is in general necessary to evaluate numerically this gradient in the presence of non linear constitutive laws.

The search direction $\underline{p}_i$ for the current iteration is determined on the basis of the search direction of the preceding iteration $\underline{p}_{i-1}$ and of gradients $\underline{g}_i$ and $\underline{g}_{i-1}$.

$$\underline{p}_i = -\underline{g}_i + [(\underline{g}_i^T \underline{g}_i)/(\underline{g}_{i-1}^T \underline{g}_{i-1})] \; \underline{p}_{i-1} \qquad (4)$$

Finally, the following equation leads to a new approximation $\underline{x}_{i+1}$ of the minimum,

$$\underline{x}_{i+1} = \underline{x}_i + \delta_i \, \underline{p}_i \qquad (5)$$

where the scalar parameter δ_i is evaluated by a one dimensional mini-mization along the direction $\underline{p}_i$.

Various procedures can be used for this linear search, like e.g. the trial-and-error one adopted in Rosenbrock's method. In this study the so called Powell's method [Powell, 1964] was adopted, that performs the one dimensional minimization through iterations each of which requires the minimization of a quadratic function locally approximating, in the current direction, the non linear objective function.

3. AN AD HOC PROCEDURE FOR BACK ANALYSIS OF ELASTICITY CONSTANTS

This method for the calibration of the elastic constants of finite ele-ment models was proposed by Kavanagh and Clough [1971] in a structural engineering context and was subsequently applied to geomechanics prob-lems, see e.g. [Cividini et al., 1981]. In order to apply this proce-dure it is necessary to establish a linear relationship between the stiffness matrix of each finite element and the unknown material para-meters p_i. Consequently, the stiffness matrix of the assembled finite element model can be written in the following form,

$$\underline{K} = \Sigma_i \, p_i \, \underline{K}_i \qquad (6)$$

where $\underline{K}_i$ is the assembled stiffness matrix obtained by setting all ma-terial parameters to zero except for the i-th parameter set equal to 1.

It is also necessary to write the final system of linear equations in the following form,

$$\begin{bmatrix} \underline{K}_{11} & \underline{K}_{12} \\ \underline{K}_{21} & \underline{K}_{22} \end{bmatrix} \begin{Bmatrix} \underline{u}^* \\ \underline{u}_2 \end{Bmatrix} = \begin{Bmatrix} \underline{f}_1 \\ \underline{f}_2 \end{Bmatrix} \qquad (7)$$

where vector $\underline{u}^*$ collects all the measured displacement components, and $\underline{f}_1$ and $\underline{f}_2$ are known nodal force vectors.

On the basis of eqs.(6) and (7), through some algebraic manipula-tions, it is possible to reach the following non linear equation sys-tem, the solution of which leads to the optimal parameter vector.

$$\underline{R}^T \underline{R} \, \underline{p} = \underline{R}^T \, (\underline{f}_1 - \underline{Q} \, \underline{f}_2) \qquad (8)$$

where

$$\underline{Q} = \underline{K}_{12} \, \underline{K}_{22}^{-1} \qquad (9)$$

and

$$\underline{R} = [\underline{r}_1 \mid \underline{r}_2 \mid \ldots] \quad ; \quad \underline{r}_i = (\underline{K}_{11,i} - Q \, \underline{K}_{21,i}) \, \underline{u}^* \qquad (10a,b)$$

The stiffness matrices in eq.(10b) are obtained by partitioning matrices $\underline{K}_i$ (eq.6) with the same criteria used in eq.(7).

The solution of the above system is reached with a simple iterative procedure requiring at each step the inversion of a part $(\underline{K}_{22})$ of the assembled stiffness matrix calculated on the basis of the parameters determined at the end of the preceding iteration.

It can be easily shown that, when only one elastic modulus E is considered as unknown for the entire rock mass, the non linear equation system (8) is replaced by the following linear relationship.

$$\overline{E} = (\underline{u}_{(1)}^T \, \underline{u}^*)/(\underline{u}_{(1)}^T \, \underline{u}_{(1)}) \qquad (11)$$

Here, $\underline{u}_{(1)}$ is the vector of nodal displacement components, corresponding to the measured ones $\underline{u}^*$, computed assuming a unit value for the elastic modulus.

4 A COMPARATIVE EXAMPLE

The example solved by [Arai et al. 1983], and concerning an excavation in a layered, linearly elastic soil deposit, was chosen as a benchmark test for comparing the performance of the three previously described solution algorithms.

A preliminary stress analysis, based on the finite element mesh and load distribution shown in fig.1, was carried out in order to obtain the complete displacement distribution within the rock mass. Subsequently, only the horizontal and vertical displacements of the surface nodes (1,6,5,...14 in fig.1) were used as input data in the calibration analyses. The results of these calculations are summarized in fig.2 where the variation of the back calculated bulk B and shear G moduli of the rock mass is shown with increasing number of iterations. For briefness sake only the elastic parameter of the first and third layers are considered.

The diagrams in fig.2 apparently indicate that the rates of converge of Kavanagh (curve 1) and conjugate gradient (curve 2) methods are superior to those of Rosenbrock (curve 3) and Arai (curve 4) procedures. It can be observed that the different results obtained with the present conjugate gradient procedure and with the one used by Arai et al. are probably due to the different characteristics of the adopted linear search techniques.

It has to be considered, however, that the implemented methods are likely to require different computational efforts for performing the same number of iterations, consequently a correct comparison among them should be based on the total CPU time required for solution, rather than on the number of iterations.

The result of this further comparison are reported in fig.3 where the decrease of the following non dimensional error is shown with increasing CPU time.

$$ERR_p = \{\Sigma_i \, [(p_i - p^*_i)/p^*_i]^2\}^{1/2} \qquad (12)$$

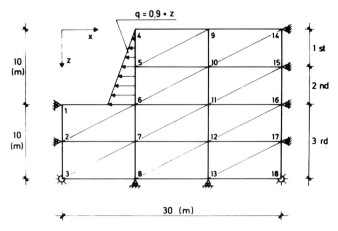

Fig. 1   Finite element mesh for the benchmark problem (after Arai et al. [1983]).

In eq.(12) p_i^* denotes the "real" value of the i-th elastic parameter.

In order to make the results independent from the computer adopted, the CPU time is divided by the time, CPU*, required for a linear stress analysis of the same problem. Note that Arai's analyses are not intro- duced in fig.3 since the corresponding CPU time was not indicated in the original publication.

Once again it turns out that the most convenient solution procedure for the considered benchmark problem is the one proposed by Kavanagh and Clough [1971]. In fact, Kavanagh's method required a non dimension- al CPU time equal to 4.2 for reaching an error of 0.05 (see line (a) in fig.3), while the times necessary with conjugate gradient and Rosen- brock's methods were, respectively, 13.7 and 27.6.

In order to check whether the above conclusion holds true also for problems of different size, some analyses were performed by varying, as indicated in the Table I, the number of unknown parameters N_p, of de- grees of freedom N_f of the entire finite element mesh and of "measured" displacements N_u. Other analyses were also performed by keeping constant, and equal to 4, the number of unknown parameters and increasing only the number of nodes in the finite element mesh. Each analysis was terminated when the non dimensional error expressed by eq.(12) reached a value equal to, or slightly smaller than, 0.05.

The results of these additional calculations, presented in fig.4, clearly show that Kavanagh's method has in most cases the fastest rate of convergence, and that it is always faster than the conjugate gradi- ent method. Only for problems involving a large number of nodal displa- cements Kavanagh's approach becomes slightly slower than Rosenbrock's method. This is due to the non negligible computer time necessary for inverting the partitioned stiffness matrix (cf. eq.9), which is not required by the direct search procedures.

It can be concluded that, among the tested solution algorithms, Kavanagh's method is particularly suitable when a large number of elasticity parameters has to be calibrated using a relatively coarse

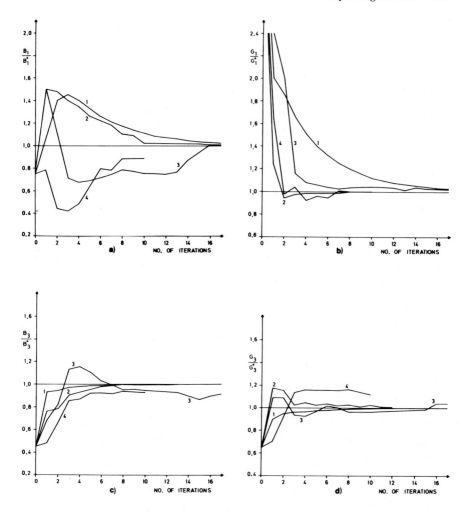

Fig. 2: Variation of bulk B and shear G moduli of layers 1 and 3 (see fig. 1) with increasing number of iterations. A star denotes the "real" parameters.
1) Kavanagh's method.
2) Conjugate gradient method.
3) Rosenbrock's method.
4) After Arai and al. (1983).

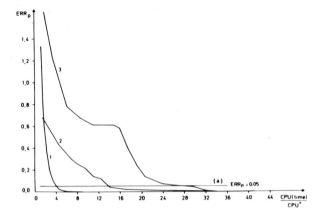

Fig. 3: Variation of the non dimensional error (cf. eq. 12) with increasing CPU time for the problem depicted in fig. 1 (CPU* = time required by an elastic stress analysis).
1) Kavanagh's method, 2) Conjugate gradient method, 3) Rosenbrock's method.

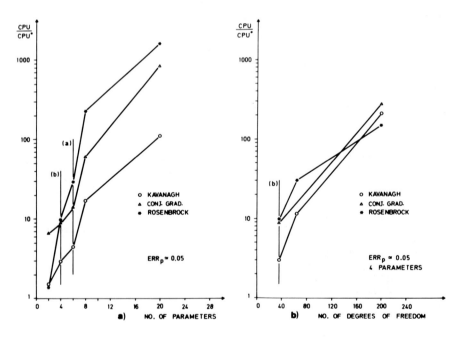

Fig. 4: Variation of the CPU time with increasing size of the calibration problem.

TABLE I

	N_p	N_f	N_u
1)	2	36	10
2)	4	36	10
3)	6	36	10
4)	8	54	14
5)	22	76	35

finite element mesh. On the contrary, Rosenbrock's procedure is convenient if a refined mesh is adopted for calibrating a small number of
elasticity parameters.

5 PLATE LOAD TEST ON LAYERED ROCKS

The plate load test represents one of the most popular in situ tests
for determining the elastic characteristics of rock masses. It seems
therefore of potential interest to check the applicability of the
previously discussed techniques to the interpretation of its results.

The hypothetical situation here examined is depicted in fig.5. It
concerns a test carried out on layered rock formations formed by three
main rock types. Two different sequences of rock layers have been considered. They are represented in fig.5 by the variations (a) and (b) of
the in situ Young modulus E* with depth. The in situ Poisson ratio is
assumed constant for all rock types and equal to 0.3. Also for this
problem the "real" displacements were determined by a preliminary finite element analysis, and only some of them were used as input data in
the calibration problems.

Based on eq.(11), a first set of back analyses was carried out considering as unknown only one "equivalent" elasticity modulus for the
entire rock mass. The results of these calculations are summarized, for
the two previously mentioned layered rocks, in fig.6. Here the diagrams
are shown relating the back calculated elastic modulus $\overline{E}$ to the number
of vertical displacements used as input data in the calibration analyses. The curves in these diagrams correspond to various measurement programs, i.e. to vertical displacements measured along different vertical
lines (lines 1,2,3 in fig.5), or along the ground surface (line 4 in
fig.5).

The results of calculations show that increasing the number of input
data the back calculated elastic modulus $\overline{E}$ converges toward a value
which depends on the adopted measurement program and that is approximately situated in a region bounded by two limit moduli. The first one,
$\overline{E}_{ALL}$, represents the modulus obtained by a back calculation in which
all the nodal displacements are introduced as input data; the second
one, E_H^*, is the "harmonic" mean of the moduli of all rock layers.

$$E_H^* = (\Sigma_i\ h_i)/(\Sigma_i\ h_i/E_i^*) \tag{13}$$

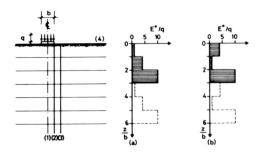

Fig. 5: Plate load problem. Variation of the in situ Young modulus E^* with depth for two hypothetical layered rock masses.

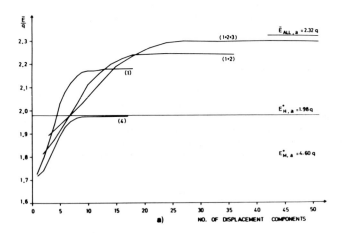

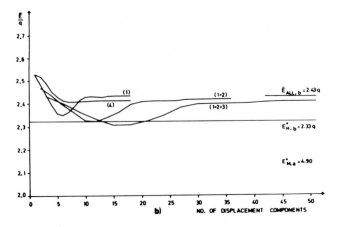

Fig. 6: Plate load problem. Variation of the back calculated Young modulus $\bar{E}$ with increasing number of vertical displacements. Figures (a) and (b) refer to the two rock masses in fig. 5.

In the above equation E_i^* and h_i represent, respectively, the modulus and the thickness of the i-th rock layer.

It is interesting to observe that the customary definition of mean elastic modulus, E_M^*, for the layered rock mass

$$E_M^* = (\Sigma_i \; E_i^* \; h_i)/(\Sigma_i \; h_i) \qquad (14)$$

leads to values much larger than those obtained with the back analysis of the measured displacements.

In order to check whether the above observation holds true also when dealing with the calibration of more than one elastic parameter, the same plate load problem was solved considering as unknowns both shear and bulk moduli for the entire rock mass. The calculations were repeated for the two previously mentioned layered rocks.

In choosing the solution algorithm for these analyses two points have been considered:
a) even discretizing only one half of the problem, because of its symmetry, a relatively large finite element problem is obtained, having 256 nodal displacement components;
b) only two unknown parameters have to be back calculated.

On the basis of these observations, and taking into account the conclusions reached in the preceding Section, Rosenbrock's method was chosen as the most convenient solution technique for the problem at hand.

It is important to note, however, that the back analysis of two parameters, requiring the minimization of a nonlinear function, has a cost much larger (more than 100 times) than that of the simple linear problem governing the calibration of only one elasticity constant.

Some preliminary tests showed that reliable values of the rock mass shear and bulk moduli could not be obtained solely on the basis of vertical displacements. Consequently, the input data for these analyses consist of the vertical displacement components of points along line (1) in fig.5, and of both vertical and horizontal components of points along lines (2) and (3).

The results of these analyses are summarized in fig.7, where the variation of the back calculated moduli with increasing number of input displacements are shown for both rock types (a) and (b). Also for this last example it turns out that the numerically evaluated elasticity constants are close to the harmonic mean of the in situ parameters, but are quite different from their "conventional" mean value.

6 CONCLUSIONS

On the basis of the numerical results discussed in the present work, concerning the back analysis of elasticity constants by means of the finite element method, the following main conclusions can be drawn:
- Even though a specific criterion for choosing the most convenient algorithm for the back analysis of elasticity parameters was not worked out, it appears that the techniques developed "ad hoc" for this purpose, like the one proposed by Kavanagh and Clough [1971], are particularly convenient when dealing with a relatively large number of unknown parameters and when the finite element mesh has a small number of nodal variables. On the contrary, general purpose direct search procedures are preferable when a few parameters are back analyzed using large finite element meshes.

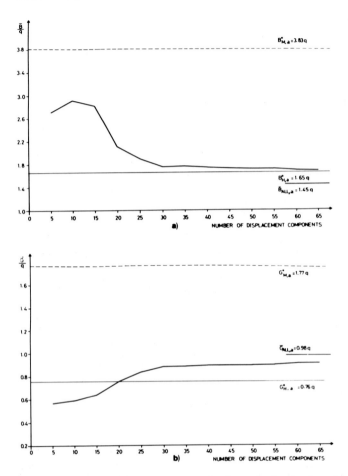

Fig. 7: Plate load problem. Variation of the back calculated bulk $\overline{B}$ and shear $\overline{G}$ moduli with increasing number of input displacements for rock mass (a) in fig. 5.

- The application of the mentioned solution algorithms to the interpretation of hypothetical plate load tests on layered rock masses
showed that the back calculated "equivalent" elasticity constants for
the entire mass are markedly smaller than the mean values, determined
according to the customary definition, of the in situ parameters. It
was also observed that the back calculated constants represent a
reasonable approximation, from the engineering view point, of the
"harmonic" mean values (cf.eq.11) of the in situ moduli.
- While one equivalent elastic modulus for the entire (layered) rock
mass can be evaluated considering only vertical displacements as input data, the determination af two independent elasticity constants
(i.e. shear and bulk moduli) require the knowledge of both vertical
and horizontal displacements at some locations.

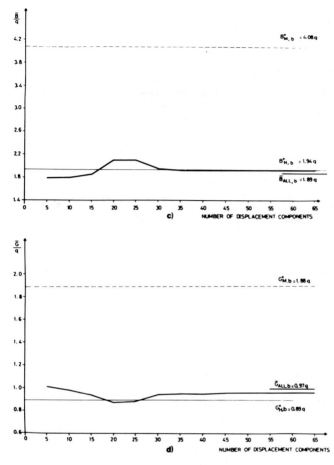

Fig. 7 cont.: Plate load problem. Variation of the back calculated bulk $\bar{B}$ and shear $\bar{G}$ moduli with increasing number of input displacements for rock mass (b) in fig. 5.

ACKNOWLEDGEMENTS

The authors are indebted to Prof. S. Sakurai for his suggestions and stimulating comments. Part of this study was prepared during a stage of the first and third author at the Department of Civil Engineering of Kobe University. The financial support of Fukada Foundation and of the Ministry of Education (MPI) of the Italian Government is gratefully acknowledged.

REFERENCES

Arai K., Otha H., Yasui T., "Simple optimization techniques for evaluating deformation moduli from field observations", Soils and Foundations, Vol.23, No.1, 1983.

Cividini A., Jurina L., Gioda G., "Some aspects of 'Characterization' problems in geomechanics", Int.J.Rock Mechanics and Mining Sciences, Vol.18, 1981.

Fletcher R., Reeves C.M., "Function minimization by conjugate gradient", The Computer Journal, Vol.7, 1964.

Kavanagh K.T., Clough R.W., "Finite element application in the characterization of elastic solids", Int.J.Solids Structures, Vol.7, 1971.

Powell M.J.D., "An efficient method of finding the minimum of a function of several variables without calculating derivatives", The Computer Journal, Vol.7, 1964.

Rosenbrock H.H., "An automatic method for finding the greatest or least value of a function", The Computer Journal, Vol.3, 1960.

Sakurai S., Takeuchi K., "Back analysis of measured displacements of tunnels", Rock Mechanics and Rock Engineering, Vol.16, 1983.

Sakurai S., Shimizu N., Matsumuro K., "Evaluation of plastic zone around underground openings by means of displacement measurements", Proc.5th.Int.Conf.on Numerical Methods in Geomechanics, Nagoya, 1985.

2nd International Symposium on Field Measurements in Geomechanics, Sakurai (ed.)
© 1988 Balkema, Rotterdam. ISBN 90 6191 778 6

Observational procedure of earth wall structures

Kazuo Furuya & Yoshihiro Ito
Sato Kogyo Co. Ltd, Engineering Research Institute, Japan

1. INTRODUCTION

In recent years, large-scale, complicated excavation works are being done in narrow spaces where adjacent structures exist. Due to the progress in high level utilization of underground space or highly intensified use of urban space, the soil is weak and the geological conditions are poor.

Under these circumstances, a so-called observational procedure that rationalizes construction based on the information obtained by site instrumentation is gradually being introduced.

In order to implement information construction of sheathing works, the "back analyzing system for earth retaining structures", is used to feed the measured data back into the construction process. It has already been applied to the performance of several sheathing works.

This report describes the instrumentation and prediction analyses which were performed in the construction of building foundations adjacent to existing structures.

2. SHEATHING WORK AND SUMMARY OF GEOLOGY

The work in question was the foundation work for an RC building with 5 storeys above ground and 2 storeys underground. The scope of excavation was about 36 x 33 m on the surface and about 12 m in depth, as shown in Fig. 2-1. The struts were set in four stages using the preload system. The diaphragm wall constituted an insitu reinforced concrete continuous underground wall 60 cm thick. The ground at the construction site consisted from the ground surface of filled soil, an alluvial soil layer (silt) with N-value of 2 - 5, and a mudstone layer with N-value of more than 50, as shown in Fig. 2-2.

3. INSTRUMENTATION

3.1 Summary of instrumentation

The stability of the diaphragm wall was tentatively confirmed through design calculations, in advance. However, to retain the profile of

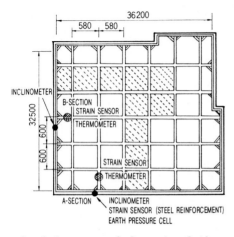

Fig.2-1 general layout of the instrumented excavatior

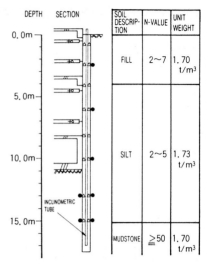

● EARTH PRESSURE CELL
△ STRAIN SENSOR (STEEL REINFORCEMENT BAR)
+ STRAIN SENSOR (STRUT)
○ THERMOMETER

Fig.2-2 measuring cross section and soil profile

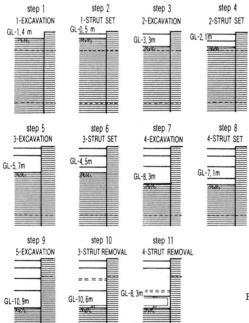

Fig.2-3 steps of construction work

Table 3-1 measuring items and instruments

OBJECT	ITEM	INSTRUMENTS		
		TRANSDUCER	CAPACITY	ACCURACY
WALL	SIDE PRESSURE	EARTH PRESSURE CELL	100~300KPa	±0.5%
	STEEL REINFORCEMENT STRESS	STRAIN SENSOR	300MPa	±0.5%
	DEFLECTION	INCLINOMETER	300'	0.1mm/1m
STRUT	LOAD	STRAIN SENSOR	±100MPa	±0.5%
	TEMPERATURE	THERMOMETER	-10~60℃	±0.5%

excavation for a long period of time and to perform the construction work both safely and smoothly, it was necessary for the instrumentation to be capable of quickly predicting the occurrence of abnormal conditions that could not be foreseen at the time of design and construction planning, and to take adequate measures thereagainst. In order to achieve safety in the excavation and sheathing works during this construction period, two sections were provided for measurement as shown in Fig. 2-1, and measurement of deformation etc., of the wall was performed.

Table 3-1 shows a table of items for measurement and the instruments used. An earth pressure cell, strain sensor (steel reinforcement), inclinometer, strain gauge and thermometer were located at section A, a main measurement section. And at section B, an inclinometer, strain gauge and thermometer were installed.

3.2 Results of measurements

3.2.1 Side pressure

The change in side pressure with the passing of time was as shown in Fig. 3-1. The following were indicated:
1) As construction progressed, wall deflection occurred in the direction of excavation and consequently the side pressure decreased, recording the minimum at the 5th stage of excavation. The tendency toward a decrease became more significant at depths larger than 10 m.
2) When the third strut was removed the side pressure increased, but showed almost no change thereafter.
3) When struts were set at the various stages, the side pressure increased temporarily due to the effect of preload. Particularly, as the wall near the strut where the preload is applied displaced toward the natural ground, the earth pressure increased locally.
4) From the above results, it can be seen that the side pressure coefficient was not constant during the construction and increased or decreased depending on the deflection of the diaphragm.

3.2.2 Horizontal deflection of diaphragm

The wall deflection at each construction step is shown in Fig. 3-2. The following were indicated:
1) As construction progressed, the deflection in the direction of excavation increased. However, when the strut was preloaded, deflection occurred in the direction of the natural ground.
2) In the initial step, the maximum deflection was observed near the head, gradually moving toward the lower part of the wall and after the installation of the second strut it moved deeper to below the bottom of the excavation.
3) At a depth shallower than 5 m, deflection tended toward the natural ground after the installation of the second strut, and after 4th stage of excavation, deflection was observed toward the natural ground before the start of excavation.
4) The maximum deflection occurred when 4th stage of excavation was performed, and was about 9 mm.

3.2.3 Bending moment

Fig. 3-3 shows the distribution of the bending moment at each excavation step. The following were indicated:
1) As construction progressed, the maximum bending moment increased showing the maximum value at the 5th stage of excavation, which was about 300 KN·m/m. Thereafter, it decreased to about 200 KN·m/m when the third strut was removed, showing almost no change afterward.
2) The location of occurrence of the maximum bending moment was approximately the bottom of excavation for each step.

3.2.4 Axial force of strut

Fig. 3-4 shows the change in the axial force of struts at each step of excavation. The following were indicated:
1) The first strut
The axial force decreased until 5th stage of excavation. When the third and fourth struts were removed, the axial force tended to increase due to the increase in the shared load as a result of their removal.
2) The second strut
This showed a tendency similar to that of the first strut. However, after the increase at the 3rd stage of excavation it decreased significantly due to the effect of preload in the third strut. Thereafter, the axial force tended to decrease until the 5th stage of excavation and increased significantly when the third and fourth struts were removed.
3) The third strut
After the axial force increased at the 4th stage of excavation, it decreased by 60 KN/m due to the effect of the preload in the fourth strut, decreasing somewhat at the 5th stage of excavation.
4) The fourth strut
The axial force decreased by about 30 KN/m at the 5th stage of excavation. When the third strut was removed it increased by about 50 KN/m.

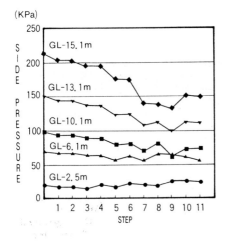

Fig.3-1　side pressure
　　　　　 at steps

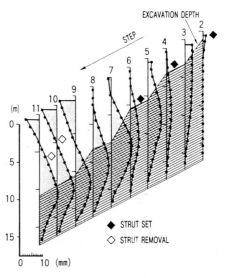

Fig.3-2　distribution of
　　　　　 deflection of wall
　　　　　 at steps

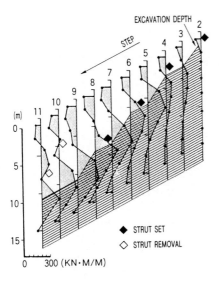

Fig.3-3　distribution of
　　　　　 bending moment of
　　　　　 wall at steps

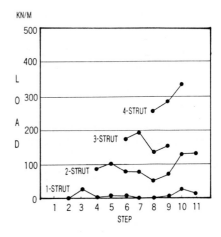

Fig.3-4　load of struts
　　　　　 at steps

4. PREDICTION ANALYSIS

4.1 Summary of prediction analysis

For the purpose of predicting the behavior of the wall quantitatively, back analysis is performed to obtain the optimum combination
of the ground parameters which induce the same results as the measured
wall deflection.
The fundamental method of calculation used for back analysis is so
called "elasto-plastic method" which is one of the retaining wall
design methods. Fig. 4-1 shows a conceptual drawing of this analysis
model. And back analysis uses the simplex method as an optimization
method.
In this analysis, wall deflection data and axial force data of
struts are used as input data. And the ground parameters to be presumed are the coefficient of active earth pressure of fill, that of
silt, the coefficient of passive earth pressure of silt, the coefficient of subgrade reaction of silt and that of mudstone. Other ground
parameters which do not affect the analysis greatly are fixed as the
constants used in the previous design.

4.2 Results of analysis

In this paper, back analysis performed at the 5th stage of excavation and the analysis to predict the movements of the wall after that
time are described. Table 4-1 shows the ground parameters assumed
from back analysis at the 5th stage of excavation. The coefficient of
active earth pressure of fill is very large. This corresponds to the
fact that the wall moved toward the back ground and it is considered
that earth pressure shows passive behavior. Conversely, the coefficient of active earth pressure of silt is rather small.
The prediction analysis up to the removal of the fourth strut is
performed using the parameter obtained. Fig. 4-2 shows comparison
between the calculated results and the measured data of the wall
deflection and the bending moment. Table 4-2 shows the axial force of
the struts. The following could be deduced from this figure and
table:
1) With regard to wall deflection, the predicted values approximately matched the measured data.

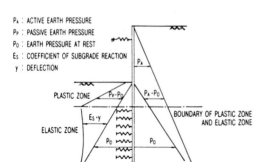

P_A : ACTIVE EARTH PRESSURE
P_P : PASSIVE EARTH PRESSURE
P_O : EARTH PRESSURE AT REST
E_S : COEFFICIENT OF SUBGRADE REACTION
y : DEFLECTION

Fig.4-1   analysis model

Table 4-1   ground parameters

PARAMETER SOIL	K_O	K_A	K_P	E_S MN/m³
FILL	—	1.27	—	—
SILT	0.66	0.41	10.00	4.91
MUDSTONE	0.50	0.33	3.00	97.10

UNDERLINED PARAMETERS ARE ESTIMATED

2) The bending moments approximately matched the bottom of excavation where maximum bending moment occurred. However, near the installation location of the struts, they did not match very well. The measured data did not clearly show the effect of the axial force of the strut as in the case of predicted value. This may be due to the fact that the side pressure generated by the preload of the strut is not considered in the analysis, and that while the axial force of the strut acts on the back ground as a distributed load through the wale and the wall, in this analysis, it is simply given as a concentrated load.

3) The axial force of the struts approximately matched at Step 10. However, the predicted value of the second strut exceeded the measured data by about 40% at Step 11. The precision of the predicted value of the axial force of struts generally depends on the accuracy of the predicted value of wall deflection. Even in this case, as the decrease in the side pressure due to the wall deflection cannot be taken into account, the predicted value is somewhat larger than the measured data.

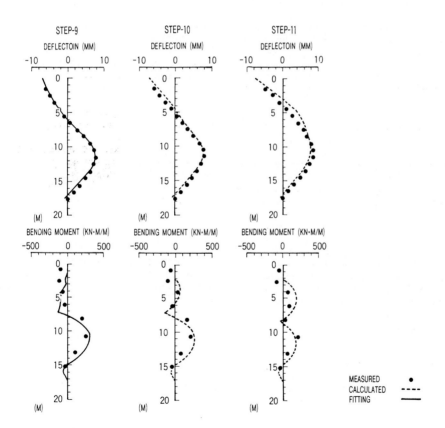

Fig. 4-2   comparison of wall deflection and bending moment between measured data and calculated results

Table 4-2   comparison of axial force of struts
          between measured data and calculated
          results

STRUT	STEP	STEP-9 5-EXCAVA-TION	STEP-10 3-STRUT REMOVAL	STEP-11 4-STRUT REMOVAL
1-STRUT	MEASURED	4.6	27.0	15.7
	CALCULATED	4.6	16.1	15.8
2-STRUT	MEASURED	70.2	130.0	132.9
	CALCULATED	70.2	132.7	185.6
3-STRUT	MEASURED	153.5	—	—
	CALCULATED	153.5	—	—
4-STRUT	MEASURED	280.5	331.1	—
	CALCULATED	280.5	359.3	—

UNIT:KN/m

5. EXAMINATION OF ANALYSIS METHOD

In the conventional method described above, the side pressure is given by means of the unit weight and the coefficient of active earth pressure. For this reason, the side pressure takes a certain fixed value regardless of the increase or decrease in the wall deflection. In other words, the active earth pressure obtained using the conventional method is nothing but the coefficient of side pressure when the back analysis is performed.

However, as it is apparent that the side pressure varies with the wall deflection, the calculation assuming that the coefficient of earth pressure would not change in the process causes a lowering in the accuracy of the prediction.

In recent years, the preload of struts has been employed to control wall deflection. It may cause a side pressure larger than the initial earth pressure at rest. In such cases, problems like this will occur simply by representing the side pressure as a certain constant value.

5.1 Improvement of analysis method

To solve the problems involved with the conventional method, a model is considered which supports the entire wall with springs whose characteristics are elasto-plastic, as shown in Fig. 5-1. With these, the earth pressure that acts upon the back of the wall either increases or decreases with the wall deflection.

5.2 Application to measured data

From the above point of view and based on the measured data at the 5th stage of excavation, prediction analysis is performed. The following results are obtained:

1) Fig. 5-2 shows the comparison between the measured data and the calculated results. From the figure, it can be seen that the accuracy of the wall deflection has been improved as compared with the conventional method.

The bending moments are also seen to match well near the location of the strut installation at Step 11.

2) Fig. 5-3 shows the results of a comparison of the axial force of the struts. From the figure, it can be seen that the axial force which was predicted largely becomes very near the measured data.

As mentioned above, by solving the problem of the side pressure that could not be considered in the conventional method, the accuracy of the analysis has been well improved.

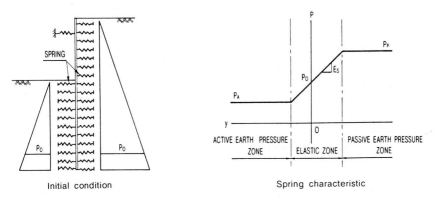

Initial condition   Spring characteristic

Fig.5-1   analysis model of proposed method

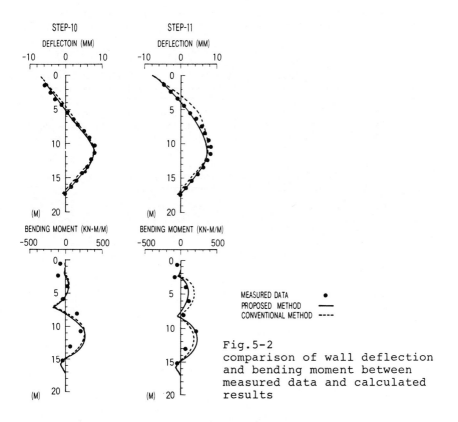

MEASURED DATA   •
PROPOSED METHOD   ——
CONVENTIONAL METHOD ----

Fig.5-2
comparison of wall deflection
and bending moment between
measured data and calculated
results

1154 *Interpretation*

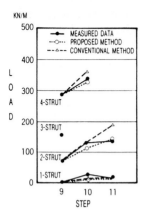

Fig. 5-3 comparison of axial force of struts
between measured data and calculated
results

6. CONCLUSION

In this work, the instrumentation of deflection, stress of the
retaining wall, axial force of the struts, and earth pressure near the
wall were performed, and prediction analysis was carried out to control the performance and to maintain safety. As a result, the construction period was shortened by changing the schedule for the
removal of the struts. Also, by comparing the measured data and the
results of the prediction analysis, the analysis technique was improved.

The method of analysis proved to be effective in these construction
works. To evaluate the instrumentation data using an analysis method
such as this one also makes it possible to derive the references for
examination of the adequacy of the analysis method itself.

In the future, it will be possible to realize a more rational and
higher quality method for design and construction, making effective
use of the information obtained by instrumentation.

Interpretation of the results of displacement measurements in cut slopes

S.Sakurai
Department of Civil Engineering, Kobe University, Japan

1 INTRODUCTION

Even when using a sophisticated numerical analysis method such as FEM,
the predicted deformational behaviour of cut slopes is often different from the real one observed during the cutting. This is mainly due to
the fact that the geological conditions and mechanical properties of
geomaterials, which are necessary for predicting the slope behaviour, are
too complex to evaluate correctly. In order to overcome this difficulty,
field measurements are performed, and slope stability is monitored by
interpreting the field measurement results.

This paper deals with a method of interpreting the results of displacement measurements performed by inclinometers. The method is based on a
back analysis of measured displacements, and is capable of assessing the
mechanism of deformational behaviour of cut slopes.

2 MECHANICAL MODEL OF CUT SLOPES

The ground in which a slope is cut has been classified into three
groups, i.e., (a) continuous, (b) discontinuous, and (c) pseudo continuous, as shown in Fig. 1. Type (a) is for ground consisting of soils,
Type (b) represents jointed rock masses, and Type (c) is for highly
fractured and/or weathered rock masses, thus, for this type, the global behaviour of the ground seems to be a continuous body. We call
this the pseudo continuous type of ground.

The stability of slopes cut in Type (a) ground can be analyzed by means
of a mechanical model based on continuum mechanics, while a discontinuous
model such as those proposed by Cundall (1975) and Kawai (1980) is used
for analyzing Type (b), where the joint elements in finite element
analysis are also useful.

Concerning the ground of Type (c), one can of course adopt a discontinuous model similar to that for Type (b). In engineering practices,
however, it is almost impossible to explore all the joint systems, or
to investigate all their mechanical characteristics. Moreover, it seems
that this type of ground behaves in a global sense, just like a continuous body. Considering these circumstances, therefore, the Author
recommends a continuous mechanics model for analyzing the stability of
slopes cut in the ground of this (c) Type. It should be noted, however,
that this continuous model must be one capable of taking into account
the effect of discontinuities.

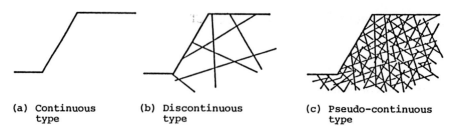

(a) **Continuous**
    type

(b) **Discontinuous**
    type

(c) **Pseudo-continuous**
    type

Fig. 1 Classification of the ground

   In this paper, only the continuous and pseudo continuous types of
ground are discussed, and a method is proposed for interpreting the
results of displacement measurements taken during the cutting of slopes
in these types of ground.   In the interpretation, one can first determine
the initial stress and mechanical characteristics of the ground materials
by analyzing the measurement results.   This process is just the reverse
of ordinary stress analysis, and thus, it is called "back analysis".
When determining the initial stress and mechanical characteristics by
back analysis, one can re-evaluate the input data used at the design
phase, and can assess the stability of slopes.
   The most difficult and important task in back analysis is the choosing
of a mechanical model to represent the mechanical behaviour of the ground.
However, the mechanical model is not input for the back analysis, but
rather output, which must be obtained through consideration of the results
of field measurements.
   Concerning the deformation of slopes cut in the continuous and pseudo
continuous types of ground, the deformational mode is classified into
three different groups; namely, (a) elastic, (b) sliding, and (c) toppling,
as shown in Fig. 2.   This means that the mechanical model for back analysis must be one which includes all three deformational modes as a potential, and by changing the material constants of the model, one or more of
the modes will be derived.   Elastic deformation, of course, can be
analyzed by ordinary elasticity.   The finite element method with joint
elements is effective for sliding deformation, while the distinct element
method (Cundall, 1975) may be the most promising for analyzing the
toppling of slopes.   Regarding the design work of cut slopes, however,
a single model suitable for analyzing all those deformational modes is
preferable.
   In the following section, a back analysis method is described, which
is based on continuum mechanics, and is capable of analyzing all three
deformational behaviours, that is, elastic, sliding, and toppling by
using a single mechanical model.

3 CONSTITUTIVE EQUATION

It has been found from careful investigations of the results of laboratory tests on cut slopes (Deeswasmongkol and Sakurai, 1984), that shear
deformation of the ground above the potential sliding surface occurs
easily, and that the largest shear deformation takes place parallel to
the sliding surface.   Considering these deformational behaviours of
slopes, an anisotropic continuous material is assumed to have the smallest shear rigidity parallel to the sliding surface.

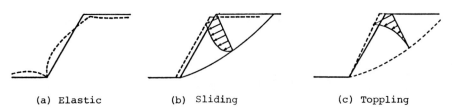

(a) Elastic   (b) Sliding   (c) Toppling

Fig. 2   Deformational modes of cut slopes

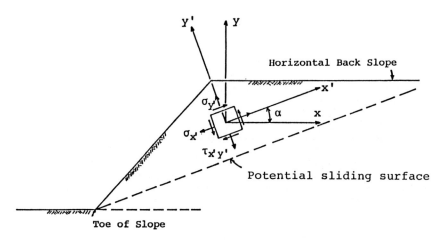

Fig. 3   Coordinate system

Now let the local coordinate system x'-y' be taken as shown in
Fig. 3, where the x' axis is parallel to the sliding surface. The
proposed constitutive equation is then given as follows;

$$\left\{ \begin{array}{c} \sigma_{x'} \\ \sigma_{y'} \\ \tau_{x'y'} \end{array} \right\} = [\,D'\,] \left\{ \begin{array}{c} \varepsilon_{x'} \\ \varepsilon_{y'} \\ \gamma_{x'y'} \end{array} \right\} \tag{1}$$

where

$$[\,D'\,] = \frac{E}{1-\nu-2\nu^2} \left[\begin{array}{ccc} 1-\nu & \nu & 0 \\ \nu & 1-\nu & 0 \\ 0 & 0 & m(1-\nu-2\nu^2) \end{array} \right] \tag{2}$$

$\{\ \sigma_{x'}\ \sigma_{y'}\ \tau_{x'y'}\ \}^T$ and $\{\ \varepsilon_{x'}\ \varepsilon_{y'}\ \gamma_{x'y'}\ \}^T$ are stress and strain
in the x'-y' coordinate system, respectively. E is Young's modulus,
ν is Poisson's ratio, and m is a mechanical constant representing
anisotropy of material characteristics. If $m = 1/2\,(1+\nu\,)$, the equation
becomes identical to that of ordinary isotropic elastic material. We
define the constant m as an "anisotropic parameter". Once the constitu-

tive equation in the local coordinate system is known, it is easy to extend it to the x - y global coordinate system, that is,

$$
\left\{
\begin{array}{c}
\sigma_x \\
\sigma_y \\
\tau_{xy}
\end{array}
\right\}
= [\, D\,]
\left\{
\begin{array}{c}
\epsilon_x \\
\epsilon_y \\
\gamma_{xy}
\end{array}
\right\}
\tag{3}
$$

where

$$
[\, D\,] = [\, T\,]\, [\, D'\,]\, [\, T\,]^T
\tag{4}
$$

$$
[\, T\,] =
\begin{bmatrix}
\cos^2\alpha & \sin^2\alpha & -2\sin\alpha\,\cos\alpha \\
\sin^2\alpha & \cos^2\alpha & 2\sin\alpha\,\cos\alpha \\
\sin\alpha\,\cos\alpha & -\sin\alpha\,\cos\alpha & \cos^2\alpha-\sin^2\alpha
\end{bmatrix}
\tag{5}
$$

where α is the angle between the x and x' axies and is called the "direction of anisotropy".

4 DETERMINATION OF MECHANICAL CONSTANTS AND INITIAL STRESS

The potential sliding surface is firstly assumed after careful investigation of field observations and measurements, thus, the direction of anisotropy α is a given value. All the mechanical constants and initial stress are then determined so as to minimize the following value δ

$$
\delta = \sum_{i=1}^{N} (u_i{}^c - u_i{}^m)^2 \qquad \longrightarrow \quad \text{min.}
\tag{6}
$$

where

$u_i{}^c$: calculated displacement at measurement point i,
$u_i{}^m$: measured displacement at measurement point i, and
N : total number of measurement points.

For this minimization analysis, computer programs such as Simplex, Rosenbrock, etc., supplied in the computer center, can easily be used.

The computer program referred to as BAPSS has been developed for determining all the mechanical constants and initial stress in cut slope problems.

5 PRACTICAL APPLICATION

5.1 Case study (A)

A cut slope appeared adjacent to the portal of a highway tunnel. The stability of the slope became a serious problem, and therefore, field measurements were performed for monitoring the behaviour of the slope during excavation.

The casing tube for a borehole inclinometer was installed prior to excavation 2m apart from the slope surface and 9m below the floor of excavation, as shown in Fig. 4. The geological formation of the ground consists of horizontal layers of sand and gravel, as shown in Fig. 5.

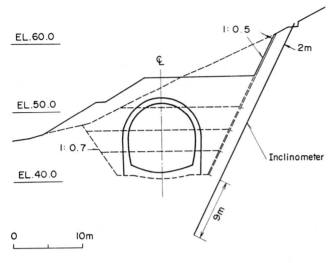

Fig. 4 Configuration of slope and location of inclinometer

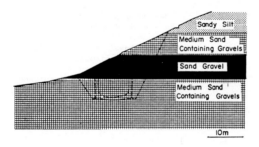

Fig. 5 Geological condition

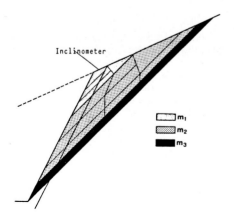

Fig. 6 A family of potential sliding surface

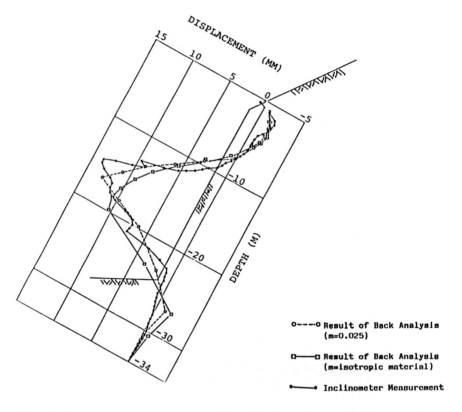

Fig. 7   Comparison between measured and back analyzed displacements

The displacements due to the cutting of the slopes were measured by the inclinometer, and the results were back analyzed to obtain the mechanical constants and initial stress. As mentioned earlier, back analysis requires the configuration of the potential sliding surface. It should be achieved from careful investigation of geological conditions, as well as field measurement results. In this case study, the straight line is assumed, and the ground above this potential sliding surface is divided into three layers as shown in Fig. 6. Each layer may have a different value for the anisotropic parameters. All the material constants, including the anisotropic parameters, as well as the initial stress existing before excavation are then back analyzed from measured displacements. The results are as follows (Kondoh and Shinji, 1986);

$$\sigma_{x0}/E = -0.274 \times 10^{-2} \qquad m_1 = 0.385 \text{ (isotropic)}$$
$$\sigma_{y0}/E = -0.478 \times 10^{-2} \qquad m_2 = 0.385 \text{ (isotropic)}$$
$$\tau_{xy0}/E = -0.113 \times 10^{-2} \qquad m_3 = 0.025$$
$$\nu = 0.3 \text{ (assumed)}$$

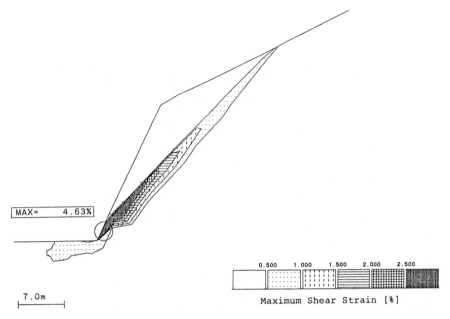

MAX=   4.63%

0.500   1.000   1.500   2.000   2.500

Maximum Shear Strain [%]

7.0m

Fig. 8   Maximum shear strain distribution

where σ_{x0} , σ_{y0} , τ_{xy0} are the components of initial stress acting at
the toe of the slope. Assuming that the vertical component of initial
stress is equal to the overburden pressure, the other components of
initial stress are determined.

Once all these values are known, one can calculate the displacements
caused by the cutting of slopes by means of the finite element method,
and compare them with the measured values to verify the accuracy of the
back analysis. Fig. 7 illustrates this comparison and shows a good
agreement between the measured and calculated values. In this figure,
the isotropic case is also shown for reference. It is seen from this
figure that only a small discrepancy appears between the linear isotropic
and proposed anisotropic calculations. This means that the behaviour of
this cut slope is similar to that of elastic materials. Thus, the slope
is classified as an elastic type. However, the maximum shear strain
distribution shown in Fig. 8 demonstrates that the potential sliding
surface seemingly starts to occur, although it is not too serious.

5.2 Case study (B)

When constructing a tunnel by the cut-and-cover method, much attention
must be paid to the slope stability problem. The case study shown here
is for a railway tunnel constructed in ground consisting of layers of
sedimentary soft rock and detritus deposits, as shown in Fig. 9. The
displacement measurements were carried out through use of inclinometers.
The first readings were taken before the excavation, so that the total
displacements due to excavation could be measured. Using these measurements, back analysis was performed to assess the stability of the slope.

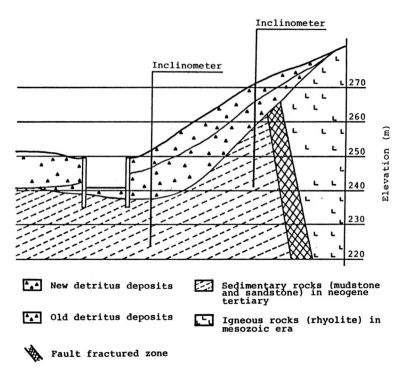

New detritus deposits

Old detritus deposits

Fault fractured zone

Sedimentary rocks (mudstone and sandstone) in neogene tertiary

Igneous rocks (rhyolite) in mesozoic era

Fig. 9   Geological condition

Results of one back analysis are as follows;

$$\sigma_{x0}/E \; = \; -0.871 \times 10^{-2}$$
$$\sigma_{y0}/E \; = \; -0.148 \times 10^{-2}$$
$$\tau_{xy0}/E \; = \; 0.473 \times 10^{-3}$$
$$\nu \; = \; 0.3 \text{ (assumed)}$$

$$m_1 = 0.01$$
$$m_2 = 0.05$$
$$m_3 = 0.1$$
$$m_4 = 0.004$$

$$m_5 \; = \; 0.006$$
$$E_L/E \; = \; 400$$
$$E_B/E \; = \; 0.01$$

This is for the final phase of excavation and is used as input data for the analysis of displacements by the ordinary finite element method.

The back analyzed displacements are compared with the measured values. Fig. 10 shows this comparison and indicates a good agreement between the two.

The maximum shear strain distribution is given in Fig. 11, and it is obvious that there is a clear potential sliding surface occurring along the layer of sedimentary rocks. Fig. 11 also reveals the fact that a large strain occurs at the floor of excavation. Thus, it has been recommended that floor concrete be placed, and that rock anchors be installed in a downward direction so as to prevent a heave of the ground. These support measures will increase the stability of the slope tremendously.

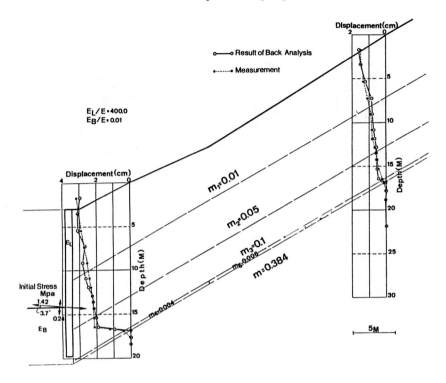

Fig. 10 Comparison between measured and back analyzed displacements

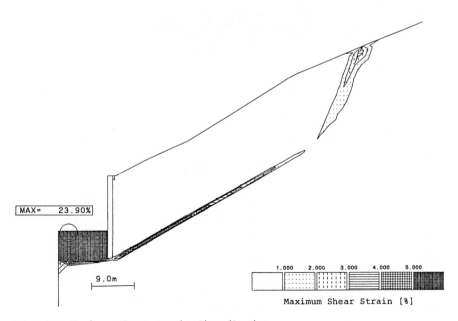

Fig. 11 Maximum shear strain distribution

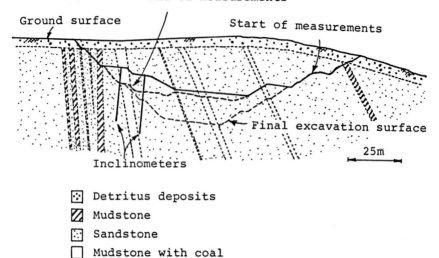

Fig. 12   Configuration of slope with geological condition

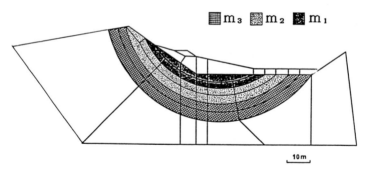

Fig. 13   Potential sliding surfaces

5.3   Case study (C)

The cut slope in concern is one appearing due to the construction of a
highway. The ground in which the slope is cut, consists mainly of sandstone with mudstone layers. Some of the layers contain coal. The geological formation of the ground is shown in Fig. 12.

   In order to assess the stability of the slope, two inclinometers were
installed to measure the displacements due to excavation. The configuration of the cut slope and the location of the inclinometers are also shown
in Fig. 12.

   The potential sliding surface is assumed to be of a circular shape.
Considering the inclinometer measurement results, however, it is not a
single surface, but constitutes a family of surfaces as shown in Fig. 13.

One of the results of back analysis performed during excavation is given as follows;

$$\sigma_{x0}/E = -0.467 \times 10^{-2} \qquad m_1 = 0.018$$
$$\sigma_{y0}/E = -0.144 \times 10^{-2} \qquad m_2 = 0.012$$
$$\tau_{xy0}/E = 0.268 \times 10^{-3} \qquad m_3 = 0.047$$
$$\nu = 0.3 \text{ (assumed)}$$

Fig. 14 indicates the comparison between the back analyzed and measured displacements, and shows an extremely good agreement between the two. It is obvious from these results that the behaviour of this cut slope is of a toppling type. The maximum shear strain distribution is then determined and shown in Fig. 15.

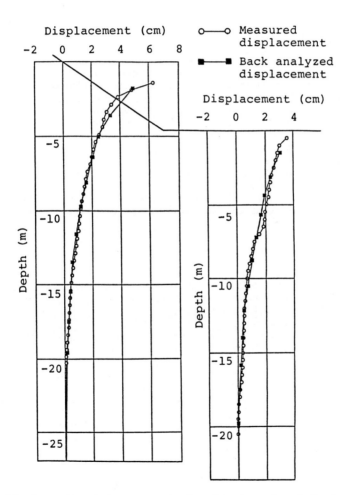

Fig. 14 Comparison between measured and back analyzed displacements

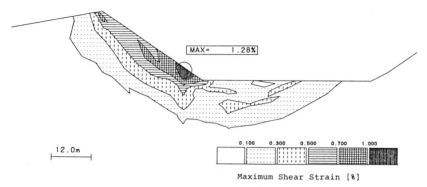

Maximum Shear Strain [%]

Fig. 15   Maximum shear strain distribution

6   CONCLUDING REMARKS

In this paper, a back analysis method is proposed for determining the mechanical constants and initial stress from the displacement measurements conducted during the cutting of slopes. The method is based on continuum mechanics, and thus, it is easily applicable to practical engineering problems.

According to the proposed method, three different types of deformations in cut slopes, that is, elastic, sliding, and toppling behaviours can be analyzed by using a single mechanical model and changing the anisotropic parameters. This means that the type of deformation can be assessed by this back analysis method.

When determining mechanical constants and initial stress, one can calculate a strain distribution which reveals the mechanism of deformation for the cut slopes.

ACKNOWLEDGEMENTS

The author is grateful to Dr. M. Shinji and Mr. Maekawa for their help on back analysis in the case studies. Thanks are also due to Ms. H. Griswold for her proofreading and typing.

REFERENCES

Cundall, P.A., M. Voegele and C. Fairhurst, 1975. Computerized design of rock slopes using interactive graphics for the input and output of geometrical data. 16th US Symposium on rock mechanics, University of Minnesota, Pub. ASCE.
Kawai, T. 1980. Some considerations on the finite element method. Int. J. Num. Meth. Engng., 16, pp. 81-120.
Deeswasmongkol, N. and S. Sakurai, 1984. Study on rock slope protection of toppling failure by physical modelings. 26th US Symposium on rock mechanics, Rapid City, vol. 1, pp. 11-18.
Kondoh, T. and M. Shinji, 1986. Back analysis of assessing for slope stability based on displacement measurments. Proc. Int. Sympo. Engineering in complex rock formations, pp. 809-815.

2nd International Symposium on Field Measurements in Geomechanics, Sakurai (ed.)
© *1988 Balkema, Rotterdam. ISBN 90 6191 778 6*

Open excavation with the aid of rockbolts and shotcrete

Shunsuke Sakurai
Kobe University, Japan

Toshihiko Okamoto & Akira Nakano
Kobe Local Road Public Corporation, Japan

Koichi Ono
Konoike Construction Co. Ltd, Osaka, Japan

1. INTRODUCTION

It would be possible to excavate a section of the ground by applying
rockbolts and shotcrete on the newly excavated face. Excavation using
this type of earth retaining wall (called an RS wall) would have the
following merits;
(1) Total excavation period could be reduced
(2) Wide working space could be secured
(3) Cost of construction could be reduced
(4) Excavation of ground with a gravel layer could also be done easily.
However, experiences with excavations using an RS wall are very insufficient and almost non-existent here in Japan.
Therefore, a rational design and safety control techniques during
excavation have to be established in order to get the RS wall into general use.
As the first step, a test excavation using an RS wall was performed.
This paper deals with the test excavation, including field measurements, and proposes a method for representing the deformational behavior of the wall.

2. FIELD TEST OF AN RS WALL

2.1 Excavation

Fig. 1 shows the size of the area to be excavated, which is 5 m in
width, about 20 m in length, and 4.1 m in depth.
The rear ground surface on the right hand side of the excavation
area (side B) is paved with asphalt and the other side (side A) is
unpaved. Fig. 2 shows the soil condition of the excavation area.
From the surface, the ground over the clay layer consists of layer of
surface soil, a layer of gravel, and a layer of silt. Fig. 2 also
shows each excavation level.
Construction of the RS wall was conducted in the order of excavation,
shotcreting over laths of steel mesh, and then rockbolting.
It took about 4 to 5 hours to complete each level of excavation,
about 8 hours for each stage of shotcreting, and about 8 hours for
each stage of rockbolting.
The thickness of the shotcrete is 10 cm, and the length of the rock-

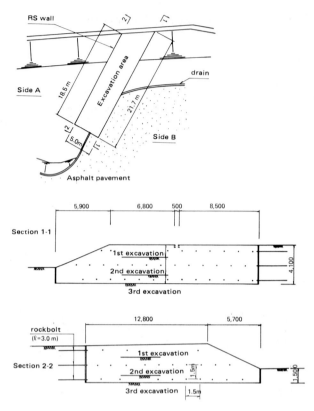

Fig. 1   The size of excavation and the RS wall

bolt is 3 m.   Rockbolts were placed 1.5 m apart from each other.   Fig. 3 shows the construction site of the test excavation.

2.2 Loading test.

A loading test was done by constructing a gravel embankment on side A as shown in Fig. 4.
   The embankment was made in 2 days, and was divided into 4 levels. The first level was up to 0.5 m, the second up to 1.0 m, the third up to 1.5 m, and the fourth up to 2.4 m.
   The first level of the embankment was begun 36 days after completion of the excavation.
Fig. 5 shows the completed bank.

2.3 Field measurements

Field measurements on settlement and lateral displacements of the rear ground surface, lateral displacements of the wall, and axial force developed in the rockbolts were performed during the excavation and banking.   Fig. 6 shows the measurement points and Table 1 shows the measured items and the instruments used.

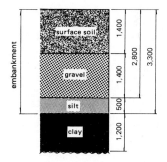
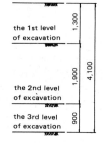

Fig. 2   Soil condition

Fig. 3   Construction site of the excavation

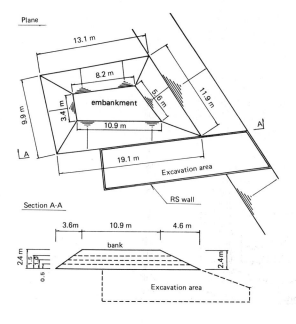

Fig. 4   Loading test

Table 1   Items of the measurements and the instruments used

item	instrument	No. of points
settlement of the ground level	level	36 points
lateral displacement of the ground level	steel tape	36 points
lateral displacement of the wall	inclinometer	2 bored holes
axial force of rockbolt	wire strain gauge	3 rockbolts

Fig. 5 After completing of the last stage of construction

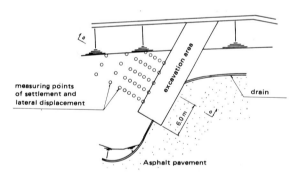

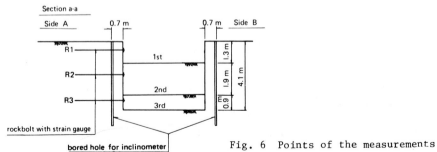

Fig. 6 Points of the measurements

2.4 Test results

Fig. 7 shows the lateral displacements of the wall and the settlement of the rear ground surface with respect to each excavation stage. Fig. 8 shows the difference in lateral displacements at the wall top between sides A and B.

Table 2 shows the maximum value of the settlement and the lateral displacements of the rear ground surface, and the settled area at each stage of excavation.

Table 2 also includes the average value of lateral strain of the ground surface within points a and b shown in Fig. 7.

Fig. 9 shows the axial force developed by rockbolts during the excavation.

Fig. 10 shows the lateral displacement of the wall occuring at each stage of banking.

Fig. 11 shows the variation in axial forces of the rockbolts by loading.

These observations indicate the following points:

(1) The lateral displacements of the RS wall of side A reached 6.1 mm at the third stage of excavation, and had extended to 9.4 mm by 36 days after completion of the excavation without a clear conversion. On the other hand, that of side B reached only 2.8 mm at the third stage of excavation and converged to 4.6 mm at about 7 days thereafter.
      The difference in the magnitude of lateral displacement as well as the mode of the distribution between sides A and B might be due to the presence or absence of an asphalt pavement.

(2) The lateral displacement due to the formation of the embankment was forced to increase by another 6.0 mm. However, the displacement converged within 13 days after the entire loading by only increasing another 2.0 mm, and thus, the RS wall remained stable.

(3) The lateral strain of the ground surface near the wall top reached a tension level of about 0.4 % after completion of the excavation. However, no cracks were observed on the surface of the rear ground.

(4) Settlement of the rear ground surface due to excavation occurred up until the region whose width was almost the same as the excavation depth, and whose mode of settlement was almost the same as that of the lateral displacement of the wall.

(5) Axial force introduced in the rockbolt by the excavation was only 1.5 tons in tension, although its allowable value is 12 tons. Moreover, its variation due to the banking remained without a limited value although the axial force in the rockbolts tended to compress when the banking was increased.
      These observations, including the axial force distribution, infer that the allocation and length of the rockbolts adopted here were suitable. However, the depth of the excavation was not large enough to cause an increase in the axial force of the rockbolts.

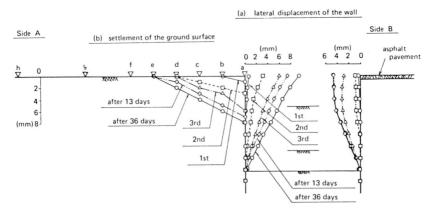

Fig. 7 Lateral displacement of the wall and
settlement of the rear ground surface

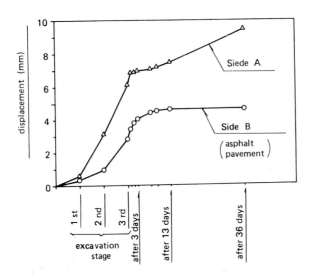

Fig. 8 Lateral displacement at the wall top

3. BACK ANALYSIS OF MEASURED DISPLACEMENTS

3.1 Method

Mechanical properties and initial stress of the ground can be evaluated by analyzing the deformational behaviour of the ground during excavation. This analysis is just the reverse calculation of ordinary stress analysis. Thus, it is called "back analysis". Sakurai, et al have proposed a back analysis method for determining the mechanical

Table 2   Ground movement and the maximum value

Excavation stage	1st		2nd		3rd		36 days after 3rd excavation	
	max.	x	max.	x	max.	x	max.	x
settlement of the ground surface	1 mm	1 m	4 mm	2 m	6 mm	4 m	8mm	5 m
lateral displacement of the ground surface	1 mm	1 m	4 mm	3 m	7 mm	4 m	10mm	4 m
lateral strain of the ground surface	0.1 %		0.3 %		0.4 %		0.4 %	
depth of excavation z	1.3 m		3.1 m		4.1 m		4.1 m	

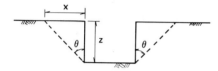

x : settled region in the ground surface by excavation

z : depth of excavation

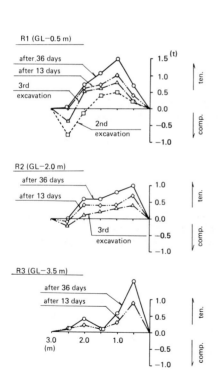

Fig. 9 Axial force of rockbolts generated by excavation

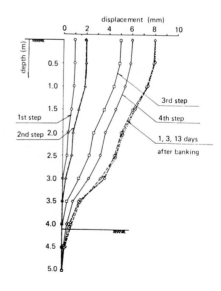

Fig. 10 Lateral displacement of the wall by loading

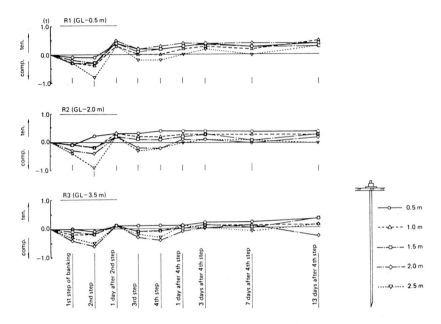

Fig. 11 Variation of axial force in rockbolts due to loading

constants and initial stress from measured displacements in cut slopes
(Sakurai, Deeswasmongkol and Shinji 1986, Sakurai, 1987).

Since a detailed description of the method has been presented elsewhere, only a brief summary is given here.

The back analysis method is based on continuum mechanics and is formulated with the finite element method. Discontinuous behaviour of
the ground is modelled by proposing an anisotropic constitutive equation. In this constitutive equation, the anisotropic parameter m and
the direction of anisotropy α are introduced to represent the discontinuous behaviour of the ground.

If $m = 1/2(1+\nu)$, the constitutive equation becomes identical to that
for ordinary isotropic elastic materials, and if $m < 1/2(1+\nu)$, the
shear rigidity of the ground is decreased along the particular direction represented by the orientation angle α.

It has been demonstrated that the proposed constitutive equation covers most of the aspects of deformational behaviour of the rear ground;
that is, elastic, sliding, and toppling types of behaviours, by changing the magnitude of the material constants (Sakurai, 1987). Non-homogeneity of the ground can be taken into account by subdividing it
into a number of regions, of which mechanical constants are different
from each other.

In the process of determining the material constants and initial stress,
stress and strain occurring in the ground due to excavation can also be
calculated.

This stress and strain can provide sufficient information for monitoring the stability of slopes in the rear ground.

3.2 Results of back analysis

The lateral displacement of the wall, as well as the settlement of the ground surface of side A at Section a-a, are used as input data for the back analysis described in the previous section to determine the initial stress and mechanical constants.

The results obtained by the back analysis are as follows,

$$\sigma_{xo}/E = -0.201 \times 10^{-3}$$
$$\sigma_{yo}/E = -0.362 \times 10^{-4}$$
$$\tau_{xyo}/E = 0.593 \times 10^{-4}$$

Anisotropic parameter $m = 0.05$
Direction of anisotropy $\alpha = 45°$
Poisson's ratio $\nu = 0.35$

where σ_{xo}, σ_{yo}, and τ_{xyo} are the horizontal, vertical, and shear stress, respectively which existed initially at the bottom corner of the exca- vated level before excavation. E denotes Young's modulus of the ground below the potential sliding plane as shown in Fig. 12. Assuming the horizontal component of initial stress as,

$$\sigma_{xo} = 0.5 \ \gamma H \ (\gamma: \text{unit weight of the ground,}$$
$$\text{H: overburden height)},$$

Young's modulus is obtained as

$$E = 1990 \ \text{kg/cm}^2,$$

and the shear modulus along the potential sliding plane as

$$G = mE = 100 \ \text{kg/cm}^2$$

These back analyzed initial stress and mechanical constants are used as input data in ordinary finite element analysis to calculate the dis- placement and strain caused by the excavation.

The calculated lateral displacement of the wall and the settlement of the ground surface are shown in Fig. 12 together with the measured val- ues. The figure shows that there is a good coincidence between them and that the deformational behaviour of the wall is of the toppling type.

Fig. 13 shows the maximum shear strain distribution. The maximum value is about 0.55 % occurring near the bottom corner of the excavated level of Side A.

The stability of the wall can be assessed by comparing the maximum shear strain with the critical strain of the ground, which is approxi- mately 0.75 % of that obtained in laboratory tests.

4. CONCLUDING REMARKS

Open excavation of the foundation with the aid of rockbolts and shot- crete was found to be possible through this test excavation and the loading test, although the excavation was relatively small in scale.

The difference in the mode and magnitude of the wall deformation with or without pavement on the rear ground surface indicates that the strengthening of the rear ground by, for example rockbolts and shot- crete, would increase the stability of the RS wall.

A good correlation between the measured value of the wall deforma- tion and the calculated one indicates that the proposed method of de-

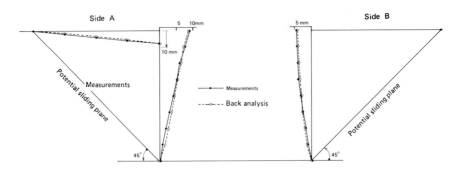

Fig. 12   Comparison between back analyzed and measured displacements

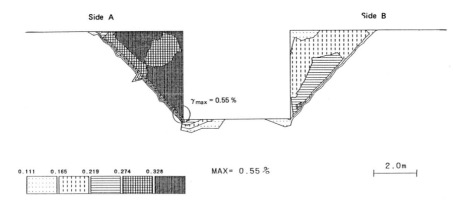

Fig. 13   Maximum shear strain distribution

formational analysis could be extended to evaluate the stability of
the RS wall.
   More field tests under various conditions are desired to establih
both a rational design technique of the RS wall and safety control during excavations.

References.

Sakurai, S., N. Deeswasmongkol and M. Shinji 1986.   Back analysis for
determing material characteristics in cut slopes. Proc. of International Symposium on Engineering in Complex Rock Formations. pp. 770 ∿
776.

Sakurai, S. 1987. Interpretation of the results of displacement measurements in cut slopes. 2nd International Symposium on Field Measurements in Geomechanics, Kobe.

2nd International Symposium on Field Measurements in Geomechanics, Sakurai (ed.)
© 1988 Balkema, Rotterdam. ISBN 90 6191 778 6

Measurement and interpretation of loading tests of concrete top blocks on soft ground

Katsuhiko Arai
Fukui University, Japan
Yuzo Ohnishi & Masakuni Horita
Kyoto University, Japan
Ikuo Yasukawa
Fushimi Technical High School, Kyoto, Japan

1 INTRODUCTION

Most of the flat plain where the large cities in Japan are located in Japan is covered with soft or very soft alluvial deposits. We often face problems in civil engineering when constructing structures on such very soft ground. When a structure is not very heavy, it can be built safely with only proper improvements (without piles) to the nearsurface and/or subsurface ground. Engineers must decide what improvement method is the best for construction, judging by cost performance, reliability and handling capability.

Recently a new method has been invented which assembles the top-shaped concrete blocks and places them on the ground. A group of these top blocks is used as a shallow foundation replacing the short piles.

A schematic diagram of the top base method is shown in Fig. 1, where the top shaped blocks are placed on crusher-run, which is spread over the soft ground. To locate the position of an individual top block and to reinforce a group of top blocks, a lattice of iron rods is placed on the crusher-run. It has been reported that several real structures constructed with this new method show the drastic improvements of reducing settlement and increasing bearing capacity of the structures. However, no one has investigated why this happens.

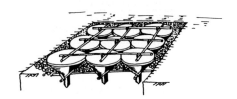

Fig.1 A View of Top Base Foundation

This paper investigates the behavior of the top base foundation on a soft ground. The top base method is believed to modify and improve the settlement and the bearing capacity of the foundation ground. In-situ plate loading tests have been performed to varify the differences between several foundation methods. In addition, laboratory model tests have been conducted to study the behavior of the foundation ground when subjected to the load applied on the top base blocks.

2 FIELD TESTS

2.1 Test site

In-situ plate loading tests have been conducted at two sites in Chiba Prefecture near Tokyo. One (No.1) is a short term ordinary plate loading test, and the other one (No.2) is the long term plate loading test, in which the test continued for two years at a certain load level and the settlement of the foundation due to consolidation was measured.

The properties of soils sampled at the sites is shown in Table 1. The geologic conditions are very similar at both sites where organic clay

Table 1 Properties of Soil

	Items	In Situ Test	Plate Loading Test	Long Term Plate Loading Test
Physical Properties	w_n (%)		137.0	125.6
	w_l (%)		128.2	120.6
	w_p (%)		63.3	55.8
	I_p		64.9	64.3
	Sand (%)		17.0	10.0
	Silt (%)		41.0	58.0
	Clay (%)		42.0	32.0
	G_s		2.538	2.579
Wet Density			1.333	1.358
Japanese Unified Soil Classification System			OH	OH
N Value			0	0
Uniaxial Compression Test	Compressive Strength q_u (kgf/cm^2)		0.11 0.09 (Average 0.10)	0.115 0.09 (Average 0.10)
	Modulus of Deformation E_{50} (kgf/cm^2)		3.8 3.3 (Average 3.55)	3.0 2.8 (Average 2.90)

is deposited at the backside of a river pond. The soft ground has uniform arealextend and the soft clay is deposited up to 3m below the ground surface. Undistributed samples were taken from the soil layer at between 0.8m and 1.3m below the surface. The physical properties may be closely related to the behavior of the foundation at the plate loading test. At the No.2 test site, in addition to physical tests, oedometer tests were performed. The results of these tests are as follows:

Initial void ratio   e = 3.270
Coefficient of consolidation Cc = 1.26
Pre-consolidation pressure   pc = 0.245 kgf/cm2

2.2 Experimental methods

At both sites, a 1m x 1m square concrete block (10cm thick) was set on the foundation. 7 types of foundations used in this series of experiments are shown in Fig.2.

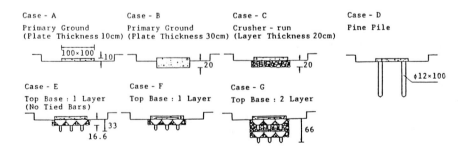

Case - A
Primary Ground
(Plate Thickness 10cm)

Case - B
Primary Ground
(Plate Thickness 30cm)

Case - C
Crusher - run
(Layer Thickness 20cm)

Case - D
Pine Pile

Case - E
Top Base : 1 Layer
(No Tied Bars)

Case - F
Top Base : 1 Layer

Case - G
Top Base : 2 Layer

Fig.2 Foundation Types for Loading Tests

(1) Plate loading tests

All 7 tests shown in Fig.2 were performed. A hydraulic jack was used to load on the foundation, starting from 0.4 tf/m2 to 2.4 tf/m2 with an increment of 0.4 tf/m2. At each loading step, settlement was recorded for 30 minutes and thereafter, the test proceeded to the next load step.

(2) Long term consolidation tests

5 among the 7 tests shown in Fig.2 were performed without Case-B and Case-E. 1m x 1m concrete blocks of 0.25tf each were put one on top of another to attain the prescribed consolidation load. The loading was done at 5-minute intervals and an initial settlement was measured at each step. A standard consolidation pressure was set to be 2.75tf/m2. However, in Case-A (without any ground improvement) the loading pressure was limited up to 2.25tf/m2 in order to avoid a sudden failure. Similarly in Case-C (with crusher-run only), the load of 2.5tf/m2 was applied. The total settlement was divided into the immediate and consolidation settlements. The consolidation set-tlement is obtained by subtracting the settlement measured at the loading stages from the total (final) settlement caused by long term loading.

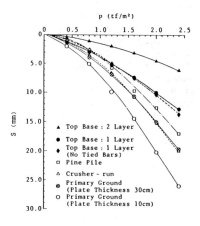

Fig.3 Applied Pressure-Settlement Relationships in Plate Loading Tests

3 TEST RESULTS

3.1 Plate loading tests

The purpose of a series of plate loading tests was to compare the bearing capacities of the ground with different types of foundations. Fig.2 shows 7 different foundations on which plate loading tests were conducted. Fig.3 shows the load-settlement relationships of these foundations. At the same load, the top base foundations show less set-tlement than the others.

The load-settlement curves in Fig.3 show the tendency of a soft clay ground which has local shear failure. However, the load-settle-ment curves for the top base founda-tion show possibilities of having general shear failure under higher pressure because they display only a small settlement. Generally, an ultimate bearing capacity is pro-posed to be 1.5 times that of the yield stress if the yield stress is determined from the load-settlement curve. The logarismic plots of the applied load versus settlement are tried in Fig.4 because it was very difficult to determine the yield

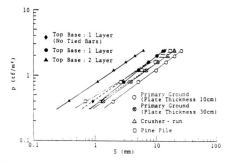

Fig.4 Logarismic Plot of Applied Pressure-Settlement Curves of Plat Loading Tests

stresses from the curves in Fig.3. A yield stress is determined at the crossing point of two characteristic lines with different slopes in a logarismic plot of applied load versus settlement. Unfortunately, Fig.4 does not show any clear yield stress because the applied stress was too small.

Practically speaking, on the other hand, an applied load-settlement curve can be fitted by a hyperbolic curve which has two asymptotes. One of the asymptotes is $p=p_f$, where p_f is the ultimate bearing capacity. The hyperbolic fitting is applied to cases B, C, E, and F shown in Fig.2. The results are presented in Fig.5, and show fair fitting. From this fitting method, the ultimate bearing capacity of the top base foundation is larger than that of the other

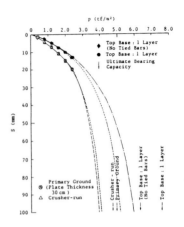

Fig.5 Hyperboric curve Fitting to Applied Pressure-Settlement Curves

foundations. Since the allowable bearing capacity is generally one third of the ultimate bearing capacity, a higher allowable bearing capacity will be expected in the installation of the top base. Judging from Fig.5, the allowable bearing capacity of the top base foundation is expected to be 1.5 times larger than that of the primary ground without any treatment. The allowable bearing capacity of the top base foundation estimated from the curve in Fig.5 is twice as large as that of this primary ground calculated based on the Terzaghi's equation with the material properties shown in Table 1. Such kinds of comparison will require more studies under different conditions.

3.2 In-situ consolidation tests

Fig.6 shows the time-settlement curves of the foundation. Cases A, C, D, F, and G shown in Fig.2 are under a constant load. The applied load is 2.2 tf/m2 on the primary ground, 2.5 tf/m2 on the crusher gravel, and 2.75 tf/m2 on the pine pile foundation and on the top base foundations. The top base foundations show less settlement than the other foundations. In particular,, the two-layer top base foundation exhibits the most effect in reducing settlement. The arrows in Fig.6 show the points at the end of the primary consolidation. They show that the top base foundations finish primary consolidation much faster than the other foundations, and that the top base foundations have smaller secondary compression and smaller coefficients of secondary compression than the other foundations.

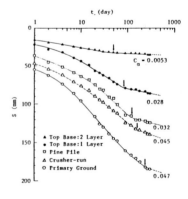

Fig.6 Settlement-Time Curves for In-situ Consolidation Tests

In summary, the top base foundations exhibit the effect of preventing the settlement of the soft clay ground. They have smaller primary consolidation and secondary compression than the other foundations. Also the top base foundations help to finish the primary consolidation faster.

4 DISCUSSION ON SETTLEMENT CONTROL

4.1 Settlement control by top base foundation

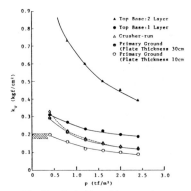

Fig.7 Relationships between Coefficient of Subgrade Reaction and Applied Pressure

The coefficient of subgrade reaction, Kv, is an index to characterize a load-settlement relationship in a plate loading test. Fig.7 summarizes kv of the foundations shown in Fig.2. The area hatched in Fig.7 (Kv=0.178 - 0.250 kgf/cm2) is the range calculated from the data of unconfined compression tests listed in Table 1. Generally, the coefficient of subgrade reaction is determined from the data under a pressure of one third that of the ultimate bearing capacity. Fig.7 shows a good matching between the calculated Kv from the data of unconfined compression test and the Kv of the primary ground determined from the plate loading test. Fig.7 also shows that the top base foundation decreases ground settlement clearly.

The comparisons in settlement of different foundations are shown in Table 2. In this table the settlement values are interpolated at an applied pressure of 2.75 tf/m2 with the hyperbolic fitting method. Because the in-situ consolidation test of the primary ground was conducted under a pressure of 2.25 tf/m2, the settlement value of the primary ground in the consolidation test is estimated at an applied pressure of 2.75 tf/m2 using the coefficient of volume compressibility, mv, obtained from a laboratory consolidation test. The comparison of settlements in Table 2 shows that the immediate settlement of the one-layer top base foundation is 1/2-1/3 , and the consolidation settlement is 1/3 of those of the primary ground. The two-layer top base foundation shows more effective settlement control, 1/3-1/4 in immediate settlement, and 1/9 in consolidation settlement of the primary ground.

Table 2 Comparison in Settlement of Different Foundations based on Primary Ground

In Situ Test / Type of Foundation	Plate Loading Test		Long Term (Consolidation) Plate Loading Test			
	Immediate Settlement		Immediate Settlement		Consolidation Settlement	
	Settlement	Rate to Primary Ground	Settlement	Rate to Primary Ground	Primary Consolidation	Rate to Primary Ground
Primary Ground (Plate Thickness 10cm)	3.35 cm	1	3.95 cm	1	19.30 cm	1
Primary Ground (Plate Thickness 30cm)	2.75	0.82 (1/1.2)	———	———	———	———
Crusher - run	2.75	0.82 (1/1.2)	4.11	1.04	9.66	0.50 (1/2.0)
Pine Pile	2.25	0.67 (1/1.5)	2.31	0.58 (1/1.7)	9.56	0.50 (1/2.0)
Top Base : 1 Layer (No Tied Bars)	1.75	0.52 (1/1.9)	———	———	———	———
Top Base : 1 Layer	1.60	0.48 (1/2.1)	1.33	0.34 (1/3.0)	6.57	0.34 (1/2.9)
Top Base : 2 Layer	0.80	0.24 (1/4.2)	1.26	0.32 (1/3.1)	2.14	0.11 (1/9.1)

(Each Settlement is at an applied pressure of 2.75 tf/m²)

4.2 Mechanism of settlement control by top base foundation

In order to study the mechanism of settlement control by top base foundations, some model tests in a laboratory were conducted. The model ground was prepared with clay in a cylindrical container as shown in Fig.8. The clay had a liquid limit of 40.4 % and a plastic index of 18.4. The clay was mixed in high water content, twice as high as the liquid limit, and was consolidated in the container under a pressure of 0.5 tf/m2 after filled in the container. For the foundation, miniature top shaped blocks (ϕ =6 cm) and crusher-run were set on the ground. Pressure transducers and pore pressure transducers were installed in the clay as shown in Fig.9. The load increment on the foundations was 0.1 kgf/cm2 up to 0.4 kgf/cm2 at a loading interval of 5 minutes, and the load was kept at 0.4 kgf/cm2 for a week.

The surface settled with time as shown in Fig.10. Fig.11 shows the change in pore water pressure at each point. The settlement at t=0 minutes in Fig.10 represents the immediate settlement. The top base foundation shows more immediate settlement than the crusher-run foundation because of the compaction of pores between the top-shaped blocks, filling sand, and the ground. Fig.10 also shows that the consolidation settlement of the top base foundation is

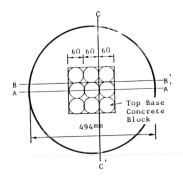

Fig.8 Plan of Top Base Foundation for Laboratory Loading Tests

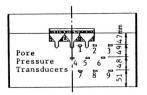

Fig.9 Location of Pore Pressure Transducers in Laboratory Loading Tests

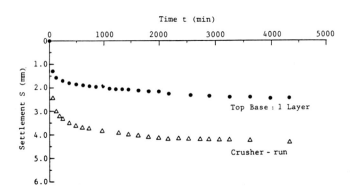

Fig.10 Comparison in Surface Settlement of Laboratory Consolidation Tests

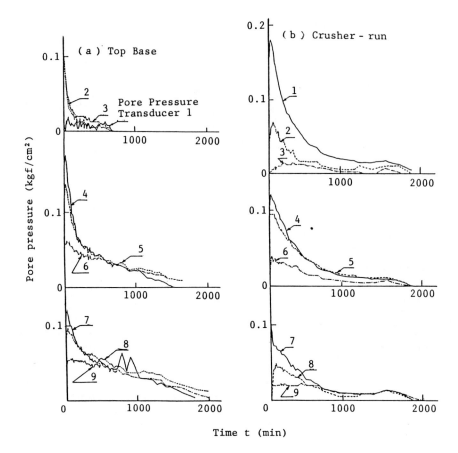

Fig.11 Pore Water Pressure Measured
in Laboratory Loading Tests

half that of the crusher-run foundation. The pressure transducer measurement shows only a small difference between the top base foundation and the crusher-run foundation. In contrast, Fig.11 clearly shows the difference in distribution of pore pressure between both foundations. The difference in the pore pressure at measurement point No.1 in Fig.9 is caused by the lateral deformation under the foundations due to dilatancy. The top base foundation prevents the lateral deformation under it, with a good interaction between the axes of top-shaped blocks and the filling sand. In short, the top base foundation causes little dilatancy under the base, generates little amount of excess pore water pressure due to lateral deformation, and the surface settlement is prevented in small. On the contrary, the crusher-run foundation causes a large amount of lateral deformation, and more excess pore water pressure generation is expected.

Fig.12 shows a comparison in consolidation settlements analyzed by a finite element method with the constitutive model of soil proposed by Sekiguchi and Ohta. The constitutive model can take dilatancy of soil in consideration. In this figure, the consolidation settlement of the top base foundation is almost half that of the primary ground. Therefore, it is very important to consider the effect of dilatancy due to lateral deformation in the mechanism of settlement control by top base foundations. The top base foundations in a good interaction

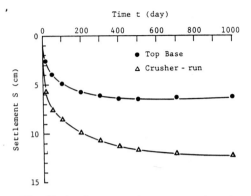

Fig.12 Surface Settlements Analyzed by F.E.M.

with filling gravel, restrain the deformation of ground in preventing the lateral deformation of clay under it, and control the settlement of the ground.

5 CONCLUSION

The top base foundations installed in soft ground are very effective in settlement control of the ground. In-situ plate loading tests show that the one-layer top base foundation has 1/2-1/3 the settlement of the primary ground in the immediate settlement, and 1/3 in the consolidation settlement. Further improvements are observed in the two-layer top base foundation. The top base foundations cause little dilatancy under the base, generate little amount of excess pore water pressure due to lateral deformation, and prevent the surface settlement.

ACKNOWLEDGEMENT

We would like to thank the MAIKOMA Manufacturers Association for the In-situ tests.

REFERENCES

Arai, K. et al. 1986. Settlement Control of Soft Ground By Top Base Foundation(in Japanese). Symposium on Lateral Deformation of Ground. JSSMFE: pp.111-114.
Japan Highway Association 1984. Manual on Highway Bridges-foundations (in Japanese). JHA: pp.193-194.
The Japanese Society of Soil Mechanics and Foundation Engineering, 1983. Manual of Plate Loading Tests (in Japanese). JSSMFE.
The Japanese Society of Architectual Engineering, 1974. Design Manual of Architectual Engineering(in Japanese). JSAE: PP.228-234.
Sekiguchi, H. & Ohta, H. 1984. Induced Anisotropy and Time Dependency in Clay. Proceedings of 9th ICSMFE, Specially Session No.9, Constitutive Equation of Soils: pp.229-238.
Yamaguchi, H. 1984. Soil Mechanics-reviced edition(in Japanese). Gyhodo: pp.254-255.

The normalization and compatibility of the micro-macro parameters in the laboratory- in situ comprehensive measurements

Tsui Zheng Chuan
Institute of Geotechnique and Survey, Yangtze, WREP, People's Republic of China

INTRODUCTION

It is a basic and important problems for geotechnique, though not solved well, that how to judge geological condition rightly in order to choose proper design parameter for reality of engineering.

As the history of development of Rock Mechanics tells us,from the end of fifties to seventies, the design parameter for engineering had been mainly obtained by both Lab. and In-situ Rock Mechanical test.

During the period, the changes had been lacking place gradually in following aspects: from small sample Lab. test to large scale in-situ test the main test became;from Yield and Peak strength to the study of whole course of Stress-Strain including Residual Strength the attention was directed; from instant Strength to Duration Strength concerned with course of Creep Deformation and from static strength to Dynamic strength the study scope expanded; and the result of Multisample tests was sub-stituded by that of Mono-sample Multi-Stage test which undergoes several periods of loads.

With the concideration of that the representative of sample and the reliabilities of the result can not be ensured, there is no final conclusion yet about how to choose the parameter conforming with the reality of structure, though a great number of both Lab. and In-situ tests have arranged and completed.

In the end of seventies, Kovari and SAKURAI took a reprsentative role to present so called Direct Back Analysis Method (DBAM) wich estimates the Macro Mechanical parameter Er and Initial stress component $\{\sigma_g\}$ based on the Displacement of Rocks $\{\delta_r\}$ as a main input information.

Since this method is developed, a new way has been created to choose design parameter for engineering. Moreover, it has more importance that the evaluation of mechanical parameters(mainly strain $\{\varepsilon f\}$) during the construction of underground openings is just the safety management process of the construction.

Thereby, the Back Analysis methods has been a great push to the development of Rock Mechanics.

But on the other hand, taking practice of Dam technique for an example, the result of preliminary design is the only basis of to work out a correct and feasible plan of construction, and therefore should be as conformable as possible to reality.

In the point of view of design, what are required are mechanical and hydraulical parameters of Dam foundation.

As everyone knows, the field of Additive Stress concerned in stability analysis is so vast as more than $100,000m^2$ in plan problem. In such a giant geological domain, it is very difficult to either choose representative samples or arrange displacement measurements.

Concerning the problems mentioned above, this paper applies the principle of Systems Engineering Geology(SEG for short) to discuss the method to choose design parameters for engineering from results of both Lab. and In-situ comprehensive measurements, and the principle of decision of foundation surface.

1. Brief philosophy of SEG

SEG consists mainly of Geological Mathematical Model GMM and Optimizing Design OPD.

GMM is the base of OPD, and the core of preliminary Design. In order to establish GMM, it is necessary at first to produce geological unit GLU and object parameter units OPU.

Therefore, the engineering exploration and Rock mechanical test and In-situ comprehensive measurements are carried out mainly for establishing GLU and OPU.

2. The Method to Establish of GLU

In tact, GIU means Rock type, Faults, Fissures, Karst cave and various geological interface located in additive stress field or in seepage field. They are discovered and determined by multiplex ways such as: the surface survey, geophysical prospecting, drilling and aditting.

However, because exploration points are limited, the geological units between boreholes can be deduced our only from statistical analysis. Consequently, in rigorous word, the geological structure model GSM is merely a sort of probability model ?BM.

To set up a connection between GFM and OPU, a geophysical parameter model MCM should be worked out on the basis of geophysical characters (called physical parameter) of all geological units.

The relationship between GSM and GPM and that between GSM and MCM are shown as figure 1.

3. The Method to compose Object Parameter Unit OPU

For the sake of convenience in SEG, the parameters from both Lab. and In-situ measurements are all called Micro Parameters MIP, while the Parameter units in GMM are called Object parameter or Macro Parameter units OPU.

The Object Parameter can be sorced into Object Parameter for Design OPUD and Object Parameter for prediction OPUP.

The contents of OPUD mainly are Elastic Modulus of Rock Mass E_R, Stiffness Coefficient of Joint unit K_n and K_s Parameters of Shear Strength C_R and ϕ_R and permeability Coefficient K_R While OPUP is the parameter for Optimizing Design and Safety inspection while mainly are real Displacement $\{\delta_R\}$, Critical Strain $\{\varepsilon_0\}$ or Failure Strain $\{\varepsilon_f\}$ and Variation of groundwater Head $\{\Delta H\}$.

In this section of the chapter, the main discussion is on the method to compose OPUD.

Traditionally, the Design parameter of geotechnique is the Static Elastic Modulus obtained by the In-situ Statical Loading test in which the loading plate diametre is D=30cm.

But what are easier to gain are the Elastic Modulus of Borehole wall, and Dynamic Elastic Modulus obtained by means of seismic prospecting.

More will occure a problems that how to Normalize parameters from different sources (E_S, E_B, E_D, ETC.) into Design parameter E_{30}.

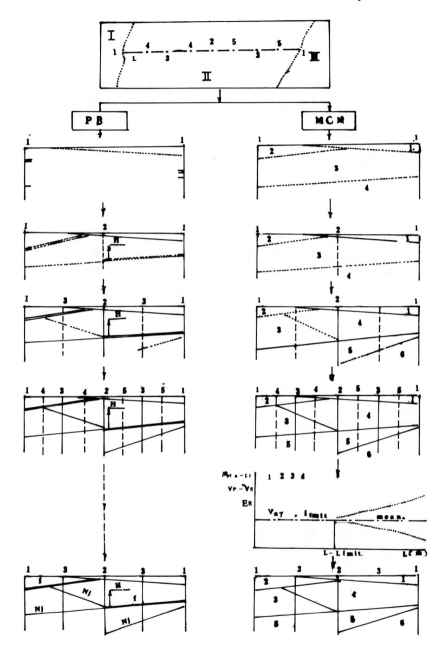

fig 1. The process of composing PBM and MCM in SGM

(1). The Compatibility between Micro and Macro Parameters

As shown in figur 2,all sort of Rock are classified,regarding their initial Elastic Modulus as basic parameter, into following four kinds ($E_s . 10^4 kg/cm^2$):

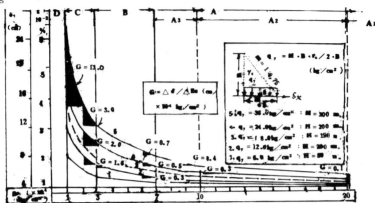

fig 2. Classification of Rock based on the Parameter E_s (simulate gravity Dam's loading conditions)

A   A1(E_s>20.0); A2 (E_s=20.0-10.0); A3(E_s=10.0-7.0):
B   (E_s=7.0-3.0):
C   (E_s=3.0-1.0):
D   (E_s<1.0).

Where;

Type A and B (called Strong Rock):
Stress Level; $\sigma_1 < \sigma_2 < .. < \sigma_i < .. < \sigma_y < \sigma_p > \sigma_r$
Strain Level; $\varepsilon_1 = \varepsilon_2 = .. = \varepsilon_i = .. = \varepsilon_y = \varepsilon_p = \varepsilon_r$
Modulus Level; E1>E2>..>E1>..>Ey>EP>Er
Type C (called Sort Rock):
Stress Level; $\sigma_1 < \sigma_2 < . < \sigma_i < .. < \sigma_y < \sigma_p > \sigma_r$
Strain Level; $\varepsilon_1 < \varepsilon_2 < .. < \varepsilon_i < .. < \varepsilon_y < \varepsilon_p < \varepsilon_r$
Modulus Level;E1=E2=..=Ei=..=Ey=Ep=Er

(here foot-notation Y,P,R are indicate of Yield, Peak and Residual separately).

Type C, between Type A,B and Type D, is called Transition Rock.
The meaning derived from formulas 1 and 2 is that the Strains gained from intact of A,B, $\{\varepsilon_s\}$ or $\{\varepsilon_{os}\}$ and $\{\varepsilon_{fs}\}$ can be directly used as strain of site Rock Mass $\{\varepsilon_{fR}\}$ or $\{\varepsilon_{OR}\}$ and $\{\varepsilon_{FR}\}$; and that the Deformations E_s gained from intact Rock of D can be used as Deformation Modulus of site Rock Mass E_R.

These relationships are called Compatibilities betweet Micro and Macro Parameters, and that is an important conclusion.

(2). Normalization for Object Parameters
In formulas 1 and 2 , E_s and E_R for Rock or A,B, $\{\varepsilon_s\}$ and $\{\varepsilon_R\}$ for Rock of D are not compatible to each other.

Hence, it is necessary to seek a new way to normalize them and the normalization may only be realized by means of third parameters called Medium Parameter, according to the following path:

a. Velocities of wave V_p and V_S

In the Softening region of Stress-Strain Histories curve are carried out the measurement of modulus E_i ($i=1,2,...n$), Compressional Velocity V_{pi} and Shear Wave V_{si} ($i=1,2,...n$) corresponding to each Stress Level C_i p_i ($i=1,2,...n$).

Then the following formula is established;

$$E_i/E_S = f(V_{pi}/V_{ps}) \qquad (i=1,2,...,n) \qquad\qquad 3$$

$$E_i/F_s = f(V_{si}/V_{ss}) \qquad (i=1,2,...,n) \qquad\qquad 4$$

Then, the mean Velocities V_{PR} and V_{SR} are measured In-situ of Engineering Geological area which is treated as one Unit. The fitting relationships of V_{PR} versus V_{pi} and that of V_{sr} versus V_{si} are analysed.

The Object Parameter E_R is offered by formulas:

$$E_R = f\left(\frac{E_s \cdot V_{PR}}{V_{ps}}\right) = f\left(\frac{E_s \cdot V_{pi}}{V_{ps}}\right) \qquad\qquad 5$$

or

$$E_R = f\left(\frac{E_s \cdot V_{sr}}{V_{ss}}\right) = f\left(\frac{E_s \cdot V_{si}}{V_{ss}}\right) \qquad\qquad 6$$

b. joint Parameter N_j

When Elastic Modulus of E_i is obtained, their Equivalent Modulus E_e ($=E_R$) can be produced by numerical analysis method, as:

$$E_e = E_R = (\alpha, N_j, E_s)^{\beta} \qquad\qquad 7$$

and

$$= f(E_j, N_j) \qquad\qquad 8$$

where α is Coefficient of Strength reduction, β serves as a index of curve. They can be gained through more than two grups of numerical solutions.

c. Parameters of Physical Properties ρ_0, and ρ_R

The filling Resistivity ρ_0 and apparent Resistivity of Rock Masses ρ_R are measured before:

$$E_R = 1.5874\left(\frac{1 - \frac{\rho_0}{\rho_R}}{2 + \frac{\rho_0}{\rho_R}}\right)^{2/3 \cdot C'} \cdot E_s \qquad\qquad 9$$

where C' is a seeker constant which can be obtained by method of parallel measurements.

d. Coefficients of Stiffness K_n and K_s

The parameters K_n and K_s can be obtained through compression tests and shear tests. They also, however, may be calculated by Commom Methods.

1. For Stress Problems in two Dimensions:

$$K_n = \frac{E_J}{1-\dfrac{\nu_j^2}{E_J}}$$

$$K_s = \frac{E_J}{2(1+\nu_j)} = G_j \qquad 10$$

2. For Strain problems in two Dimensions:

$$K_n = \frac{(1-\nu_j)E_j}{(1+\nu_j)(1-2\nu_j)}$$

$$K_s = G_j \qquad 11$$

where ν_i is Poisson's ratio of Joint, G_j is Shear Modulus of Joint.

e. The parameters of C_R and ϕ_R :

A common formula can given for the Rock of A, B and C:

$$C_R = \frac{1-Sin_R}{2-Cos_R} \cdot E_R \cdot \epsilon_{\phi R} \qquad 12$$

While for the Rock of D, there can be:

$$C_R = C_S$$
$$\phi_R = \phi_S \qquad 13$$

The Undulation and the Roughness for the wavy interstratified layer can be treated on the basis of the principle described in reference 1.

f. Permeatibility Coefficient K_R and K_j:
The Equivalent Parameter $K_e (=K_R)$ can be determined by association of following two path:

a. To find out K_e corresponding Lugeon Value Lu. on the relational curve of K_e versus unit Lu. as shown in the figures 3; and

b. To calculate it according to following formulas:

$$K_e = \frac{1}{\dfrac{2\rho_R}{\rho_w}} \cdot \frac{(P_R)^3 \cdot 10^8}{S_V^2(1-P_R)} \qquad 14$$

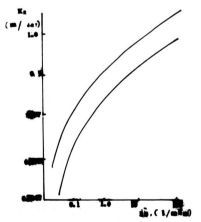

Where P_R is Porosity of Rock Masses, ρ_w and ρ_R are the Resistivity of water and Rock Masses separetely and S_v is the Relative area of grains.

When flow volum of aquifer Q_j is measured, the parameter K_j may be obtained based on following formulas:

$$K_j = \frac{Q_j}{2\pi \cdot \Delta H} Ln \frac{R}{r} \qquad 15$$

and

$$R=300(H-h)\sqrt{K_j} \qquad 16$$

fig 3. An example of relativity of K_e and Lu.

in which, ΔH is the fall head of water, r is the diametre of borehole, R is the radius of influence of which the effect to result can be omitted because it is not so large in seepage flow field consisting of fissures except for the Karst regions.

To obstract what has been presented above, the method of composing object parameter unit can be summerized in following flow chart graph:

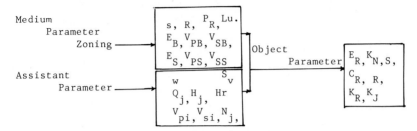

fig 4. The method of composing OPU

4. Design of GMM Program

Since the methods of composing geological unit and parameter unit are settled in application, the GMM of additive Stress-Seepage field can be established, and this procedure can be outlined as a Systems Program.

GMM editing program consists mainly of following five general parts:

(1). Program for drawing Geological Sections:SUB GPMM

This subroutine is used to draw ordinary Geological Map presenting the first-hand information truthfully. The Data to be input mainly are:

1. Borehole number, Coordinates (X,Y,Z) and Depth;

2. All kinds of geological interfaces, such as Soft-Intercalations, Faults, Joints and Caverns. etc. which are gathered from field survey and Borehole drillings.

3. Elevation of Aquifer, Water Head H or its Interval ΔH, Flow Volume{Q}or its Variation{ΔQ}, Flow Velocity V_w.

4. Other sorts of Data such as: Upper ———— Lower Water Level of river; Reservoir Water Level; the design Data of Dam and so on.

(2). Program for Probability Model of Geological Structure: SUB-GBM

This work synchronized with exploration. As shown in fig 1, the probabilities study of Geological Structure between Boreholes are continuously carried on with the increase of the numbers of Boreholes until the deduction of main Geological problems meet the probability demanded. Following are the necessary Data to be input:

1. Elevations of main Geological interfaces, sach as Soft Intercalations; Weathered Zone; Low dip fissures and Faults.

2. The average value of Direction; Angle; Length; Interval; Thickness; Frequency and so forth of those Interfaces mentioned above.

3. The Density of Joints N_j.

Its output is the Geological Structure Probability Model of Additive Stress-seepage field with the same scale as GPMM.

(3). Program for Chart of Medium Parameters Zoning: MCM

MCM is also called Physical Parameter Model PPM. It synchronize with exploration as GBM dose, and its imput Data are mainly Physical Parameter (see fig 1). which are:

1. Apparent Resistivity of Rock Mass and Groundwater ρ_R and ρ_w;

2. Filling Resistivity ρ_p ;
4. Wave Velocity V_{pR} and V_{sR} measured both in and across Boreholes;
5. Elastic Modulus of Borehole Wall E_R;
and so on.

Process in application starts from Surface, regarding each Stratum or every Five-Metre Interval Rock as an unit, to calculate average of each parameter mentioned above $\overline{X}_{\rho_R}$, $\overline{X}_{\rho_s}$, $\overline{X}_{pR}$, $\overline{X}_{VsP}$, $\overline{X}_{EB}$ and their Variance Value $V_{ar.\rho_R}$, $V_{ar.\rho_s}$, $V_{ar.PR}$, $V_{ar.VsR}$, $V_{ar.EB}$, etc.

The output result is shown as figure 5.
(4). Program for Object Parameter Unit: SUB-DPU

The assistant parameters shown in figure 4 are accepted to calculated Object Parameters: E_R, C_R, ϕ_R, K_n, K_s, K_e, K_j of every parameter zone in MCM.

(5). Program for Geological Mathematical Model; SUB-GMM

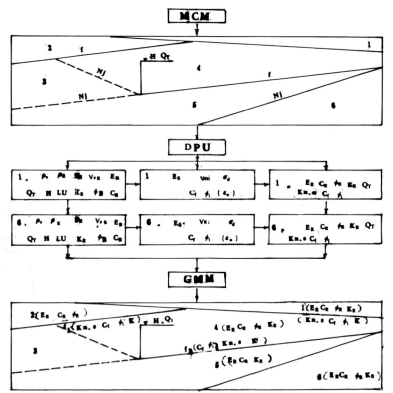

fig 5. The relationship between SUB-MCM, SUB-DPU AND SUB-GMM
The output result are:
1.Section with the same scale as GPMM, GBM, and MCM.
2.The table of Object parameters corresponding with GMM section (table 1.);

where f1, f2, ... , are the Fault Unit, IB1, ... ,are the Intercalation Unit.
This Program can be expanded into Tree-Dimension GMM program.

Table 1. Objective Parameter of GMM

Geological Unit	Objective Parameter							
	E_R	C_R	R	K_n	K_s	H or H	K_e	K_j
1	*	*	*				*	
2	*	*	*				*	
3	*	*	*				*	
4	*	*	*				*	
.	*	*	*				*	
	*	*	*				*	
f1	*	*	*	*	*	*		*
f2	*	*	*	*	*	*		*
.	*	*	*	*	*	*		*
.	*	*	*	*	*	*		*
IB1	*	*	*	*	*		*	
.	*	*	*	*	*		*	
.	*	*	*	*	*		*	

5. Principle of Determing Foundational Surface and Optimizing Design
   According to the principle of SEG, the aim of GMM is to realize the Optimizing Design.
   Optimizing Design is of two aspects: One is that the choice of Foundation Surface (or called as Base Surface Level) is accurate, another is that the Design of Super-structure such as Dam is the most optimized one for the local Geological conditions.
   For determination of foundation surface, the traditional method utiliges slightly weathered zone and fresh Rock Mass beneath as a supporting bed, and treats completely and Weakly Weathered zone all as what have to be excavated away Such method of deterination may be called Geological method or Geological principle.
   But in SEG, all Gelogical Units of GMM are expressed by Objective Parameter Units without special dividing Weathered zone.

   Such, the procedure to choose foundation surface starts down from Ground surface with reduction of the surface elevation to try design on different elevations until the standards of stability and antiseepage are met, as shown in figure 6.

   fig 6. Procedure of determining foundational surface based on the GMM
   Herein, the stability condition is checked on following two aspects:The first is that the upper part of Dam Foundation never

produces any probability of Man-Caused slip or the demand of anti-slip is satisfied .

The second is that the strain of each point $\{\varepsilon_R\}$ Inside Dam foundation must fit the requirement $\{\varepsilon_R\}\leqslant\{\varepsilon_{OR}\}$.

Here $\{\varepsilon_{OR}\}$ is the permission strain decided in advance corresponding with the extent and importance of Dam.

And the antiseepage condition means that the seepage is controlled under a demand value $\{Q_{con}\}$ or parameters fit $\{Q_{TR}\}\leq\{Q_{TO}\}$ with conventional grouting measure. Also, pervious-Deformation is prevented away.

To realize the Optimum Design, it is necessary to pay attention to following two points; one is that the profile of monitoring for Displacement must to comform to that of stability analysis profiles; another is that it is required to be apt to tell out the deformation caused by deformation of Dam foundation from whole Dam Bending flexure which can be expressed by formula;

$$\delta_B = A + B + C + D \qquad\qquad 17$$

where, A is the item due to temperature stress field of the Dam-Concrete; B is that due to stress field caused by variation of reservoir water level; C is that due to comprehensive stress field through long operation of reservoir; and D is that due to stress field of Dam foundation.

The formula 17 may be rewritten as following:

$$\delta_B = \{A,B,C\} + D \qquad\qquad 18$$

where $\{A,B,C\} = A+B+C$, called external strain value of Dam, can be determined through analysis of multivariant-regressin analysis of Dam's Bending Flexure δ_B versus temperature; Water Level and Time.

$\{D\}$, called Strain of Dam Foundation, should theoretically be as following:

$$\{D\} = \delta_B - \{A,B,C\}$$

Therefore, with $\{D\}$ as a givem value, the population-Equivalent Young's Modulus of foundation can be obtained.

Since deep Borehole strain gauge came into application, it has become reality to measure displacment of each point $\{\delta_{Ri}\}$ of Dam foundation directly.

Consequently, with $\{D_R\}$ can directly calculated out with $\{\delta_{Ri}\}$. And with the help of error analysis:

$$\{D\}=\{D_R\}-\{D\} \qquad\qquad 20$$

whether Dam is or not on safe operation will be judged correctly.

AFTERWORD

This paper briefly introduces some of what the author has been studing on SEG for recent years.

Although for some reason the study has not yet been applied as a whole for engineering, the author is confident that this study must be the sole path of Ehgineering Geology and Rock Mass Mechanics.

The author would like to have a close cooperation with experts at home and abroad in order to advance this study. And that is why the paper was written.

REFERENCES

1). CHUI Zheng Chuan: Systems Engineering Geology and Geotechnique.
              Procedings of the International Symposium on
              Engineering in Complex Rick Formations (ECRF).
              November 3-7, 1986 Beijing, China
2). Shunsuke SAKURAI and Kunifumi TAKEUCHI:
              Back analysis of Displacement measurements in
              Tunnelling. Report of the society of civil
              engineers of japan, No. 337.Sept. 1983.
3). K.Kovari, Ch. Amstad and P. Fritz:
              Integrated Measuring Technique for Rock Pressure
              Determination.Proceeding of the International
              Symposium on field Measurements in Rock Mechanics,
              Zurich. April 1977.
4). ISRM Commission on Standardization of Laboratory and field
     Tests: Suggested Methods for the quantative description
              of discontinuities in Rock Masses. Int j.Rock
              Mech. Sci. & Geomech.
              Abstr. Vol, 15, 1978.
5). Koichi AKAI, Yuzo OHNISHI and Del Ho Lee:
              Multiple stage Triaxial Test and its application
              to fully saturated Soft Rock.
              Japan Society of Civil Engineers, Report No.311
              1981.
6). L.Müllor: Rock Mass Behaviour-Determination and application in
              Engineering Practice.
              Proc 3rd ISRM Conf., 1-A, 1977.
7). Toshiaki TAKEUCHI, Tateo SUZUKI and Naoshi KUNIMATSU:
              A Consideration of the In-Situ C and ϕ as Deter-
              mination by Borehole Load Test.
              OYO Technical Report No.3 1981.
8). Hiromichi MANAME, Shoji UENO and Michio MORINO:
              Use of a Micro-Flow Meter in Investigating of
              Underground Water.
              OYO Technical Report No.1. 1979.

Supervision of a tunnel using FEM back-analysis based on field measurements

Koichi Fujino
Electric Power Development Co. Ltd, Hokkaido, Japan
Kouichi Suzuki
Shingijutsu Keikaku Co. Ltd, Hokkaido, Japan

1 INTRODUCTION

The Finite Element Method is often used in design prior to construction, or for carrying out reviews after construction. However, this method has seldom been used for supervision during construction. The reason is that much time and expense would be required to develop the hardware for handling such enormous amounts of calculations in the field and the software for supervision. The Electric Power Development Co., Ltd. (EPDC) resolved the problems involved, those of hardware by utilizing the EPDC microwave network and connecting the terminal equipment at the construction office with the host computer at the headquarters in Tokyo more than 1,000 km away, and those of software by establishing the processes described below.

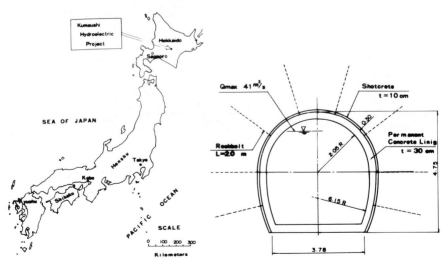

Fig. 1 Key map

Fig. 2 Typical cross section of the Kumaushi headrace tunnel

EPDC is presently constructing the Kumaushi Hydroelectric Power Station in Hokkaido which is scheduled to start operation in November 1987, and EPDC itself is carrying out supervision of the construction. The headrace tunnel of the power station is a non—pressure water conduit with a horseshoe—shape cross section 4.1 m in inside diameter and 6,200m in length, the entire length having been driven by the so—called New Austrian Tunneling Method (NATM) from August 1984 to September 1986.

With the objective of utilizing the features of NATM to the maximum, a detailed design of the support system for the tunnel was not done prior to construction, and it was clearly stated in the specifications that the support system was to be optimized according to the measure—ments made during the construction and/or the geological conditions. In this case, a process for rapidly and accurately evaluating the results of field measurements and feeding them back in to the construction execution was necessary. This process is shown in Fig. 3 and described below.

① Using the FEM back—analysis technique, char—acteristics of rock are estimated from the values of convergence measure—ments obtained during construction.

② Inputting the above values to FEM forestep analysis, the stress condition of the present support system is calculated.

③ If this stress con—dition is not in the allow—able range, the support system is optimized through trial and error numerical simulation using FEM fore—step analysis.

④ In case the geologi—cal conditions change, this process is to be repeated aiming at all times for the optimum support system.

In this paper, the var—ious calculation processes are introduced including examples of application as well as the scope of the technique.

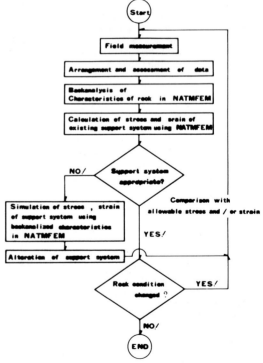

Fig. 3 Flow chart of supervision

2. APPLICATION OF FEM ANALYSIS TO SUPERVISION

2.1 NATMFEM program

The NATMFEM program was developed by the Railway Technical Research Institute of the Japan National Railways (JNR) who have abundant experience in tunnel construction in Japan. This program has the following features:
① The nonlinear properties and creep phenomena of rock and support members can be represented.
② Any configuration can be represented including very thin elements.
③ The condition of the tunnel construction stages can be represented.
④ The phenomenon such as the bond of rock bolts becoming lost can be represented.
⑤ It is fundamentally a two-dimensional analysis, but quasi-three-dimensional analysis accompanying face progression is possible.
The main input and output items of NATMFEM are given in Table 1.

Table 1. I/O items of NATMFEM

Input data
1. Initial ground pressure (horizontal, vertical)
2. Characteristics of rock (modulus of deformation both initial and in failure, limit of elasticity)
3. Strength of rock   (C, ϕ , tensile strength)
4. Characteristics of creep
5. Speed of face progression
6. Procedure of excavation
7. Characteristics of support members (shotcrete, rockbolt etc.)

Output data
1. Displacement of each node ($\delta x, \delta y$)
2. Stress of each element   ($\sigma x, \sigma y, \tau xy, \sigma_1 , \sigma_2 , \tau \max$)
3. Proximity to failure of each element

2.2 Addition of back-analysis function to NATMFEM

The NATMFEM program developed by JNR involves forestep analysis where calculations are possible only when the characteristics of the individual elements are known. However, information obtained during construction normally consists only of convergence values, and in order to find the stress condition of the support system, it is necessary to estimate the characteristics of the individual elements, especially ground pressure and the modulus of rock deformation, using the back-analysis function.
    Before preparation of the program for back-analysis, a parameter study was made, running the NATMFEM program many times. As a result, it was found that the dominant input parameters affecting deformation of the tunnel were the initial modulus of deformation (Do) and the initial factor of horizontal ground pressure (Ko).

Fig. 4 shows the calculation results arranged into the relationship between Ko and the displacedement ratio λ , for each value of Do.

The larger is Ko which represents the horizontal component of external force, the larger will be λ which represents the horizontal component of deformation. The relationships of Ko and λ are roughly linear where Do> 245 MPa (2,500 kgf/cm²), and become independent of Do where Do≧ 490 MPa (5,000 kgf/cm²). This indicates the possibility of Ko to be back-analyzed from the measured value of λ .

Fig. 5 shows the relationships of Do and 1/Δ c (inverse of horizontal convergence) for each value of Ko. The value of 1/Δ c becomes large, in effect Δ c is calculated to be smaller, as Do increases.   If Ko≧ 0.5, the relationships of Do and 1/Δ c will be those of straight lines connecting with the origin. This indicates that Do will be back-analyzed using both Ko estimated in Fig. 4 and Δ c obtained by measurements.

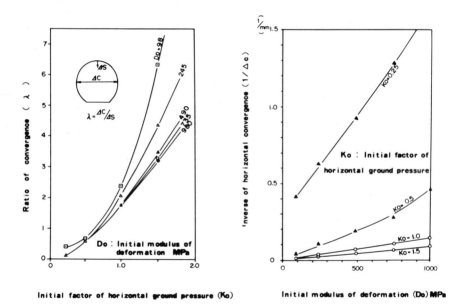

Fig. 4 Parameter study of
　　　　　　　　NATMFEM (1)

Fig. 5 Parameter study of
　　　　　　　　NATMFEM (2)

Based on the results of the abovementioned parameter study, a program for back-analyzing Do and Ko which are the main characteristics of the ground was compiled according to the flow chart of Fig. 6. This is the modification of NATMFEM.

Fig. 7 gives examples of back-analysis using this program and forestep analysis using estimated values. Here the situation that convergence and stress of both shotcrete and rockbolts are changing with time in step with face progression of the tunnel, is shown in the form of comparisons between calculated and measured values. It is considered that the stress conditions of the extremely complex tunnel support system are simulated with ample precision for practical purposes. The running time of this program for back-analysis is comparatively short since it does not include a great number of repeated calculations, nor calculations of large-scale inverse matrices. In other words, this technique differs from the normal FEM back-analysis and does not have a mathematical generality, being an approximation method which is valid for practical purposes within a certain scope of application.

2.3 Calculation of the safety factor for the support system

The support system consists mainly of shotcrete and rockbolts. The safety factor of the support system for a tunnel driven in soft rock such as the Kumaushi site is generally dominated by the safety factor of shotcrete.

As is well known, the load increases with the tunnel face progression, while the modulus of deformation and strength of the shotcrete also increase with age. Therefore, it is necessary for the safety factor to be calculated from the ratio between analyzed stress of shotcrete and the corresponding strength.

As a result of laboratory tests, the characteristics of shotcrete were obtained as follows:

$$Do(t) = 7,800 \ln(t) + 18,000 \ (kgf/cm^2) = 765 \ln(t) + 1,765 \ (MPa)$$
$$\sigma c(t) = 55 \ln(t) + 130 \ (kgf/cm^2) = 5.4 \ln(t) + 12.75 \ (MPa)$$

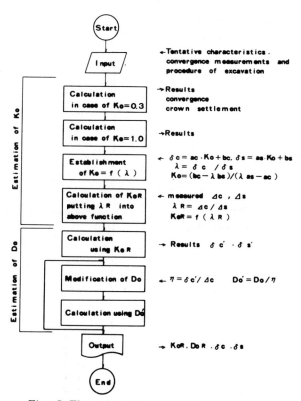

Fig. 6 Flow chart of NATMFEM program

where
　　Do(t)　: modulus of deformtion of concrete at age t (day)
　　σ c(t) : unconfined compressive strength at age t (day)

The safety factor of shotcrete which represents the safety factor of
the support system is defined by the following equation:

　　S.F.=Min.(S.F.(t)) =Min.(σ c(t) /Sc(t))
where
　　S.F.　　: safety factor of shotcrete
　　Sc(t)　: compressive stress of shotcrete calculated by NATMFEM for
　　　　　　　　age t

The required value of the safety factor (S.F.) was established as
more than 3.0 in consideration of the following,

①　The allowable stress of concrete in Japan is to be one-third of
the compressive strength.
②　The support system, including shotcrete, is a temporary work until
the permanent lining is placed, and therefore, it is comparatively
unimportant.

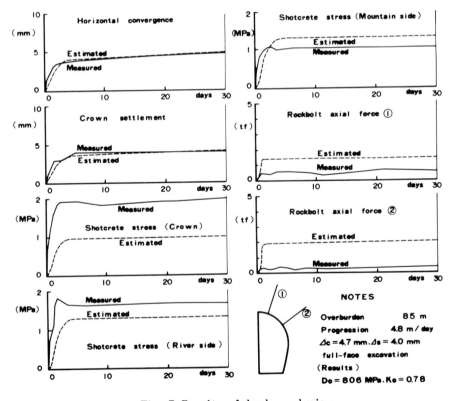

Fig. 7 Results of back-analysis

③ Some errors are unavoidable in this method because of both scatter-ing in the measurements of Do(t) as well as σ c(t), and differences be-tween stress calculation results and measured values.

3 CASE OF APPLICATION IN THE KUMAUSHI HEADRACE TUNNEL

3.1 Geology

The geology of the Kumaushi headrace tunnel can be divided into three formations as shown in Table 2. They have small unconfined compressive strength (qu) and comparatively large angles of internal friction, and are non-to-medium-consolidated typical soft rocks. Full length of the tunnel was driven by excavating machines so that blasting was not used at all. In addition, springing of ground water during construction amounted to as much as 9,500 1/min at maximum, and the excavation work was extremely difficult.

Table 2. Geological condition of the Kumaushi tunnel

Name of stratum	Component material	Unit weight (t/m³)	Modulus of deformation in borehole (MPa)	C (MPa)	ϕ (deg)	Unconfined compressive strength (MPa)
Iwamatsu forma-tion	gravel 80% sandy silt 20%	1.8–2.2 (dry) 2.1–2.3 (wet)	390–750	–	–	–
Kuttari forma-tion	welded tuff	1.2–2.4 (dry) 1.7–1.8 (wet)	300–590	0.09–0.13	40–43	0.42–1.07
Shibusan forma-tion	silt 20% sand 50% gravel 30%	1.1–1.8 (dry) 1.6–1.8 (wet)	160–300	0.07–0.09	36–40	0.25–0.72

Rockbolts are not effective under such geological conditions, and it was unavoidable that the support system was dependent on shotcrete. When this was inadequate, steel ribs and/or partial excavation— excava-tion of a part of the cross section and application of shotcrete without delay — were provided.

3.2 Application to the Kumaushi tunnel

Two cases are introduced below where the support systems were altered during construction using the abovementioned technique.

Case 1

In this case, in view of the small convergence and the favorable condition of the original ground, a study was made for omitting rockbolts using this technique.

The geology, results of measurements and supports in Case 1 are shown in Fig. 8. Estimation A gives the results of numerical simulation for safety factors of shotcrete before alteration of the support system which includes rockbolts. It is estimated that the minimum value of the safety factor is 5, which indicates that support members can be reduced. Estimation B (Fig. 9) gives the simulation results of the safety factor of shotcrete employing back-analyzed input parameters in case of shotcrete only and no rockbolt. The minimum value of the safety factor in B is 4.5, and it is shown that even though rockbolts are not used, the stress of the shotcrete has ample allowance against compressive strength.

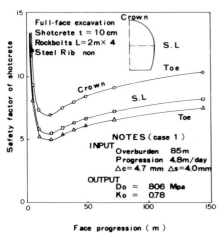

Fig. 8 Transition of safety factors of shotcrete (case 1-A planned support system)

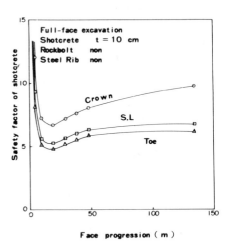

Fig. 9 Transition of safety factors of shotcrete (case 1-B alternative support system)

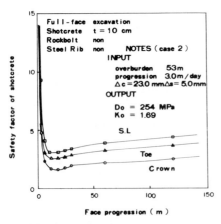

Fig.10 Transition of safety factors of shotcrete (case 2-A planned support system)

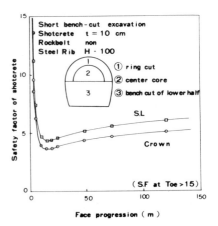

Fig.11 Transition of safety factors of shotcrete (case 2-B alternative support system)

Case 2

This is a case of improvement of the excavation procedure and the size of supports.

The ground in Case 2 consists of non-consolidated tuff. As the tunnel approached the ground over which a gully existed, the tunnel face gradually losed its self-standing ability. Thus, although the overburden below the gully was only 27 m in depth, its displacement increased to a value of more than that for the 90 m overburden.

As a result of the back-analysis of the safety factor of shotcrete using the field measurements at a point where overburden was 53 m, the minimum value was 1.6 (Fig. 10). An alternation of the support system was contemplated since the safety factor would become less than 1.0 directly below the gully if this trend were to continue.

As a result of forestep analyses of several cases of different construction procedures and support sizes, it was judged that the excavation could be done suitably by the short bench-cut method, while providing lining of shotcrete with steel ribs.

It was estimated that the minimum value of the safety factor of shotcrete would be raised to 3.5 through this alteration (Fig. 11). As a result of this application to the actual construction, the excavation speed was lowered to 80 %, but due to a reduction in the deformation of the tunnel, the excavation was performed safely.

4. DISCUSSION

The key points of this technique are that the comparatively simple relationships of ($K_o - \lambda$) and ($D_o - \Delta c, \Delta s$) can be found as shown in Figs. 4 and 5.

The scopes of application for K_o and D_o must be grasped from these relationships. According to experiences at the Kumaushi tunnel, it is possible for input parameters to be back-analyzed so that the calculation results will have errors in a range of about ± 10 % in relation to measurment results. Further, at $D_o \geqq 440$ MPa and $K_o \geqq 0.5$, the errors will be in a range of ± 3 % so that a very high degree of accuracy is attained.

However, there are some cases that exceed the scope of application of this technique other than the abovementioned problems. In fact there exist differences between the continuities of the component elements assumed in FEM and the actual geological conditions. Examples are when the tunnel face is washed away by springing water, or when sliding occurs due to the low of groundwater.

Such conditions were encountered several times during execution of the Kumaushi tunnel, an example being as follows.

Numerous cracks were formed in shotcrete 50 to 80 m behind the face while excavating through the Kuttari welded tuff formation, and the inner section of the tunnel was displaced 30 to 50 cm. The convergence of this place had once been stabilized prior to the occurrence of this accident, and the safety factor of the shotcrete was higher than 3.0 according to back-analysis at that time. As a result of investigations, it was found that there was a flow of groundwater behind the side wall and sliding occurred. It was conceivable that fine grained materials in the crack had been washed away by the path of groundwater as excavation progressed. Strainers were installed in the shotcrete in subsequent excavations to prevent development of any groundwater path. Such a case is out of the scope of this technique, and the judgment of the supervisor in the field is very important.

5. CONCLUSION

This "Supervision of a tunnel using FEM back-analysis based on field measurements" was developed and put into practice for the first time in Japan by EPDC at the Kumaushi tunnel.

   The greatest objective of this technique was to establish a procedure for rapid and quantitative feedback of measurement results. It appears that this technique has achieved the objective for a comparatively broad range of geological conditions.

   Taking into account the performance at Kumaushi, the technique is summarized as follows:

(1) Back-analysis of FEM input parameters based on measurements

This technique uses a nonlinear, two-dimensional FEM (NATMFEM) with which it is possible to express stress release accompanying face progression. The initial factor of horizontal ground pressure Ko is back-analyzed from the displacement ratio λ of the measurement results. Next, using the estimated Ko, the initial modulus of deformation Do is back-analyzed so that the calculation results and measurement results of convergence (Δ c) coincide.

(2) Evaluation of support system

The safety of the support system can be evaluated with the results of forestep simulations employing the back-analyzed input parameters. Where the geology consists of soft rock such as the Kumaushi Tunnel, the safety factor of stress of shotcrete is chosen for the evaluation of the safety of the support system. The calculations take into consideration the time dependent changes in strength and the modulus of elasticity of shotcrete.

(3) Scope of application

Unlike ordinary FEM back-analysis, this technique does not have a mathematical generality, and is an approximation method with which initial factors of horizontal ground pressure Ko and modulus of deformation Do can be obtained for practical purposes within a certain limit. The scope of application at the Kumaushi tunnel was determined as Do$\geqq$ 245 MPa and Ko$\geqq$ 0.5.

REFERENCES

Tsuchiya, T. 1984. Design program for a new tunneling method using bolts and shotcrete. Proc. of Japan Society of Civil Engineers. No. 346, III -1 (in Japanese)

Fujino, K. & Suzuki, K. 1986. Back-analysis of input parameters of NATMFEM using convergence data. Proc. 41th Annual Symposium of JSCE. III -386 (in Japanese)

2nd International Symposium on Field Measurements in Geomechanics, Sakurai (ed.)
© 1988 Balkema, Rotterdam. ISBN 90 6191 778 6

Analytical estimation of tunnel deformation based on field measurements

H.Kunimi & H.Takasaki
Civil Engineering Division, Shimizu Construction Co. Ltd, Tokyo, Japan

1 INTRODUCTION

Although the New Austrian Tunnelling Method (NATM) is based upon a more scientific concept than the conventional tunnelling method, it is still difficult under the present state of the art of technology to deterministically and precisely predict the behavior of the ground and tunnel system in the planning and design state, thereby resulting in an extensive difference between the predicted values and the actual behavior after excavation. This means that we have not established the design analysis models for adequately describing:
  1. the method for clarifying the detailed stratum structure of intended ground;
  2. the method for evaluating the physical properties and engineering properties not as rock but as bedrock;
  3. combined behaviors of ground-support interaction system and execution process.
Moreover, this means that we have not necessarily been able to clarify the dynamic behavior of the ground and tunnel system from just our experience alone. In weak rock and soil stratum where the ground strength is small and the tunnel is likely to become unstable, precise prediction of the deformation and stress is greatly required. This requirement can be readily recognized judging from the fact that greater deformation and stress than expected cause substantial effect upon the structural stability, excavation period and schedule of a tunnel.
To ensure precise prediction of the deformation and stress, an examination of the results of measurement in detail and stress is needed. And upon fully recognizing the results of the examination, it is essential to establish a method for applying the finite element method (FEM), theoretical analysis method and comparatively practicable analysis techniques available to us today, for predicting the deformation behavior of the ground and tunnel system. More concretely, miscellaneous problems concerning:
  1'. supporting mechanism and modelling of rock bolt, shotcrete and other support members;
  2'. establishment of the modulus of deformation and other constants of ground and support members to be input to FEM and theoretical analysis;
can be considered. On the basis of the above-mentioned background, modelling is performed from the in-situ observation data regarding the theme of 1', and the observation data is examined according to the back analysis method regarding the theme of 2' in this paper. Furthermore, it is tried herein to establish numerical estimation models of deformation and stress by summarizing this study data.

2 MODELLING OF GROUND-SUPPORT INTERACTION

2.1 Rock bolt

According to the measurement data of the axial force of rock bolt and strain in the ground adjacent thereto, it has been reported in many papers that there arises relative deformation between both of them (K. Hayashi, et al 1979, H. Takasaki, et al 1982). Therefore, the model presented in Figure 1 is considered to be adequate for expressing the actual phenomena (H. Takasaki, et al 1984). In other words, although both the adhesion and friction act between the rock bolt and ground up to a certain relative deformation δ, the adhesion no longer acts and only the friction acts where δ exceeds a certain limit.

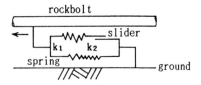

Figure 1. Adhesion-slide model of ground-rockbolt interaction

Now, let us examine the effect of rock bolts upon the restriction of deformation of ground and tunnel using numerical analysis. For this analysis, the elastic FEM model which takes into consideration the preceeding deformation in front of tunnel face and excavation steps are used (T. Tsuchiya 1981). The tunnel adopted herein is of a horse-shoe shape with an approximate diameter of 9 m, while theground condition is of two types, namely, the sand layer in Kokubu Drainage Tunnel with a ground modulus of deformation D = 350 kgf/cm² and mudstone (D = 2,200 kgf/cm²). The rock bolt, excluding its inverted part, is pattern-bolted per one bolt/m² × 3m length, with the thickness of shotcrete being 20 and 10 cm, respectively. As an interaction model between the ground and rock bolt, moreover, the fixed model which makes it possible to attain the effect of the rock bolt upon restriction of deformation has been adopted. This is intended to disregard the relative deformation between both of them.

The results of analysis indicated that the effect of the rock bolt upon the restriction of tunnel wall deformation was only 0 to 10% when compared with that in the case where no rock bolt was provided. The reasons that only such a small value was indicated are mainly considered to be due to the facts that

1. the equivalent force caused on an excavation was substantially larger than the stiffness of the rock bolt,

2. because of continous ground conditions and the shape of the tunnel being similar to a circular form, the majority of the equivalent force caused on an excavation was distributed in a tangential direction of the tunnel and the tunnel was stabilized due to perfect formation of the so-called ground arch ring, and

3. analysis was carried out by setting the rock bolt after the preceeding deformation occupying 30 to 40% of the total amount of deformation had been released.

From the above-mentioned results, there would be no problem if the model of ground and rock bolt were analyzed while disregarding the rock bolt, as long as it is intended to examine only the deformation of a tunnel. Should it be intended to obtain the stress of rock bolt as well by numerical analysis, however, it is concluded to be appropriate to adopt the model in Figure 1. As another new problem will arise on how to obtain the constants of spring and slider in case the model in Figure 1 is to be applied, a fixed model (this means that K_1 and K_2 = ∞) is considered to be applied in advance.

Meanwhile, in light of the fact that the flexural rigidity of rock bolt is smaller than the total value of the entire ground, modelling of the rock bolt itself should be performed by replacement with a rod element which takes only axial force into account.

2.2 Shotcrete

In light of the fact that shotcrete is in close contact with the ground, it can be easily imagined that the force between the ground and shotcrete is transmitted by means of adhension. This phenomenon can be observed also in the sand layer where the adhension is weakest (C.Uchiyama, et al 1985). In view of the fact that the tunnel is of an arch shape, moreover, it is evident that the external force acting in the radius direction will cause the axial force

to initiate in a tangential direction to the shotcrete. Although the former and the latter can be called an adhesion model and an axial model (T.Konda 1983), respectively, it is concluded to be suffi-cient if a combination of the adhesion model and axial model is taken into account as the interaction modelling of the ground and shotcrete systems. This is expressed by the model presented in Figure 2.

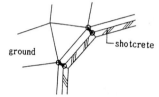

Figure 2. Adhesion+axial model of ground-shotcrete interaction

Meanwhile, since shotcrete lacks in flexural rigidity, the shotcrete proper may be considered as a rod element which transmits only the axial force.

3. SETTING OF INPUT CONSTANTS

3.1 Equivalent elastic modulus of shotcrete member

The shotcrete constitutes a major support member of a tunnel according to NATM, and it is essential to set its elastic modulus in order to estimate the stress arising in the shotcrete by design calculation. In many cases, $100,000$ kgf/cm^2 or greater is generally adopted as the elastic modulus of shotcrete in the design calculation stage. This value is equivalent to that in the old age. According to a great deal of stress measurement data of shotcrete in the past, however, it has been found that the actual measurements are substantially smaller than the calculated estimation values. Therefore, it can be said that a considerably smaller value should be adopted as the elastic modulus of shotcrete for use in design.

Now, let us define the equivalent elastic modulus Eeq of shotcrete based upon the wall deformation of a tunnel and stress measurement data of shotcrete as follows:

(1) $$Eeq = \frac{r}{u} \sigma_s$$

where r: radius of tunnel; u: radial deformation of tunnel wall;and σ_s: stress in shotcrete. All of these values are available from actual measure-ments. The reciprocal of r/u on the right side of equation (1) refers to the strain in the tangential direction of the tunnel wall. As a result of applying this concept for the previously mentioned Kokubu Drainage Tunnel, Eeq = $5,100$ kgf/cm^2 was obtained (M.Akada, et al 1984, T.Fujimori, et al 1985, T.Chishaki, et al 1986), while a maximum of 28,000 kgf/cm² was obtained from the other application example. Both of these values are equivalent to the smaller values corresponding to those in the young age.

The reasons that the equivalent elastic modulus is small as mentioned above are considered to be due to the following two points:

1. The face has advanced and an equivalent force caused on an excavation has been released when the elastic modulus is small because of the young age of shotcrete; and

2. The tunnel has been settled or deformed, and together with an increase in the deformation of the tunnel, the stress in shotcrete has been lowered at the same time in the top heading excavation stage and immediately after excavation of the lower section.

In other words, it might well be recognized that the higher the tunnel excavation velocity, the smaller the equivalent elastic modulus becomes due to reason 1, and the poorer the ground, the smaller the elastic modulus becomes due to reason 2. Therefore, it can be said to be appropriate to adopt a smaller value based upon the concept of equivalent elastic modulus as the elastic modulus of shotcrete to be input for design calculation.

3.2 Ground modulus of deformation and coefficient of lateral pressure according to back analysis

Presented in Figure 3 is a conceptual diagram indicating fluctuation of deformation at the point adjacent to the face of a tunnel associated with the advance of excavation. In this diagram, $U_{o'}$ is the preceeding deformation initiating before the face has been reached, $(U_N - U_{o'})$ is the amount of deformation on the tunnel wall in case a support has been set after excavation, and $(U_{N'} - U_{o'})$ is the amount of deformation in case the tunnel has been excavated without any support. The respective curves indicate the transition of

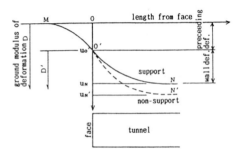

Figure 3.   Deformation distribution around tunnel face

deformation associated with the length from a face, and the support is set at the point $0'$. In this diagram, D must be adopted as a ground modulus in case the preceeding deformation is combinedely taken into account, but if only the deformation after excavation is taken into consideration, D' must be adopted as the ground modulus of deformation. In other words, it should be sufficiently known in advance whether the ground modulus of deformation required from the structural characteristics of design and analysis models is according to D or D', or some other concept. In the case of a model which searches for the execution step including preceeding deformation, D should be input. When analysis is carried out using D should be the instantaneous excavation and instantaneous installation of support model, on the other hand, the ground modulus of deformation D' should be applied, and the amount of deformation and support member stress after excavation will be obtained as a result of the analysis.

In many cases, the values of deformation and stress in the ground and tunnel system predicted in the design and planning stage are quite different from those observed during excavation. As a means of solving this problem, it is possible to apply the concept of back analysis to obtain the ground modulus of deformation by using the actual measured deformation. Proposed in this paper is a new back analysis method based on a trial and error method (M.Akada, et al 1984, T.Fujimori, et al 1985, T.Chishaki, et al 1986). In proposing this method, the following four points have been taken into consideration:

1. It is possible to take into consideration an absolute amount of deformation prior to excavation of a tunnel;

2. It is possible to take into account the setting period and reaction force of the tunnel support members;

3. In light of the requirement that the variables other than the ground modulus of deformation be as small as possible, the analysis method to be used is simplified; and

4. The ground modulus of deformation obtained has a theoretical invariance applicable also for executing analysis according to the methods other than the analysis method used herein.

Along with the progress of the tunnel face, the absolute amount of deformation from the initial stage of a ground generally changes as shown by the solid line in Figure 3, while the amount changes as shown by the dotted line after reaching the face unless the tunnel support member has been set. When an instantaneous excavation and an instantaneous installation of the support model in the case of a full section excavation is taken into consideration, therefore, the ground modulus of deformation can be back-analyzed according to the following procedures, provided, however, that the coefficient of lateral pressure is discussed here with respect to known cases:

Step 1. From the measurement data of incremental strain in the tangential direction of the tunnel wall and stress initiating in the support member after the setting of a support member, the equivalent elastic modulus of a support member is back-estimated as referred in 3.1.

Step 2. By using the equivalent elastic modulus of a support member and amount of deformation $(U_N - U_{o'})$, the ground modulus of deformation D' is subtained according to the trial and error method based upon the concept of instantaneous excavation and instantaneous installation of the support model.

Step 3. By using the ground modulus of defomation obtained above, the amount of deformation $(U_{N'} - U_{o'})$ under a non-support state is obtained. In this case, it is sufficient if analysis by the instantaneous excavation and non-support model is carried out.

Step 4. The total amount of deformation $U_{N'}$ is obtained by adding the amount of preceeding deformation $U_{o'}$ (known from measurement) until reaching the face to the amount of deformation $(U_{N'} - U_{o'})$, obtained in Step 3. The procedures in Steps 2, 3 and 4 have been taken into account as it was intended to express the preceeding deformation and support setting time by using instantaneous excavation and instantaneous installation of the support model.

Step 5. The ground modulus of deformation D, presumably equivalent to the total amount of deformation $U_{N'}$, is obtained.

Although the back analysis method presented above is used to obtain the ground modulus of deformation conforming to the amount of deformation at a specific point, it is required to expand the above-mentioned method where the amount of deformation and deformation mode (horizontal deformation / vertical deformation) are to be complied with each other at the same time. The amount and mode of deformation of a tunnel are determined according to the cor-relationship between the ground modulus of deformation and the coefficient of lateral pressure of ground so that it is impossible to independently deal with both of the respective values. Now, the ground modulus of deformation and

1212 *Interpretation*

coefficient of lateral pressure are obtained by using a diagram solution method. Firstly, the co-efficient of lateral pressure and ground modulus of deformation D', conforming to the deformation mode and amount of deformation measured after the face has been reached, are obtained. Secondly, the required ground modulus of deformation D is obtained according to the procedures in Steps 3, 4 and 5 presented above, while making the coefficient of lateral pressure obtained above constant.

Now, let us present an example of applying the method proposed above to the Kokubu Drainage Tunnel. As shown in Figure 4, several cases of the combination of ground modulus of deformation and coefficient of lateral pressure are assumed and calculated at first. Then, a graph indicating the relationship between these moduli and coefficients, and mode and amount of deformation is prepared. From this graph, the ground modulus of deformation D' and coefficient of lateral pressure which would simultaneously satisfy the actually measured mode and amount of deformation are found out. According to this instance, since the curve of each ground modulus of deformation happened to intersect at the point of 0.52 of the actually measured deforma-tion mode, the coefficient of lateral pressure was readily obtained to be 0.75. Moreover, as D' = 250 in the curve at the intersecting point between the actually measured vertical deformation $(U_N - U_{0'}) = 30.7$ mm and coefficient of lateral pressure K = 0.75, D' was obtained to be 250 kgf/cm². As mentioned above, since such values could not be discovered in most cases, finding out the ground modulus of deformation and coefficient of lateral pressure locating on a straight line of actually measured mode and amount of deformation by using a propor-tional allotment technique is required.

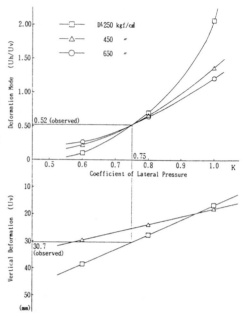

Figure 4. Correlation between ground modulus of deformation and coefficient of lateral pressure, and deformation and mode in the Kokubu Drainage Tunnel

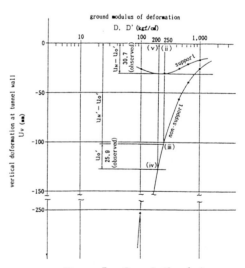

Figure 5. Correlation between ground modulus of deformation and deformation in the Kokubu Drainage Tunnel case

Next, the vertical deformation of a tunnel is calculated by changing the ground modulus of deformation while fixing the coefficient of lateral pressure at 0.75, and a relationship diagram presented in Figure 5 is obtained. From this diagram, the ground modulus of deformation D is obtained at 200 kgf /cm² according to the procedures in Steps 3, 4 and 5 presented above.

4 STRUCTURE OF ANALYTICAL ESTIMATION MODEL

4.1 System of deformation estimation

Based on the various factors described above, the procedures for studying and clarifying the dynamic behavior in measurement and management during excavation work can be summarized as presented in Figure 6. That is, by replacing the rock bolt and shotcrete constituting the support members into an elastic rod element as well as in to the interacton model with ground, respectively, into a fixed model and (axial force + adhesion) model, the models for analyzing the ground-tunnel support member system are at first prepared mainly on the basis of the FEM. Secondly, the equivalent elastic modulus of a shotcrete is obtained from the deformation, stress and other data obtained from investigation and measurement at the time of excavation, and the coefficient of lateral pressure and ground modulus of deformation are further estimated by back-analysis. These moduli and constants are applied in the analysis for the purpose of evaluating the structural safety of the section not yet excavated in case design changes must be made.

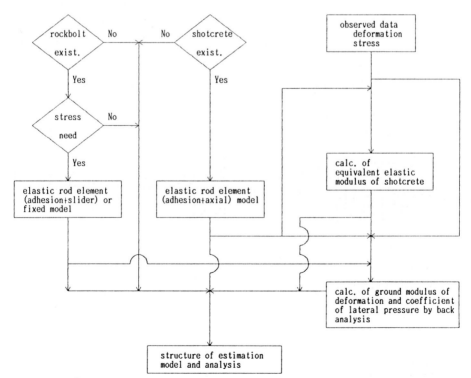

Figure 6.   System-flow for analytical estimation of tunnel deformation

4.2 Inspection of analytical estimation method

The design and analysis models which have been established up to the previous section will be appied to the Kokubu Drainage Tunnel in diluvial sand stratum, and the prediction accuracy and practicability of these models will be clarified.

The FEM method to be used is the instantaneous excavation and instantaneous installation method of the support model, and because of comparatively small deformation as a result of measurement, elastic analysis will be conducted. While disregarding the rock bolts, the shotcrete will be replaced into an elastic rod element, and the interaction with the ground will be converted into a model of (axial force + adhesion). With regard to the stiffness of shotcrete and ground, and coefficient of lateral pressure, the equivalent elastic modulus Eeq = 5,100 kgf/cm² described in Clause 3.1 will be used for the former, while the value obtained according to the back analysis method described in Clause 3.2 will be used for the latter. Figure 7 indicates the input constants related to the ground and the finite element mesh.

Presented in Table 1 is a list of actual measurement values and results of analysis. According to this table, both the deformation of the tunnel wall, and stress in support members obtained from the analysis, comply excellently with the actual measurement values. The predicted value of ground surface settlement varies extensively from the actual measurement value, and can be

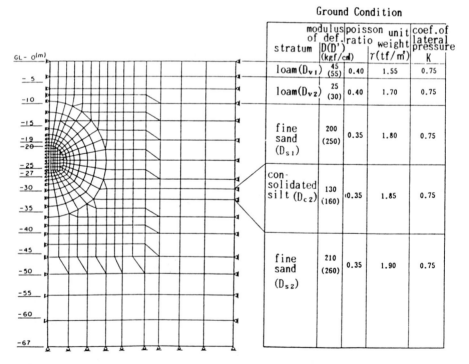

Ground Condition

stratum	modulus of def. $D(D')$ (kgf/cm²)	poisson ratio	unit weight γ(tf/m³)	coef.of lateral pressure K
loam(D_{v1})	45 (55)	0.40	1.55	0.75
loam(D_{v2})	25 (30)	0.40	1.70	0.75
fine sand (D_{s1})	200 (250)	0.35	1.80	0.75
con-solidated silt (D_{c2})	130 (160)	(0.35	1.85	0.75
fine sand (D_{s2})	210 (260)	0.35	1.90	0.75

Figure 7. Finite element mesh and input constants of ground in the Kokubu Drainage Tunnel case

said to lack in reliability. This means that although the stiffness of support member and ground, and coefficient of of lateral pressure were obtained and input by back analysis based upon actual measurement values of tunnel wall deformation and support members, it is difficult to maintain the prediction accuracy even at remote measurement points beyond the objective range.

As mentioned above, although there are some problems yet to be solved, it can be said to be possible to obtain results substantially close to actual phenomena as far as

Table 1. Comparison between observed and and estimated behavior of Kokubu Drainage Tunnel

		observed	estimated
ground modulus of deformation (D_{si}) D (D') kgf/cm^2		350~600	200 (250)
settlement of ground surface mm		29	9
tunnel wall def. mm	vertical	57	56
	horizontal	18	17
axial force of shotcrete (ave.) tf/m		50	53
note		preceeding deformation involved	proposed analytical model

deformation of the tunnel wall and stress in support members are concerned, provided that the input constants are appropriately evaluated and selected according to the analysis models presented in this paper. Therefore, the justifiability and practicability of the proposed prediction method can be confirmed.

5 CONCLUSION

Presented in this paper is the modelling of interaction between ground and support members, and the property setting method pertaining to shotcrete and ground on the basis of the back analysis concept. Moreover, the analysis model for predicting and calculating the deformation and stress of a tunnel by incorporating these into the 2-dimensional FEM or theoretical analysis has been established as summarized in the previous section. This method is mainly applied for predicting deformation and stress in cases where various input constants are determined from the results of measurements obtained from an already excavated section at a specific work site, and the support system of the section not yet to be excavated is to be changed. Considering that a comparatively simple concept and easy analysis techniques are applied, and that it is possible to process data sufficiently with a personal-computer installed at the site, this method is said to be highly useful.

Although the analysis model proposed in this paper is intended for use in observational procedures at the time of the excavation work, it is mentioned here that the model can be applied even in the design stage. In this case, however, since the actual measurement data of the concerned tunnel has not yet been obtained, the input constants must be determined by referring to data from other experiences. In such cases, the constants, namely, the equivalent elastic modulus of support members, the ground modulus of deformation, and the coefficient of lateral pressure, must be determined by back analysis.

Acknowledgement

The authors would like to extend their gratitude to Prof. Dr. T. Chishaki of the Engineering Department at Kyushu University for his backup and instruction,

as well as to Messrs. T. Kurata, F. Kusumoto, H. Kumasaka and M. Akada who
are the joint researchers at Shimizu Construction Co., Ltd.

REFERENCES

Hayashi K., Y. Nishida, Y. Koga & H. Takasaki 1979. Tunnel Investigation
   Sustained Expansive Earth Pressure. The 12th Symposium on Rock Mechanics.
   Japan Soc. Civ. Engrs: 51-55. (in Japanese)
Takasaki H., T. Kurata & F. Kusumoto 1982. Peripheral Behavior of Tunnel in
   Kamuikotan Metamorphic-zone. The 14th Symposium on Rock Mechanics. Japan
   Soc. Civ. Engrs: 101-105. (in Japanese)
Takasaki H., A. Fukushima & T. Kitagawa 1984. Interaction between Ground and
   Support in Small Tunnel by NATM. In Proc. 39th Ann. Conf. Japan Soc. Civ.
   Engrs.3: 455-456. (in Japanese)
Tsuchiya T. 1981 Research for design of system rock-bolt tunneling method.
   In Weak Rock: soft, fractured and weathered rock, proceedings of the inter-
   national symposium, Tokyo. vol.2, P 951-956. Rotterdam: Balkema.
Uchiyama C., T. Kurata, T. Kondoh & H. Tsuchiya 1985. Examples of Measurement
   and Analysis of Stress on Shotcrete. The 17th Symposium on Rock Mechanics.
   Japan Soc. Civ. Engrs: 271-275. (in Japanese)
Konda T. 1983. Action effect of shotcrete. The 3rd Tunnelling Symposium —
   Shotcrete in NATM — . Japan Tunnelling Association: 1-9. (in Japanese)
Akada M., T. Yamaguchi, Y. Kijima & H. Takasaki 1984. Estimation of ground
   modulus of deformation based on tunnel behavior observation. In Proc. 39th
   Ann. Conf. Japan. Soc. Civ. Engrs: 431-432. (in Japanese)
Fujimori T., C. Uchiyama, H. Kunimi & H. Takasaki 1985. Use of the NATM in
   soft ground near Tokyo, Japan. Proceeding of the fourth international
   symposium on Tunnelling '85. The Institute of Mining and Metallurgy, P.93-
   102. England: Stephan Austin / Hertford.
Chishaki T., H. Takasaki & M. Akada 1986. Study on Presumption of Ground
   Modulus of Deformation by Back-Analysis in Tunnelling. Technology Reports
   of the Kyushu Univ. Vol.59, No.4: 489-493. (in Japanese)

2nd International Symposium on Field Measurements in Geomechanics, Sakurai (ed.)
© 1988 Balkema, Rotterdam. ISBN 90 6191 778 6

Assessment of tunnel face stability by back analysis

M.Hisatake
Kinki University, Osaka, Japan

1. INTRODUCTION

Mechanical constants and initial stresses of the ground are the most important parameters affecting tunnel movements in squeezing ground. From a practical viewpoint in tunnel construction, it is very useful to evaluate the equivalent mechanical constants which reflect not only the properties of the sample, but also the effects of joints and non-homogeneity of the material on tunnel movements. Recently research on back analysis methods has started to estimate the equivalent mechanical properties of rocks from measured displacements [1].

The most important problem in tunneling is how to assess the stability of the tunnel face. Back analysis is expected to give answers to this problem. In the assessment of tunnel face stability, geometrical characteristics around the face should be taken into account. But most back analysis methods presented previously are not three dimensional but two dimensional. By following two dimensional back analysis, it is difficult to get a reasonable solution to this problem.

In the assessment of tunnel face stability, stresses and strains ahead of the face should be known. For this purpose, the equivalent mechanical properties and the excavation stresses released at the face must be determined beforehand.

In this study, in order to evaluate stresses and strains ahead of a tunnel face from displacements measured in the tunnel, a three dimensional back analysis method is proposed, in which equivalent mechanical constants of the ground and excavation nodal forces at the tunnel face are estimated from equilibrium conditions of the excavation nodal forces.

2. BACK ANALYSIS METHOD

The excavation of a tunnel face can be expressed by the following stiffness equation in the finite element method,

$$\begin{Bmatrix} g \\ g_1 \\ g^*_2 \\ g^*_3 \end{Bmatrix} = \begin{bmatrix} k_{11} & k_{12} & k_{13} & k_{14} \\ k_{21} & k_{22} & k_{23} & k_{24} \\ k_{31} & k_{32} & k_{33} & k_{34} \\ k_{41} & k_{42} & k_{43} & k_{44} \end{bmatrix} \begin{Bmatrix} u \\ u^*_1 \\ u_2 \\ u^*_3 \end{Bmatrix} \qquad (1)$$

where, k_{ij} is a component of a stiffness matrix, g is the excavation nodal force released, g_1 is the force at the fixed points where displacements are constrained. g^*_3 is the force at the points of inner boundary of the lining where displacements are measured. g^*_2 equals the forces at the inner boundary of the lining and at the interior of both the ground and the lining where displacements are not measured. u, u_1, u_2 and u^*_3 are the displacements at the points corresponding to the points of g, g_1, g_2 and g^*_3, where known values have the symbol of *. Because g^*_2, g^*_3 and u^*_1 are all 0, g can be expressed with measured displacements(u^*_3) from Eq.(1),

$$\{g\} = [C] \{u^*_3\} \qquad (2)$$

where,

$$[C] = -[k_{11} \ k_{13}] \begin{bmatrix} k_{31} & k_{33} \\ k_{41} & k_{43} \end{bmatrix}^{-1} \begin{bmatrix} k_{34} \\ k_{44} \end{bmatrix} + [k_{14}]$$

When the number of components of u^*_3 is greater than that of g, a similar equation like Eq.(2) can be obtained by using the least squares method. For example, from Eq.(2), the equilibrium relationship of the excavation nodal forces(g_v) in the vertical direction is

$$\sum_i g_{vi} - W = \sum_i \sum_j C_{ij} u^*_{3j} - W \qquad (3)$$

In the above equation, C_{ij} contains the unknowns of the modulus of elasticity(E) and Poisson's ratio.

By the way, in ordinary back analysis of the direct formulation method, the following objective function(J) is made minimum to find out material constants,

$$J = \sum_{j=1}^{N} (u^*_j - u_j)^2 \qquad (4)$$
(N :Number of measured displacements(u^*_j))

This method, however, has a serious shortcoming because it is impossible to calculate u_j without guessing the value of the excavation nodal forces(g) beforehand. The values of "g" are strongly affected by the initial stresses of the ground and by the external stresses acting on the lining behind the tunnel face, and which are impossible to guess. On the other hand, if real values of E and Poisson's ratio are given in Eq.(3), the right hand side of Eq.(3) becomes 0. Therefore, if the right hand side of Eq.(3) is chosen as objective function(J_1), the

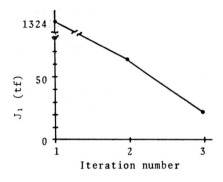 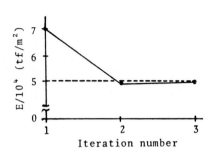

Fig.1 Convergency characteristics

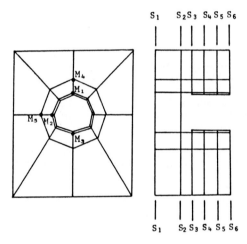

Fig.2 FE meshes used

values of the material constants may be found out without guessing the values of the excavation nodal forces(g),

$$J_1 = \sum_i \sum_j C_{ij} u^*_{3j} - W \qquad (5)$$

Eq.(5) does not contain "g", but contains the unknowns E and Poisson's ratio. The value of Poisson's ratio does not have much effect on the estimation of "E", so it is possible to give a value for Poisson's ratio before doing back analysis.

In order to check this theory, convergency characteristics "J_1" and "E" are shown in Fig.1, in which displacements of the nodal points on the inner boundary of the lining are given and the simplex method is used. Also, Fig.2 shows the finite element meshes used, in which tunnel excavation width=5m, one excavation length=1.25m, lining thickness=10cm, initial stress ratio of horizontal and vertical direction=0.5, initial stress of vertical direction=200tf/m^2, E=50,000tf/m^2, Poisson's ratio of the ground=0.3, modulus of elasticity of the lining=1,000,000tf/m^2 and Poisson's ratio of the lining=0.15.   And the value of W(weight of rock masses which are taken out at the excavation) is neglected because W is very small compared to the excavation stresses in this case. Fig.1 indicates that the back analyzed value E(49,605 tf/m^2) is very close to the real value(50,000 tf/m^2). Namely, the validity of using Eq.(5) in the back analysis is confirmed with these results.

But in the above analysis, the simplex method, which is one of the optimization techniques, is used, and which is based on an iteration method. Therefore, the three dimensional finite element analysis has to be carried out many times(9 times in the above analysis), and this increases computation time and costs.
In order to overcome this shortcoming, Hisatake[2] showed the linear relationship between "J_1" and "E". By using these results, Eq.(5) can be expressed in such a modified form as

$$J_1 = aE + b \tag{6}$$

where, "a" and "b" are constants determined later.
From Eqs.(5) and (6), the value of E can be calculated easily without using the iteration method.
Firstly, calculate two J_1-values(JX1 and JX2) by giving two arbitrary values of "E1" and "E2" into Eq.(5).   Then, the constants "a" and "b" in Eq.(6) are calculated by giving the two sets of (JX1,E1) and (JX2,E2),

$$a=(JX1 - JX2)/(E1 - E2)$$
$$\tag{7}$$
$$b=JX1 - E1(JX1 - JX2)/(E1 - E2)$$

The value of the left hand side of Eq.(6) becomes 0 when the real value of E(=E_r) is given. By using this fact, E_r can be calculated with "a" and "b" as follows,

$$E_r = -b/a \tag{8}$$

After determination of the modulus of elasticity from Eq.(8), the excavation nodal forces released are calculated from Eq.(2).

3. ESTIMATION OF STRESSES AND STRAINS

Fig.3 shows the maximum shearing stresses in the lining caused by one excavation of the tunnel face.   In Fig.3, SA means sequence analysis with real input data and SA(Back) shows sequence analysis with back-analyzed values of "E" and nodal forces released.   In the back analysis, all displacements on the inner boundary of the lining are given except those in tunnel axial direction, and the incorrect Poisson's ratio(=0.4) is given which has an error of 33%.   From Fig.3, it may be understood that the lining stresses can be estimated with high accuracy by SA(Back).   But in this method, many input displacements have to be given, and this is a shortcoming of this method.   In the following, a method to decrease the number of input displacements is shown.

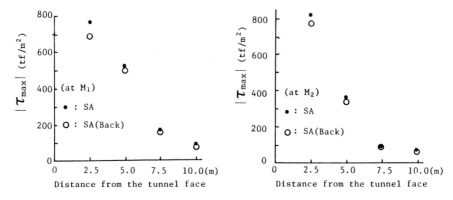

Fig.3 Comparison of the maximum shearing stresses calculated by SA and SA(Back)

Fig.4 shows lining displacements in the tunnel axial sections through crown(M_1), side wall(M_2) and invert(M_3) when the nodal forces(g_a and g_b) are given, respectively, at sections S_2 and S_3 in Fig.2. The displacements caused by g_a are, of course, much smaller than those by g_b. This result shows that the estimation of g_a from measured displacements is very difficult. Fig.5 shows a comparison between real g_a and g_a estimated from g_b directly($g_a=g_b$). From this result, it may be possible to estimate g_a by g_b from an engineering viewpoint. By following this method, the number of unknown nodal forces(g_a) can be decreased. This means that the same number of input displacements measured can be decreased.

Figs.6 and 7 show comparisons of the maximum shearing stress of the lining and the maximum shearing stress ahead of the tunnel face, respectively, between SA and SA(Back). M_1 and M_2 are shown in Fig.2. From these results, not only the stresses of the lining behind the tunnel face, but also the strains ahead of the tunnel face can be estimated from the displacements measured behind the tunnel face. In other words, the stability of the tunnel face may be assessed and reasonable support systems may be determined with the help of the back analysis results.

4. CONCLUSIONS

In order to assess the stability of the tunnel face, equivalent mechanical properties of the ground and excavation stresses released at the tunnel face are determined by the three dimensional back analysis method proposed, in which input displacements are measured behind the tunnel face. Through numerical analyses, the validity of this method has been confirmed.

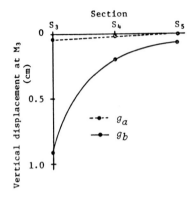

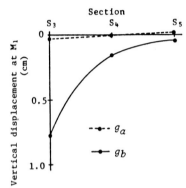

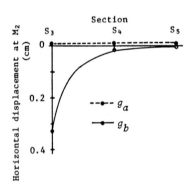

Fig.4 Displacements on the
inner boundary of the
lining when excavation
nodal forces g_a and g_b
are given respectively

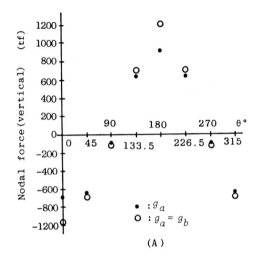

(A)

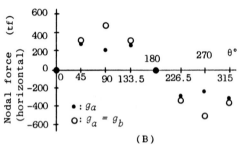

(B)

Fig.5 Comparison of the maximum
shearing stresses by SA and
SA(Back)

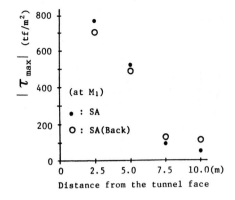

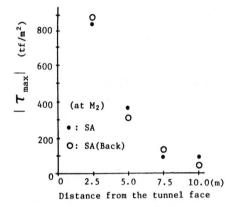

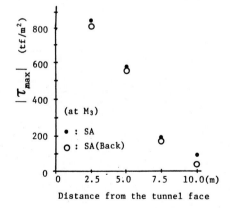

Fig.6 Comparison of the maximum shearing stresses by SA and SA(Back) in which g_a is determined by g_b

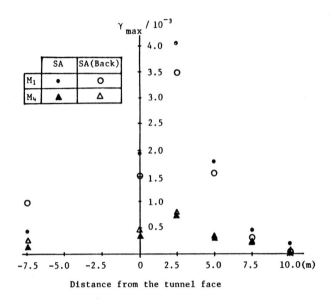

Fig.7 Comparison of the maximum shearing strain calculated by SA and SA(Back)

REFERENCES

[1] HISATAKE, M. and ITO, T.: Back analysis for tunnels by optimization method, Proc. 5th Int. Conf. on Numerical Methods in Geomech., pp.1301–1307, Vol.2, 1985.
[2] HISATAKE, M.: Three dimensional back analysis for tunnel, Proc. Int. Symp. Eng. Complex Rock Formations, pp.791–797, 1986.

2nd International Symposium on Field Measurements in Geomechanics, Sakurai (ed.)
© 1988 Balkema, Rotterdam. ISBN 90 6191 778 6

Technique of field measurement interpretation in erecting underground openings of round section

N.S.Bulychev & I.I.Savin
Tula Polytechnical Institute, USSR

INTRODUCTION

Design of multi-layer lining of round cross-section underground working by the results of stress measurements (normal and tangential) and deformations upon an arbitrary contact of a monolayer lining or upon a contact of the lining
with the rock body undergoes investigation in work presented. A specific feature of the design is the possibility of
fully reestiblishing stresses in lining by the results of
measuring some arbitrary kind of stresses or deformations.

The design is made applying the elastic model of contact
interraction of the lining with the rock body. The rock body appears a linearly deformed medium. Deep embedding wor-
kings $\mathcal{H} \gg R$ ($\mathcal{H}$ – the depth of bedding, R – the radius of
working) is under investigation. The inverse problem un-
dergoes its solution: stress initial field specifications
are estimated by the normal radial or tangential stress
known, deformation or displacements: (coefficient of lateral pressure λ and the coefficient α^* which takes into
account the lining erection lagging behind rock exposure).
    The following expressions have been received in general
for coefficients stated above :

$$\lambda = \frac{\mathcal{H}_n (D_1 A_{22} - D_2 A_{12}) - 2(D_2 A_{11} - D_1 A_{21})}{\mathcal{H}_n (D_1 A_{22} - D_2 A_{12}) + 2(D_2 A_{11} - D_1 A_{21})};$$

$$\alpha^* = \frac{(D_1 A_{22} - D_2 A_{12})(1 + \mathcal{H}_n)}{\gamma^{\mathcal{H}} (A_{11} A_{22} - A_{12} A_{21})(1 + \lambda)},$$

where $\mathcal{H}_n$ – is the coefficient of the type of rock body
               stressed state

$$\mathcal{H}_n = 3 - 4\mu_n$$

μ_n – is the rock Poisson coefficient ;

γ – is the weight of the rock in a volume unit, MH/m^3;
H – is the depth the working is sunk at, m ;
$A_{11}, A_{12}, A_{21}, A_{22}, D_1, D_2$ – the coefficients which are esti-
         mafor every particular case of measurements.
 Order of estimating unknown coefficients A_{11}, A_{12}, A_{21},
A_{22}, D_1, D_2 :
1. In case if there N measurements of normal radial stresses upon an external outline of an arbitrary i – th layer,
the coefficients stated above are estimated by the formulae:

$$A_{11} = N\,W_{o(i+1)} \;;\qquad A_{12} = W_{2(i+1)}\sum_{j=1}^{N}\cos 2\theta_j \;;$$

$$A_{21} = W_{o(i+1)}\sum_{j=1}^{N}\cos 2\theta_j \;:$$

$$A_{22} = W_{2(i+1)}\sum_{j=1}^{N}\cos^2 2\theta_j \;;$$

$$D_1 = \sum_{j=1}^{N} P_{(i)}^j \;;\qquad D_2 = \sum_{j=1}^{N} P_{(i)}^j \cos 2\theta_j \;,$$

where

$$W_{o(n)} = K_{o(n)} \;;\qquad W_{2(n)} = K_{11(n)} \;;\qquad Z_{2(n)} = K_{21(n)} \;;$$

$$\{W_i\} = [K_i]\{W_{i+1}\} \;;$$

$$\{W_i\} = \begin{Bmatrix} W_{o(i)} \\ W_{2(i)} \\ Z_{2(i)} \end{Bmatrix} \;;\qquad [K_i] = \begin{bmatrix} K_{o(i)} & 0 & 0 \\ 0 & K_{11(i)} & K_{12(i)} \\ 0 & K_{21(i)} & K_{22(i)} \end{bmatrix}$$

Here K_o, K_{11}, K_{12}, K_{21}, K_{22} – are the coefficients of external loads transfer / 1 / ;
 θ_j is the angle of coordinate of the point of the j -th
measurement ($j = 1,2,\ldots, N$)
 $P_{(i)}^j$ is the value of measured normal radial stresses upon an
external outline of the i -th layer.
 Shaft lining design of the "Centralnaya" coal mine elaborated by the results of radial stresses measurements upon
the contact of lining with the rock body. The results of
the design are given in fig. 1.
2. If there are N measurements of normal tangential stresses $\sigma_{\theta(i)}^j$ upon the internal outline of the arbitrary i -th
layer, the unknown coefficients A_{11}, A_{12}, A_{21}, A_{22}, D_1,
D_2 are estimated by the formulae :

$$A_{11} = N\left(W_{o(i+1)}m_1 - W_{o(i)}m_2\right);$$

$$A_{12} = -\left(W_{2(i+1)}n_1 - Z_{2(i+1)}n_2 - W_{2(i)}n_3 + Z_{2(i)}n_4\right)\sum_{j=1}^{N}\cos 2\theta_j \;;$$

$$A_{21} = -\left(W_{o(i+1)}m_1 - W_{o(i)}m_2\right)\sum_{j=1}^{N}\cos 2\theta_j \;;$$

$$A_{22} = \left(W_{2(i+1)}n_1 - Z_{2(i+1)}n_2 - W_{2(i)}n_3 + Z_{2(i)}n_4\right)\sum_{j=1}^{N}\cos^2 2\theta_j \;;$$

$$D_1 = \sum_{j=1}^{N} \sigma_{\theta(i)}^{j} \; ;$$

$$D_2 = -\sum_{j=1}^{N} \sigma_{\theta(i)}^{j} \cos 2\theta_j \; ,$$

where coefficients m_1 , m_2 , n_1 , n_2 , n_3 , n_4 are estimated for i-*th* layer coming from the geometric spe-cifications of the layer in question / 1 /.

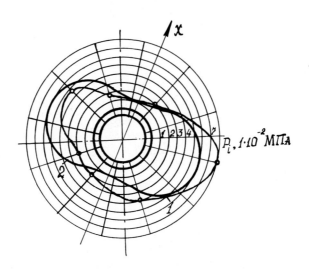

Fig. 1
1. is the epure of measured stresses
2 is the epure of designed stresses

Epures of measured and designed normal tangential stres-ses upon the cage shaft concrete lining internal outline of the "Centralnaya" coal-mine of the Donskoy are given in fig. 2.
3. In the case if the normal tangential stresses $\sigma_{\theta(i)}^{j}$ upon an external outline of the arbitrary i -th layer of multi-layer lining are measured, the coefficients A_{11} , A_{12} , A_{21} , A_{22} , D_1 , D_2 are estimated by the formulae :

$$A_{11} = N(W_{0(i+1)} m_1' - W_{0(i)} m_2') \; ;$$

$$A_{12} = (W_{2(i+1)} n_1' - W_{2(i)} n_3' - Z_{2(i+1)} n_2' + Z_{2(i)} n_4') \sum_{j=1}^{N} \cos 2\theta_j \; ;$$

$$A_{21} = (W_{0(i+1)} m_1' - W_{0(i)} m_2') \sum_{j=1}^{N} \cos 2\theta_j \; ;$$

$$A_{22} = (W_{2(i+1)} n_1' - W_{2(i)} n_3' - Z_{2(i+1)} n_2' + Z_{2(i)} n_4') \sum_{j=1}^{N} \cos^2 2\theta_j \; ;$$

$$D_1 = \sum_{j=1}^{N} \sigma_{\theta(i)}^{j} \; ;$$

$$D_2 = \sum_{j=1}^{N} \mathfrak{S}_{\theta(i)}^{j} \cos 2\theta_j \; .$$

Here the coefficients m_1' , m_2' , n_1' , n_2' , n_3' , n_4' are estimated in accordance with /1/.

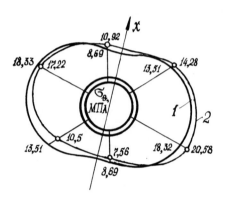

Fig. 2
1 - is the epure of measured stresses
2 - is the epure of designed stresses

4. The case of measuring normal tangential stresses in ring lining fitting is spread. The specific feature of the design in this particular case is that the mean values of normal tangential stresses $\bar{\mathfrak{S}}_{\theta(i)}^{d}$ in ring fitting sections are measured as a rule. In this case coefficients A_{11} , A_{12} , A_{21} , A_{22} , D_1 , D_2 are estimated by the formulae :

$$A_{11} = N \frac{E_2}{2\overline{E}_i} \left[W_{0(i+1)}(m_1 - m_1') - W_{0(i)}(m_2 - m_2') \right];$$

$$A_{12} = -\frac{E_2}{2\overline{E}_i} \left[W_{2(i+1)}(n_1 - n_1') - Z_{2(i+1)}(n_2 - n_2') - W_{2(i)}(n_3 - n_3') + Z_{2(i)}(n_4 - n_4') \right] \sum_{j=1}^{N} \cos 2\theta_j \; ;$$

$$A_{21} = -\frac{E_2}{2\overline{E}_i} \left[W_{0(i+1)}(m_1 - m_1') - W_{0(i)}(m_2 - m_2') \right] \sum_{j=1}^{N} \cos 2\theta_j \; ;$$

$$A_{22} = \frac{E_2}{2\overline{E}_i} \left[W_{2(i+1)}(n_1 - n_1') - Z_{2(i+1)}(n_2 - n_2') - W_{2(i)}(n_3 - n_3') + Z_{2(i)}(n_4 - n_4') \right] \sum_{j=1}^{N} \cos^2 2\theta_j \; ;$$

$$D_1 = \sum_{j=1}^{N} \bar{\mathfrak{S}}_{\theta(i)}^{d} \; ;$$

$$D_2 = \sum_{j=1}^{N} \bar{\mathfrak{S}}_{\theta(i)}^{d} \cos 2\theta_j \; ,$$

where E_2 is the elastic modulus of the fitting material, MPa;

$\bar{E}_i$ is the given elastic modulus of the lining layer in question, MPa ;

E_1 is the elastic modulus of the filler material, MPa;

F_i' is the degree of reinforcement ;

$$\bar{E}_i = E_1(1-F_i') + E_2 F_i' \; ;$$

$$F_i' = \frac{F_2}{F_1 + F_2} \; ;$$

F_1 is the area occupied by the filler, m^2;

F_2 is the area occupied by the fitting, m^2;

5. If radial displacements (convergences) $U_{(i)}^d$ of the internal outline points of the arbitrary i-th layer of a multilayer lining have been measured, the coefficients A_{11}, A_{12} , A_{21} , A_{22} , D_1 , D_2 are estimated by the formulae :

$$A_{11} = N\frac{R_{i-1}}{4\,G_i(C_i^2-1)} \left(W_{0(i+1)}\,d_1 - W_{0(i)}\,d_2\right);$$

$$A_{12} = \frac{R_{i-1}}{12\,G_i\,D_i}\left[W_{2(i+1)}(a_1+3a_1')-Z_{2(i+1)}(a_2+3a_2')-W_{2(i)}(a_3+3a_3')+Z_{2(i)}(a_4+3a_4')\right]\sum_{j=1}^{N}\cos2\theta_j ;$$

$$A_{21} = \frac{R_{i-1}}{4\,G_i(C_i^2-1)}\left(W_{0(i+1)}\,d_1 - W_{0(i)}\,d_2\right)\sum_{j=1}^{N}\cos2\theta_j \; ;$$

$$A_{22} = \frac{R_{i-1}}{12\,G_i\,D_i}\left[W_{2(i+1)}(a_1+3a_1')-Z_{2(i+1)}(a_2+3a_2')-W_{2(i)}(a_3+3a_3')+Z_{2(i)}(a_4+3a_4')\right]\sum_{j=1}^{N}\cos^2 2\theta_j ;$$

$$D_1 = \sum_{j=1}^{N} U_{(i)}^d \; ; \qquad D_2 = \sum_{j=1}^{N} U_{(i)}^d \cos 2\theta_j \, .$$

where R_{i-1} is the internal radius of the lining layer being designed m ;

G_i is the Young's modulus of the lining layer material being designed, MPa ;

coefficients D_i , c_i , d_1 , d_2 , a_1 , a_1' , a_2 , a_2' , a_3 , a_3' , a_4 , a_4' are estimated for lining layer being designed in accordance with /1/.

6. If measurements of tangential deformations $\varepsilon_{\theta(i)}^i$ upon the internal outline arbitrary i-th layer are conducted, the coefficients A_{11} , A_{12} , A_{21} , A_{22} , D_1 , D_2 estimated by the formulae :

$$A_{11} = \frac{N}{2\,G_i}\left\{W_{0(i+1)}\,m_1(1-\mu_i)-W_{0(i)}\left[m_2(1-\mu_i)+\mu_i\right]\right\} ;$$

$$A_{12} = \frac{1}{2\,G_i}\left\{W_{2(i+1)}\,n_1(1-\mu_i)-Z_{2(i+1)}\,n_2(1-\mu_i)-W_{2(i)}\left[n_3(1-\mu_i)-\mu_i\right]+ Z_{2(i)}(1-\mu_i)n_4\right\}\sum_{j=1}^{N}\cos2\theta_j \; ;$$

$$A_{21} = \frac{1}{2\,G_i}\left\{W_{0(i+1)}\,m_1(1-\mu_i)-W_{0(i)}\left[m_2(1-\mu_i)+\mu_i\right]\right\}\sum_{j=1}^{N}\cos2\theta_j \; ;$$

$$A_{22} = \frac{1}{2G_i} \{ W_{2(i+1)} n_1 (1-\mu_i) - Z_{2(i+1)} n_2 (1-\mu_i) - W_{2(i)} [n_3 (1-\mu_i) - \mu_i] +$$
$$+ Z_{2(i)} (1-\mu_i) n_4 \} \sum_{j=1}^{N} \cos^2 2\theta_j \ ;$$

$$D_1 = \sum_{j=1}^{N} \varepsilon'_{\theta(i)} \ ;$$

$$D_2 = \sum_{j=1}^{N} \varepsilon'_{\theta(i)} \cos 2\theta_j \ .$$

where μ_i is the Poisson coefficient of material of the lining layer in question.

CONCLUSION

The specific feature of the design technique given is that specifications of the initial stressed state of the body λ and α^* estimated as a result of the design do not depend on the geometric and deformation specifications of the lining in which the measurements were conducted and as a result of the independence mentioned can be used for designing other construction of lining, operating in similar conditions.

REFERENCES

Bulychev, N.S. Mechanics of underground construction.
  Moscow , "Nedra", 1982 , 270 p.
Fotieva, N.N. & Bulychev, N.S. 1979, Using data of full-
  scale measurement in lining design for underground struc-
  tures. Proc, Fourth Congr. of the Intern. Soc, for Rock
  Mech; Montreus, Switzerland, Sept. 02.08, vol. 1:387-392.

Analysis of plastic zone around underground openings

Xu Wen-Huang & Huang Zheng
Department of Civil Engineering, Southwestern Jiaotong University, Emei, Sichuan, People's Republic of China

ABSTRACT

This paper provides a group of ϕ-C curves which may be used
to determine the maximal radius of plastic zone around underground openings.These curves are drawn by nonlinear FEM analy-
sis.Once the cohesion C and the internal friction angle ϕ are
given,we may get the radius of plastic zone directly from the
ϕ-C curves.It is conveninet to use for engineers.

1 INTRODUCTION

As is well known,the analysis of plastic zone around underground openings is important for engineers in geomechanics
field.The stability of the surrounding rock around an underground opening depends upon the stress state of the rock,and
especialy upon the range of the plastic zone around the ope-
ning.During the excavation of an opening,it is very important
to know the extent of the plastic zone occurring around the
opening whenever necessary.
    The methods mostly used in the analysis of the plastic zone
are numerical methods,such as finite element method(FEM),boun-
ary element method(BEM),etc..S.Sakurai and others (1985) present a method to evaluate the plastic zone around underground
openings,which is applicable and effective to engineering practice,and think that a method,in which the field measurements,
laboratory experiments,coputing and analysis are more closely
combined,may be more effective.The application of the curves
of cohesin C versus internal friction angle ϕ proposed in this
paper is such a method.
    If the internal friction angle and the cohesion of the sur-
rounding medium around an underground opening are determined
by laboratory experiments,the maximal extent of the plastic
zone of the medium can be get from the ϕ-C curves under the
corresponding initial stresses which may be obtained by back
analysis of the displacements measured during the excavation
of the opening (Sakurai and Takeuchi 1983).Obviously,it is
very helpful for monitoring the stability of underground openings.

2 DESCRIPTION OF   C CURVES

To a given underground opening,we call all the factors inf-
luncing the range of the plastic zone around the opening,ex-
cept the cohesion and the internal friction angle,an external
condition.

If an external condition and a pair of ϕ,C are given,we can
obtain,by mechanical analysis,a ratio R,that is

$$R=Rmax/Ro \qquad (1)$$

where Rmax is the maximal radius of the plastic zone and Ro
the radius of the opening.

To a given external condition, a ratio R is obviously a function of the parameters ϕ,C only.Thus the following representation appears

$$R=R(\phi,C). \qquad (2)$$

Let R equal to a constant R1,then due to Eg.(2)

$$R1=R(\phi,C). \qquad (3)$$

We should keep in mind that R1 is the specified maximal radius of the plastic zone per unit radius of the opening.Giving
a C,therefore,a corresponding ϕ,which satisfies the Eq.(3)
together with the given C,may be get.If n pairs of and C have
been found by the same way,we can make n points in a ϕ-C plane,with which the curve of R=R1 may be drawn in the plane.

As a result,m such curves in the -C plane may be drawn by
making R=R1,R2,...,Rm respectively and the equation of the ith
curve with R=Ri reads

$$Ri=R(\phi,C), \qquad i=1,2,...,m \qquad (4)$$

where Ri, i=1,2,...,Rm are distinct constants.

The curves of the Eq.(4) are shown in Fig.1.

Eq.(2) is the basic equation of drawing the ϕ-C curves.But
generally speaking,the analytical representation of the function R(ϕ,C) is difficult to get because of the complexity of
underground opening problems.We take,consequently,the nonlinear finite element analysis as an approximation of Eq.(2).

With the finite element analysis,the following assumptions
are made:

1.The surrounding medium around an underground opening is
isotropic,homogeneous and continuous.

2.The medium is elasto-ideal-plastic and obeys Mohr-Coulomb
yielding criterion.

3.No linings and their effects are taken into account.

4.Only plane-strain condition is considered.

In this paper,all the ϕ-C curves are drawn by means of the
FEM.

After having adquate ϕ-C curves,under a given external condition,the maximal radius of the plastic zone may be determined directly from the curves once the cohesion and the internal friction angle of the surrounding medium are known.

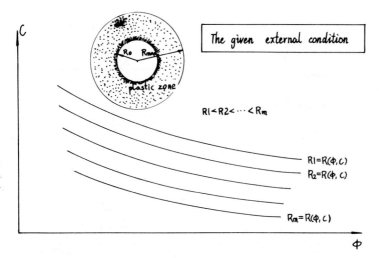

$Rl < R2 < \cdots < R_m$

$Rl = R(\phi, c)$
$R_2 = R(\phi, c)$

$R_m = R(\phi, c)$

ϕ

Fig.1. The $\phi - C$ curves under a given external condition

3 APPLICATION OF THE ϕ-C CURVES

Set R=1.0,1.1,1.2,1.3 respectively,then the corresponding ϕ-C curves,under the given external condition,are drawn in Fig.2 by the method described in previous section.

The curve of R=1.0 may be called "critical line",that is say ,when a point (ϕ,C) is above the line,the surrounding medium whose parameters ϕ and C make up the point is in elastic sta- te.And the medium is in elasto-plastic state if the point is below the line.If the point is on the line,we have Rmax=Ro, and the medium is in "critical state".

As examples,now we give three different points Pk(ϕk,Ck), k=1,2,3.Every one of the three points should correspond an su- rrounding medium.The positions of the three points in the ϕ-C plane are shown in Fig.2.

The point P1($\phi 1$,C1) is above the curve of R=1.0 (critical line,cf. Fig.2),that is, the surrounding medium with 1 and C1 is in elastic state.

The point P2($\phi 2$,C2) falls on the critical line,then Rmax=Ro for the surrounding medium with $\phi 2$ and C2.

The point P3($\phi 3$,C3) is below the critical line and on the curve of R=1.2,therefore the surrounding medium with $\phi 3$ and C3 is in elasto-plastic state and has Rmax=1.2Ro.

With the illustration above,we have seen that,only if ade- quate ϕ-C curves are made and the values of ϕ and C of the su- rrounding medium are known,the maximal radius the plastic zone of the medium may be easily determined from appropriate ϕ-C curves with the accuracy required in engineering practice and no further evalution is demanded.

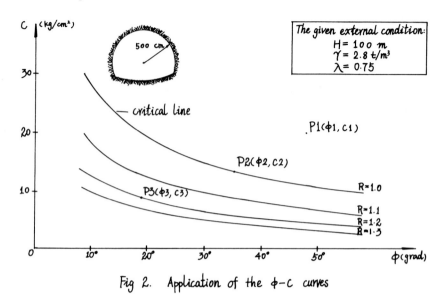

Fig 2.   Application of the ϕ-c curves

4 INFLUENCE OF THE EXTERNAL CONDITION

As mention in previous sections,a group of ϕ-C curves can be
used only under a corresponding external condition.Different
groups of the curves should be made for different external
conditions.But it is impossible to draw a group of the curves
for every one of all external conditions which are possible
to exist in engineering problems.
    We hope to make finite groups of the curves suitable for
changeful external conditions.

The main influencing factors
included in an external condition are the depth H of the
opening embeded underground,
the unit weight γ of the surrounding medium and the ratio
λ of horizontal component of
the initial stresses of the
medium to vertical one of the
initial stresses.The radius
of the opening has obviously
no influence upon the ratio
R.
    An idealized external con-
dition is given in Fig.3.

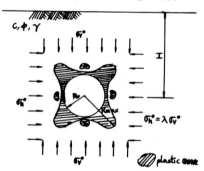

Fig.3. An idealized external condition

We take γH as an influencing factor instead of considering
γ and H respectively.Let σh and σv represent the horizontal
and vretical components of initial stresses respectively,and
$\sigma h=\lambda\sigma v$.Thus,an external condition may be expressed only by
the initial stresses if let $\sigma v=\gamma H$.
    A few groups of the ϕ-C curves have been made for some diff-
rent initial stress states (Xu Wen-Huan et al 1986).

5 CLOSURE

Summarizing the discussions above,the ϕ-C curves proposed in this paper is simple and efficient in determining the maximal radius of the plastic zone of the surrounding medium around an underground opening.The curves required can be finished in doors and used in engineering practice with the initial stresses measured in field or back-calculated by the displacements measured in field.

The cohesion and the internal friction angle may be obtained by experiments in laboratory,but it is also possible to determine the two parameters by back analysis of the measurements in field (Gioda & Maier 1980).

REFERENCES

Sakurai,S. & Shimizu,N. 1985. Evaluation of plastic zone around underground openings by means of displacement measurements. 5th Int. Conf. Numerical Method in Geomechanics. 1:111-118.
Sakurai,S. & Takeuchi,K. 1983. Back analysis of measured displacements of tunnels. Rock Mechanics and Rock Engineering. 16(3):173-180.
Gioda,G. & Maier,G. 1980. Direct search solution of an inverse problem in elastoplasticity:identification of cohesion,friction angle and in situ stress by pressure tunnel tests. Int. J.Numer.Meth.Engng. 15:1823-1843.
Xu,W.H. & Liu,G.X. 1986. Relation between rock classification and parameters of rock (in chinese). Research Report. Department of Civil Engineering, Southwestern Jioatong University,China.

2nd International Symposium on Field Measurements in Geomechanics, Sakurai (ed.)
© 1988 Balkema, Rotterdam. ISBN 90 6191 778 6

Assessment of the stability of the ground surrounding a shallow tunnel by means of a back analysis method of measured displacements

Kaiei Sakaguchi
Ishikawa Prefecture, Wajima-shi, Japan
Tatsutoshi Kondoh, Yukihiko Okabe & Masato Shinji
OYO Corporation, Tokyo, Japan

1 INTRODUCTION

It is very important to obtain a continuous feedback of information for the estimation of the stability of the ground surrounding a tunnel, and for use in the design of underground structures such as tunnels, underground powerhouses and nuclear waste disposal stations. The feedback of information in the form of field measurements is called the "Observational Procedure", and is especially vital for providing the information on underground structures in shallow and soft ground scheduled for excavation.

The Shin-Ushizu Tunnel, located in Ishikawa Prefecture in the northern part of Honshu-Island, was under construction as part of an improvement project on National Road No. 249. Fig. 1 shows a geological section of this tunnel. In this figure, the interval between Stations 25 and 37 consists of a shallow overburden and weak rock media. There are a lot of houses and a public school located on the surface of this ground. Consequently, it was imperative to protect against settlement caused by tunnel excavation. At first, excavation of the tunnel using H type steel ribs with the wooden sheet pile method was planned. However, the design was modified for that interval, and another method, aimed at avoiding settlement as much as possible, was adopted. The interval to which this method was applied was 240m long. Firstly, side pilot tunnels were constructed prior to the upper section progress. To support the upper section, a combination of back filling concrete, rock bolting and fore-piling was used. Using Finite Element analysis, it was predicted that settlement of the ground surface would be 10mm maximum, so that we decided that

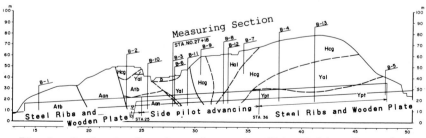

Fig.1 Geologic Section along the Shin-Ushizu Tunnel

the control value for feedback of information for the estimation of the stability of the ground surrounding the tunnel was 0.6% as a maximum shear strain. Ultimately, settlement was held down to 8.4mm, a satisfactory amount for safe tunnelling and protection of the surface.

In this paper, we deal with a method of excavation and a system for monitoring the excavation of tunnels. Also, we discuss the usefulness of the system of continuous feedback by means of a back analysis procedure.

2 GEOLOGY OF SHIN-USHIZU TUNNEL

In the area along the Shin-Ushizu Tunnel, the strata mainly consists of Miocene Anamizu formation, Yanagida formation and Higashi-Innai alternation. The Anamizu formation consists mostly of andesite and tuff breccia. The andesite has few cracks, and is very strong. A geological section of the ground along the tunnel is shown in Fig. 1.

The Yanagida formation consists mostly of sandstone and tuff, including a small pumice. This formation is very weak. The Higashi-Innai alternation consists mostly of conglomerate, sandstone and silt stone. This alternation is unconsolidated and soft. The Yanagida formation irregularly covers the anamizu formation. The conglomerate, including andesite breccia (ancient detritus), is distributed near the area at Station 28 (see Fig. 1). The Higashi-Innai alternation formation irregularly covers the Yanagida formation.

No major fault is present in this area, but a minor fault (2 cm in width) exists. This fault extends across the tunnel, and dips 40 degrees near Station 30. At Station 28, Andesite and the ancient detritus are partially sheared, and contain clay. A syncline extends across the tunnel at Station 32.

3 DESIGN OF THE TUNNEL TO PROTECT SETTLEMENT OF SURFACE

Table.1 shows one of the results of laboratory testing of sandstone from the Yanagida formation (Yal) in the area from Stations 28 to 32, and the Higashi-Innai alternation (Hal) in the area from Stations 32 to 33. It is very clear that the modulus of elasticity in these formations is very low. Therefore, it was expected that excavation of the alternation would be accompanied by an unstable behavior of the rock, tunnel face, failures, etc.

Table.1   Results of Laboratory Test

Tertiary Miocene		Density (g/cm^3)	Modulus of Elasticity (kg/cm^2)	Poisson's Ratio	Cohesion (Kg/cm^2)	Internal Friction Angle
Anamizu Formation	Andesite	2.0	9000.0	0.3	10.0	45°
Yanagida Formation	Sandstone	1.9	1300.0	0.45	1.0	30°
	Pumicetuff	1.5	2000.0	0.4	1.0	30°
Higashi-Innai Alternation	Sandstone Mudstone	1.8	800.0	0.45	3.0	25°
	Conglomerate	2.0	2000.0	0.4	1.5	40°

We used numerical simulation of the finite element method to investigate the excavation techniques, with a view to protecting against ground settlement. As a result, we adopted the following excavation technique (see Fig.2). Firstly, two sets of side pilot tunnels were excavated in the unstable formation. Secondly, we constructed a temporary side wall to support H type steel ribs in the side pilot tunnels. Thirdly, the upper section was constructed, using H type steel ribs, expand metals, fore-piling and rock bolting. After that, a temporary concrete lining instead of shotcrete lining was constructed. The effect of temporary lining is guaranteed by simulation of the finite element method. Finally, we excavated the lower section of the tunnel. Fig. 3 shows the design of this tunnel.

Fig. 4 shows the surface settlement that was forecast by the numerical simulation for this excavation technique. Fig. 4 also indicates the forecast surface settlement when there is no temporary concrete lining.

From these results and the results of laboratory tests, we predicted that the amount of surface settlement would be less than 10mm, and the critical strain ε_0 (Sakurai,1981) would be 1.0%.

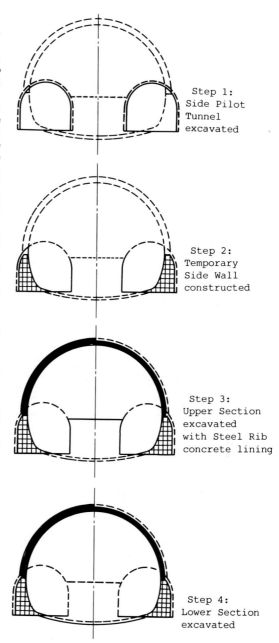

Step 1:
Side Pilot
Tunnel
excavated

Step 2:
Temporary
Side Wall
constructed

Step 3:
Upper Section
excavated
with Steel Rib
concrete lining

Step 4:
Lower Section
excavated

*Fig.2   Excavation Procedure of
the Shin-Ushizu Tunnel*

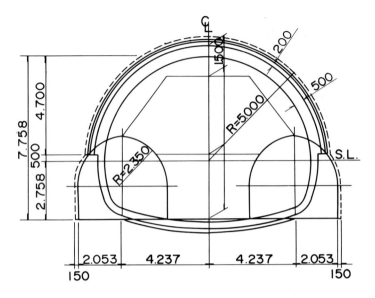

Fig.3   *General Section of the Shin-Ushizu Tunnel*

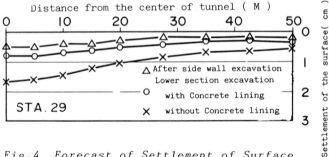

Fig.4   *Forecast of Settlement of Surface*
        by means of Numerical Simulation

4 FIELD MEASUREMENT

Field measurements were taken in the area between Stations 27 and 38. In this area, we measured the displacement of the ground during excavation. Fig. 5 shows the arrangement of a typical section of the measurement systems. In this figure, SM is the omission of the SLIDING MICROMETER-ISETH, and INC is the omission of OYO Q-Tilt.

   Fig.6 shows the final results of vertical displacement distribution by means of the SLIDING MICROMETER when the upper section was excavated. The final settlement of the surface is about 8mm. Fig.6 also shows the distribution of horizontal displacement of each borehole in the upper section that was excavated. It is clear from these figures that horizontal and vertical displacement appear similarly. However, horizontal displacement occurs immediately around the tunnel section,

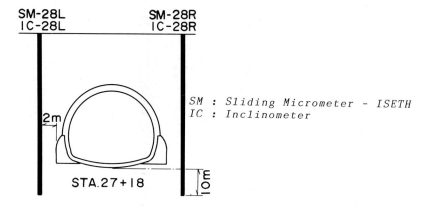

SM : Sliding Micrometer - ISETH
IC : Inclinometer

Fig.5 Arrangement of Measuring Instruments

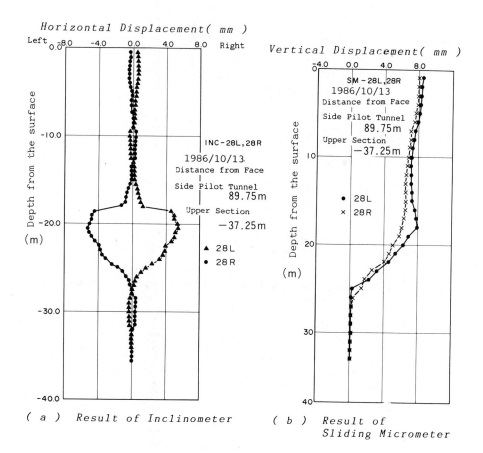

(a) Result of Inclinometer

(b) Result of Sliding Micrometer

Fig.6 Results of Field Measurements

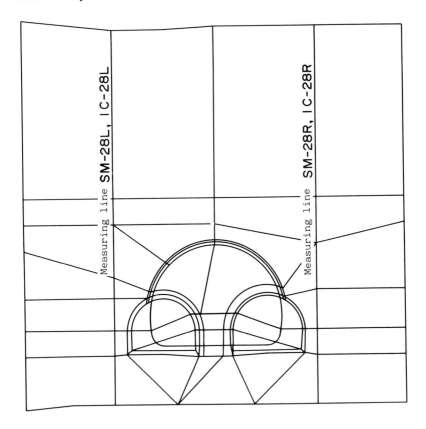

Fig.7   Finite Element Mesh by using this Study

and vertical displacement occurs continuously from the bottom of the tunnel section to the surface.

5 BACK ANALYSIS AND THE FORECAST OF SETTLEMENT BY MEANS OF FIELD MEASUREMENTS

We conducted back analysis using an NEC personal computer, the PC-9801 VM2, to estimate the stability of tunnel excavation and the surface settlement. This computer system located near the tunnelling site. The back analysis program is "DBAP/M (Direct Back Analysis Program for Microcomputer)" proposed by one of the authors. Fig.7 shows a part of the finite element mesh of this study. The process of back analysis can be done in time together with graphic display.

Fig.8 compares the actual displacement, as measured by the sliding micrometer and inclinometer, with the displacement determined by back analysis.

The results of back analysis using the data in Figs.7 and 8 are as follows;

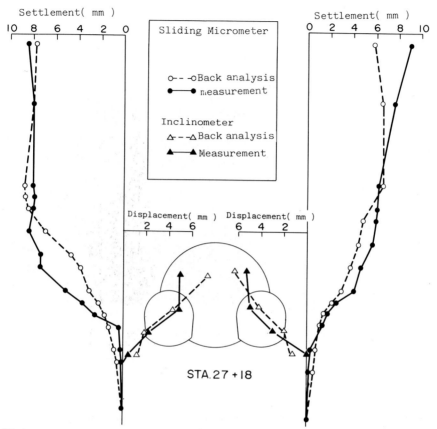

Distance from the face(Upper Section): 19.75M

Fig.8   Comparison between Measured and Calcuclated displacements

$$\sigma_{x0} / E = -0.4845 \times 10^{-2} \qquad (1)$$

$$\sigma_{y0} / E = -0.2612 \times 10^{-2} \qquad (2)$$

$$\tau_{xy0} / E = 0.6041 \times 10^{-4} \qquad (3)$$

where, σ_{x0} /E, σ_{y0} /E, τ_{xy0} /E are the initial state of stress existing on the section being excavated, and E is the modulus of elasticity of the ground.

If we assume the σ_{y0} as the overburden pressure, the modulus of elasticity of the ground is as follows;

$$\sigma_{y0} = \gamma \times H = 2.0 \text{ gf/cm}^3 \times -2300 \text{ cm}$$
$$= -4.6 \text{ kgf/cm}^2 \ (-0.45 \text{ Mpa}) \qquad (4)$$

Substitute Eq.(4) into Eqs.(1) and (3), we can obtain the other initial state of stress as follows;

$$\sigma_{x0} = -8.53 \ \text{kgf/cm}^2 \ (\ 0.84 \ \text{Mpa} \ )$$
$$\tau_{xy0} = 0.106 \ \text{kgf/cm}^2 \ (\ 0.01 \ \text{Mpa} \ )$$
$$E = 1760 \ \text{kgf/cm}^2 \ (\ 172 \ \text{Mpa} \ )$$

Fig.9 shows one of the results of back analysis. This computer graphics is the maximum shear strain distribution of the ground when the side pilot tunnels are excavated. Fig.10 also shows one of the results of back analysis when the upper section is excavated.

It is very clear from these figures that the maximum value of the maximum shear strain is enough smaller than the critical strain (0.6%) that the stability of this tunnel is guaranteed.

Fig.11 shows the distribution of settlement in the upper section during excavation. When there was enough distance between the upper section and the measuring section, the settlements of the surface of each sliding micrometer are almost 4mm respectively. When the upper section reached the measuring section, the surface settlement became almost 5mm. Finally, the surface settlement was 8.3mm.

At this tunnelling site, we calculated the surface settlement by the back analysis method. Fig.11 indicates the amount of the surface settlement forecast. In this figure, it is very clear that the forecast of settlement by numerical simulation before the tunnel excavation coincides with the actual settlement.

In addition, it is very clear that the feedback system and design of the

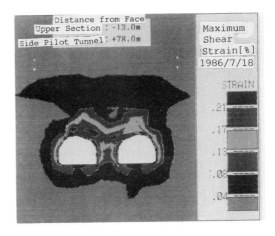

Fig.9   Final Maximum Shear Strain Distribution by means of Excavations of Side Pilot Tunnels

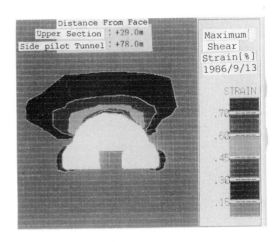

Fig.10   Final Maximum Shear Strain Distribution by means of Excavations of Upper Section

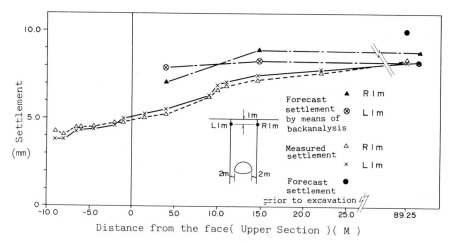

Distance from the face(Upper Section)(M)

Fig.11 *Comparison between Measured and Forecasted*
Settlements

tunnel proposed in this study are useful for protecting the settlement
of the surface during tunnel excavation.

6 CONCLUSION

Tunnel excavation in this section was successfully conducted in a
shallow overburden by applying a system of the continuous feedback of
information obtained by back analysis and precise field measurements.
The outcome was that the actual surface settlement was within the range
allowable for safety.
  In conclusion, the authors would like to express their gratitude to
Mr. Shinzaburo ODA of Hokuto-Gumi Corporation for his corporation in
allowing us to present the field measurement data on the Shin-Ushizu
Tunnel.

REFERENCES

SAKURAI, S., Direct Strain Evaluation Technique in Construction of
    Underground Openings., Proc. 22nd U.S. Rock Mechanics Symposium, MIT,
    pp. 278-282, 1981.

SAKURAI, S. and M. SHINJI, A Monitoring System for the Excavation of
    Underground Openings based on Microcomputers., Proc. Design and
    Performance of Underground Excavations, Cambridge, pp.471-476, 1984.

2nd International Symposium on Field Measurements in Geomechanics, Sakurai (ed.)
© 1988 Balkema, Rotterdam. ISBN 90 6191 778 6

Stability assessment of the ground around a shallow tunnel

K.Kimura
Japan Railway Construction Public Corporation

1 INTRODUCTION

For relieving commuter congestion, a new traffic network of railway lines has been planned and constructed in the Tokyo Metropolitan Area. The Japan Railway Construction Public Corporation has been serving an important role in accomplishing such projects.

For the past ten years, six shallow tunnels (except the Narita Airport Tunnel and the Kokubugawa Tunnel nearby it) and three other tunnels built by two other authorities have been excavated in diluvial sand. They are located in a north-east direction from the center of Tokyo and are listed in Table 1. Although the shield tunneling method is usually applied to the construction of tunnels in this district, the demand for many shallow tunnels allows us to develop an economic and safe tunneling method. For this purpose, we have attempted to use the new tunneling method called NATM. However, the new tunneling method has shortcomings, namely, that an unsupported area appears around the face, and it takes a longer period of time than the shield tunneling method to complete the cross-sectional tunnel supports. Therefore, much care should be taken due to the fear of trouble or the collapse of the ground.

We carried out one trial test at the Kuriyama Tunnel to assess the ground stability around the face, which was based on the characteristics of ground behavior derived from other shallow tunnelings.

This paper deals with the following matters,
1. Typical ground behavior around a shallow tunnel
2. The way to assess ground stability around a shallow tunnel with three indexes termed as Shear Index, Simple Shear Strain, and Maximum Surface Settlement
3. Its applications to the Kuriyama Tunnel

2 GEOLOGICAL AND ENVIRONMENTAL CONDITIONS

The north-east district of the Tokyo Metropolitan Area is a hilly district covered with thick sand layers and thin loam. The above-mentioned tunnels lie in diluvial sand layers called the Narita Layers, which occupy the greater part of it. A ground water table lies near the ground surface. The representative characteristics of diluvial sand are shown in Table 2,which gives us its low uniformity coefficient and rich deformability.

The material with low water content above the ground water table and the saturated sand run and ravel rapidly after being exposed by excavation.

Table 1. Tunnel list

TUNNEL NAME	GEOLOGY	EXCAVATION AREA (m²)	TUNNEL DIA.(m)	CROWN DEPTH /DIA. RATIO	SURFACE & GROUND WATER CONDITION	EXCAVATION METHOD
HORINOUCHI T.	Diluvial sand & Loam	85	10.7	0.39-1.01	Slope & Flat, A little water	Short bench with Supporting body
TOKKO T.	Diluvial sand & Loam	85	10.7	0.35-0.84	Almost flat, Little water	Short bench with Supporting body
KOMAINO T. (No.1)	Diluvial sand & Loam	85	10.7	0.24-0.86	Slope,Little water	Short bench with Supporting body
KOMAINO T. (No.2)	Diluvial sand & Loam	85	10.7	0.19-0.89	Slope,Little water	Short bench with Supporting body
OHNUKI T. SOUTHERN SEC.	Diluvial sand & Loam	36	5.96	0.35-1.54	Almost flat, Little water	Short bench with Supporting body
NARITA AIR-PORT(8th SEC)	Diluvial sand & Loam	136	14.3	0.34-0.57	Almost flat, A little water	With side drifts
NARITA AIR-PORT(9th SEC)	Diluvial sand & Loam	122	13.4	0.54-0.61	Almost flat, A little water	Bench with Temporary invert
KOKUBUGAWA	Diluvial fine sand & Loam	58.6	8.6	0.81-2.21	Flat, Much water drained off by wells	Short bench with Supporting body
KURIYAMA T. (YAGIRI SEC.)	Diluvial fine sand & Loam	71.8-90	10.23-12.39	0.91-1.03	Almost flat, A little water	Short bench or Side drifts

Table 2. Characteristics of
diluvial sand in
Narita Layers.

Items	Samples at the tunnel face		Samples from boring cores before tunnel excavation
Grain Composition	Sand	99%	94%
	Silt	1%	6%
Uniformity Coefficient	1.7		2.0
Water Content	15%		25 - 35%
Cohision (Cd)			0.35kg/cm²
Friction Angle (φd)			31 degree
Void ratio			0.84 - 1.05

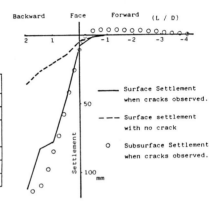

Fig. 1. Comparison between longitudinal settlement curves.

Table 3. Phenomena observed on the ground surface & in the ground.

GROUND SURFACE			SUBSURFACE		
NORMAL PHENOMENON	FAILURE	TUNNEL NAME	NORMAL PHENOMENON	FAILURE	TUNNEL NAME
1,Pre-upheaval or relative pre-upheaval (L -1m)		HT,TT, OT,			
	2,Transversal crack	KT,			
3,Pre-settlement		All.	3,Pre-settlement & Relative upheaval		All.
4,Enlargement of settlement trough		HT,		5,Local collapse	KT,HT, KWT,
6,Accelerative settlement (-3 L 3m)		All.	6,Accelerative settlement & Relative settlement		All.
	7,Failure of groung	HT,KT		8,Collapse	HT,TT, KT,KWT
9,Convergence of settlement (1m L)		All.	9,Convergence of settlement 10,Decrease of relative settlement		All. HT,
	11,Longitudinal crack	HT,KT			

Remarks; Definition
1) Pre-upheaval means the upheaval before the tunnel face reaches the measuring point.
2) Relative pre-upheaval means the phenomenon that the smaller settlement in the area can be seen comparing to the surroundings.
3)Transversal crack is the crack which occurs before the tunnel face reaches.
4) Pre-settlement means the settlement which occurs before the tunnel face reaches.
5) Relative upheaval means the phenomenon that the surface settlement is larger than the subsurface settlement.
6) Relative settlement means the phenomenon that the surface settlement is smaller than the subsurface settlement.
7) Longitudinal crack is the crack which can be seen along the tunnel axis apart from the point above the tunnel wall.
8) L ; Distance from the tunnel face, Minus means the condition before the tunnel face reaches.
9) HT :Horinouchi T., TT :Tokko T., OT :Ohnuki T., KT :Kuriyama T., KWT :Kokubugawa T.

This caused the serious problem of how to keep the face and crown stable.
The tunnel crown depth-tunnel diameter ratio varies from 0.19 to 2.21. The
Kuriyama Tunnel is located partly below a residential area and partly below
agricultural land.

3 CONSTRUCTION PROCEDURES

As shown in Table 1, the 8th section of the Narita Airport Tunnel and a
part of the Yagiri Section of the Kuriyama Tunnel were excavated with two
side drifts. The other tunnels were excavated with short bench. The bench
length was kept at a value nearly equal to each tunnel diameter.
   A dewatering system by deep wells from the ground surface and shallow ones
from the inside of the tunnels was applied to the saturated ground with
much water. This system worked successfully and didn't cause any problems
even in the rather densely populated district above the Kuriyama Tunnel.
In rather stable diluvium, the excavator with a rotating head was used, but
in unstable ground, power shovels and hand-held spades and knives were
applied. Besides these, many kinds of supplementary aids were used to keep
the crown and face stable because of loose sand and largeness of the
excavation area. Detailes are reported by Yokoyama et al. (1983) and
Horiuchi et al. (1986).

4 GROUND SETTLEMENT BEHAVIOR

Eleven typical settlement behaviors of the ground have been observed
through the tunnelings, which are shown in Table 3. Associated with tunnel
advance, we observed the following phenomena along the tunnel's center line.
   1. Pre-upheaval or relative pre-upheaval on the ground surface
Pre-upheaval means the upheaval before the tunnel face reaches the
instrument line, and relative pre-upheaval is the phenomenon that the
smaller settlement can be seen in the area relative to the surroundings.
   The phenomena are considered to be observed when the rigidity of the
ground and the geometry of the tunnel and ground surface satisfy a certain
condition.
   Table 4 shows the list of pre-upheaval and relative pre-upheaval observed
in four tunnels and their magnitudes. The phenomena appears along the
tunnel's center line when the face was rather far away from the instrument
line, and continued until the face came to about 3-5m from the line.
Along the transversal direction of tunnel advance, upheaval was also
observed after the face passed the line in the same tunnels. Interestingly
in three tunnels where pre-upheaval could not be seen on the ground
surface, pre-upheaval was measured by horizontal inclinometers along the
casing installed just above the tunnel crown. The magnitude varies from
2mm to 4mm.
   2. Transversal crack on the ground surface
The crack occurred before the face had reached the instrument line. we
found it near the entrance of the Kuriyama Tunnel. The appearance started
at the entrance slope and continued to the level section of 25m. Relations
between the transversal crack locations and ground surface conditions are
shown in Table 5. In the slope section, the phenomenon was found far away
from the tunnel face. The distance decreased as the ground surface varied
from slope to level.
   Fig. 1 shows a comparison between longitudinal settlement curves in two
cases. In one case the apparent crack on the surface was observed, and
in the other, there was no crack. A vast difference is found in the

Table 4. Pre-upheaval or relative pre-upheaval on the ground surface.

TUNNEL NAME	LONGITUDINAL DIRECTION				TRANSVERSAL DIRECTION			
	MEASURED POINT	VALUE (mm)	L / D	REMARKS	MEASURED POINT	VALUE (mm)	L / D	REMARKS
OHNUKI D=5.96m	13k482.5m	4	-3.10~-0.92	H=7.98m	13k492.5m	3	0.68-2.36	Upheaval & Relative up-
	13k492.5m	5	-1.43~-0.08	H=8.22m	13k424m	3	0.68-2.36	heaval
	13k605m	3	-3.02~-0.50	H=7.25m	13k605m		0.68-2.36	Relative upheaval
TOKKO	62k755m	3	-1.30~-0.37	H=7.2m , Slope	62k780m	3	0.00-2.73	L/D becomes larger accord-
	62k775m	3	-2.72~-0.65	H=8.9m				ing to the tunnel advance-
	62k780m	4	-4.20~-0.56	H=8.9m				ment.
	62k785m	3	-3.46~-1.03	H=8.9m				
	62k790m	3	-4.49~-0.65	H=8.9m				
	62k795m	3	-4.86~-0.75	H=8.9m				
NARITA 8th Sec.	63k282m	(3)		3mm upheaval after pilot excavation could be observed before the center portion was excavated.	63k252m	5		
					63k310m	4		
NARITA 9th Sec.	63k084m	3		Upheaval could be seen 5 to 19days before the tunnel face reached.	63k084m	4		Upheaval could be seen for 44days before and after tunnel excavation.

NOTICE : L ; Distance from the tunnel face or the tunnel axis. Minus means the condition before the tunnel face reaches.
D ; Tunnel diameter
H ; Crown depth
L/D means the tunnel face location when upheaval or relative upheaval could be observed.

Table 5. Relationship between transversal crack locations and ground conditions.

SURFACE CONDITION	L / D	CROWN DEPTH /DIA. RATIO
LONGITUDINAL SLOPE	1.08~1.37	0.25~0.93
FLAT (1)	0.73~1.03	0.92~0.93
FLAT (2)	0.39~0.49	0.92~0.97

REMARKS 1)FLAT (1) means the section between the slope and the level section.
2)FLAT (2) means the level section.
3)L/D means the distance between the tunnel face and the cracks on the ground surface.

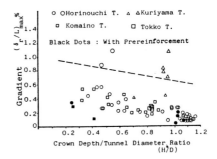

Fig. 2. Maximum relative surface settlement between two points along the tunnel axis.

shape. The magnitude of settlement in the area already excavated is supposed to influence whether the crack appears or not.
  3. Pre-settlement on and in the ground and relative upheaval between surface and subsurface
Pre-settlement means the settlement which has occurred before the face reaches the instrument line, and relative upheaval means the phenomenon that the surface settlement is smaller than the subsurface settlement.
  Table 6 shows the magnitude of maximum relative upheaval in four tunnels. It varies from 0.17mm to 6.3mm, and the phenomenon occurred just before the face passed through the instrument line.
  4. Enlargement of settlement trough
  5. Local collapse at the face and the periphery of the excavation area
  6. Accelerative surface and subsurface settlements and accelerative relative settlement between surface and subsurface
When the face passes through the instrument line, pre-settlement in the ground and relative settlement between the surface and subsurface grow acceleratively. Accelerative surface settlement which occurs in the period of each 3m before and behind the face occupies from 21 to 56 per cent of the total surface settlement. The phenomenon causes relative surface settlement between two points on the ground surface. Fig. 2 shows maximum relative surface settlement termed as Gradient by 5m against the tunnel crown

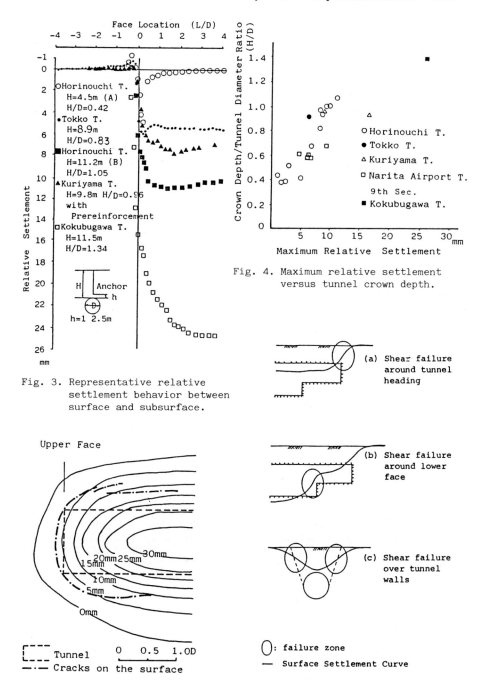

Fig. 3. Representative relative settlement behavior between surface and subsurface.

Fig. 4. Maximum relative settlement versus tunnel crown depth.

Fig. 5. Surface settlement contours (Tokko T.) and cracks (Kuriyama T.).

Fig. 6. Imaginary shear failure modes in the ground above a shallow tunnel.

Table 6. Maximum relative upheaval observed in 4 tunnels.

Tunnel	HORINOUCHI T.												TOKKO T.	NARITA AIRPORT (9th Sec.) T.	KURIYAM. T.
H / D	0.42	1.05	1.01	1.01	0.94	0.98	0.83	0.68	0.59	0.52	0.45	0.39	0.92	0.60	0.89
-L / D	0.26	0.51	0.19	0.19	0.19	0.19	0.19	0.19	0.19	0.19	0.19 / 0.09	0.19	0.28	1 day before	0.29
U_{rmax} (mm)	1.38	0.17	3.17	1.40	0.36	1.10	1.11	1.26	0.85	1.64	0.89	0.73	0.68	6.3	1.45

REMARKS : Measured point ; HORINOUCHI : 1~1.25m above the crown
   TOKKO   : 1.25m   above
   NARITA AIRPORT (9th Sec.) : about 1.2m above the crown
   KURIYAMA : 1.0m above the crown

depth-tunnel diameter ratio. It decreases in proportion to an increase in the ratio. Above the broken line some kind of ground failure has happened.
 On the other hand, Fig.3 shows relative settlement between ground surface and the point 1-2.5m above the crown in the ground against the face location. Here, a positive value reflects the increase in distance between the two points. The greater part of the settlement also appeared in the period of each 3m before and behind the face.
 Although the maximum relative settlement increases according to the increase of tunnel crown depth, the surface and subsurface settle down together when the tunnel crown depth- tunnel diameter ratio becomes smaller than 0.3, as shown in Fig. 4. In addition, when the ratio is smaller than about 0.5, relative settlement between the two points disappears immediately after the face passes (see Fig. 3).
   7. Failure of the ground
   8. Collapse
As mentioned above, if surface settlement, in a word Gradient, exceed a certain value which is shown by the broken line in Fig. 2, and doesn't converge, some kind of ground failure appears . We found transversal cracks on the surface ahead of the face and longitudinal cracks there, along both side walls of the tunnel. They have a good agreement with the surface settlement contour lines as shown in Fig. 5.
 In the case of ground failure, if the tunnel support members cannot endure the weight of ground over the tunnel, the tunnel will collapse.
   9. Convergence of settlement
   10. Decrease in relative settlement (see in 6.)
   11. Longitudinal crack (see in 8.)
The settlement behavior of the above-mentioned ground reflects not only the ground's continuity, but also its discontinuity. Therefore, for a shallow tunnel, the assessment of the ground or tunnel stability is regarded as finding or forcasting the turning point where the ground over the tunnel changes from a continuous body to a discontinuous one.

5 GROUND STABILITY ASSESSMENT

5.1 Monitoring

The settlement behavior as stated previously teaches us that failure or the instability of the ground is not only occupied by the stress state in the assessed area, but is also strongly influenced by the magnitude of settlement in the previously exacavated section, and that failure of the ground happens in the modes shown in Fig. 6.
 On the other hand model tests by Atkinson et al.(1974,1975) and Kimura et al.(1981) and practical reports by Müller(1978) and Distelmier(1983) also

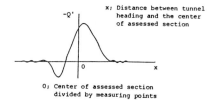

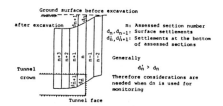

Fig. 7.(above) General conception
of Q'-value behavior

Fig. 8.(right) General conception
of Simple Shear Strain
(θ-value)

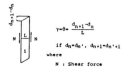

show that the collapse of a shallow tunnel in weak soil is supposed to happen in a similar manner. And the place where the failure happens coincides with the position where the shape of the surface or subsurface settlement curve changes dramatically. Therefore, the main purpose of monitoring is determined as gasping and assessing changes in shape. For this purpose, we used three indexes derived from the surface or subsurface settlement survey.

5.2 Indexes for monitoring

1. Shear Index (Q'-value)
As previously stated by authors(1986), Shear Index in the section between stations n and n+1 is defined in Eq. (1) by the magnitudes of surface or subsurface settlement measured at stations which were provided at the same interval. It can be applied to assess the failure shown in Fig. 6 (a) and (b).

$$Q' = \frac{3(d_{n+1}-d_n)-d_{n+2}+d_{n-1}}{L^3} \qquad (1)$$

where d_n is surface or subsurface settlement at station n, and L is the regular interval of each station.

As the index shows the divided difference of the third order, it reflects the elastic shear stress state of the ground in the assessed section. Therefore, the index behaves radically when the face passes through the center of the assessed section.

Fig. 7 shows the general behavior of Q'-value against the distance from the face. But the index increases or decreases dramatically when the unstable condition or some kind of failure in the mode of Fig 6 (a) or (b) happens.

On the other hand, assuming that the ground over a tunnel is regarded as a continuous beam in the longitudinal direction, Q'-value may be converted to shear stress by the beam theory in structural mechanics. Then, the ultimate Q'-value will be determined by Eq. (2) when deformation modulus (E), shear strength (τ), and sectional area (A) of the ground as a beam are chosen properly.

$$Q' = -\frac{\tau A}{E I} \qquad (2)$$

2. Simple Shear Strain (θ-value)
Simple shear strain is defined as the divided difference of the first
order as shown in Eq. (3).

$$\theta = \frac{d_{n+1} - d_n}{L} \qquad (3)$$

It then corresponds to a gradient of the settlement curve. Here, as shown
in Fig. 8, a different assumption from Shear Index is introduced. That is,
a soil beam over a tunnel longitudinally consists of rectangular blocks
divided by each instrument line. Moreover, a tight vertical connection and
no horizontal transmission of force on the common boundary are supposed,
and the blocks become sheared.
Simple Shear Strain thus reflects the discontinuous characteristics of
the ground as soil beam. However, it should be considered that the index
from surface settlements is smaller than that from settlements on the
bottom level of the blocks since the magnitude of the surface settlement
differs from that of settlement on the bottom.

3. Maximum surface settlement
This index is applied to forecast   the failure mode as shown in Fig. 6
(c). In general, it is difficult to set up many transversal instrument
lines with several stations on the ground surface. But, based on the same
assumption as Simple Shear Strain, the shear stress state along tunnel
walls is appreciated by Maximum surface settlement.
According to studies by Peck(1968) and Schmidt(1978), a transversal
surface settlement curve is generally represented by an error function
curve. Therefore, the maximum gradient as the maximum Simple Shear Strain
appears at the point of inflection and is given in Eq. (4).

$$\theta_{max} = \frac{0.61}{i} S_{max} \qquad (4)$$

where θ_{max} is the maximum gradient, i is the distance from the centerline
to the point of inflection, and S_{max} is the maximum surface settlement.
Though Peck(1968), Atkinson et al.(1977), and Schmidt(1968) also showed
that the point of inflection was changeable   according to ground conditions
and the tunnel's center line depth- tunnel diameter ratio, our experiences
tell us that i-value is nearly equal to one half the tunnel width or tunnel
diameter, when the tunnel's center line depth-tunnel diameter ratio is less
than 1.5.

6 RESULTS OF APPLICATION

Criteria for monitoring were determined with imaginary factors of safety
as shown in Table 7. Here, the effective height of soil beam and its cross-
sectional area were supposed to be 5m and rectangular respectively. But
effective height is equal to the overburden where relative settlement
occurred in positive between surface and subsurface doesn't maintain until
tunnel is completed.
At the point of 40m from the portal, unstable conditions in the ground
happened, and a crack like a tongue appeared, as forecasted in Fig. 6.
Here, values Q' and θ suddenly changed as shown in Figs. 9 and 10,
respectively. Moreover, Table 8 represents the comparison between the
magnitude of indexes observed before the unstable condition and the

Table 7. Criteria set up & input data for calculations in Kuriyama T.

Sections	Section with horizontal borehole inclinometer			Second section without horizontal borehole inclinometer			Section with side drifts			
Indexes / Criterion	Q'-value $\times 10^{-12}$ $1/\text{mm}^2$	θ-value %	Smax mm	θ'-value $\times 10^{-12}$ $1/\text{mm}^2$	θ-value %	Smax mm	θ'-value $\times 10^{-12}$ $1/\text{mm}^2$	θ-value %	Smax	
									For drifts	For center portion
First stage (Fs* = 1.5)	2320	0.87	73	271	0.67	45	271	0.67	26	26
Second stage (Fs* = 1.2)	2900	1.08	91	339	0.83	56	339	0.83	33	33
Third stage (Fs* = 1.0)	3480	1.30	109	407	1.00	67	407	1.00	39	39
Input data for calculations	H = 10m H' = 5m E = 150kgf/cm² A = BH' $I = \frac{BH'^3}{12}$			C = 0.3kg/cm² ϕ = 31 degrees γs = 1.75tf/m³ $\tau = C + \frac{\nu}{1-\nu}\gamma s(H - \frac{H'}{2})\tan\phi$			H ; Overburden H'; Effective height of soil beam			

*) Fs ; Imaginary factor of safety

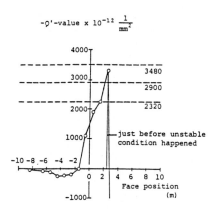

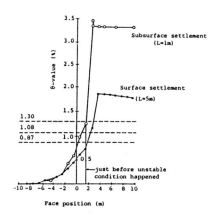

Fig. 9. Shear index in the assessed section whose center is located 36.5m from the portal, which is given by subsurface settlements (L=1.0m)

Fig. 10. Simple Shear Strain in the assessed section whose center is located 37.5m from the portal.

criteria previously determined. Their good agreement strongly encouraged us to continue assessing the ground stability thereafter.

7 CONCLUSION

In the construction of shallow tunnels in diluvial sand, the following matters were understood,

1256 *Interpretation*

Table 8   Maximum magnitude of indexes in the assessed
sections before unstable conditions occurred.

Indexes		Measurement results	Criterion
Smax	(Surface)	79 ～ 99.8 mm	
	(Subsurface)	79.27 ～ 93.55 mm	73 ～ 109 mm
Q'-value	(Surface)	254×10^{-12} 1/mm^2	
	(Subsurface)	$2130 \sim 3170 \times 10^{-12}$ 1/mm^2	$2320 \sim 3,480 \times 10^{-12}$
θ-value	(Surface)	0.734%	
	(Subsurface)	0.752 ～ 1.269%	0.87 ～ 1.30%

1. Eleven typical settlement behaviors observed here reflect the
   characteristics of sandy ground changing from a continuous material to
   a discontinuous one.
2. Three kinds of failure modes in ground over a shallow tunnel are
   assumed to correspond to the actual failure.
3. Three indexes to assess ground stability against the failure modes
   were assumed effectively, and reflect the variation in ground
   conditions. The criteria can thus be determined before excavation,
   based on geometry of the tunnel and a geological survey.

Atkinson, J.H., A.M. Cairncross & R.G. James. 1974. Model tests on shallow
   tunnels in sand and clay. Tunnels & Tunnelling. July:28–32.
Atkinson, J.H., E.T. Brown & P.M. Potts. 1975. Collapse of shallow unlined
   tunnels in dense sand. Tunnels & Tunnellings. May:81–87.
Atkinson, J.H. & M.P. Davis. 1977. Subsidence above shallow tunnels in
   soft ground. Proc. ASCE. GT4. April:307–325.
Distelmier, H. 1983. NATM usage in difficult ground condition. Proc. of
   Rapid Excavation and Tunneling Conference. Chicago:893–909.
Horiuchi, Y., T. Kudo, M. Tashiro & K. Kimura. 1986. A shallow tunnel
   enlarged in diluvial sand. Proc. of International Congress on Large
   Underground Openings. ITA. Firenze.
Kimura, T. & R.J. Mair. 1981. Centrifugal testing of model tunnels in soft
   ground. Proc. 10th ICSMFE. Stockholm:319–322.
Müller, L. 1978. Removing misconceptions on the New Austrian Tunneling
   Method. Tunnels & tunnelling. October:29–32.
Peck, R.B. 1968. Deep excavations and tunneling in soft ground. Proc. 7th
   ICSMFE. Mexico City. State-of-the-art Volume:225–290.
Schmidt, B. 1968. Settlements and ground movements associated with
   tunneling in soil. Ph. D Thesis. University of Illinois.
Schmidt, B. & G.W. Clough. 1977. Design and performance of excavations
   and tunnels in soft clay. A-State-Of-The-Art. A paper prepared for the
   International Symposium on soft clay. Bangkok. July.
Yokoyama, A., C. Tanimoto & K. Kimura. 1983. Relations between settlement
   of ground and deformability obtained by borehole tests in alluvial
   layers. Proc. International Symposium on Soil and Rock Investigations by
   In-Situ Testing. Paris:593–600.

2nd International Symposium on Field Measurements in Geomechanics, Sakurai (ed.)
© 1988 Balkema, Rotterdam. ISBN 90 6191 778 6

Development of a computer aided design and analytical system for geotechnical characterization problems using coupled FE-BE program

John L.Meek & Shinichi Akutagawa
Department of Civil Engineering, University of Queensland, Australia

GENERAL

The development approach for the construction of a computer aided
design and analytical system for geotechnical characterization
problems is discussed. In order to satisfy the fundamental needs of
geomechanics engineers, the proposed system suggests, as standard
equipment, the installation of three numerical units. They are (1)
the numerical analysis unit, (2) the field measurement data processing
unit and (3) the characterization analysis unit. Standardized
expressions are employed to identify the numerical processes and their
executions in a CAD system. Utilization of advanced software
development tools is emphasized. Introduced here are the efficient
management schemes for in-core and out-of-core data as well as an
error detection approach. Conclusions will be drawn showing the
direction of system development for various purposes.

(1) INTRODUCTION

(1.1) Problem identification
Successful application of theory into practical geotechnical
engineernig problems has been a major concern for many years. Along
with the development of computer hardware, a major proportion of
research resources has been expended on the development of computer
programs in an effort to build numerical tools that can simulate the
behaviour of the real world (R*) as closely as possible.

$$R^* \ (x,t) \qquad\qquad\qquad (1.1)$$

Here, (R*) is the general description of geotechnical quantities
(such as displacement, stress, strain, etc.) that usually depends
upon its location and time of observation. In the empirical approach,
(R*) is the first quantity to be obtained through various forms of
field observations while in the numerical approach it is the last
quantity to be computed through numerical analysis.

(1.2) Numerical analysis
In the simplest form, the process of numerical analysis parallels the
logic of formulating and solving a highly complicated non-linear
equation of N-input parameters. In this process, once the equation is

explicitly described, namely once the proposed theory is expressed in the form of a computer program, the theoretical solution of the equation depends entirely on the assumed values of N-input parameters.

$$R = F(P) \qquad\qquad (1.2)$$

where: R is an output quantity obtained by the execution of F.
         F is a numerical process (the computer program) which produces R.
         $P = P_1, P_2, \ldots\ldots\ldots, P_n$ is a set of input quantities that are necessary for the execution of F.

   The process of solving this equation can be regarded as the fundamental unit of numerical analysis and for many years, the major concern of many studies has been the construction of the function F which produces the theoretical results on the assumptions of N-input parameters.

(1.3)   Field measurement
The highly sophisticated techniques of field measurements have been developed and are serving the practical engineering needs in both direct and indirect manners.
   Field measurement offers us the following information:

$$R_m \qquad\qquad (1.3)$$

$$P_m \qquad\qquad (1.4)$$

where: R_m is the geomechanics quantity obtained at a finite number of measurement points. This relates to output quantity in expression (1.2).
         P_m is the geomechanics quantity obtained at a finite number of measurement points. This relates to input quantities in expression (1.2).

   The number of measurement points is limited by economic reasons. Therefore, measurement strategies (type of data to be measured, measurement locations, measurement timing, etc) must be carefully planned so that the data obtained could be of the best use.

(1.4)   Characterization analysis
Quantitative information which is obtained from field measurements can best be utilized when they are incorporated into the comparative study with the theoretical prediction through characterization (parameter identification or back analysis) procedures (Gioda, Cividini, Sakurai). Characterization procedures are uniquely designed numerical algorithms which fall into the field of non-linear optimization. They are based on the concept that assuming the numerical process is fixed during the analysis, the best choice of P $(= P_1 \ldots\ldots P_n)$ results in the minimum discrepancy between the theoretical prediction and the practical observation.
   This is expressed by the following expression:

$$P' = C(R_m, P_m, R, P) \qquad\qquad (1.5)$$

where: C is a characterization process which produces the improved quantity P' when theoretical and measurement information is given.
        P' is an improved input quantity that can be used for the next numerical process as $R' = F(P')$.

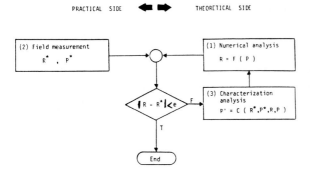

Figure(1) Fundamental flow chart of the proposed CAD system

They take the form of either the direct (non-iterative) or indirect (iterative) numerical algorithms and are designed to determine the best choice of P in order to satisfy the following criteria.

$$|R - R_m| < \text{Epsilon} \hspace{4cm} (1.6)$$

Hereby the applied form of the fundamental numerical unit for an integrated geotechnical software package is indicated in Figure (1).

This is the fundamental flow chart of interactive characterization analysis which is made of (1) the numerical analysis unit, (2) the field measurement data processing unit, and (3) the characterization analysis unit. This logical loop will be the central conceptual base of the proposed CAD system where the analysts can actually participate in the analysis environment in an interactive manner through intensive use of computer graphics.

(2)   STRUCTURAL DESCRIPTION OF CAD SYSTEM

(2.1)   Structure of the numerical process

An attempt is made here in order to broaden the expression (1.2). Shown in Figure (2) is the general flow chart of the hierarchically structured computer program. Each rectangular box represents various forms of an executable numerical process. It could be a program itself, subroutines, functions, or even a single executable statement. Dotted lines and solid lines show the transitions of the controller between two numerical processes of different levels and the same levels, respectively.

According to this chart, an executable program is regarded as a numerical process of level = 1. As the controller goes deeper into the program structure, the level indicator increases. Numbers indicated in circles are check point numbers. A check point is defined as a point in the program structure which the operational controller passes through ONCE per one program execution.

(2.2)   Execution of numerical process

A random component (indicated as a rectangular box) is now sampled and defined as a numberical process F with additional attributes. These are si (structural status identifier) and vi (validity identifier) and

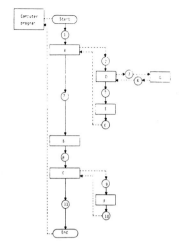

Figure(2) General flow chart of a computer program

are introduced here in order to explicitly identify structural and executional status of the numerical process under consideration.

If the input and output quantities of F are defined as P and R with similarly defined attributes, the execution of this process is stated by the following expression:

$$R_{vi,si} = F_{vi,si} (P_{vi,si}) \qquad\qquad (2.1)$$

with:
vi = 1 or 0 and si = CB:CA:LV

where: vi is a validity identifier. If the numerical object (either data or software) is regarded as being valid at the moment of execution, vi = 1, otherwise 0. In order to obtain correct results (R with vi = 1), vi must be 1 for all numerical objects that are related to that particular procedure.
 si is a structural status identifier that represents the following three attributes:
 CB: check point number located just before the entrance of the numerical process F.
 CA: check point number located just after the exit of the numerical process F.
 LV: level indicator of the process F.
 R is a set of output quantities which are defined as a result of the execution of F.
This expression is versatile. Firstly, if LV = 1 the expression (2.1) automatically means the execution of the program itself. P and R mean the input data and result output, respectively. They are usually stored in files. Sampling numerical process of level = 1, its successful execution is stated as follows:

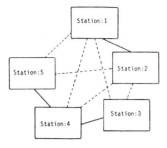

Figure(3) Master program as a network of STATIONs

Example (1) program execution
$$R_{1,si} = F_{1,si}(P_{1,si}) : si = 1:11:1 \qquad (2.2)$$

Secondly, if LV is greater than 1, (2.1) means the execution of sub-program (subroutines, functions, etc.) of level = LV. P and R mean input and output arguments that are input into and output from the numerical process F. For the process indicated as D, the following expression applies for its successful execution:

Example (2) sub-program execution
$$R_{1,si} = F_{1,si}(P_{1,si}) : si = 2:5:3 \qquad (2.3)$$

Thirdly, as an extreme case with LV being greater than 1, (2.1) could be the expression of a single executable statement in a computer program. In this case, P means the variable that appears on the right hand side of the equal sign, and R on the left hand side. For the process indicated as G, the following expression applies for its successful execution:

Example (3) single statement execution
$$R_{1,si} = F_{1,si}(P_{1,si}) : si = 3:4:4 \qquad (2.4)$$

(2.3)   Structure of the master program
The master program is the integrated software network (shown in Figure (3)) that is comprised of singly or multiply connected components which are defined as STATIONS. Complexity and structural characteristics of this network depend on the language and hardware to be used.
These stations are built into the system on a parallel structure basis. Therefore, any transition between two different stations is performed on the same conceptual level. Any station can be chosen as the first station to be activated in the total CAD system. Further transition to other stations (which can be performed without the deactivation of the system) is limited only to the ones which are connected to the first by the solid lines. This limitation is also language and hardware dependent.

(2.4)   Operation in a station
In Figure (4), a typical example of a station is illustrated schematically with its corresponding program library and data base. A station is a software environment in which the executable program is

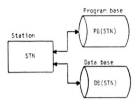

Figure(4) A station with its associated program and data bases

to be chosen from the program library to enter the interactive analysis
environment.

When the station = STN is the current active station, the current
operational status is defined by the following expression:

$$OPERATION = STN : PB(STN) : DB(STN) \qquad (6)$$

where: STN   is the station code name.
        PB(STN)   is the list of the executable program modules that
                  must be accessible from station = STN.
        DB(STN)   is the list of data sets that must be accessible from
                  station = STN.

This expression uniquely identifies in which station the current
operation is being undertaken and which program and data libraries
must be in the current storage device for the successful operation.
Use of this expression helps install the system on different machines
with various storage sizes.

(3)   ADVANCED SOFTWARE TOOLS

(3.1)   General
In the development and maintenance of an integrated software package,
the use of advanced software tools is essential. These tools are in
the form of subroutine utilities and help programmers write
sophisticated products in various ways. In these utilities, there are
groups of subroutine libraries that deal with free field input,
in-core and out-of-core data management, error detection, matrix and
vector operations, etc. In this paper, discussion is limited to those
that are associated with the management of data and the error
detection system.

(3.2)   In-core data management
In the advanced form of the in-core data management system (Hoit,
Arora, et al), all the dynamic partitioning of arrays is performed
automatically by the in-core data manager (ICDM).

In this method, one big array (IA : super array) is installed in a
blank common block. Any array to be used in the program can be
registered in this super array IA. On registration, each array
consumes two portions of the super array's memory. The first portion
is used to store the array data. The second portion is used to store
the directory information, comprised of the array identifier name,
size, type and other important attributes. This is illustrated in
Figure (5).

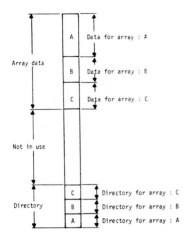

Figure(5) Internal structure of the SUPER ARRAY

Information to be registered in the super array does not always have to be in the form of an array and can be just a single variable of any type. Creation (space reservation), duplication, size adjustment, deletion or any other operations can be performed by single subroutine calls. Any address change accompanied with these operations is automatically calculated by the ICDM.

The use of ICDM not only helps programmers achieve flexible data management, but also makes programs transparent to all potential users. It is even possible, by the use of the ICDM, to allocate all the in-core information in a single super array. This technique is specially appealing for standardized modularization, flexible control of data flow, and easier debugging procedures.

Consider a program structure of level = 1 in Figure (2). If all the significant information is registered in a super array, the state of the program execution can be identified simply by:

[super array]   (3.1)

at any check point marked on level = 1 in Figure (2).

(3.3) Out-of-core data management

On the other hand, the out-of-core data management system (OCDM) is a set of subroutines that takes care of organizing and managing information stored on an out-of-core storage device. Use of OCDM helps achieve the efficient utilization of out-of-core storage available. It also plays a crucial part in data transmission in conjunction with the ICDM when several modules are executed interactively in one particular station.

(3.4) Error treatment

Errors have to be interrupted and when they occur, appropriate treatment must be taken. In the proposed CAD system, the systematic approach shall be taken for remedial treatment through the combined use of ICDM, OCDM and EDMS (error detection management system).

Consider a numerical process Fvi,si (Pvi,si) where si = Cl:C2:LV.
Assume that all the data sets P_1......P_n are registered in the super
array.

Firstly, just before entering the numerical process F, validity of
the data sets is checked at a check point Cl using the validity
information which is given to the software preliminary. If vi = 1 for
all the data sets, the controller proceeds to the numerical process.
Should any error be detected, the appropriate error message is given
and corrective treatment is made either automatically by the program
or interactively by the user. This approach is adapted in an intension
to allow only the admissible data sets to enter the numerical process.
Similar actions are taken against errors that might occur during the
execution of process or at the exit check point C2.

Along with the error detection performances, the content of super
array is saved into the file at the most significant check points for
security purposes. Points are usually on structural level = 1. Therefore, if the execution is terminated due to the occurrence of a fatal
error, users can not only read the error message, but also fully examine
the state of the execution, which is retained in the content of the
super array which was saved before the error occurrence.

Saving the contents of the super array at the most significant check
points is advantageous in another sense. Suppose the program of 10
serial processes was terminated due to the fatal error at step 10 in
the first execution. If the user finds the cause of the error only in
the 10th numerical process, steps 1 through 9 do not have to be
repeated and can be skipped for the second execution by using a simple
command. This is possible because repeating steps 1 through 9 is
equivalent to recovering the super array stored in a file at check
point 9 in the last execution.

(4)   DIRECTION OF SYSTEM DEVELOPMENT

As was stated earlier, the underlying concept of this system is the
installation of three fundamental numerical units as the standard
equipment for the geotechnical characterization software package.

In this package, several options are available in the analytical and
characterization stations and users are responsible for deciding which
option is to be adapted in each session. The measurement data
processing station will be equipped to process the measurement data
obtained from the site into acceptable form for the characterization
analysis. Other components are added to enhance the efficiency of the
CAD operation through the introduction of a flexible graphic routine
package. The structure of the master program environment of the
proposed system is schematically illustrated in Figure (6) and the
numerical processes to be installed in each station are summarized in
Table (1).

All the software units employed in this system are written following
a standardized coding method using advanced software tools. Therefore,
any software developed elsewhere in the same manner can directly be
registered as another member of the system's software family. Should
any new software be developed to be added to the system, it is
essential to follow the standardized coding method to make it transparent to potential users.

Interactability of the system operation is also carefully designed
with and without the help of graphic software. Therefore, the system
can be installed on a machine without a graphic device. This will also

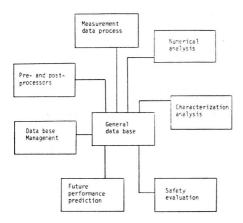

Figure(6) Master program environment of the proposed CAD system

Table(1) Numerical processes to be installed in the proposed CAD system
(*) is optional.

STATION CODE NAME	NUMERICAL PROCESSES TO BE INSTALLED
Numerical analysis	Linear and non-linear FEM (2 and 3D) BEM (2 and 3D) Coupled FE-BE (2 and 3D)
Field measurement data process	Displacement data processor Stress data processor Strain data processor
Characterization analysis	Least square method Direct search method Statistical identification method(*)
Pre- and post-processor	Graphic pre-processor Graphic post-processor Non-graphic pre-processor Non-graphic post-processor
Safety evaluation	Safety evl. by displacement criteria Safety evl. by stress criteria Safety evl. by strain criteria
Future performance prediction	Time-dependent analysis
Help information center	Information services for CAD operation
Data base management	Data base manager
General data base	

help the system to be adapted for various applications.

Lastly, a new attempt will be made to develop the intelligent software environment in which the system can achieve the best numerical approximation of geomechanical quantity $R(x,t)$ not only through the iterative algorithm to find the best P, but also to find the best F, using the artificial intelligence concept of the learning processes.

REFERENCES

Gioda. 1985. Some remarks on back analysis and characterization problems in geomechanics. 5th Int. Conf. on Numerical Methods in Geomechanics, Nagoya, 47-61.

Cividini, A., Jurina, L., Gioda, G. 1981. Some aspects of characterization problems in geomechanics. Int. Jnl. Rock. Mech. Sci. & Geomech. Abstr. Vol. 18, 487-503.

Sakurai, S., Shinji, M. 1984. A monitoring system of excavation of underground opening based on micro-computers. Int. Symp. on Design and Performance of Underground Excavations, Cambridge, U.K.

Hoit, M.I. 1983. New computer programming techniques for structural engineering. Ph.D. thesis, University of California, Berkeley, 1983.

Arora, J.S., Lee, H.H., Jao, Y. SMART: Scientific database management and engineering analysis routines and tools. Advances in Engineering Software, Vol. 8, No. 4, 194-199.

2nd International Symposium on Field Measurements in Geomechanics, Sakurai (ed.)
© 1988 Balkema, Rotterdam. ISBN 90 6191 778 6

Ground movements caused by tunnelling

E.W.Brand
Geotechnical Control Office, Hong Kong

Throughout this excellent Symposium, we have heard a large number of
accounts of field measurements on a wide range of projects under construction. Of particular note, is the vast amount of data which has
been amassed during the construction of tunnels, particularly in Japan.
In some countries, notably Japan, it has clearly become standard practice to install sophisticated means of monitoring deformations caused
by tunnelling operations in all types of ground. The data obtained
relates mainly to movements at the tunnel face and to deformations of
the lining, but a great deal of information is also being collected on
ground surface settlements. In a few cases, these settlements have
been correlated with the resulting damage to existing structures.

As a result of the extensive tunnelling monitoring programmes undertaken in recent years, the published literature now contains a wealth
of first class data for many different geological conditions and tunnelling methods. In some geographical locations, enough measurements would
seem to have been made to form a sound basis for the reliable prediction
of the approximate deformations which are likely to be caused by future
tunnelling activities. In my view, therefore, more effort should be
expended on establishing meaningful correlations between measured deformations and geology, tunnel size/depth and tunnelling method. The amount
of data available is adequate, I feel, to enable useful design guidance
charts to be produced for a wide range of situations.

The data correlation exercise that I envisage as necessary need be no
more than an extension of the approach originally adopted by Peck (1969)
and Schmidt (1969) for predicting the ground movements associated with
tunnelling in soft clay. With the availability of further data, others
have tried to improve upon those early correlations for soft clay, notably Hanya (1977), Morton & Dodds (1979), Clough & Schmidt (1981), Reséndiz & Romo (1981), Fujita (1982) and Hurrell (1984). Other state-of-
the-art papers which provide predictions of deformations caused by tunnelling in firmer ground conditions are those by Cording & Hansmire
(1975), Attewell (1977), O'Reilley & New (1982), Attewell & Yeates (1984),
Attewell et al (1986), Lo et al (1984) and Eisenstein & Negro (1985).
It may well be that there are important reviews on ground movements
caused by tunnelling published in Japanese and other languages of which
I am not aware. There is no doubt, however, that recently published
field data, including that contained in the volumes of this Symposium
Proceedings, merits a further comprehensive state-of-the-art review.

REFERENCES

Attewell, P.B. 1977. Ground movement caused by tunnelling in soil.
  (State-of-the-art Paper). Proc. Conf. Large Ground Movements and
  Structures, Cardiff, 812-948. (Discussion, 999-1009).
Attewell, P.B. & Yeates, J. 1984. Tunnelling in soil. Ground Movements
  and Their Effects on Structures, ed. P. Attewell & R.K. Taylor, 132-
  215. Surrey Univ. Press, London.
Attewell, P.B., Yeates, J. & Selby, A.R. 1986. Soil Movements Induced
  by Tunnelling and Their Effects on Pipelines and Structures. Blackie,
  Glasgow, 325 p.
Clough, G.W. & Schmidt, B. 1981. Design and performance of excavations
  and tunnels in soft clay. Soft Clay Engineering, ed. E.W. Brand &
  R.P. Brenner, 569-634. Elsevier, Amsterdam.
Cording, E.J. & Hansmire, W.H. 1975. Displacements around soft ground
  tunnels. (General Report). Proc. 5th Panam. Conf. SMFE, Buenos Aires,
  4:571-633. (Errata, 5:46-47).
Eisenstein, Z. & Negro, A. 1985. Excavations and tunnels in tropical
  lateritic and saprolitic soils. Proc. 1st Int. Conf. Geomechanics
  in Tropical Lateritic and Saprolitic Soils, Brasilia, 4:299-333.
Fujita, K. 1982. Prediction of surface settlements caused by shield
  tunnelling. Proc. Int. Conf. Soil Mechanics, Mexico City, 1:239-246.
Hanya, T. 1977. Ground movements due to construction of shield-driven
  tunnel. Proc. 9th Int. Conf. SMFE, Tokyo, case history vol.:759-790.
Hurrell, M.R. 1984. The empirical prediction of long-term surface
  settlements above shield-driven tunnels in soils. Proc. 3rd Int. Conf.
  Ground Movements and Structures, Cardiff, 161-170. (Discussion, 820-
  825).
Lo, K.Y., Ng, M.C. & Rowe, R.K. 1984. Predicting settlement due to
  tunnelling in clays. Tunnelling in Soil and Rock (Proc. two sessions,
  Geotech '84, ASCE, Atlanta, Georgia), 46-76.
Morton, J.D. & Dodds, R.B. 1979. Ground subsidence associated with
  machine tunnelling in fluvio-deltaic sediments. Tunnels & Tunnelling,
  11(8):13-17 and 11(9):23-28.
O'Reilley, M.P. & New, B.M. 1982. Settlements above tunnels in the
  United Kingdom - their magnitude and prediction. Proc. 3rd Int. Symp.
  Tunnelling (Tunnelling '82), Brighton, England, 173-181.
Peck, R.B. 1969. Deep excavations and tunnelling in soft ground.
  (State-of-the-art Report). Proc. 7th Int. Conf. SMFE, Mexico City,
  state-of-the-art vol.:225-290. (Discussion, 4:320-330).
Reséndiz, D. & Romo, M.P. 1981. Settlements upon soft-ground tunneling:
  Theoretical solution. Soft Ground Tunneling: Failures and Displace-
  ments, ed. D. Reséndiz & M.P. Romo, 65-74. Balkema, Rotterdam.
Schmidt, B. 1969. Settlements and Ground Movements Associated with
  Tunneling in Soil. PhD Thesis, Univ. Illinois, Urbana, 234 p.

Author index